Handbook of Fruit Wastes and By-Products

Handbook of Fruit Wastes and By-Products: Chemistry, Processing Technology, and Utilization addresses the chemistry and valorization of wastes and by-products generated during the processing of fruits. It provides an overview of the recovery of bio-functional components from fruit processing residues and their utilization. Besides, this book is a valuable resource for scientists, researchers, professionals, and enterprises that aspire in the management of fruit processing wastes and by-products, and their utilization.

Key features

- Provides comprehensive information about the chemistry of wastes and by-products obtained during fruit processing
- Provides in-depth information about the bioactive potential of fruit processing wastes and by-products
- Explores new strategies used for the proper valorization of fruit processing residues
- Describes the utilization of nutraceutical components derived from fruit processing residues in the fabrication of novel functional foods

Handbook of Fruit Wastes and By-Products

Chemistry, Processing Technology, and Utilization

Edited by
Khalid Muzaffar, Sajad Ahmad Sofi,
and Shabir Ahmad Mir

CRC Press
Taylor & Francis Group
Boca Raton London New York

CRC Press is an imprint of the
Taylor & Francis Group, an **informa** business

First edition published 2023
by CRC Press
6000 Broken Sound Parkway NW, Suite 300, Boca Raton, FL 33487-2742

and by CRC Press
4 Park Square, Milton Park, Abingdon, Oxon, OX14 4RN

CRC Press is an imprint of Taylor & Francis Group, LLC

© 2023 selection and editorial matter, Khalid Muzaffar, Sajad Ahmad Sofi, and Shabir Ahmad Mir, individual chapters, the contributors

ISBN: 9780367724740 (hbk)
ISBN: 9780367758950 (pbk)
ISBN: 9781003164463 (ebk)

DOI: 10.1201/9781003164463

Typeset in Times
by Deanta Global Publishing Services, Chennai, India

Contents

Suheela Bhat, Priyanka Suthar, Shafiya Rafiq, Asmat Farooq, and Touseef Sheikh

Fozia Hameed, Neeraj Gupta and Rukhsana Rehman

Nadia Bashir, Beena Munaza, Shafiya Rafiq, Monika Sood, and Sushil Sharma

**Shiv Kumar, Poonam Baniwal, Harpreet Kaur, Rekha Kaushik,
Sugandha Sharma and Naseer Ahmed**

Preface

Processing of fruits produces large volumes of wastes and by-products, which can create environmental problems. However, these fruit processing residues have amazing nutritional composition, containing good amounts of sugars, vitamins, minerals, and health-promoting bioactive components (polyphenols, carotenoids, dietary fibers, peptides, ω-fatty acids, and so on). Due to the presence of bioactive components, these can be used for a number of purposes in the food, pharmaceutical, and cosmetic industries. The present aim is to efficiently utilize these fruit wastes and by-products and minimize their impact on the environment. Proper utilization of fruit processing wastes and by-products will not only be a source of profit to the fruit processing industry, but will also help lessen environmental pollution.

Handbook of Fruit Wastes and By-Products: Chemistry, Processing Technology, and Utilization is the first book devoted to fruit processing wastes and by-products, exploring a wide range of important fruits including tropical, subtropical, and temperate fruits. This book provides in-depth information about fruit processing wastes and by-products, their nutritional composition, biochemistry, processing technology of by-products, and the utilization of these residues in various food applications. This book will deal with the transformation of fruit processing residues into value-added products. This book also offers in-depth information about the potential of these fruit processing residues as a source of bioactive ingredients and their utilization in the development of novel functional foods. In addition, the various novel technologies useful in the extraction of functional components from these fruit processing residues are discussed.

Although there are some general books on by-products of the food processing industry, they are limited in context, discussing only some particular fruits. The unique quality of this book is that it provides a full-length study of the different developments, from the basic technologies involved in the management of fruit wastes and by-products to the recent advancements and future areas of research to be done on this subject. This book includes the valorization of fruit processing wastes and by-products of important fruits grown in different climatic conditions of the world. The book comprises 22 chapters, with each chapter providing a detailed study of the nutritional profile, biochemical composition, nutraceutical potential, and utilization of wastes and by-products of a particular fruit.

The book will be of interest in almost all regions of the world, with considerable market appeal. This book is a comprehensive reference written for teachers, scientists, researchers, students, and others who have an interest in the management of fruit processing wastes and by-products, and their utilization. This book is a firsthand resource to policy developers and personnel working in the food industry regarding the valorization of fruit processing residues generated in the food industry.

Khalid Muzaffar

Sajad Ahmad Sofi

Shabir Ahmad Mir

Editors

Dr. Khalid Muzaffar (Ph.D.) currently working as Postdoc Research Associate fellow at Department of Food Technology, Islamic university of Science and Technology, UT of Kashmir, India. Worked as a guest faculty at Government Boys Degree College Baramulla, J & K in the Department of Food Technology. Also worked as a Junior Research Fellow at university of Kashmir, Srinagar, J & K, India. Qualified National Eligibility Test conducted by Indian Council of Agricultural Research (ICAR) India. Got Maulana Azad National Fellowship from University Grant Commission, New Delhi, India. Has written one book, number of book chapters and has more than 40 publications published in reputed international journals. Attended and presented papers in many national and international conferences. Potential peer reviewer of reputed international journals related to Food Science and Technology that belong to popular publishing house viz. Elsevier, Taylor-Francis, Wiley, Springer etc. Recipient of Reviewer Recognition certificates from Elsevier, Taylor-Francis, Springer and Wiley Online.

Sajad Ahmad Sofi (Ph.D.) is a Food Analyst at the Food Testing Laboratory, Department of Food Technology, IUST, Awantipora, India. He received an M.Sc. (Food Technology) degree from Islamic University of Science & Technology, Awantipora, India, and a Ph.D. (Food Science & Technology) degree from SKUAST, Jammu, India. Dr. Sofi has also qualified National Eligibility Test in Food Technology conducted by the Indian Council of Agricultural Research. The author has been awarded a gold medal for outstanding performance during his M.Sc. degree program and UGC award for Ph.D. fellowship. Dr. Sofi has more than 30 publications in International reputed journals and 10 book chapters. He is an active reviewer for many scientific journals of repute. He has attended several national and international conferences, workshops, and seminars.

Shabir Ahmad Mir (Ph.D) obtained his PhD in Food Technology from Pondicherry University, Puducherry, India. At present, he is an Assistant Professor at the Government College for Women, Srinagar, India. He has received the Best PhD Thesis Award 2016 for outstanding research work by the Whole Grain Research Foundation. He has organized several conferences and workshops in Food Science and Technology. Dr. Mir has published numerous international papers, book chapters, and edited six books.

Contributors

Madiha Abdel-Hay
Home Economic Department
Faculty of Specific Education
Mansoura University
Mansoura, Egypt

Naseer Ahmed
Department of Food Technology
DKSG Akal College of Agriculture
Eternal University
Baru Sahib, India

Nadira Anjum
Division of Food Science and Technology
Sher-e-Kashmir University of Agricultural
    Science & Technology
Chatha, India

Nazmin Ansari
Department of Food Technology and Nutrition
Lovely Professional University
Phagwara, India

Poonam Baniwal
Food Corporation of India
India

Rahul Islam Barbhuiya
Department of Food Process Engineering
National Institute of Technology
Rourkela, India

Nadia Bashir
Food and Drug Administration
J&K
Jammu, India

Suheela Bhat
Department of Food Engineering and
    Technology
Sant Longowal Institute of Engineering and
    Technology
Longowal, India

Aparna Bisht
Department of Food Technology and Nutrition
Lovely Professional University
Phagwara, India

Divya Chauhan
Department of Food Technology
DKSG Akal College of Agriculture
Eternal University
Baru Sahib, India

Chhaya
Department of Food Technology
DKSG Akal College of Agriculture
Eternal University
Baru Sahib, India

B.N. Dar
Department of Food Technology
Islamic University of Science & Technology
Awantipora, India

Saptashish Deb
Department of Food Engineering and
    Technology
Sant Longowal Institute of Engineering and
    Technology
Longowal, India

Vildan Eyiz
Department of Food Engineering
Faculty of Engineering and Architecture
Necmettin Erbakan University
Konya, Turkey

Asmat Farooq
Division of Biochemistry
Sher-e-Kashmir University of Agricultural
    Science & Technology
Chatha, India

Seherpreet Kaur Flora
Department of Food Technology and Nutrition
Lovely Professional University
Phagwara, India

Yogesh Gat
Department of Food Technology
Institute of Chemical Technology
Marathwada Campus
Jalna, India
and

Department of Food Technology and Nutrition
Lovely Professional University
Phagwara, India

Falak Grover
Department of Food Technology and Nutrition
Lovely Professional University
Phagwara, India

Neeraj Gupta
Division of Food Science and Technology
Sher-e-Kashmir University of Agricultural
    Science & Technology
Chatha, India

Fozia Hameed
Division of Food Science and Technology
Sher-e-Kashmir University of Agricultural
    Science & Technology
Chatha, India

Sumaira Jan
Department of Food Technology
DKSG Akal College of Agriculture
Eternal University
Baru Sahib, India

Fozia Kamran
School of Science
Western Sydney University
Penrith, Australia

Gurwinder Kaur
IK Gujral Punjab Technical University
Kapurthala, India

Harpreet Kaur
Department of Biotechnology
Deenbandhu Chhotu Ram University of
    Science and Technology
Murthal, India

Jaspreet Kaur
Department of Food Science and Technology
RIMT University
Mandi Gobindgarh, India

Manpreet Kaur
IK Gujral Punjab Technical University
Kapurthala, India

Samandeep Kaur
Department of Food Engineering and
    Technology
Sant Longowal Institute of Engineering and
    Technology
Longowal, India

Rekha Kaushik
Department of Food Science (Hotel
    Management)
Maharishi Markandeshwar
Mullana, India

Bababode Adesegun Kehinde
Department of Food Technology and Nutrition
Lovely Professional University
Phagwara, India

Gülşah Çalışkan Koç
Food Technology Program
Esme Vocational High School
Usak University
Usak, Turkey

Ashwani Kumar
Department of Food Technology and Nutrition
Lovely Professional University
Phagwara, India

Harish Kumar
Department of Food Technology
Amity Institute of Biotechnology
Amity University Rajasthan
Jaipur, India

Krishan Kumar
Department of Food Technology
DKSG Akal College of Agriculture
Eternal University
Baru Sahib, India

Shiv Kumar
Department of Food Technology
DKSG Akal College of Agriculture
Eternal University
Baru Sahib, India
and
Department of Food Science (Hotel
    Management)
Maharishi Markandeshwar
Mullana, India

Yogesh Kumar
Department of Food Engineering and
 Technology
Sant Longowal Institute of Engineering and
 Technology
Longowal, India

Arshied Manzoor
Department of Post-Harvest Engineering and
 Technology
Faculty of Agricultural Sciences
Aligarh Muslim University
Aligarh, India

Shabir Ahmad Mir
Department of Food Technology
Government College for Women
Srinagar, India

Beena Munaza
Sher-e-Kashmir University of Agricultural
 Science & Technology
Chatha, India

Khalid Muzaffar
Department of Food Technology
Islamic University of Science & Technology
Awantipora, India

Haroon Naik
Division of Food Science and Technology
Sher-e-Kashmir University of Agricultural
 Science & Technology
Shalimar, India

Saadiya Naqash
Division of Food Science and Technology
Sher-e-Kashmir University of Agricultural
 Science & Technology
Shalimar, India

Razieh Niazmand
Research Institute of Food Science and
 Technology
Mashhad, Iran

Shafiya Rafiq
Department of Food Technology and Nutrition
Lovely Professional University
Phagwara, India
and

Department of Food Technology
Amity Institute of Biotechnology
Amity University Rajasthan
Jaipur, India

Rafia Rashid
Department of Food Science and Technology
Sher-e-Kashmir University of Agricultural
 Science & Technology
Chatha, India

Bibi Marzieh Razavizadeh
Research Institute of Food Science and
 Technology
Mashhad, Iran

Reetu
Maharaja Ranjit Singh Punjab Technical
 University
Bathinda, India

Rukhsana Rehman
Division of Food Science and Technology
Sher-e-Kashmir University of Agricultural
 Science & Technology
Chatha, India

Qurat-Ul-Eain Hyder Rizvi
Department of Food Technology
DKSG Akal College of Agriculture
Eternal University
Baru Sahib, India

Charanjiv Singh Saini
Department of Food Engineering and
 Technology
Sant Longowal Institute of Engineering &
 Technology
Longowal, India

D.C. Saxena
Department of Food Engineering and
 Technology
Sant Longowal Institute of Engineering and
 Technology
Longowal, India

Harish Kumar Sharma
Department of Chemical Engineering
National Institute of Technology
Agartala, India

Renu Sharma
Department of Food Technology and Nutrition
Lovely Professional University
Jalandhar, India

Sugandha Sharma
Department of Nutrition and Dietetics
Chandigarh University
Mohali, India

Sushil Sharma
Sher-e-Kashmir University of Agricultural
    Science & Technology
Chatha, India

Mohd Aaqib Sheikh
Department of Food Engineering and Technology
Sant Longowal Institute of Engineering &
    Technology
Longowal, India

Touseef Sheikh
Department of Biochemistry
Govt Degree College
Anantnag, India

Barinderjit Singh
Department of Food Technology
IK Gujral Punjab Technical University
Kapurthala, India

Sushil Kumar Singh
Department of Food Process Engineering
National Institute of Technology
Rourkela, India

Tajendra Pal Singh
Department of Food Technology
DKSG Akal College of Agriculture
Eternal University
Baru Sahib, India

Poonam Singha
Department of Food Process Engineering
National Institute of Technology
Rourkela, India

Sajad Ahmad Sofi
Department of Food Technology
Islamic University of Science & Technology
Awantipora, India

Monika Sood
Sher-e-Kashmir University of Agricultural
    Science & Technology
Chatha, India

Yashi Srivastava
Central University of Punjab
Bathinda, India

Priyanka Suthar
Department of Food Technology and
    Nutrition
Lovely Professional University
Phagwara, India

Barbara E. Teixeira-Costa
Faculty of Agrarian Sciences
Federal University of Amazonas,
Manaus, Brazil

Priyanka Thakur
Department of Food Technology
DKSG Akal College of Agriculture
Eternal University
Baru Sahib, India

Vidisha Tomer
VIT School of Agricultural Innovations and
    Advanced Learning
Vellore Institute of Technology
Vellore, India

Ismail Tontul
Department of Food Engineering
Faculty of Engineering and Architecture
Necmettin Erbakan University
Konya, Turkey

Mohammad Ubaid
Department of Food Engineering and
    Technology
Sant Longowal Institute of Engineering &
    Technology
Longowal, India

CHAPTER 1

Fruit Processing Wastes and By-Products

Khalid Muzaffar, Sajad Ahmad Sofi, Fozia Kamran,
Barbara E. Teixeira-Costa, Shabir Ahmad Mir and B.N. Dar

CONTENTS

1.1 INTRODUCTION TO FRUIT PROCESSING BY-PRODUCTS AND WASTES

Fruits are considered to be widely used food items from horticultural produce that are consumed either fresh or processed into different products. With changes in dietary habits and increasing population size, the production and processing of fruits has increased exponentially to meet consumer demands (Sagar et al., 2018). As a consequence, high volumes of waste and by-products are generated during production and processing of fruits. The increasing popularity of processed fruit products such as juices, nectars, and minimally processed fruit products has amplified the production of fruit-based wastes and by-products. Many fruits, such as oranges, pineapples, apples, peaches, and pomegranates are utilized for juice extraction, jams, and frozen pulp, and produce significant amounts of waste and by-products. Fruit waste and by-products include pomace, peels, rind, seeds, stones, and trimmings (Rodriguez et al., 2006; Torres-Leon et al., 2018). Many fruits generate about 25% to 30% of their fresh weight as waste, which is not subsequently used (Ajila et al., 2007, 2010). During slicing of apple, 10.91% of the fruit is lost as seed waste, leaving 89.09% of final product. Processing of pineapple yields waste materials such as peel (14%), core (9%), pulp (15%), and top (15%), with a total final product of 48%. Mangos upon processing yield about 11% peel, 13.5% seed, 18% unusable pulp, and 58% of finished product (Ayala-Zavala et al., 2010; Joshi et al., 2012). Waste streams of 8.5% peel, 6.5% seed, and 32% unusable pulp leaves a final product of 53% during dicing of papaya. In addition, about 5.5 million metric tonnes of pomace waste is produced during juice production from fruits and vegetables (Sagar et al., 2018).

By-products and waste obtained during the processing of fruits can have a bad impact on the environment due to its high biodegradability. These residues contribute to emission of greenhouse gases to the atmosphere (Girotto et al., 2015). As most of these biomaterials are currently not utilized, they find their way into municipal landfills where they create serious environmental issues because of microbial decomposition to carbon dioxide and methane and production of leachate. The cost of handling and disposing of these fruit residues has an adverse impact on the economy, while the management of large amounts of residues poses an environmental challenge (Torres-Leon et al., 2018). The European Union (EU) action plan for the circular economy could be an effective strategy to reduce the level of waste and by-products generated during fruit processing. This plan is based on the reduction, reuse, recovery, and recycling of materials and energy, so as to enhance the value and consequently the useful life of products, materials, and resources in the economy. The utilization of agro-industrial wastes and by-products can represent a renewable source for food additives already used in the food system or even the generation of new additives with enhanced functional properties to be used in the development of novel, functional food products (Faustino et al., 2019). Currently, the food industries are concerned with innovations to achieve zero waste, where the residual wastes generated are used as raw material for the development of new products. Such types of action have a direct impact on the Millennium Development Goals, the forthcoming goals of sustainable development, and the Zero Hunger Challenge.

The wastes and by-products of fruit processing are a rich source of valuable bioactive compounds that can potentially be used in the food industry as economical sources of food additives. Although horticultural waste and by-products have not been taken seriously as very valuable materials in the past, the scenario has now changed. These fruit processing residues are potential sources of phenolic compounds, dietary fibers, pigments, sugar derivatives, minerals, organic acids, and other components. Most of these biomolecules possess various health benefits including antiviral, antibacterial, antimutagenic, anti-inflammatory, antitumor, gastroprotective, immunomodulatory, antioxidative, and cardioprotective activities. Fruit residues can be used to extract and isolate valuable biomolecules that can be used in the food, cosmetics, pharmaceutical, and textile industries (Dilas et al., 2009; Yahia, 2017). Extraction of bioactive compounds from fruit residues is the most critical step. The type and quantity of bioactive components that can be derived from fruit processing wastes and by-products depend on the extraction method adopted. Furthermore, the extraction method may vary with respect to the targeted bioactive molecules. Extraction process parameters such as the type and quantity of solvent, temperature, extraction time, and method of extraction could have a detrimental effect on the extraction of bioactive compounds (Sagar et al., 2018).

Therefore, although wastes and by-products generated during the processing of fruits can be considered to be unavoidable, the proper utilization of these residual materials generated from horticultural produce may provide an initiative for sustainable development to reduce the environmental issues. Human health can be improved through novel foods enriched with functional components derived from these residues. The derived functional ingredients may be regarded as nutraceutical ingredients, allowing for the fabrication of products with enhanced nutritional value, potential health benefits, longer shelf-life, and acceptable sensory properties. Thus, appropriate utilization of these fruit residues could generate economic gains for the industry, help to reduce nutritional problems, provide beneficial health effects, and lessen the environmental implications.

1.2 MANAGEMENT OF FRUIT PROCESSING WASTES AND BY-PRODUCTS

Large amounts of wastes and by-products are generated from fruits during processing and thus constitute a matter of great concern (Banerjee et al., 2017). Because of high moisture content and microbial load, these fruit residues are highly biodegradable and may cause high levels of environmental problems. The anaerobic biological decomposition of organic matter is considered to be the third largest anthropogenic source of atmospheric methane emissions (Breeze, 2018). Waste derived from fruits and vegetables takes place throughout the food supply chain, although the amount of residual waste generated at each stage differs significantly from one country to another, depending on the postharvest technologies adopted in the areas where crops are cultivated. Processing techniques adopted in developing countries are not up to the mark and contribute to waste generation during fruit processing. Reduction in the amount of these residues during processing is complicated and requires substantial investment (Jiménez-Moreno et al., 2020). Meanwhile, in today's world, food security is a growing problem due to the exhaustion of natural food resources combined with increasing demand from growing populations, creating a gap between the production of food and its consumption. It is thus of the greatest importance that the food material produced and its by-products are used efficiently with minimal generation of food waste (Villacís-Chiriboga et al., 2020). To cope with the nutritional challenges of today's society, we require more sustainable nutritional sources. In this regard, fruit wastes and by-products are of paramount importance due to the presence of valuable quantities of starch, proteins, lipids, bioactive compounds, micronutrients, and dietary fibers in these processing residues.

Waste management is defined as the collection, transport, recovery, and disposal of waste, along with the supervision of such operations, and the waste management system consists of the whole set of activities related to handling, disposing, or recycling of waste materials (Plazzotta et al., 2017). To counterbalance the waste problem, managing environmental sustainability and overcoming the economic development model of "take, make, and dispose", the circular economy approach has been presented to utilize the waste and by-products, using sustainable and profitable technologies (Maina et al., 2017). Transforming food waste and by-products into a resource provides a major opportunity to shift from a linear to a closed loop economic system. In addition, it is one of the most important strategies to understand the concept of circular economy. In this modern approach, food waste management involves the valorization of wastes and by-products by recapturing their functional components and/or developing novel products with a market value (Otles et al., 2015). Nowadays, food industries have realized the efficient use of by-products and waste so as to minimize the effect of these residues on the environment (Villacís-Chiriboga et al., 2020) A holistic concept of food production was introduced by Laufenberg et al. (2003), which includes the interactions between the different goals pursued by food production: maximum product quality and safety, maximum production efficiency, and protection of the environment. This concept is meant for reduction of waste generation and recycling of valuable components in the short term, to add value to food by-products in the medium term, and to apply environmentally friendly manufacturing processes and the development of novel food products in the long term (Jiménez-Moreno et al., 2020).

1.3 PROCESSING OF FRUIT WASTES AND BY-PRODUCTS

Wastes and by-products generated during fruit processing can present very interesting chemical compositions which act as a driving force for the design of various biorefinery processes for their valorization. These fruit residues can be processed into a number of valuable components and have a potential role in the food industry. Dietary fibers, protein, pectin, starch, oils, phenolic compounds, organic acids, pigments, flavors, enzymes, etc. can be derived from these fruit processing residues (Sagar et al., 2018). Bioactive components, such as phenolic compounds, dietary fiber, proteins, etc., have health benefits and can be used in the development of functional foods and in the preparation of pharmaceuticals. Bioactivity of these functional components depends on the source, the nature of the compound, and the extraction method (Azmir et al., 2013; Pagano et al., 2021).

Fruit processing by-products and wastes obtained from grapes, mangos, bananas, citrus fruits, avocados, apples, pears, oranges, and dates are important sources of dietary fiber. Residues of these fruits are used as dietary fiber supplements (gelling and thickening agents) in refined foods (Wadhwa et al., 2015). Apple peel has a dietary fiber content of 0.91% fresh weight (FW) consisting of 0.46% FW insoluble and 0.45% FW soluble (Gorinstein et al., 2001). Apple pomace also contains substantial amounts of dietary fiber (Li et al., 2014). Grape pomace is a rich source of dietary fibers consisting of cellulose, hemicelluloses, and small amounts of pectins (Kammerer et al., 2005). Mango peel and kernel have been shown to possess appreciable amounts of dietary fibers. Total dietary fiber of mango peel is 51.2% of dry matter (DM), of which 32% DM consists of insoluble fibers while 19% DM is soluble fibers (Ajila et al., 2007, 2008). Mango peel fiber shows high hydration capacities and is considered to have great potential in the preparation of dietary fiber-rich foods (Wadhwa et al., 2015). Mango kernel has about 3.96% DM of crude fiber although less than that of peel. Lemon peel was reported to have a dietary fiber content of 14% DM consisting of about 9.04% DM insoluble and 4.93% DM soluble fiber (Sagar et al., 2018). Peel of "Liucheng" oranges was found to contain 57% DM total dietary fiber, of which the soluble portion was 9.41% DM and the insoluble portion was 47.6% DM. Cellulose and pectin were observed to be the main components of the fibers (Chau and Huang, 2003). Date pits obtained during processing of dates could be used as a potential source of dietary fiber. Date pit fiber is used for the production of high-fiber biscuits, cakes, and bread. Compared with wheat bran, it provides similar sensory properties and thus could be used as an alternative to wheat bran (Wadhwa et al., 2015)

Pectin is another biomolecule that can be obtained from fruit wastes and by-products, which play a role as a polymeric matrix for active packaging. Pectin extraction from fruit processing residues mostly involves the microwave-assisted extraction (MAE) technique. This kind of technique has the potential to extract high-quality pectin from peel of oranges, mangos, citrus fruits, and bananas, and from apple pomace. Pectin content of about 5% can be extracted from pineapple, musk melon, and papaya peel. (Basri et al., 2021). About 10–15% and 20–30% pectin are present in apple pomace and citrus peel, respectively (Baiano, 2014), whereas 10–20% pectin is found in mango peel (Rehman et al., 2004).

Considerable starch content is also present in fruit processing wastes and by-products that can be further explored. These starches have unique properties and can find potential applications in industry. Fruit processing residues obtained from mangos, kiwifruits, litchis, longans, pineapples, tamarinds, apples, jackfruits, loquats, bananas, and avocados, act as new, non-conventional, sustainable sources of starch (Kringel et al., 2020).

Wastes and by-products of several fruits have been reported to act as a potential source of protein. Appreciable protein contents are found in the peel of papayas, kiwifruits, and avocados (Chitturi et al., 2013). Protein contents of 2.5% to 9.0% has been found in citrus peels (Mamma and Christakopoulos, 2014). Fruit residues, including apple pomace, and peel of mangos, mosambis, oranges, bananas, and pineapples, were found to have protein contents of 4.45, 9.5, 5.4, 5.97, 6.02, and 8.7%, respectively (Sagar et al., 2018).

Citrus fruit-processing residues, mango and apricot kernels, and pomegranate seeds contain an appreciable amount of oil (10–50% by weight, dry basis) (Banerjee et al., 2017). Banana peel, blackcurrant, rambutan seed, passion fruit seed, and date pits are other good sources of oils rich in polyunsaturated fatty acids. Banana peel has an oil content of about 2.2–10.9%, which is rich in polyunsaturated fatty acids, primarily α-linolenic acid, and linoleic acid. Depending on the variety, rambutan (*Nephelium lappaceum*) seeds varies between 14 and 41% oil. Guava seeds contain about 5–13% oil, which is rich in essential fatty acids. Oil from passion fruit seed exhibits antioxidant activity and predominantly contains unsaturated fatty acids (87.6%), primarily oleic acid (13.8%) and linoleic acid (73.1%). Peach seed also acts as a good source of edible oil and contains palmitic acid, stearic acid, oleic acid, and linoleic acid as the main fatty acids (Wadhwa et al., 2015).

The rind, peel, and seeds of fruits contain high amounts of bioactive, phenolic compounds. Fruit peels (lemons, oranges, and grapes) and seeds (avocados, longans, jackfruits, and mangos) have a higher phenolic content than fruit pulp (Sagar et al., 2018). Citrus waste (particularly peel) is considered to be a rich source of phenolics and has a higher phenolic content than the edible part of the fruit (Balasundram et al., 2006). Peel of other fruits also has a higher phenolic content than the edible part of the fruit. The phenolic content of the peel of apples, pears, and peaches is twice that of the peeled fruits (Gorinstein et al., 2001). Banana (*Musa acuminata*) pulp has been found to contain 232 mg/100 g DM of phenolics, which is only about one-quarter that of the peel (Someya et al., 2002). Peel of pomegranate was reported to contain 249.4 mg phenolic compounds/g, while 24.4 mg/g was recorded for the fruit pulp (Li et al., 2006). Grape pomace also acts as a rich source of phenolic compounds, particularly proanthocyanidins (mono, oligo, and polymeric forms) (Torres and Bobet, 2001). Fruit seeds are considered to be nutrient-dense functional ingredients rich in phytochemicals (Pelegrini et al., 2006; Udenigwe and Aluko, 2012). Seeds contain polyphenols including phenolic acids, flavonoids, catechins, hydrolyzable tannins, xanthanoids, and other secondary metabolites (Ballesteros-Vivas et al., 2019; Torres-León et al., 2016). Phytochemicals are more highly concentrated in the seed coat part of fruit seeds (Villacís-Chiriboga et al., 2020). Bioactive phenolic compounds, including phenolic acids, flavanols, flavonols, anthocyanins, and dihydrochalcones, are also present in fruit pomace. Recovery of phenolic compounds from fruit residues depends on the extraction technique used. Several thermal (heating, radiofrequency, microwave, and infrared heating) and nonthermal methods (ultrasound, high hydrostatic pressure, irradiation, pulsed electric field, and pulsed light) are used for phenolic extraction (Sagar et al., 2018).

Wastes and by-products from fruits can serve as a source of flavors, aromas, organic acids, enzymes, pigments, and alcohols. Vanillin can be synthesized from pineapple peel waste, which contains a precursor for vanillic acid called ferulic acid. A three-step microbial biotransformation is used for the synthesis of vanillin from pineapple waste. Citrus fruit waste is used for commercial production of rhamnose which is the basic raw material for the synthesis of strawberry-flavored furaneol (2,5-dimethyl-4-hydroxy-3(2H)-furanone). Production of pineapple flavoring from apple pomace is carried out with the help of the fungus, *Ceratocystis fimbriata*. A number of volatile compounds like esters, alcohols, aldehydes, ketones, and acids have been extracted from pineapple residues left behind after extraction of juice from the fruit (Sagar et al., 2018). Fruit wastes and by-products are rich sources of natural pigments like anthocyanins, betalains, chlorophylls, and carotenoids. Extraction of pigments from these fruit processing residues can meet the demand of natural pigment production at an industrial scale which find a potential application in the food, cosmeceutical, or pharmaceutical industries. Various novel methods like pulsed electric field, high-pressure processing, pulsed light, and ionizing radiation could be used for the extraction of these natural colorants from fruit wastes and by-products (Jimenez-Moreno et al., 2020). Organic acids derived from fruit residues are regarded as important biomolecules with potential application in the food, chemical, or cosmetic industries. Citric acid can be produced by the fungus *Aspergillus niger* using apple pomace as a substrate material for the microbe (Dhillon et al., 2011). Wastes from other fruits such as pineapple, mandarins, and mixed-fruit wastes can also be used for the production of citric acid (Prabha and Rangaiah, 2014). Large amounts of citric acid can be produced from pineapple waste using *Yarrowia lipolytica* and *Aspergillus niger* as the fermenting microbes (Imandi et al., 2008). Lactic acid, an industrially important organic acid, which is used as an acidulant and preservative in food products, can be produced by several microbes using fruit by-products as the substrate (Rodriguez Couto, 2008). Microorganisms like *Lactobacillus casei*, *Lactobacillus plantarum*, and *Lactobacillus delbrueckii* have been used for production of lactic acid using mango and orange residue as substrates (Panda and Ray, 2015).

Various studies have also been carried out with regard to the production of bioethanol from fruit wastes and byproducts by yeast (*Saccharomyces cerevisiae*) fermentation. Fruit residues with high pectin, hemicellulose, and cellulose content can be used as suitable substrates for bioethanol production. Amongst the various fruit processing residues, apple pomace, banana peel, beet pomace, kinnow (a high-yield mandarin hybrid) and peach wastes have shown promising results with respect to bioethanol production. Several studies related to bioethanol production from fruit peel (sweet lime, pineapple, and orange), and avocado seed have also been carried out. *S. cerevisiae* has been used in the production of bioethanol from date waste (Wadhwa et al., 2015).

Many fruit residues are used as substrates for the production of enzymes. Wastes from bananas, dates, citrus fruits, loquats, and mangos are used for enzyme production. Orange waste can be used for the production of α-amylase under the action of *A. niger* (Djekrif-Dakhmouche et al., 2006). Enzyme filter-paperase (FPase) can be generated by *Trichoderma reesei* from kinnow waste (Oberoi et al., 2010). Banana solid waste and a mixed bacterial culture (*Bacillus megaterium*, *Cellulomonas carte*, *Pseudomonas putida*, and *Pseudomonas fluorescens*) was used to produce cellulase and β-D-glucosidase (Dabhi et al., 2014). Solid-state fermentation (SSF) of grape pomace can be used for the production of pectinase from the fungus *Aspergillus awamori* (Botella et al., 2005), while pineapple peel can be used as a substrate for the synthesis of pectinase by *Penicillium chrysogenum*. Peel of orange, lemon, and banana can be used for production of pectinase using *A. niger* (Mrudula and Anitharaj, 2011). Under the action of the fungus *Aspergillus foetidus*, wastes obtained during the processing of banana, grapes, cashew apple, and pineapple can be used for the production of pectinase (Venkatesh et al., 2009).

1.4 NUTRACEUTICAL POTENTIAL

Fruit wastes and by-products possess health benefits due to the presence of bioactive components. The processing industries of fruit with nutraceutical potential have started to exploit the bioactive ingredients in wastes and by-products in the preparation of innovative functional foods. The peel, pomace and seed are rich in polyphenols, dietary fiber, proteins, and secondary metabolites which showed various bioactive properties, such as antitumor, antioxidant, antimicrobial, anti-inflammatory, antiallergenic, and antithrombotic, as well as in the treatment of cardiovascular diseases (Coman et al., 2020). The applications of advanced technology for the isolation and purification of bioactive compounds from fruit wastes and by-products have focused on their utilization in nutraceutical foods to reduce the risk of chronic diseases (Kumar, 2020). Bioactive components isolated from fruit processing residues act as strong antioxidants and provide a defense system for the body against harmful reactive oxygen species (ROS) and can thus mitigate various diseases associated with elevated ROS levels. Citrus-based peel, pomace, and seeds are rich sources of antioxidant pigments with high antioxidant action (Rafiq et al., 2018). The fruit peel, pomace, and seeds from mangos, grapes, avocados, apples, melons, citrus, pineapples, plums, apricots, pears, and pomegranates are rich sources of polyphenols and exhibit high antioxidant potential (Suleria et al., 2020; Torres-León et al., 2016; Sójka et al., 2015). The bioactives present in fruit pomace compounds have strong ROS scavenging activity and protein glycation inhibitory properties, as well as anti-tumor activities (Fernández-Fernández et al., 2021; Coman et al., 2020; Shao et al., 2015). Several *in-vitro* and *in-vivo* studies have revealed that bioactive compounds present in fruit wastes and by-products exhibit activity against cancers such as breast cancer, prostate cancer, colorectal cancer, and leukemia (Panth et al., 2017). Bioactive compounds like naringin, hesperidin, bromalin, limonin, and anthocyanins present in fruit processing residues have anticancer properties (Gualdani et al., 2016). Klavins et al. (2018), Majerska et al. (2019), Villacís-Chiriboga et al. (2020), Sojka et al. (2015), Mahato et al. (2018) have reported the anticancer properties of berry pomace, banana peel, apple pomace, avocado seeds, plum pomace, and citrus pomace, respectively. The peel, pomace, and seed from different fruits (citrus, mangos, bananas, watermelons, berries, apples, and avocados) have been used as functional ingredients with nutraceutical potential against cardiovascular diseases, inflammation, atherosclerosis, allergy, and obesity (Baeeri et al., 2018; Ramdhonee and Jeetah, 2017; Chavan et al., 2018; Uchenna et al., 2017; Wen et al., 2019). Fruit peel contains bioactive compounds with potent antioxidant, antitumor, antiviral, and immunomodulatory activities (Lau et al., 2021). A protein, which was one of the ingredients present in fruit seeds, was reported to possess various bioactive properties including antihypertensive, anti-inflammatory, and antifungal properties (Udenigwe and Aluko, 2012). The oil from fruit seeds is rich in health-beneficial components, such as omega fatty acids and phytochemicals that exhibit health benefits like anticancer, anti-inflammatory, and antiallergenic activities (Banerjee et al., 2017).

1.5 EXTRACTION OF BIOACTIVE COMPONENTS

Extraction of bioactive materials from fruit processing wastes and by-products can be an effective approach for the valorization of these fruit processing residues. The most common extraction techniques for biologically active compounds are solid-liquid extraction *via* utilization of Soxhlet apparatus, maceration and hydro-distillation using different organic solvents as well as combinations of them with heating and mixing procedures (Azmir et al., 2013; Wen et al., 2020), although recently many green methods such as ultrasound-assisted extraction, enzyme-assisted extraction, microwave-assisted extraction, supercritical CO_2, subcritical water, gas-expanded liquids, use of deep eutectic solvents or ionic liquids, and others has been used for the isolation of bioactive compounds (Amran et al., 2021; Wen et al., 2020; Zuin and Ramin, 2018)

1.5.1 Conventional Extraction Method

Conventional methods of extraction have been used for a long time and are still considered for extraction of bioactive compounds from fruit wastes and by-products. Conventional methods are easy-to-operate and readily available techniques for the extraction and isolation of bioactive compounds. The details of the technique used depend upon the nature of the fruit matrix, solvent type, purity and concentration (Rifna et al., 2021). The conventional techniques used for extraction of bioactive compounds from fruit waste and by-products (peel, pomace, and seeds) are solid-liquid extraction, maceration, and hydro-distillation (Azmir et al., 2013; Wen et al., 2020). The use of particular conventional methods for extraction of bioactive compounds depends upon the fruit matrix properties and the target bioactive compound (Zuin and Ramin, 2018).

Solid-liquid or solvent extraction is a common conventional method used in the form of Soxhlet apparatus, maceration, decoction, infusion, and percolation (Rifna et al., 2021). The solvents used for solid-liquid extraction have different polarities (hexane, benzene, cyclohexane, acetone, and chloroform) for rupturing the fruit matrix and releasing targeted bioactive molecules (Adeoti and Hawboldt, 2014). The Soxhlet method of extraction is simple and has shown good reproducibility (Sagar et al., 2018). The solvent extraction method is generally applied for extraction of bioactive compounds which are lipophilic in nature (total phenolics, anthocyanins, antioxidants, and essential fatty acids) (Corona et al., 2015; Machado et al., 2015; Pereira et al., 2019b; Santos et al., 2021; Lucarini et al., 2021). Maceration has been extensively used for the extraction of thermolabile bioactive compounds from fruit wastes and by-products which are similar to those isolated by Soxhlet extraction (Rifna et al., 2021; Sagar et al., 2018). Maceration is an economical, widely used technique which does not require sophisticated equipment and is an efficient technique due to its continuous extraction nature (Rifna et al., 2021). The maceration process involves reduction of sample size, extraction with a solvent to achieve greater contact with the fruit matrix for extraction of bioactive compounds, and then fractionation to obtain the desired bioactive from solvent obtained (Azmir et al., 2013; Sagar et al., 2018). The maceration process can be associated with novel techniques (ultrasound, microwave, ohmic heating) to increase the yield and bioactivity of the extracted bioactive compounds (Safdar et al., 2017). Hydro-distillation is another conventional technique similar to solvent extraction for isolation of bioactive compounds without the use of organic solvents (Azmir et al., 2013; Sagar et al., 2018). Hydro-distillation methods used for extraction of bioactive compounds from fruit wastes and by-products uses different methods such as water and steam distillation, water distillation, and direct steam distillation (Azmir et al., 2013; Sagar et al.,

2018). Hydro-distillation methods use hot water and steam as carrier for extraction of active agents with bioactive properties. Hydro-distillation is a simple and easy technique but it is not suitable for extraction of heat-labile, bioactive compounds which are degraded or oxidized (Sagar et al., 2018).

1.5.2 Green Extraction Methods

The bioactive compounds extracted by conventional methods are characterized by low yields, environmental problems, limitations in the use of different food matrixes, and low activities of the extracted bioactive compounds. Although conventional extraction techniques are well established and frequently used as standard methods of extraction, these limitations have encouraged industries to use novel methods of extraction. The limitations associated with conventional extraction methods are the costly use of organic reagents, as in the case of solvent extractions; prolonged processing time; low selectivity of targeted compounds; potential degradation or oxidation of biological substances, and the need for purification (Figure 1.1). These limitations have stimulated the use of emerging, innovative techniques for extraction of bioactive compounds, such as microwave-assisted extraction (MAE), ultrasound-assisted extraction (UAE), deep eutectic solvent extraction (DES), supercritical fluid extraction (SFE), pulsed electric field extraction (PEF), and others.

Green extraction techniques use less hazardous chemicals by replacing them with safer, alternative solvents, with efficient energy consumption and reduced analysis time, using renewable feedstock, with low toxic residue production and being environment friendly (Azmir et al., 2013). In addition, replacing toxic or non-renewable organic solvents is a crucial approach when using greener extraction alternatives (Zuin and Ramin, 2018). Green extraction methods exhibit good reproducibility, high efficiency and high selectivity robustness, with efficient processes and the use of safer products without affecting the health (Zuin and Ramin, 2018).

1.5.2.1 Microwave-Assisted Extraction (MAE)

MAE is considered to be an advanced technique for the extraction of bioactive compounds using microwave energy, resulting in enhanced bioactivities and yields from fruit matrixes (Azmir et al., 2013; Sagar et al., 2018; Vinatoru et al., 2017). Microwaves (MW) are radiation on the

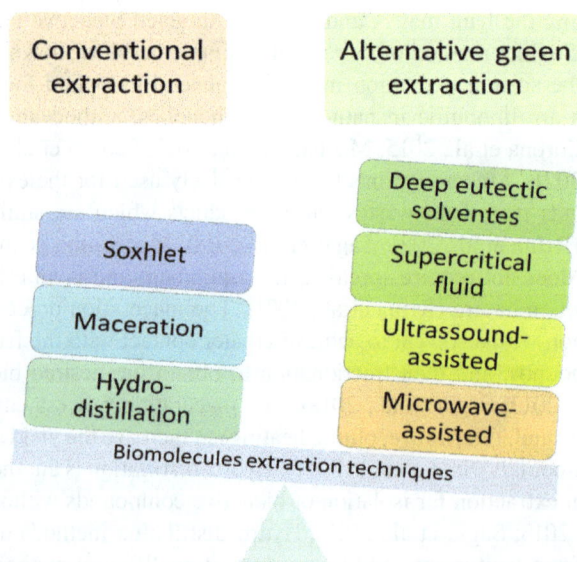

Figure 1.1   Graphical balance between conventional and green extraction methods for bioactive compounds.

electromagnetic spectrum that vary from 300 MHz (radio radiation) to 300 GHz (infrared radiation) in frequency (Vinatoru et al., 2017). MW extraction occurs due to the transfer of electromagnetic energy to heat, affecting molecules by dipole polarization and ionic conduction (Flórez et al., 2015). The dipole rotation occurs due to the alignment of dipolar molecules on the alternating electric field, making them oscillate and collide with each other and the surrounding molecules, releasing thermal energy (heat) (Flórez et al., 2015; Vinatoru et al., 2017). The MW used for extraction of bioactive compounds from fruit wastes and byproducts uses three types, based on solvent usage, namely microwave-assisted solvent extraction (MASE), solvent-free microwave extraction (SFME), and microwave extraction combining hydro-diffusion and gravity (MHG) (Rombaut et al., 2014). The extraction of bioactive compounds from fruit wastes and by-products occurs by mass transfer with increased yield, uniformity, and activities (Flórez et al., 2015). The MAE has potential application for extraction from fruit wastes and by-products of various bioactive compounds such as phenolic compounds, essential oils, aromas, pigments, antioxidants, and organic compounds (Rombaut et al., 2014; Vinatoru et al., 2017; Flórez et al., 2015; Rombaut et al., 2014; Zhang et al., 2011).

1.5.2.2 Ultrasound-Assisted Extraction (UAE)

UAE is one of the innovative and non-thermal methods for extraction of bioactive compounds with high yield, short processing time, high activity, and low solvent volumes as compared with conventional extraction methods (Alexandre et al., 2018). The UAE used for extraction of bioactive compounds from fruit matrixes is based upon hydrodynamic cavitations without affecting the extracted bioactive compounds (Vinatoru et al., 2017). The most used equipment for extraction of bioactive compounds are ultrasonic baths and ultrasound probes, depending on the nature of the fruit wastes and by-products (Rombaut et al., 2014). UAE has been extensively used in treatment of fruit wastes and by-products for extraction of various bioactive compounds, such as phenolic compounds, polysaccharides, volatile compounds, and enzymes with increased yield and reduced extraction time (Alexandre et al., 2018; Pan et al., 2012; Tabaraki et al., 2012; Rombaut et al., 2014; Chemat et al., 2017; Alexandre et al., 2018).

1.5.2.3 Supercritical Fluid Extraction (SFE)

SFE is a novel and modern method of extraction and used for the isolation of various bioactive compounds from diverse food matrixes. The SFE uses critical solvents, such as carbon dioxide, water, etc., in the supercritical state with gas and liquid properties, such as diffusion, viscosity, surface tension, and solvation power, with increased extraction rate and high mass transfer and bioactivities of extracted compounds (Azmir et al., 2013; Zuin and Ramin, 2018). SFE have been widely used as an environmentally friendly technique for the extraction of diverse bioactive compounds, such as plant lipids, free fatty acids, essential oils, phenolics (such as resveratrol), flavonoids, carotenoids, chlorophylls, and alkaloids from fruit wastes and by-products (Alexandre et al., 2018; Costa et al., 2016; Rai et al., 2016; Durante et al., 2014). Table 1.1 shows various extraction methods and the bioactive compounds which can be extracted from fruit wastes and byproducts.

1.5.2.4 Pressurized Liquid Extraction

Pressurized Liquid Extraction (PLE) involves the use of liquid solvents in their subcritical state under controlled conditions of temperature and pressure. PLE is an efficient technique for the extraction of bioactive phenolic compounds from plant materials in less time using solvents (organic or water) (Pereira et al., 2019a). Pressurized liquids have the ability to recover phenolic compounds faster than conventional extraction methods. The nature of the extract obtained after PLE is significantly affected by both the temperature and solvent system used. High temperatures recover more phenolics (Machado et al., 2015). Due to the involvement of high pressure, it is possible to reach higher

Table 1.1 Bioactive Components and Their Methods of Extraction

Extraction Methods	Fruit Wastes and By-Products	Target Compounds	Reference
Hydro-distillation	Pineapple fruit (*Ananas comosus*) processing residues	Volatile compounds	Barretto et al. (2013)
Subcritical water extraction	Papaya (*Carica papaya*) seeds agroindustrial residue	Phenolic substances	Rodrigues et al. (2019)
	Defatted orange peel	Flavanones	Lachos-Perez et al. (2018)
	Pomegranate (*Punica granatum*) seed residues	Phenolic substances	He et al. (2012)
Microwave-assisted extraction	Watermelon (*Citrullus lanatus*) fruit rinds wastes	Pectin	Prakash Maran et al. (2014)
	Sweet orange (*Citrus sinensis*) peel waste	Essential oil	Boukroufa et al. (2015)
Ultrasound extraction	Passion fruit (*Passiflora* ssp.) peel	Pectin	Oliveira et al. (2016)
	Jackfruit (*Artocarpus heterophyllus*) fruit peel waste	Pectin	Moorthy et al. (2017)
CO_2 supercritical fluid extraction	Sea almond (*Terminalia catappa*) fruits	Vegetable oil	Santos et al. (2021)
	Papaya (*Carica papaya* L.) agroindustrial waste	Phenolic substances	Castro-Vargas et al. (2019)
Pressurized liquid extraction	Blackberry (*Rubus fruticosus*) residues	Phenolic substances	Machado et al. (2015)
Deep eutectic solvent extraction	Bambangan (*Mangifera pajang*) fruit waste	Phenolic substances	Ling et al. (2020)
	Olive (*Olea europaea*) wastes from oil processing	Phenolic compounds	Bonacci et al. (2020)
	Citrus peel waste	Flavonoids	Xu et al. (2019)

temperatures than the normal boiling point of the solvent used. Due to this, processing times and volumes of solvent required are less as compared to conventional extraction techniques. Due to increase in temperature, the dielectric constant of the liquid solvent decreases thereby lowering its polarity up to the required values so as to extract the compounds of interest (Joana Gil-Chavez et al., 2013).

1.5.2.5 *High Intensity Pulsed Electric Field Extraction*

High Intensity Pulsed Electric Field (HIPEF) extraction is an emerging non-thermal extraction technology that has shown promising results in extracting bioactive phenolic compounds from agro food wastes. This extraction technique is considered a high efficiency technology and involves low temperatures for the extraction of natural bioactive ingredients. HIPEF extraction involves the application of high voltage pulses between 20–80 kV/cm on the material between two electrodes. HIPEF induces a synthetic effect which includes strong physical and chemical reactions. This increases cell membrane permeability causing inactivation of cells and increasing the release of intracellular compounds through cell membrane thus achieving higher efficiency and low extraction time (Yan et al., 2017). This technique has been effective in the extraction of various compounds like anthocyanins from blueberry by-products, b-carotene from carrot pomace, and polyphenols from fruit peels (for example, grape, orange, and plum peel) (Jimenez-Moreno et al., 2019).

1.6 FOOD APPLICATIONS OF FRUIT WASTES AND BY-PRODUCTS

Due to their amazing diversity of chemical composition, renewability, non-toxicity, versatility, biocompatibility, and biodegradability, fruit wastes and by-products have a wide range of food applications. The bioactive compounds extracted from fruit wastes and by-products have potential applications as functional ingredients due to their techno-functional and nutraceutical properties. Polysaccharides, especially cellulose, lignin, pectin, and hemicelluloses, present in fruit wastes and by-products are among the major natural polymers which act as dietary fibers (Rivas et al., 2021). In the food industries, dietary fiber from these fruit processing wastes is used for production of functional bread, pasta, biscuits, and extruded cereals (Duţa et al., 2018). Cellulose from fruit wastes and by-products is widely used in food products (Harini et al., 2018). Cellulose from fruit wastes and by-products has a versatile application in the food industry as a fat substitute, to design biodegradable composite packaging, and to encapsulate and deliver bioactive substances (Baghaei and Skrifvars, 2020; Rivas et al., 2021). Oils rich in bioactive compounds extracted from the processing residues of different fruit (apples, citrons, grapes, guavas, kumquats, mangos, melons, oranges, papaya, passion fruits, and strawberries) have wide applications in food products. These bioactive oils display good fatty acid profiles, indicating their potential as a source of essential fatty acids (da Silva and Jorge, 2017). Bioactive polyphenols derived from fruit residues can be used as natural preservative agents in food products, replacing synthetic additives. Fruits pomaces are used in the development of functional bakery products (biscuits and cereal bars) with good nutritional profiles (Ferreira et al., 2015). Food packaging obtained from fruit processing residues is of great interest due to its unique qualities. Fruit processing residues are used for preparation of edible films as potential food packaging for shelf-life extension of food products (Riaz et al., 2018). Packaging materials with added polyphenolic extracts from fruit wastes and byproducts have been prepared for development of antimicrobial packing materials (Goulas et al., 2019). When these wastes and by-products are used in food packaging systems, they offer several advantages, including increased antioxidant activity, improved mechanical properties and antimicrobial activity, and improved quality of protected food products (Bayram et al., 2021). Extracts obtained from fruit by-products and wastes are used in high-fat dairy foods like butter and cheese, in order to prevent lipid oxidation and enhance the microbial safety of these food products. Fruit beverages can also be enriched with these extracts to increase their phenolic content, with a subsequent increase in antioxidant potential. Utilization of extracts derived from fruit processing residues helps to delay lipid and/or protein oxidation of meat and meat products (Trigo et al., 2020). Fruit wastes and by-products (apple pomace, papaya peel, and watermelon rinds) are used in the development of low-glycemic index foods. Fruit pomace (particularly apple pomace) is utilized in the production of gluten-free food formulations (Kowalska et al., 2017).

REFERENCES

Adeoti, I. A., & Hawboldt, K. (2014). A review of lipid extraction from fish processing by-product for use as a biofuel. *Biomass and Bioenergy*, *63*, 330–340.

Ajila, C. M., Aalami, M., Leelavathi, K., & Rao, U. P. (2010). Mango peel powder: A potential source of antioxidant and dietary fiber in macaroni preparations. *Innovative Food Science & Emerging Technologies*, *11*(1), 219–224.

Ajila, C. M., Leelavathi, K. U. J. S., & Rao, U. P. (2008). Improvement of dietary fiber content and antioxidant properties in soft dough biscuits with the incorporation of mango peel powder. *Journal of Cereal Science*, *48*(2), 319–326.

Ajila, C. M., Naidu, K. A., Bhat, S. G., & Rao, U. P. (2007). Bioactive compounds and antioxidant potential of mango peel extract. *Food Chemistry*, *105*(3), 982–988.

Alexandre, E. M. C., Moreira, S. A., Castro, L. M. G., Pintado, M., & Saraiva, J. A. (2018). Emerging technologies to extract high added value compounds from fruit residues: Sub/supercritical, ultrasound-, and enzyme-assisted extractions. *Food Reviews International*, *34*(6), 581–612.

Amran, M. A., Palaniveloo, K., Fauzi, R., Satar, N. M., Mohidin, T. B. M., Mohan, G., Razak, S. A., Arunasalam, M., Nagappan, T., & Seelan, J. S. S. (2021). Value-added metabolites from agricultural waste and application of green extraction techniques. *Sustainability (Switzerland)*, *13*(20), 1–28.

Ayala-Zavala, J. F., Rosas-Domínguez, C., Vega-Vega, V., & González-Aguilar, G. A. (2010). Antioxidant enrichment and antimicrobial protection of fresh-cut fruits using their own byproducts: Looking for integral exploitation. *Journal of Food Science, 75*(8), R175–R181.

Azmir, J., Zaidul, I. S. M., Rahman, M. M., Sharif, K. M., Mohamed, A., Sahena, F., Jahurul, M. H. A., Ghafoor, K., Norulaini, N. A. N., & Omar, A. K. M. (2013). Techniques for extraction of bioactive compounds from plant materials: A review. *Journal of Food Engineering, 117*(4), 426–436.

Baeeri, M., Sarkhail, P., Hashemi, G., Marefatoddin, R., & Shahabi, Z. (2018). Data showing the optimal conditions of pre-extraction and extraction of *Citrullus lanatus* (watermelon) white rind to increase the amount of bioactive compounds, DPPH radical scavenging and anti-tyrosinase activity. *Data in Brief, 20*, 1683–1685.

Baghaei, B., & Skrifvars, M. (2020). All-cellulose composites: A review of recent studies on structure, properties and applications. *Molecules, 25*(12), 2836.

Baiano, A. (2014). Recovery of biomolecules from food wastes—A review. *Molecules, 19*(9), 14821–14842.

Balasundram, N., Sundram, K., & Samman, S. (2006). Phenolic compounds in plants and agri-industrial by-products: Antioxidant activity, occurrence, and potential uses. *Food Chemistry, 99*(1), 191–203.

Ballesteros-Vivas, D., Álvarez-Rivera, G., Morantes, S. J., Sánchez-Camargo, A. d. P., Ibáñez, E., Parada-Alfonso, F., & Cifuentes, A. (2019). An integrated approach for the valorization of mango seed kernel: Efficient extraction solvent selection, phytochemical profiling and antiproliferative activity assessment. *Food Research International, 126*, 108616.

Banerjee, J., Singh, R., Vijayaraghavan, R., MacFarlane, D., Patti, A. F., & Arora, A. (2017). Bioactives from fruit processing wastes: Green approaches to valuable chemicals. *Food Chemistry, 225*, 10–22.

Barretto, L. C. de O., Moreira, J. de J. da S., Santos, J. A. B. dos, Narendra, N., & Santos, R. A. R. dos. (2013). Characterization and extraction of volatile compounds from pineapple (*Ananas comosus* L. Merril) processing residues. *Food Science and Technology (Campinas), 33*(4), 638–645.

Bayram, B., Ozkan, G., Kostka, T., Capanoglu, E., & Esatbeyoglu, T. (2021). Valorization and application of fruit and vegetable wastes and by-products for food packaging materials. *Molecules, 26*(13), 403.

Bonacci, S., Di Gioia, M. L., Costanzo, P., Maiuolo, L., Tallarico, S., & Nardi, M. (2020). Natural deep eutectic solvent as extraction media for the main phenolic compounds from olive oil processing wastes. *Antioxidants, 9*(6), 513.

Botella, C., De Ory, I., Webb, C., Cantero, D., & Blandino, A. (2005). Hydrolytic enzyme production by *Aspergillus awamori* on grape pomace. *Biochemical Engineering Journal, 26*(2–3), 100–106.

Boukroufa, M., Boutekedjiret, C., Petigny, L., Rakotomanomana, N., & Chemat, F. (2015). Bio-refinery of orange peels waste: A new concept based on integrated green and solvent free extraction processes using ultrasound and microwave techniques to obtain essential oil, polyphenols and pectin. *Ultrasonics Sonochemistry, 24*, 72–79.

Breeze, P., & Breeze, P. (2018). Landfill waste disposal, anaerobic digestion, and energy production. In P. Breeze (Ed.), *Energy from waste* (pp. 39–47). Academic Press.

Castro-Vargas, H. I., Baumann, W., Ferreira, S. R. S., & Parada-Alfonso, F. (2019). Valorization of papaya (*Carica papaya* L.) agroindustrial waste through the recovery of phenolic antioxidants by supercritical fluid extraction. *Journal of Food Science and Technology, 56*(6), 3055–3066.

Chau, C. F., & Huang, Y. L. (2003). Comparison of the chemical composition and physicochemical properties of different fibers prepared from the peel of *Citrus sinensis* L. Cv. Liucheng. *Journal of Agricultural and Food Chemistry, 51*(9), 2615–2618.

Chavan, P., Singh, A. K., & Kaur, G. (2018). Recent progress in the utilization of industrial waste and by-products of citrus fruits: A review. *Journal of Food Process Engineering, 41*(8), e12895.

Chemat, F., Rombaut, N., Sicaire, A.-G., Meullemiestre, A., Fabiano-Tixier, A.-S., & Abert-Vian, M. (2017). Ultrasound assisted extraction of food and natural products. Mechanisms, techniques, combinations, protocols and applications. A review. *Ultrasonics Sonochemistry, 34*, 540–560.

Chitturi, S., Gopichand, V., & Vuppu, S. (2013). Studies on protein content, protease activity, antioxidants potential, melanin composition, glucosinolate and pectin constitution with brief statistical analysis in some medicinally significant fruit peels. *Der Pharmacia Lettre, 5*(1), 13–23.

Coman, V., Teleky, B.-E., Mitrea, L., Martău, G. A., Szabo, K., Călinoiu, L.-F., & Vodnar, D. C. (2020). Bioactive potential of fruit and vegetable wastes. In F. Toldrá (Ed.), *Advances in food and nutrition research* (Vol. 91, pp. 157–225). Academic Press.

Corona, M. A. G., Beatriz Gómez-Patiño, M., de Jesús Perea Flores, M., Alberto Moreno Ruiz, L., Margarita Berdeja Martinez, B., & Arrieta-Baez, D. (2015). An integrated analysis of the *Musa paradisiaca* peel, using UHPLC-ESI, FTIR and confocal microscopy techniques. *Annals of Chromatography and Separation Techniques*, *1*(1), 1–5.

Costa, B. E. T., Santos, O. V. Dos, Corrêa, N. C. F., & França, L. F. De. (2016). Comparative study on the quality of oil extracted from two tucumã varieties using supercritical carbon dioxide. *Food Science and Technology*, *36*(2), 322–328.

da Silva, A. C., & Jorge, N. (2017). Bioactive compounds of oils extracted from fruits seeds obtained from agroindustrial waste. *European Journal of Lipid Science and Technology*, *119*(4), 1600024.

Dabhi, B. K., Vyas, R. V., & Shelat, H. N. (2014). Use of banana waste for the production of cellulolytic enzymes under solid substrate fermentation using bacterial consortium. *International Journal of Current Microbiology and Applied Sciences*, *3*(1), 337–346.

Dhillon, G. S., Brar, S. K., Verma, M., & Tyagi, R. D. (2011). Enhanced solid-state citric acid bio-production using apple pomace waste through surface response methodology. *Journal of Applied Microbiology*, *110*(4), 1045–1055.

Đilas, S., Čanadanović-Brunet, J., & Ćetković, G. (2009). By-products of fruits processing as a source of phytochemicals. Chemical Industry and Chemical Engineering Quarterly/CICEQ, *15*(4), 191–202.

Djekrif-Dakhmouche, S., Gheribi-Aoulmi, Z., Meraihi, Z., & Bennamoun, L. (2006). Application of a statistical design to the optimization of culture medium for α-amylase production by Aspergillus niger ATCC 16404 grown on orange waste powder. *Journal of Food Engineering*, *73*(2), 190–197.

Durante, M., Lenucci, M. S., D'Amico, L., Piro, G., & Mita, G. (2014). Effect of drying and co-matrix addition on the yield and quality of supercritical CO_2 extracted pumpkin (*Cucurbita moschata* Duch.) oil. *Food Chemistry*, *148*, 314–320.

Duță, D. E., Culețu, A., & Mohan, G. (2018). Reutilization of cereal processing by-products in bread making. In C. M. Galanakis (Ed.), *Sustainable recovery and reutilization of cereal processing by-products* (pp. 279–317). Woodhead Publishing.

Faustino, M., Veiga, M., Sousa, P., Costa, E. M., Silva, S., & Pintado, M. (2019). Agro-food byproducts as a new source of natural food additives. *Molecules*, *24*(6), 1056.

Fernández-Fernández, A. M., Dellacassa, E., Nardin, T., Larcher, R., Gámbaro, A., Medrano-Fernandez, A., & del Castillo, M. D. (2021). In vitro bioaccessibility of bioactive compounds from citrus pomaces and orange pomace biscuits. *Molecules*, *26*(12), 3480.

Ferreira, M. S. L., Santos, M. C. P., Moro, T. M. A., Basto, G. J., Andrade, R. M. S., & Gonçalves, É. C. B. A. (2015). Formulation and characterization of functional foods based on fruit and vegetable residue flour. *Journal of Food Science and Technology*, *52*(2), 822–830.

Flórez, N., Conde, E., & Domínguez, H. (2015). Microwave assisted water extraction of plant compounds. *Journal of Chemical Technology & Biotechnology*, *90*(4), 590–607.

Girotto, F., Alibardi, L., & Cossu, R. (2015). Food waste generation and industrial uses: A review. *Waste Management*, *45*, 32–41.

Gorinstein, S., Zachwieja, Z., Folta, M., Barton, H., Piotrowicz, J., Zemser, M., … & Màrtín-Belloso, O. (2001). Comparative contents of dietary fiber, total phenolics, and minerals in persimmons and apples. *Journal of Agricultural and Food Chemistry*, *49*(2), 952–957.

Goulas, V., Hadjivasileiou, L., Primikyri, A., Michael, C., Botsaris, G., Tzakos, A. G., & Gerothanassis, I. P. (2019). Valorization of carob fruit residues for the preparation of novel bi-functional polyphenolic coating for food packaging applications. *Molecules*, *24*(17), 3162.

Gualdani, R., Cavalluzzi, M. M., Lentini, G., & Habtemariam, S. (2016). The chemistry and pharmacology of citrus limonoids. *Molecules*, *21*(11), 1530.

Harini, K., Ramya, K., & Sukumar, M. (2018). Extraction of nano cellulose fibers from the banana peel and bract for production of acetyl and lauroyl cellulose. *Carbohydrate Polymers*, *201*(June), 329–339.

He, L., Zhang, X., Xu, H., Xu, C., Yuan, F., Knez, Ž., Novak, Z., & Gao, Y. (2012). Subcritical water extraction of phenolic compounds from pomegranate (*Punica granatum* L.) seed residues and investigation into their antioxidant activities with HPLC–ABTS+ assay. *Food and Bioproducts Processing*, *90*(2), 215–223.

Imandi, S. B., Bandaru, V. V. R., Somalanka, S. R., Bandaru, S. R., & Garapati, H. R. (2008). Application of statistical experimental designs for the optimization of medium constituents for the production of citric acid from pineapple waste. *Bioresource Technology*, *99*(10), 4445–4450.

Jiménez-Moreno, N., Esparza, I., Bimbela, F., Gandía, L. M., & Ancín-Azpilicueta, C. (2020). Valorization of selected fruit and vegetable wastes as bioactive compounds: Opportunities and challenges. *Critical Reviews in Environmental Science and Technology*, *50*(20), 2061–2108.

Joana Gil-Chavez, G., Villa, J. A., Fernando Ayala-Zavala, J., Basilio Heredia, J., Sepulveda, D., Yahia, E. M., & González-Aguilar, G. A. (2013). Technologies for extraction and production of bioactive compounds to be used as nutraceuticals and food ingredients: An overview. *Comprehensive Reviews in Food Science and Food Safety*, *12*(1), 5–23.

Joshi, V. K., Kumar, A., & Kumar, V. (2012). Antimicrobial, antioxidant and phyto-chemicals from fruit and vegetable wastes: A review. *International Journal of Food and Fermentation Technology*, *2*(2), 123–136.

Kammerer, D., Claus, A., Schieber, A., & Carle, R. (2005). A novel process for the recovery of polyphenols from grape (*Vitis vinifera* L.) pomace. *Journal of Food Science*, *70*(2), C157–C163.

Klavins, L., Kviesis, J., Nakurte, I., & Klavins, M. (2018). Berry press residues as a valuable source of polyphenolics: Extraction optimisation and analysis. *LWT-Food Science & Technology*, *93*, 583–591.

Kowalska, H., Czajkowska, K., Cichowska, J., & Lenart, A. (2017). What's new in biopotential of fruit and vegetable by-products applied in the food processing industry. *Trends in Food Science & Technology*, *67*, 150–159.

Kringel, D. H., Dias, A. R. G., Zavareze, E. D. R., & Gandra, E. A. (2020). Fruit wastes as promising sources of starch: Extraction, properties, and applications. *Starch-Stärke*, *72*(3–4), 1900200.

Kumar, K. (2020). Nutraceutical potential and utilization aspects of food industry by-products and wastes. In M. R. Kosseva & C. Webb (Eds.), *Food industry wastes* (2nd ed.) (pp. 89–111). Academic Press.

Lachos-Perez, D., Baseggio, A. M., Mayanga-Torres, P. C., Maróstica, M. R., Rostagno, M. A., Martínez, J., & Forster-Carneiro, T. (2018). Subcritical water extraction of flavanones from defatted orange peel. *The Journal of Supercritical Fluids*, *138*(March), 7–16.

Lau, K. Q., Sabran, M. R., & Shafie, S. R. (2021). Utilization of vegetable and fruit by-products as functional ingredient and food. *Frontiers in Nutrition*, *8*, 261.

Laufenberg, G., Kunz, B., & Nystroem, M. (2003). Transformation of vegetable waste into value added products:(A) the upgrading concept;(B) practical implementations. *Bioresource Technology*, *87*(2), 167–198.

Li, X., He, X., Lv, Y., & He, Q. (2014). Extraction and functional properties of water-soluble dietary fiber from apple pomace. *Journal of Food Process Engineering*, *37*(3), 293–298.

Li, Y., Guo, C., Yang, J., Wei, J., Xu, J., & Cheng, S. (2006). Evaluation of antioxidant properties of pomegranate peel extract in comparison with pomegranate pulp extract. *Food Chemistry*, *96*(2), 254–260.

Ling, J. K. U., Chan, Y. S., & Nandong, J. (2020). Extraction of antioxidant compounds from the wastes of *Mangifera pajang* fruit: A comparative study using aqueous ethanol and deep eutectic solvent. *SN Applied Sciences*, *2*(8), 1365.

Lucarini, M., Durazzo, A., Bernini, R., Campo, M., Vita, C., Souto, E. B., Lombardi-Boccia, G., Ramadan, M. F., Santini, A., & Romani, A. (2021). Fruit wastes as a valuable source of value-added compounds: A collaborative perspective. *Molecules*, *26*(21), 6338.

Machado, A. P. D. F., Pasquel-Reátegui, J. L., Barbero, G. F., & Martínez, J. (2015). Pressurized liquid extraction of bioactive compounds from blackberry (*Rubus fruticosus* L.) residues: A comparison with conventional methods. *Food Research International*, *77*, 675–683.

Mahato, N., Sharma, K., Sinha, M., & Cho, M. H. (2018). Citrus waste derived nutra-/pharmaceuticals for health benefits: Current trends and future perspectives. *Journal of Functional Foods*, *40*, 307–316.

Maina, S., Kachrimanidou, V., & Koutinas, A. (2017). From waste to bio-based products: A roadmap towards a circular and sustainable bioeconomy. *Current Opinion in Green and Sustainable Chemistry*, *8*, 18–23. https://doi.org/10.1016/j.cogsc.2017.07.007.

Majerska, J., Michalska, A., & Figiel, A. (2019). A review of new directions in managing fruit and vegetable processing by-products. *Trends in Food Science & Technology*, *88*, 207–219.

Mamma, D., & Christakopoulos, P. (2014). Biotransformation of citrus by-products into value added products. *Waste and Biomass Valorization*, *5*(4), 529–549.

Mohd Basri, M. S., Abdul Karim Shah, N. N., Sulaiman, A., Mohamed Amin Tawakkal, I. S., Mohd Nor, M. Z., Ariffin, S. H., … & Mohd Salleh, F. S. (2021). Progress in the valorization of fruit and vegetable wastes: Active packaging, biocomposites, by-products, and innovative technologies used for bioactive compound extraction. *Polymers*, *13*(20), 3503.

Moorthy, I. G., Maran, J. P., Ilakya, S., Anitha, S. L., Sabarima, S. P., & Priya, B. (2017). Ultrasound assisted extraction of pectin from waste *Artocarpus heterophyllus* fruit peel. *Ultrasound Sonochemistry, 34*, 525–530.

Mrudula, S., & Anitharaj, R. (2011). Pectinase production in solid state fermentation by *Aspergillus niger* using orange peel as substrate. *Global Journal of Biotechnology and Biochemistry, 6*(2), 64–71.

Oberoi, H. S., Chavan, Y., Bansal, S., & Dhillon, G. S. (2010). Production of cellulases through solid state fermentation using kinnow pulp as a major substrate. *Food and Bioprocess Technology, 3*(4), 528–536.

Oliveira, C. F., Giordani, D., Lutckemier, R., Gurak, P. D., Cladera-Olivera, F., & Marczak, L. D. F. (2016). Extraction of pectin from passion fruit peel assisted by ultrasound. *LWT-Food Science and Technology, 71*, 110–115.

Otles, S., Despoudi, S., Bucatariu, C., & Kartal, C. (2015). Food waste management, valorization, and sustainability in the food industry. In C. Galanakis (Ed.), *Food waste recovery—Processing technologies and industrial techniques* (pp. 3–23). Academic Press.

Pagano, I., Campone, L., Celano, R., Piccinelli, A. L., & Rastrelli, L. (2021). Green non-conventional techniques for the extraction of polyphenols from agricultural food by-products: A review. *Journal of Chromatography A, 1651*, 462295.

Pan, Z., Qu, W., Ma, H., Atungulu, G. G., & McHugh, T. H. (2012). Continuous and pulsed ultrasound-assisted extractions of antioxidants from pomegranate peel. *Ultrasonics Sonochemistry, 19*(2), 365–372.

Panda, S. K., & Ray, R. C. (2015). Microbial processing for valorization of horticultural wastes. In L. B. Sukla, N. Pradhan, S. Panda, & B. K. Mishra (Eds.), *Environmental microbial biotechnology* (pp. 203–221). Springer.

Panth, N., Manandhar, B., & Paudel, K. R. (2017). Anticancer activity of *Punica granatum* (pomegranate): A review. *Phytotherapy Research, 31*(4), 568–578.

Pelegrini, P. B., Noronha, E. F., Muniz, M. A. R., Vasconcelos, I. M., Chiarello, M. D., Oliveira, J. T. A., & Franco, O. L. (2006). An antifungal peptide from passion fruit (*Passiflora edulis*) seeds with similarities to 2S albumin proteins. *Biochimica et Biophysica Acta (BBA) - Proteins and Proteomics, 1764*(6), 1141–1146.

Pereira, D. T. V., Tarone, A. G., Cazarin, C. B. B., Barbero, G. F., & Martinez, J. (2019a). Pressurized liquid extraction of bioactive compounds from grape marc. *Journal of Food Engineering, 240*, 105–113.

Pereira, M. G., Maciel, G. M., Haminiuk, C. W. I., Bach, F., Hamerski, F., de Paula Scheer, A., & Corazza, M. L. (2019b). Effect of extraction process on composition, antioxidant and antibacterial activity of oil from yellow passion fruit (*Passiflora edulis* var. *Flavicarpa*) seeds. *Waste and Biomass Valorization, 10*(9), 2611–2625.

Plazzotta, S., Manzocco, L., & Nicoli, M. C. (2017). Fruit and vegetable waste management and the challenge of fresh-cut salad. *Trends in Food Science & Technology, 63*, 51–59.

Prabha, M. S., & Rangaiah, G. S. (2014). Citric acid production using Ananas comosus and its waste with the effect of alcohols. *International Journal of Current Microbiology and Applied Sciences, 3*(5), 747–754.

Prakash Maran, J., Sivakumar, V., Thirugnanasambandham, K., & Sridhar, R. (2014). Microwave assisted extraction of pectin from waste *Citrullus lanatus* fruit rinds. *Carbohydrate Polymers, 101*(1), 786–791.

Rafiq, S., Kaul, R., Sofi, S. A., Bashir, N., Nazir, F., & Nayik, G. A. (2018). Citrus peel as a source of functional ingredient: A review. *Journal of the Saudi Society of Agricultural Sciences, 17*(4), 351–358.

Rai, A., Mohanty, B., & Bhargava, R. (2016). Supercritical extraction of sunflower oil: A central composite design for extraction variables. *Food Chemistry, 192*, 647–659.

Ramdhonee, A., & Jeetah, P. (2017). Production of wrapping paper from banana fibres. *Journal of Environmental Chemical Engineering, 5*(5), 4298–4306.

Rehman, Z. U., Salariya, A. M., Habib, F., & Shah, W. H. (2004). Utilization of mango peels as a source of pectin. *Journal-Chemical Society of Pakistan, 26*, 73–76.

Riaz, A., Lei, S., Akhtar, H. M. S., Wan, P., Chen, D., Jabbar, S., Abid, M., Hashim, M. M., & Zeng, X. (2018). Preparation and characterization of chitosan-based antimicrobial active food packaging film incorporated with apple peel polyphenols. *International Journal of Biological Macromolecules, 114*, 547–555.

Rifna, E. J., Misra, N. N., & Dwivedi, M. (2021). Recent advances in extraction technologies for recovery of bioactive compounds derived from fruit and vegetable waste peels: A review. *Critical Reviews in Food Science and Nutrition*, 1–34.

Rivas, M. Á., Casquete, R., Martín, A., Córdoba, M. de G., Aranda, E., & Benito, M. J. (2021). Strategies to increase the biological and biotechnological value of polysaccharides from agricultural waste for application in healthy nutrition. *International Journal of Environmental Research and Public Health*, *18*(11), 5937.

Rodrigues, L. G. G., Mazzutti, S., Vitali, L., Micke, G. A., & Ferreira, S. R. S. (2019). Recovery of bioactive phenolic compounds from papaya seeds agroindustrial residue using subcritical water extraction. *Biocatalysis and Agricultural Biotechnology*, *22*(October), 101367.

Rodríguez, R., Jimenez, A., Fernández-Bolanos, J., Guillen, R., & Heredia, A. (2006). Dietary fibre from vegetable products as source of functional ingredients. *Trends in Food Science & Technology*, *17*(1), 3–15.

Rodríguez Couto, S. (2008). Exploitation of biological wastes for the production of value-added products under solid-state fermentation conditions. *Biotechnology Journal: Healthcare Nutrition Technology*, *3*(7), 859–870.

Rombaut, N., Tixier, A.-S., Bily, A., & Chemat, F. (2014). Green extraction processes of natural products as tools for biorefinery. *Biofuels, Bioproducts and Biorefining*, *8*(4), 530–544.

Safdar, M. N., Kausar, T., & Nadeem, M. (2017). Comparison of ultrasound and maceration techniques for the extraction of polyphenols from the mango peel. *Journal of Food Processing and Preservation*, *41*(4), e13028.

Sagar, N. A., Pareek, S., Sharma, S., Yahia, E. M., & Lobo, M. G. (2018). Fruit and vegetable waste: Bioactive compounds, their extraction, and possible utilization. *Comprehensive Reviews in Food Science and Food Safety*, *17*(3), 512–531.

Santos, O. V., Lorenzo, N. D., Souza, A. L. G., Costa, C. E. F., Conceição, L. R. V., Lannes, S. C. da S., & Teixeira-Costa, B. E. (2021). CO2 supercritical fluid extraction of pulp and nut oils from Terminalia catappa fruits: Thermogravimetric behavior, spectroscopic and fatty acid profiles. *Food Research International*, *139*, 109814.

Shao, D., Venkitasamy, C., Li, X., Pan, Z., Shi, J., Wang, B., McHugh, T. (2015). Thermal and storage characteristics of tomato seed oil. *LWT-Food Science and Technology*, *63*(1), 191–197.

Sójka, M., Kołodziejczyk, K., Milala, J., Abadias, M., Viñas, I., Guyot, S., & Baron, A. (2015). Composition and properties of the polyphenolic extracts obtained from industrial plum pomaces. *Journal of Functional Foods*, *12*, 168–178.

Someya, S., Yoshiki, Y., & Okubo, K. (2002). Antioxidant compounds from bananas (Musa Cavendish). *Food Chemistry*, *79*(3), 351–354.

Suleria, H. A. R., Barrow, C. J & Dunshea, F. R. (2020). Screening and characterization of phenolic compounds and their antioxidant capacity in different fruit peels. *Foods*, *9*, 1206.

Tabaraki, R., Heidarizadi, E., & Benvidi, A. (2012). Optimization of ultrasonic-assisted extraction of pomegranate (*Punica granatum* L.) peel antioxidants by response surface methodology. *Separation and Purification Technology*, *98*, 16–23.

Torres, J. L., & Bobet, R. (2001). New flavanol derivatives from grape (Vitis vinifera) byproducts. antioxidant aminoethylthio– flavan-3-ol conjugates from a polymeric waste fraction used as a source of flavanols. *Journal of Agricultural and Food Chemistry*, *49*(10), 4627–4634.

Torres-León, C., Ramírez-Guzman, N., Londoño-Hernandez, L., Martinez-Medina, G. A., Díaz-Herrera, R., Navarro-Macias, V., … & Aguilar, C. N. (2018). Food waste and byproducts: An opportunity to minimize malnutrition and hunger in developing countries. *Frontiers in Sustainable Food Systems*, *2*, 52.

Torres-León, C., Rojas, R., Contreras-Esquivel, J. C., Serna-Cock, L., Belmares-Cerda, R. E., & Aguilar, C. N. (2016). Mango seed: Functional and nutritional properties. *Trends in Food Science & Technology*, *55*, 109–117.

Trigo, J. P., Alexandre, E. M., Saraiva, J. A., & Pintado, M. E. (2020). High value-added compounds from fruit and vegetable by-products–Characterization, bioactivities, and application in the development of novel food products. *Critical Reviews in Food Science and Nutrition*, *60*(8), 1388–1416.

Uchenna, U. E., Shori, A. B., & Baba, A. S. (2017). Inclusion of avocado (Persea americana) seeds in the diet to improve carbohydrate and lipid metabolism in rats. *Revista argentina de endocrinología y metabolismo*, *54*(3), 140–148.

Udenigwe, C. C., & Aluko, R. E. (2012). Food protein-derived bioactive peptides: Production, processing, and potential health benefits. *Journal of Food Science*, *77*(1), R11–R24.

Venkatesh, M., Pushpalatha, P. B., Sheela, K. B., & Girija, D. (2009). Microbial pectinase from tropical fruit wastes. *Journal of Tropical Agriculture*, *47*(1), 67–69.

Villacís-Chiriboga, J., Elst, K., Van Camp, J., Vera, E., & Ruales, J. (2020). Valorization of byproducts from tropical fruits: Extraction methodologies, applications, environmental, and economic assessment: A review (Part 1: General overview of the byproducts, traditional biorefinery practices, and possible applications). *Comprehensive Reviews in Food Science and Food Safety*, *19*(2), 405–447.

Vinatoru, M., Mason, T. J., & Calinescu, I. (2017). Ultrasonically assisted extraction (UAE) and microwave assisted extraction (MAE) of functional compounds from plant materials. *TrAC Trends in Analytical Chemistry*, *97*, 159–178.

Wadhwa, M., Bakshi, M. P. S., & Makkar, H. P. S. (2015). Wastes to worth: Value added products from fruit and vegetable wastes. *CAB International*, *43*, 1–25.

Wen, C., Zhang, J., Zhang, H., Duan, Y., & Ma, H. (2019). Effects of divergent ultrasound pretreatment on the structure of watermelon seed protein and the antioxidant activity of its hydrolysates. *Food Chemistry*, *299*, 125165.

Wen, L., Zhang, Z., Sun, D.-W., Sivagnanam, S. P., & Tiwari, B. K. (2020). Combination of emerging technologies for the extraction of bioactive compounds. *Critical Reviews in Food Science and Nutrition*, *60*(11), 1826–1841.

Xu, M., Ran, L., Chen, N., Fan, X., Ren, D., & Yi, L. (2019). Polarity-dependent extraction of flavonoids from citrus peel waste using a tailor-made deep eutectic solvent. *Food Chemistry*, *297*(January), 124970.

Yahia, E. M. (Ed.). (2017). *Fruit and vegetable phytochemicals: Chemistry and human health, 2 Volumes*. John Wiley & Sons.

Yan, L. G., He, L., & Xi, J. (2017). High intensity pulsed electric field as an innovative technique for extraction of bioactive compounds—A review. *Critical Reviews in Food Science and Nutrition*, *57*(13), 2877–2888.

Zhang, H. F., Yang, X. H., & Wang, Y. (2011). Microwave assisted extraction of secondary metabolites from plants: Current status and future directions. *Trends Food Science and Technology*, *22*, 672–688.

Zuin, V. G., & Ramin, L. Z. (2018). Green and sustainable separation of natural products from agro-industrial waste: Challenges, potentialities, and perspectives on emerging approaches. *Topics in Current Chemistry*, *376*(1), 3–57.

CHAPTER 2

Pomelo Wastes
Chemistry, Processing, and Utilization

Rahul Islam Barbhuiya, Poonam Singha, and Sushil Kumar Singh

CONTENTS

2.1 INTRODUCTION

Pomelo is native to Southeast Asia and is also known as the principal ancestor of the grapefruit. Pomelo has a history of consumption and cultivation dating back 4,000 years (Chang and Azrina, 2016) and is grown in Thailand, India, Indonesia, Philippines, Malaysia, China, Japan, and the USA. The pomelo tree is 4.9–15.2 m tall with irregular, low-hanging branches. Depending upon the

DOI: 10.1201/9781003164463-2

cultivar, a pomelo fruit is 15–25 cm in diameter with a weight of 1–2 kg. About 30% of the total weight of the fruit consists of the peel, which contains many phytochemicals.

There are two types of pomelo fruit: a sour kind with pinkish flesh, more likely to be used for ceremonial purposes, and a sweet kind with white flesh, mostly eaten for its juiciness, sweetness, and delicious taste (Zain et al., 2015). As a result of crossbreeding with other citrus species, numerous pomelo cultivars are cultivated in many countries (Puglisi et al., 2017).

The pomelo fruit has two parts: the pulp and the peel, which can be easily detached. The fruit's pulp (or flesh) is pink or white in colour and is uneven, with large, spindle-shaped sacs of juice. The flesh is consumed fresh or subjected to juice extraction, while the peel is discarded. Pomelo processing industries produce a large amount of waste (Rosnah et al., 2015). The waste includes different parts of pomelo peel, which covers the flesh of the fruit. Pomelo peel has three different parts; flavedo (exocarp), albedo (endocarp) and lamella (segment membrane), as shown in Figure 2.1. Pomelo also contains few seeds which are individually relatively large, although some varieties have numerous seeds. Very little research has been conducted on the pomelo seed, so this chapter will solely discuss the pomelo peel.

Pomelo-producing countries generally do not have appropriate, cost-effective ways of utilizing the peel waste. Processors eliminate the waste *via* burning or landfill disposal (Zhu et al., 2017). Landfill and burning of such waste produce greenhouse gases that will negatively impact the environment (Tonini et al., 2018). As a consequence, the demand for pomelo waste valorization into beneficial product is increasing day by day. Many published reports have shown promising outcomes for sustainable and cost-effective pomelo waste utilization.

Development of innovative processes and technologies is a promising practice of fruit waste management through production of activated carbon, biodiesel, craft paper, and value-added food

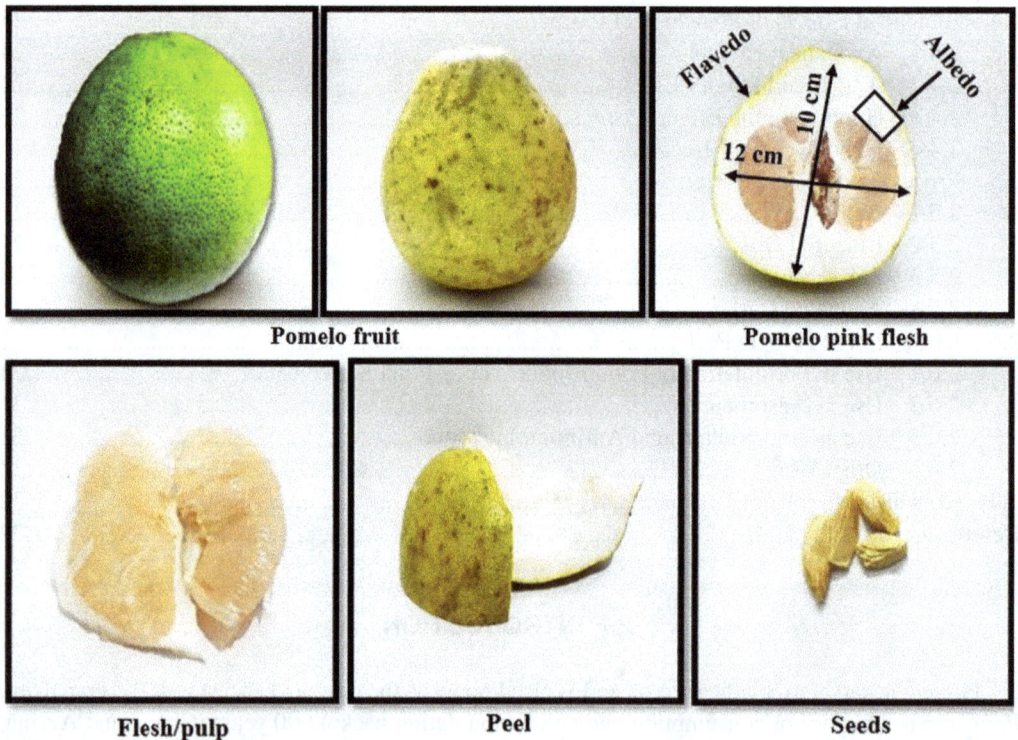

Figure 2.1 Different parts of pomelo fruit.

materials, such as essential oils, pectin, etc. This opportunity of waste utilization through product development maintains environmental stability, creates sustainable socio-economic development, generates employment, improves the total returns to the producer, and much more. In this chapter, we discuss different parts of pomelo peel, their nutritional compositions, chemistry, processing, and possible utilization opportunities along with information on value-added uses through the retrieval of potentially bioactive compounds inherent to pomelo peel waste, to deliver up-to-date insights on potential ways to take benefit of this underutilized organic source.

2.2 NUTRITIONAL AND ANTINUTRITIONAL COMPOSITION OF POMELO PEEL

Pomelo peel is an excellent source of important nutrients. Similar to most citrus fruits, the main components of pomelo peel are water, cellulose, hemicellulose, soluble sugars, and lipid components (mainly D-limonene). Integration of pomelo peel with food material can increase its nutritional and phytochemical content; it also improves the flavor and acceptability of the food for human consumption (Ani and Abel, 2018). Pomelo peel contains (a) considerable numbers and concentrations of vitamins, minerals, and carbohydrates, (b) low levels of ash, protein, and crude fiber, and (c) moderate levels of fat and moisture. Essential human bodily functions, such as metabolism, health maintenance, and growth, require essential organic compounds like vitamins (Gropper & Smith, 2012). The total vitamin concentration of pomelo peel ranges from 1.35 mg/100 g to 279.60 mg/100 g, with 1.36 mg/100g of folate, 4.45 mg/100 g of vitamin E, 11.20 mg/100 g of thiamine, 19.34 mg/100 g of vitamin C, 224.16 mg/100 g of niacin and 279.63 mg/100 g of pyridoxine (Ani and Abel, 2018). Vitamins E and C act as antioxidants and quenching free radicals to provide protection against diseases. Pyridoxine and niacin can be isolated from the peel for industrial uses. Similarly, the presence of thiamine in the diet can improve the overall metabolism of carbohydrate and its interconnectedness with amino acid metabolism *via* α-keto acids, improving the development of acetylcholine for neural function (WHO/FAO, 2001).

Other than vitamins, the macronutrient which is most abundant in pomelo peel is carbohydrate (Zahra et al., 2017). Feumba et al. (2016) studied eight different fruit peels and reported that the macronutrient found at highest concentration in fruit peel was carbohydrate. The approximate percentage of carbohydrate present in pomelo peel is 72 g per 100 g peel extract (Ani and Abel, 2018; Uraku, 2015). Carbohydrates extracted from pomelo peel could serve the purpose of nutrient fortification for various body functions.

Pomelo fruit peel contains a moderate amount of moisture (13.20%) (Ani and Abel, 2018), which is somewhat similar to the moisture content present in citrus peel (Uraku, 2015). The growth of microorganisms and shelf life is determined by a food's moisture content (Nasir et al., 2004). Processing the food by bottling or canning and pasteurization can help to eliminate microbial activities and improve the food's shelf life.

The concentration of fat and crude fiber in pomelo juice is low, making it a perfect constituent in weight-reducing diets. The concentration of fiber and fat present in pomelo peel is approximately 10% and 3%, respectively (Ani and Abel, 2018), comparable to the values in other citrus fruit peel. However, there might be differences in the fiber concentration of citrus fruits with different origins and varieties. The sample preparation method also affects the fiber concentration (Feumba et al., 2016; Uraku, 2015).

The role of minerals in metabolism and nutrition cannot be exaggerated. The composition of minerals in fruits mainly depends on geographical features, topography, cultivar, stage of maturity, and soil type. Pomelo peel and pomelo juice contain high concentrations have calcium. Ani and Abel (2018) studied pomelo peel composition, and reported that 515.7 mg calcium/100 g was present in the peel. Similarly, Okwu and Emenike (2007) studied different citrus varieties and reported that the highest concentration mineral in citrus fruits was calcium. Calcium plays a significant role in enzyme activation, muscle contraction, blood clotting, and cellular processes. It is also a structural component of teeth and bones (Gropper and Smith, 2012).

Other minerals present in pomelo peel are sodium, phosphorus, iron, magnesium, and potassium. The human body's extracellular fluid contains sodium as the primary cation, which helps maintain body fluid. Similarly, phosphorus helps maintain acid-base balance, cell membrane structure, nucleic acid formation, energy transfer and storage, and bone mineralization (Gropper and Smith, 2012). Iron is necessary for better work performance, energy metabolism, temperature regulation, cognitive development, and for proper functioning of the immune system (Baynes, 2000). Potassium is known for its role in nerve impulse transmission and fluid balance (Ani and Abel, 2018). Magnesium plays a significant role in various physiological and biochemical processes (Schrauzer, 2000). The ash content of food material helps to determine the amount and type of minerals present in food. Pomelo peel contains an appreciable concentration of ash, similar to that of *Vitis vinifera* (4.24%) and *Citrus × sinensis* fruits (1.59%) (Uraku, 2015).

Pomelo peel contains a relatively high concentration of antinutritional compounds compared with pomelo juice. The substances that decrease nutrient consumption, ingestion, and utilization, and which give rise to additional adverse effects are known as antinutrients (Aberoumand, 2012). Phytic acid, tannin, and oxalate are the main antinutritional compounds present in the pomelo peel. Phytic acid is present in the highest quantity among the three antinutrients. It binds with vital minerals, such as zinc, iron, calcium, and magnesium, to make insoluble phytates (Agte et al., 1999).

Similarly, tannin binds with amino acids, proteins, and other organic compounds, including alkaloids. Essential trace metals form complexes with oxalate which makes them inaccessible to the digestive system for their biochemical and physiological purposes (Ladeji et al., 2004). However, being an antinutrient is not an inherent property of any composite, as it depends on the ingesting animal's digestive system and the amount ingested. Additionally, heat treatment of antinutritionals renders them inactive (Iyayi et al., 2008).

2.3 BIOACTIVE COMPONENTS OF POMELO PEEL

Bioactive components of pomelo peel include phenolic acids, flavonoids, and carotenoids. These biomolecules are primarily found in the albedo and flavedo layers of the peel (Tocmo et al., 2020).

2.3.1 Phenolic Acids and Tannins

Pomelo peel is an excellent source of phenolic compounds. Some common phenolic acids present in pomelo peel linked with antioxidant and anti-inflammatory effects are caffeic acid, ferulic acid, and chlorogenic acid (Gülçin, 2006; Liang and Kitts, 2016; Lü et al., 2016). Table 2.1 presents some major phenolic acids in pomelo peel from various cultivars. Wang et al. (2008) reported that the most abundant phenolic acid in pomelo peel was chlorogenic acid compared with other common phenolic acids, viz. caffeic acid, sinapic acid, ferulic acid, ρ -coumaric acid, and chlorogenic acid. In-depth information on specific phenolic acids from various pomelo peel cultivars is still limited. Most of the researchers only reported information on total phenolic content (11 to 90 mg gallic acid equivalents/g dry weight) (Abd Rahman et al., 2018; Malleshappa et al., 2018; Toh et al., 2013). It has been reported that pomelo peel also contains a fair amount of tannins (230 mg/100 g) (Ani and Abel, 2018). Tannins can bind with proteins and precipitate them; hence, they are considered to be antinutrients (Hagerman et al., 1998). Tannins which are hydrolyzable and condensed are acknowledged for their beneficial effects on health, including antibacterial, antiviral, and antioxidant effects (Koleckar et al., 2008).

Table 2.1 Some Common Phenolic Acids Present in Pomelo Peel from Different Cultivars

Phenolic Acids	Cultivar(s)	Concentration Range (µg/g DWor as indicated)[1]	Reference(s)
Sinapic acid	Peiyou, Wendun	10 to 29	Lü et al., 2016
Caffeic acid	Peiyou, Wendun, Shatianyou, Yubei, Liangpingyou 78-8, Humiyou 1, Guanxi Miyou, Hongxinyou, Dianjiang Baiyou, Changshou Shatianyou	8.0 to 56	Lü et al., 2016; Wang et al., 2008
Chlorogenic acid	Peiyou, Wendun, Zhenlongyou 3, Yubei Shatianyou, Liangpingyou 78-8, Humiyou 1, Guanxi Miyou, Hongxinyou, Baiyou, Dianjian, Changshou Shatianyou, Wubuyou	24 to 428	Lü et al., 2016; Wang et al., 2008
ρ-Coumaric acid	Wendun, Peiyou	142 to 241	Lü et al., 2016
Ferulic acid	Peiyou, Wendun, Kuiyou, Zhenlongyou 3, Yubei Shatianyou, Liangpingyou 78-8, Humiyou 1, Guanxi Miyou, Hongxinyou, Baiyou, Dianjian, Changshou Shatianyou, Wubuyou	5.0 to 88	Lü et al., 2016; Wang et al., 2008
Gallic acid	Peiyou, Wendun, Kuiyou, Zhenlongyou 3, Yubei Shatianyou, Liangpingyou 78-8, Humiyou 1, Guanxi Miyou, Hongxinyou, Baiyou, Dianjian, Changshou Shatianyou, Wubuyou	527 to 1,097	Lü et al., 2016

[1] Range represents lowest and highest values across cultivars extracted from the indicated references.

2.3.2 Carotenoids

Pomelo peels are rich sources of carotenoids (Shan, 2016). For instance, pomelo cultivars such as 'Chuzhou Early Red' and 'Yuhuan' contain 11 carotenoids (luteoxanthin, 9-*cis*-violaxanthin, zeaxanthin, β -cryptoxanthin, lutein, β-carotene, α -carotene, lycopene, zeta-carotene, phytofluene, and phytoene) (Xu et al., 2006). Of these, phytofluene and phytoene represent 13–25% and 30–55% of total carotenoids on a dry weight basis, respectively. Other major carotenoids are zeaxanthin (0.1 to 5% dry weight), lutein (13% dry weight), and β-carotene (0.5–4.5% dry weight of total carotenoids). Similarly, in cultivars 'Wendun' and 'Peiyou', β-carotene, β-cryptoxanthin, zeaxanthin, and lutein are the major carotenoids (Wang et al., 2008).

The red/pink or golden yellow color of the fruit pulp and pomelo peel is due to carotenoids (Xu et al., 2006). Based on the cultivar, the amounts of individual carotenoids vary significantly. Various studies have shown that the total carotenoid content of different cultivars reveals an extensive range (16–63 µg/g dry weight) (Abudayeh et al., 2019; Tao et al., 2010; Wang et al., 2008). Thus, more studies are required to offer a more inclusive evaluation of individual carotenoid concentrations, using less expensive and highly efficient extraction methods that compensate for production, distribution, and material changes.

2.3.3 Flavonoids

The flavedo and albedo layers of pomelo peel contain important bioactive components known as flavonoids. Various types of flavonoids have been detected in pomelo peel. They are divided into flavanols, flavones and flavone glycosides, flavanones, and flavanone glycosides based on their chemical structures. Flavanones and flavones detected in pomelo peel are of two forms, namely glycosides and aglycones. The flavanone aglycones often recognized in pomelo peel are eriodictyol, hesperetin, and naringenin (Li et al., 2014; Lü et al., 2016). Similarly, major flavone aglycones commonly reported in pomelo peel are diosmetin, luteolin, apigenin, tangeretin, nobiletin, and sinensetin (Li et al., 2014; Lü et al., 2016; Wang et al., 2008; Yu et al., 2015). The flavanone glycosides present in pomelo peel are of two types, neohesperidosides, and rutinosides. Neohesperidosides contain a flavanone with neohesperidose (rhamnosyl-α-1,2 glucose), while rutinosides contain a disaccharide residue and a flavanone. Rutinosides include neoponcirin, hesperidin, narirutin, and eriocitrin, whereas neohesperosides include neoeriocitrin, neohesperidin, and naringin (Li et al., 2014; Ortuño et al., 1995; Yu et al., 2015). Major flavone glycosides present in pomelo peel are neodiosmin, rhoifolin, and diosmin (Li et al., 2016; Lü et al., 2016). Commonly the flavonols detected in pomelo peel are kaempferol, quercetin, and rutin (Lü et al., 2016), but they are present in much lower concentrations than the other flavonoids.

2.4 BIOLOGICAL EFFECTS OF POMELO PEEL EXTRACT

Numerous studies have revealed a wide range of bioactivities, including anticancer, hypolipidemic, anti-inflammatory, antioxidant, antimicrobial, and hypoglycemic effects, indicating that pomelo peel extracts and their isolated pure compounds have health-promoting and therapeutic potential. Some of them are discussed below.

2.4.1 Anticancer Effects

Some researchers have shown that pomelo peel extracts have anticancer properties. For instance, in four different cancer cell lines, including adenocarcinoma gastric cell line (AGS) gastric adenocarcinoma cells, HeLa cervix adenocarcinoma cells, U373MG human glioblastoma, and SNU-16 gastric cancer cells, a supercritical CO_2 (SCO_2) extract from whole pomelo fruits (peel + pulp) was found to have dose-dependent (0–400 μg/mL) antiproliferative effect (Gyawali et al., 2012). The sensitivity of the cell lines to the extracts varied, with SNU-16 cells being the most sensitive (~40 % inhibition) and AGS gastric adenocarcinoma cells being the least sensitive (~20 % inhibition). The antiproliferative effects of the extracts were attributed to their lipid-soluble contents (α-terpineol, hexadecenoic acid, auraptene, caryophyllene oxide, and limonene) and phenolic content (i.e., tangeretin and nobiletin) (Gyawali et al., 2012).

An *in vivo* investigation recently revealed evidence of pomelo peel's anticancer properties (Yu et al., 2018). The study was focused on polysaccharides generated from pomelo peel. Female Kunming mice were injected subcutaneously with S180 tumor cells as *in vivo* tumor models. Compared with mice treated with tumor cells, oral administration of polysaccharides inhibited tumor cell development, improved splenic lymphocyte proliferation, boosted natural killer (NK) cell killing activity, and increased CD4+ T cell numbers. This research revealed that pomelo peel's anticancer properties are linked to components other than phenolic chemicals.

2.4.2 Anti-Inflammatory Effects

An anti-inflammatory effect of pomelo extract was demonstrated in a mice model of carrageenan-induced paw edema, a commonly used model of acute inflammation (Ibrahim et al., 2018). Oral treatment with methanolic extract at 300 or 500 mg/kg bodyweight at 4 hours after edema induction (1 % carrageenan) inhibited paw edema by 34.47 % or 38.68 %, respectively. These findings were supported by findings from another study (Malleshappa et al., 2018), which used the same rat paw edema model. Five hours after induction, intraperitoneal dosages of 250 or 500 mg/kg pomelo peel extracts inhibited carrageenan-induced paw edema by 17 or 48 %, respectively. Qualitative examination of pomelo peel extracts revealed the presence of anti-inflammatory components, such as phenols and flavonoids, which could be the cause of the reported effects.

In another study (Kuo et al., 2017), the anti-inflammatory effects of pomelo peel-derived pure compounds were measured by inhibiting the release of the lysosomal enzyme elastase by activated neutrophils. Epoxybergamottin, isomeranzin, and 4-hydroxybenzaldehyde were reported to reduce fMLP/cytochalasin B-induced elastase release in human neutrophils with strong (IC_{50} 0.43 to 4.33 M) suppression. Although activated neutrophils are vital in the host's defense against microbial infection (Teng et al., 2017), long-term or unresolved activation of these immune cells is connected with chronic inflammation, which is linked to the pathophysiology of a variety of chronic diseases. As a result of the reported suppression of elastase release, chemicals in pomelo peel may have modulatory effects in chronic inflammation.

2.4.3 Hypolipidemic Effects

Lewis and Segal (2010) identified hyperlipidemia as a risk factor for atherosclerotic cardiovascular disease (ASCVD) and stroke (Stone et al., 2014). It is described as a cholesterol imbalance in the blood, with unusually high levels of low-density lipoprotein ("bad") cholesterol (LDL-c, $\geq$130 mg/dL) and low levels of high-density lipoprotein ("good") cholesterol (HDL-c, <40 [men] and <50 [women] mg/dL) (Stone et al., 2014). As LDL-c and HDL-c regulate body cholesterol levels, an imbalance in these lipoproteins can raise the risk of cardiovascular events including myocardial infarction (Sandhu et al., 2016). Hyperlipidemia also includes hypertriglyceridemia, a condition in which triglycerides (TG) levels are abnormally high ($\geq$150 mg/dL) (Berglund et al., 2012). Citrus fruit peels, such as pomelo (Walker et al., 2014), bergamot (Walker et al., 2014), lemon (Hashemipour et al., 2016), orange (Ashraf et al., 2017), and pomegranate (Haghighian et al., 2016), have been demonstrated to have hypolipidemic effect.

Pomelo peel extracts (1% [w/w]) from four cultivars were tested for their effect on markers of cardiometabolic disorders in mice fed with a high-fat diet (HFD) for 8 weeks (Ding et al., 2013). In comparison to HFD mice, two of the four extracts prevented HFD-induced weight gain and lowered serum total cholesterol (TC), as well as TG and TC levels in the liver. In another study (Hong et al., 2010), 1% (w/w) pomelo peel extracts supplied for 10 weeks to Sprague-Dawley rats fed an HFD-style diet reduced bodyweight, decreased TG, TC, LDL-c, and atherogenic index, and blocked systolic blood pressure elevation compared with untreated HFD-fed rats. Although the peels of numerous pomelo cultivars have been known to have a wide spectrum of chemical components, only a few cultivars have been studied for their hypolipidemic properties. Ding et al. (2013) analyzed the chemical components of pomelo peel extracts for their flavonoid content. Although pomelo peel components, particularly flavonoids, have been extensively studied, factors such as extraction procedures and varietal changes may alter chemical profiles, affecting the *in-vivo* efficacy of the extracts.

2.4.4 Hypoglycemic Effects

The healing efficacy of the pomelo extract on induced excision skin wounds in Wistar albino rats was investigated in a study (Ahmad et al., 2018). When compared with an untreated control group, doses of 400 or 600 mg/kg significantly reduced healing time (to 12 or 9 days, respectively), as compared with untreated control (15 days). Furthermore, diabetic rats given pomelo peel extract had considerably lower fasting blood glucose (FBG) levels than control diabetic rats who were not given any treatment. FBG level is a parameter that is frequently assessed in diabetes-related studies, and chemicals or extracts that reduce FBG levels in diabetic patients could be potential anti-diabetic medicines.

In another investigation, pomelo peel extract (1,200 mg/kg bodyweight) was shown to have an anti-diabetic effect in streptozotocin-induced diabetic mice (Quynh et al., 2010). When diabetic mice treated with an ethyl acetate extract of pomelo peel were compared with untreated diabetic controls, FBG levels decreased by 41%. In diabetic mice treated with pomelo peel extract, this reduction in blood sugar levels was attributed in part to considerably increased hepatic hexokinase activity and decreased hepatic glucose-6-phosphatase activity. Glycolysis and gluconeogenesis are involved in glucose modulation (John et al., 2011). Inhibiting the enzymes α-amylase and α-glucosidase, which break down carbohydrates to release glucose during digestion, can prevent a postprandial rise in blood glucose levels (Tundis et al., 2010). Several studies have shown that certain phytochemicals have anti-diabetic characteristics by their ability to block pancreatic enzymes (McDougall & Stewart, 2005). In the case of pomelo peel, it was discovered that acetone and ethyl acetate extracts (80 to 320 µg/mL) suppressed the activities of α-amylase and α-glucosidase in a dose-dependent manner (Oboh & Ademosun, 2011). In another study (Lim and Loh, 2016), phenolic extracts from the peel of the 'White Tambun' cultivar inhibited α-amylase and α-glucosidase activities by 38 and 41%, respectively. When compared with other citrus peels (i.e. lime, calamansi), pomelo peel extracts had a 41% higher inhibitory activity against α-glucosidase.

2.5 PROCESSING OF POMELO PEEL

Figure 2.2 shows the processing of pomelo peel as a raw material for production of various commodities such as pectin, essential oils, polysaccharides other than pectin, and activated carbon.

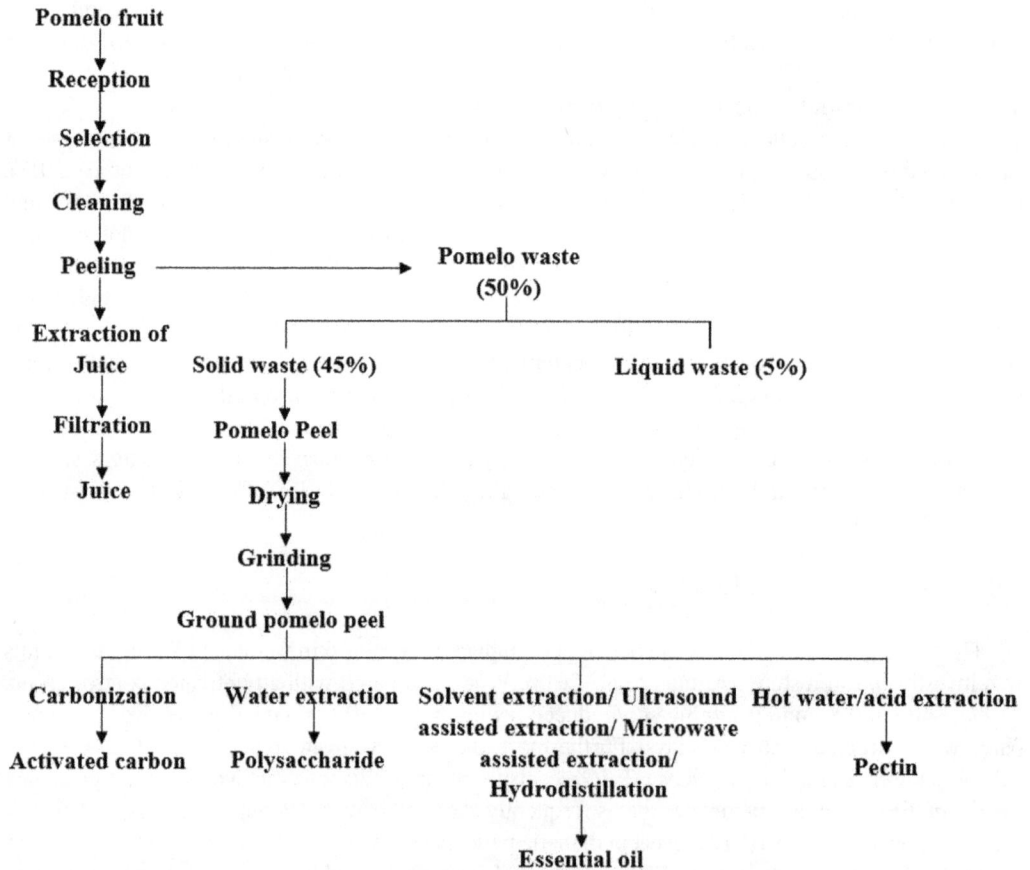

Figure 2.2 Processing of pomelo waste into industrial commodities.

2.5.1 Pectin

Pectin is a complex heteropolysaccharide used as a thickening, emulsifying, and a gelling agent (Thakur et al., 1997). Pectin has multipurpose properties that enable it to be used for different engineering practices, especially in terms of drug delivery and tissue engineering (Noreen et al., 2017). Nowadays, raw materials are being investigated to discover substitute sources of commercial pectin. The spongy white part of pomelo peel (albedo) is an excellent source of pectin (up to 20% of the fruit's total pectin) (Chavan et al., 2018).

Extraction methods used for pectin recovery are conventional water extraction (Methacanon et al., 2014), ultrasound-/microwave-assisted extraction combined with acid extraction (Liew et al., 2016; Liew et al., 2019), and solvent extraction method (Elgharbawy et al., 2019).

Depending on the extraction method, pectin yield (ratio of pectin dry weight to pomelo powder dry weight) varies from 3.6% to 94% (w/w), dry basis. Sequential ultrasound/microwave-assisted technique without acid modification had a yield of 3.4% ([w/w], dry basis) (Liew et al., 2019). However, when a sequential ultrasound–microwave process is used with citric acid, pectin yield increased to 39% (w/w), dry basis (Liew et al., 2016). On the other hand, without pH modification, the water extraction method resulted in lower pectin yield (6.5% [w/w], dry basis). However, when sodium hydroxide or hydrochloric acid are used to modify the alkalinity or acidity of the extracting medium, the yield improved to 24% (w/w), dry basis (Wandee et al., 2019). Microwave-assisted extraction with acid modification (with tartaric acid) gave 24% pectin yield ([w/w] dry basis) (Quoc et al., 2015). A novel extraction method was developed that uses natural deep eutectic solvents (NADES) to extract pectin from pomelo peel waste (Elgharbawy et al., 2019). NADES are considered "greener" options because they are composed of plant-derived primary metabolites. Examples of such natural metabolites are amino acids, amines, alcohols, sugars, and organic acids (Dai et al., 2013). When these compounds are mixed, they generate a eutectic mixture, which is a liquid salt containing a range of anionic and/or cationic species. Several combinations of NADES effectively improved pectin yield from 29% to 96% (w/w, dry basis) relative to the other extraction methods discussed earlier (Elgharbawy et al., 2019). Generally, the yield of pectin from pomelo peel was either superior to or comparable with other fruit wastes, like pomace of apple (3.5% to 90% [w/w, dry basis]) (Perussello et al., 2017) and peel of mango (9% to 26% [w/w, dry basis]) (Maran et al., 2015; Patel, 2017). Therefore, pomelo peel can be an excellent source of pectin for commercial use.

2.5.2 Essential Oils

One of the most valuable products the citrus fruit processing industry has is essential oil. Essential oils are used extensively in pharmaceutical, beverage, flavor, and cosmetic industries (Palazzolo et al., 2013; Sarkic & Stappen, 2018). They have antimicrobial and anti-inflammatory characteristics (Mancuso et al., 2019). Essential oils contain structurally varied volatile substances usually categorized into acids, ketones, esters, terpenoids, aliphatic alcohols, and aldehydes (Cheong et al., 2011; Hosni et al., 2010). These types of volatile substances cover 92% to 99% of the total essential oil (Njoroge et al., 2005).

Extraction methods used for essential oil recovery from pomelo peel are hydro-distillation (Hosni et al., 2010), steam distillation (Guo et al., 2018), microwave pre-treatment and hydro-distillation (Liu et al., 2017), solvent extraction (Cheong et al., 2011), cold-pressing (He et al., 2019), ultrasound/microwave-assisted extraction, supercritical CO_2 fluid extraction, and simultaneous distillation and extraction method (Sun et al., 2014). Depending on the extraction method, the yield of essential oils from pomelo peel varies from 1.0% to 1.7% of fresh fruit weight (Hosni et al., 2010). To date, approximately 299 volatiles have been recognized in pomelo peel, out of which 189 (63.2%) of the volatiles belong to the terpenoids. Volatiles terpenoids include tricyclic sesquiterpenoids consisting of 17 (5.7%) volatiles, bicyclic sesquiterpenoids representing 28 (9.4%) volatiles, monocyclic sesquiterpenoids with 20 (6.7%) of the volatiles, acyclic monoterpenoids representing 38 (12.7%) of the volatiles, monocyclic monoterpenoids with 41 (13.7%) volatiles, bicyclic monoterpenoids with 27 (9.0%) volatiles, diterpenoids with three (1%) volatiles, and acyclic sesquiterpenoids with 15 (5.0%) volatiles. Other volatiles found in the essential oil of pomelo peel are esters representing 26 (8.7%) of the volatiles, non-terpenoid hydrocarbons consisting of 17 (5.7%) volatiles, non-terpenoid aldehydes containing 18 (6.0%) volatiles, and non-terpenoid alcohols with 14 (4.7%) volatiles. In addition to all these volatiles, there are approximately 35 (11.7%) volatiles which are unclassified (Tocmo et al., 2020).

Out of all the volatiles, the primary volatile present is limonene regardless of extraction method or pomelo cultivar used. Other volatiles found in high concentration are nootkatone, sabinene, *cis*-linalool oxide, linalool, germacrene D, β-myrcene, α-pinene, and β-pinene (Tocmo et al., 2020). However, large variations have been observed among different pomelo cultivars in terms of the variety and concentration of volatiles (Goh et al., 2019). Furthermore, the extraction method dramatically affects the profiles of essential oils from pomelo peel. Table 2.2 shows various methods of essential oil extraction along with the chemical components isolated from pomelo peel.

2.5.3 Polysaccharides

Apart from pectin, other polysaccharides can be derived from pomelo peel. Liu et al. (2018) extracted microcrystalline cellulose and cellulose from pomelo peel by acid solution hydrolysis and mixed alkaline hydrogen peroxide liquor. The physical property of the microcrystalline cellulose extracted from pomelo peel is similar to the commercial one. It has greater oil/water-binding capacity compared with the commercial microcrystalline cellulose and has superior thermal rise stability to cellulose. These results recommend that pomelo peel-derived microcrystalline cellulose can be used for various purposes such as dietary fiber supplements, fat alternatives, and emulsion stabilizers. Polysaccharides with a molecular weight of 1.10×10^5 Da can be extracted from pomelo peel using the alcohol precipitation technique and traditional hot water extraction at 4.4% yield (Yu et al., 2018). Polysaccharides extracted from pomelo peel are also considered biocompatible, biodegradable, and non-toxic, like other natural polysaccharides (Poli et al., 2011). Hence, pomelo peel can be regarded as a vital source of polysaccharides with a wide range of applications. More research on the classification of the recovered polysaccharides, their physical and chemical properties, and optimization of extraction conditions is needed

2.5.4 Activated Carbon

Carbon materials, such as activated carbons and nanocarbons like carbon nanotubes, graphene, and other carbon materials with a large specific surface area and chemical stability, have been widely used as adsorbents for pollutant removal from wastewaters (Chen et al., 2010; Li et al., 2015). As a result, activated carbon is one of the most common materials used to clean industrial wastewaters. Carbonization, physical, and chemical modifications have frequently been used to improve the efficacy of the original pomelo peel adsorbent, giving it more adsorption sites and/or new functional groups (Wang et al., 2020).

To generate a porous biochar or activated carbon, the original pomelo peels were treated at high temperatures (250°C to 800°C) in an inert atmosphere (Wu et al., 2017). To boost the adsorbing effectiveness of the resulting biochar, certain chemical modifiers were commonly included in pomelo peel powder before carbonization. K_2CO_3, K_2FeO_4, Fe_3O_4, $FeCl_3$, H_3PO, KOH, and NaOH were among the common chemical modifiers. These substances may activate biochar by increasing surface pore size, surface area, and oxygen-containing groups, thus boosting its pollutant-binding capacity (Wang et al., 2020). For instance, Peng et al. (2014) used pyrolysis and a KOH activation technique to prepare porous activated carbon from pomelo peel, which was then heated in an argon

Table 2.2 Different Extraction Methods for Essential Oils and Volatile Components of Pomelo Peel from Various Cultivars

Extraction Method	Cultivar/Country of Origin	Volatile Components	No. of Volatiles Identified	References
Hydro-distillation	Not indicated/ Tunisia	Geranyl α-terpinene, α-cyperone, β-sinensal, l-cadinol, γ- eudesmol, Rpi-cubenol, humulene 6,7-epoxide, germacrene B, (E)-nerolidol, aromadendrene, α-humulene, β-ionone, germacrene D, cubebol, bicyclogermacrene, δ-cadinene, α-terpinyl acetate, neryl acetate, geranyl acetate, α-copaene, β-elemene, bornyl isobutyrate, β-copaene, trans-para-menth-2-ene-1-ol, geranial, (E)-(E)-2,4 decadienal, trans-pinocarvyl acetate, trans-carveol, myrtenyl acetate, α-pinene, sabinene, β-pinene, β-myrcene, α-phellandrene, limonene, (Z)-β-ocimene, (E)-β-ocimene, γ-terpinene, trans- sabinene hydrate, cis-sabinene hydrate, linalool, trans-pinocarveol, borneol, terpinen-4-ol, verbenone, β-cyclocitral, citronellol	46	Hosni et al., 2010
Steam distillation	Shatian/China	Hexanal, 2-hexanal, 3-hexen-1-ol, α-pinene, β-pinene, camphene, mycene, D-limonene, ocimene, γ-terpinene, cis-linalooxide, linalool, 1,3,8-p-menthatriene, cyclooctanone, cis-limonene oxide, trans-limonene oxide, β-terpilenol, camphor, citronellal, 3-methylenecyclohexene, terpineol, cis-carveol, (+)-carvone, citronellol, citral, nerol, perillaldehyde, 4-carene, neryl acetate, terpilene, caryophyllene, β-elemene, 2,6-nonadienal, germacrene D, α-farnesene, β-farnesene, isocarveol, hexanoic acid (hexyl ester), 4-decyne, sobrerol, β-bisabolene, patchoulane, santolina triene, nerolidol 2, caryophyllene oxide, prenyl bromide, farnesyl acetate, geranylgeraniol, cis-carvone oxide, p-α-dimethylstyrene, perillene, farnesol, methyl m-oxybenzoate, thujanol, nootkatone, β-humulene, cyclohexene (2-ethynyl-1,3,3-trimethyl), dihydrocarveol, myrcenol, α-campholenal, citronellol formate, 2-octyne, 2-acetylcyclopentanone, geranyl acetate, cis-p-mentha-2,8-dien-1-ol, D-nerolidol, (-) spathulenol, γ-selinene, cyclododecyne, longifolenaldehyde, alloaromadendrene oxide-(1), ledol, p-menth-α-3,8-diene, cryptone, 7-methyl-3-octyne, cyclopentene, 1,5,6,7. tetrahydro-4-indolene, octanoic acid (hexyl ester), β-maaliene, γ-gurjunenepoxide-(1), verbinone, limonene-1,2-diol, trans-decalone, laurine, valencen, (+)-ledene, cis-Z-α-bisabolene epoxide, isolimonene, longipinocarvone, thunbergol, cyclohexanepropanol, 1,3-cycloheptadiene, 9-dodecenol, dihydromyrcene, dihydromyrcenol, thujone, toluquinol, cis-2,8-menthadiene-1-ol, 1-(3-methylphenyl)-ethanone, 2-methyl-1octen-3-yne, α-methylcinnimal, γ-picoline 1-oxide, 2-ethoxy-pyridine, 1,4-octadiene, exestrol, sabinol, santalol	107	Guo et al., 2018

(Continued)

Table 2.2 (Continued) Different Extraction Methods for Essential Oils and Volatile Components of Pomelo Peel from Various Cultivars

Extraction Method	Cultivar/Country of Origin	Volatile Components	No. of Volatiles Identified	References
Microwave pre-treatment and hydro-distillation	Not indicated/China	α-pinene, β-pinene, β-myrcene, D-limonene, benzeneacetaldehyde, β-ocimene, cis-linaloloxide, trans-linalool oxide (furanoid), linalool, trans-p-mentha-2,8-dienol, cis-p-mentha-2,8-dien-1-ol, terpinen-4-ol, α-terpineol, decanal, nerol, carveol, D-carvone, geraniol, citral, hexyl hexanoate, butyl caprylate, β-elemene, β-gurjunene, β-cubebene, γ-elemene, γ-muurolene, germacrene D, valencene, α-farnesene, 7-epi-α-selinene, nerolidol, hexyl benzoate, isoaromadendrene epoxide, juniper camphor, trans-farnesol, farnesal, nootkatone, cyclohexadecane, squalene	39	Liu et al., 2017
Solvent extraction	Malaysian Pink and White/Malaysia	α-pinene, β-pinene, β-myrcene, limonene, trans-β-ocimene, octanal, p-cymene, terpinolene, 6-methyl-5-hepten-2-one, hexanol, trans-2-heptenal, cis-3-hexen-1-ol, nonanal, trans-2-hexen-1-ol, trans-linalool oxide, cis-linalool oxide, octyl acetate, citronellal, decanal, octanol, linalool, iso-bornyl acetate, fenchol, p-menthene-8-thiol, β-caryophellene, α-terpineol, germacrene D, neral, carvyl acetate, geranial, dodecanal, carvone, neryl acetate, trans,trans-2,4-decadienal, nerol, geraniol, trans-2-dodecenal, cis,trans-2,6 nonadienal, perillyl alcohol, α-farnesol, trans-nerolidol, elemol, carvacrol, β-sinensal, indole, nootkatone, pentanethiol, trans-epoxy-ocimene, α-copaene, camphor, 4-terpineol, citronellyl acetate, citronellol, geranyl acetate, methyl benzoate, perillyl aldehyde, carveol, benzothiazole	58	Cheong et al., 2011
Cold-pressing extraction	Guan Xi/China	α-pinene, artemisiatriene, β-pinene, β-myrcene, limonene, ocimene, propionamide, metaraminol, citral, 4-carene, nor-ephedrine, caryophyllene, cubebene, cathinone, β-copaene, bicyclogermacrene, γ-elemene, 2,6,11,15-tetramethyl-hexadeca-2,6,8,10,14-pentaene, β-farnesene, 7-methoxy-6-(3-methyl-2-oxobutyl)-2H-1-benzopyran-2-one, 2-(methylamino)-1-phenylethanol	21	He et al., 2019)
Ultrasound/microwave-assisted extraction Supercritical CO_2 fluid extraction Cold pressing, Hydrodistillation	GuanXi pomelo/China	α-pinene, sabinene, β-pinene, β-myrcene, octanal, α-phellandrene, limonene, cis-β-ocimene, trans-β-ocimene, octanol, cis-linalool oxide, terpinolene, linalool, nonanal, p-mentha-(E)-2,8(9)-dien-1-ol, cis-limonene oxide, trans-limonene oxide, citronellal, nonanol, 4-terpineol, α-terpineol, n-decenal, trans-piperitol, trans-carveol, nerol, cis-carveol, neral, carvone, geraniol, geranial, perillyl aldehyde, indole, δ-elemene, neryl acetate, α-copaene, geranyl acetate, cis-3-hexenyl hexanoate, β-elemene, dodecanal, trans- caryophyllene, γ-elemene, α-humulene, dodecenal, germacrene D, β-selinene, valencene, bicyclogermacrene, α-muurolene, α-farnesene, δ-cadinene, germacrene B, trans-nerolidol, intermedeol, farnesol, nootkatone, methyl-hexadecanoate	56	Sun et al., 2014

environment. The porous, activated carbon generated had a very high specific surface area of up to 2100 m^2 g^{-1} and a high energy density of 17.1W·h kg^{-1}.

2.6 UTILIZATION OF POMELO PEEL

The possible opportunities for utilization of pomelo peel are discussed in this section.

2.6.1 Use in Formulation of Food Products or as Food Ingredients

Pomelo peel has been utilized for food enrichment because of its bioactive compounds with significant health-promoting effects. The fiber concentration of pomelo peel is approximately 10% (Ani and Abel, 2018), which may encourage gut health by stimulating growth of beneficial micro-organisms in the gastrointestinal tract and thus can be used in formulation of fiber-rich products. Apart from having appreciable fiber content, pomelo peel has a fair amount of polyphenols, exclusively flavonoids (Zain et al., 2015). Pomelo peel is used as a tea or tea additive and in the formulation of candied peel strips and peel marmalade. However, limited scientific literature is available on such food applications across pomelo-growing countries.

2.6.2 Use as Adsorbents

Various industries, including the leather, printing, textile, and paper industries, produce harmful effluents, which cause serious environmental pollution. Removal of such effluents is a matter of great concern. Industrial procedures such as finishing and dyeing use large amounts of dyes. These dyes are generally discharged into water bodies, polluting water for industrial, agricultural, and domestic use (Lellis et al., 2019). These types of dyes and their metabolites also have carcinogenic and toxic effects on marine life (Lellis et al., 2019). Therefore, optimization of procedures for the elimination of dyes in water bodies is needed. Generally, removing dyes (especially azo dyes) includes physical, chemical, and biological methods. Amongst these methods, the cheapest and most efficient technique is the adsorption method (Katheresan et al., 2018).

Pomelo peel waste can be converted into active materials or adsorbents and has shown promising results in the removal of dyes. For example, dried pomelo peel powder can adsorb azo dyes (e.g., RB 114 dye) with the highest adsorption capacity of 16 mg/g at a temperature of 30°C and pH 2 (Argun et al., 2014). Dyes like methyl red (Nowicki et al., 2016), methyl orange (Li et al., 2016), and methylene blue (Nowicki et al., 2016) can be adsorbed by activated carbon from pomelo peel. The reason behind such adsorption mechanisms is the micropore and mesopore structure on the surface of the activated carbon from pomelo peel (Foo and Hameed, 2011), chemical bonding, and electrostatic attraction between the negatively charged surface of pomelo peel and positively charged dye cations (Argun et al., 2014).

2.6.3 Use as Antioxidants and Antimicrobials

Pomelo peel has good direct and indirect antioxidant properties due to its content of secondary metabolites, as discussed earlier. Studies on different cultivars of pomelo show varying antioxidant capacities of pomelo peel extracts due to differences in experimental methodology, solvent used for extraction, and the antioxidant assay used. The essential oils extracted from pomelo peel have natural antimicrobial properties and are used widely in various industrial applications. Numerous studies have shown antimicrobial action of pomelo peel essential oil against pathogenic bacteria such as *Micrococcus luteus*, *Penicillium chrysogenum*, *Streptococcus iniae*, *Streptococcus faecalis*, *Bacillus cereus*, *Salmonella enteritidis*, *Salmonella enterica*, *Bacillus subtilis*, *Escherichia coli*,

Pseudomonas aeruginosa, and *Staphylococcus aureus* (Ou et al., 2015). In general, antimicrobial properties of essential oils from pomelo peel offer practical applications in food safety.

2.6.4 Other Uses

Pomelo peel has total soluble sugar concentrations similar to those of other citrus fruit peel like orange peel and grapefruit peel (Marín et al., 2007). Due to its high pectin and soluble sugar concentrations, pomelo peel has become a raw material of great interest. It can be used for production of bioethanol (Doran et al., 2000; Huang et al., 2014). Pomelo peel extract can also be used to produce nanoparticles (Barbhuiya et al., 2022). Synthesis of silver nanoparticles using citrus fruit extract is an eco-friendly, cost-effective, and green technology. Pomelo peel extract can act as a reducing and capping agent. Green synthesis of silver nanoparticles using pomelo peel extract has been a topic of great interest due to their antimicrobial activity and biomedical applications such as tissue restoration or regeneration, drug delivery, etc. (Barbhuiya et al., 2022). Pomelo peel-mediated green synthesized silver nanoparticles are irregular or spherical in shape, with sizes ranging from 20 to 30 nm (Jalani et al., 2018). Such nanoparticles also exhibit inhibitory action against Gram-positive and Gram-negative bacteria (Sarvamangala et al., 2013), suggesting potential antimicrobial properties of pomelo peel-synthesized silver nanoparticles.

2.7 CONCLUSIONS

The pomelo processing industry produces a great deal of industrial waste, which is harmful to humans and the environment. To create and recover value-added products from such waste streams, numerous research studies have been undertaken to utilize pomelo wastes, produce value-added products with various applications, meet sustainable utilization, and reduce hazardous environmental impact. Pomelo peel contains large amounts/numbers of important phytochemicals and nutrients. However, very limited research has been done on pomelo seed. Chemical composition of pomelo peel has shown suitable bioactive components, such as pectin, volatile compounds (essential oils), polysaccharides, organic acids, phenolic acids, carotenoids, flavonoids, and other active volatile members. The acceptability and flavor of food products can be improved by integration with pomelo peel, which will also improve the nutrient and phytochemical densities of the food material and help reduce the risk of severe diseases and encourage biological functions. Furthermore, pomelo waste utilization can decrease environmental pollution and produce alternate fuels in the form of activated carbon, ethanol, biogas, and bio-oil. Within this chapter, we have emphasized several strategies for the valorization of pomelo waste. Still, present efforts are faced with several challenges that could be overcome by following advanced methods and comprehensive scientific research. In addition, more research on pomelo seed is needed to fully understand the potential of all pomelo fruit processing residues.

REFERENCES

Abd Rahman, N. F., Shamsudin, R., Ismail, A., Shah, N. N. A. K., & Varith, J. (2018). Effects of drying methods on total phenolic contents and antioxidant capacity of the pomelo (*Citrus grandis* (L.) Osbeck) peels. *Innovative Food Science & Emerging Technologies, 50*, 217–225.

Aberoumand, A. (2012). Screening of phytochemical compounds and toxic proteinaceous protease inhibitor in some lesser-known food based plants and their effects and potential applications in food. *International Journal of Food Science and Nutrition Engineering, 2*(3), 16–20.

Abudayeh, Z. H., Al Khalifa, I. I., Mohammed, S. M., & Ahmad, A. A. (2019). Phytochemical content and antioxidant activities of pomelo peel extract. *Pharmacognosy Research, 11*(3), 244.

Agte, V., Tarwadi, K., & Chiplonkar, S. (1999). Phytate degradation during traditional cooking: Significance of the phytic acid profile in cereal-based vegetarian meals. *Journal of Food Composition and Analysis*, *12*(3), 161–167.

Ahmad, A. A., Al Khalifa, I. I., & Abudayeh, Z. H. (2018). The role of pomelo peel extract for experimentally induced wound in diabetic rats. *Pharmacognosy Journal*, *10*(5), 885–891.

Ani, P. N., & Abel, H. C. (2018). Nutrient, phytochemical, and antinutrient composition of Citrus maxima fruit juice and peel extract. *Food Science & Nutrition*, *6*(3), 653–658.

Argun, M. E., Güclü, D., & Karatas, M. (2014). Adsorption of reactive blue 114 dye by using a new adsorbent: Pomelo peel. *Journal of Industrial and Engineering Chemistry*, *20*(3), 1079–1084.

Ashraf, H., Butt, M. S., Iqbal, M. J., & Suleria, H. A. R. (2017). Citrus peel extract and powder attenuate hypercholesterolemia and hyperglycemia using rodent experimental modeling. *Asian Pacific Journal of Tropical Biomedicine*, *7*(10), 870–880.

Barbhuiya, R. I., Singha, P., Asaithambi, N., & Singh, S. K. (2022). Ultrasound-assisted rapid biological synthesis and characterization of silver nanoparticles using pomelo peel waste. *Food Chemistry*, *385*, 132602.

Baynes, R. D. (2000). Iron. In M. Stipanuk (Ed.), *Biochemical and physiological aspects of human nutrition* (pp. 711–740). W. B. Saunders.

Berglund, L., Brunzell, J. D., Goldberg, A. C., Goldberg, I. J., Sacks, F., Murad, M. H., & Stalenhoef, A. F. (2012). Evaluation and treatment of hypertriglyceridemia: An Endocrine Society clinical practice guideline. *The Journal of Clinical Endocrinology & Metabolism*, *97*(9), 2969–2989.

Chang, S., & Azrina, A. (2016). Antioxidant content and activity in different parts of pomelo [*Citrus grandis* (L.) Osbeck] by-products. In *III International Conference on Agricultural and Food Engineering*, Kuala Lumpur, Malaysia, 1152.

Chavan, P., Singh, A. K., & Kaur, G. (2018). Recent progress in the utilization of industrial waste and by-products of citrus fruits: A review. *Journal of Food Process Engineering*, *41*(8), e12895.

Chen, S., Zhang, J., Zhang, C., Yue, Q., Li, Y., & Li, C. (2010). Equilibrium and kinetic studies of methyl orange and methyl violet adsorption on activated carbon derived from *Phragmites australis*. *Desalination*, *252*(1–3), 149–156.

Cheong, M., Liu, S., Yeo, J., Chionh, H., Pramudya, K., Curran, P., & Yu, B. (2011). Identification of aroma-active compounds in Malaysian pomelo (*Citrus grandis* (L.) Osbeck) peel by gas chromatography-olfactometry. *Journal of Essential Oil Research*, *23*(6), 34–42.

Dai, Y., van Spronsen, J., Witkamp, G.-J., Verpoorte, R., & Choi, Y. H. (2013). Natural deep eutectic solvents as new potential media for green technology. *Analytica Chimica Acta*, *766*, 61–68.

Ding, X., Guo, L., Zhang, Y., Fan, S., Gu, M., Lu, Y., Jiang, D., Li, Y., Huang, C., & Zhou, Z. (2013). Extracts of pomelo peels prevent high-fat diet-induced metabolic disorders in c57bl/6 mice through activating the PPARα and GLUT4 pathway. *PloS one*, *8*(10), e77915.

Doran, J. B., Cripe, J., Sutton, M., & Foster, B. (2000). Fermentations of pectin-rich biomass with recombinant bacteria to produce fuel ethanol. *Applied Biochemistry and Biotechnology*, *84*(1), 141–152. https://doi.org/10.1385/ABAB:84-86:1-9:141.

Elgharbawy, A. A., Hayyan, A., Hayyan, M., Mirghani, M. E., Salleh, H. M., Rashid, S. N., Ngoh, G. C., Liew, S. Q., Nor, M. R. M., & bin Mohd Yusoff, M. Y. Z. (2019). Natural deep eutectic solvent-assisted pectin extraction from pomelo peel using sonoreactor: Experimental optimization approach. *Processes*, *7*(7), 416.

Feumba, D. R., Ashwini, R. P., & Ragu, S. M. (2016). Chemical composition of some selected fruit peels. *European Journal of Food Science and Technology 4*(4), 12–21.

Foo, K., & Hameed, B. (2011). Microwave assisted preparation of activated carbon from pomelo skin for the removal of anionic and cationic dyes. *Chemical Engineering Journal*, *173*(2), 385–390.

Goh, R. M. V., Lau, H., Liu, S. Q., Lassabliere, B., Guervilly, R., Sun, J., Bian, Y., & Yu, B. (2019). Comparative analysis of pomelo volatiles using headspace-solid phase micro-extraction and solvent assisted flavour evaporation. *LWT*, *99*, 328–345.

Gropper, S. S., & Smith, J. L. (2012). *Advanced nutrition and human metabolism*. Cengage Learning.

Gülçin, İ. (2006). Antioxidant activity of caffeic acid (3, 4-dihydroxycinnamic acid). *Toxicology*, *217*(2–3), 213–220.

Guo, J.-j., Gao, Z.-p., Xia, J.-l., Ritenour, M. A., Li, G.-y., & Shan, Y. (2018). Comparative analysis of chemical composition, antimicrobial and antioxidant activity of citrus essential oils from the main cultivated varieties in China. *LWT*, *97*, 825–839.

Gyawali, R., Jeon, D. H., Moon, J., Kim, H., Song, Y. W., Hyun, H. B., Jeong, D., & Cho, S. K. (2012). Chemical composition and antiproliferative activity of supercritical extract of *Citrus grandis* (L.) Osbeck fruits from Korea. *Journal of Essential Oil Bearing Plants, 15*(6), 915–925.

Hagerman, A. E., Rice, M. E., & Ritchard, N. T. (1998). Mechanisms of protein precipitation for two tannins, pentagalloyl glucose and epicatechin16 (4→ 8) catechin (procyanidin). *Journal of Agricultural and Food Chemistry, 46*(7), 2590–2595.

Haghighian, M. K., Rafraf, M., Moghaddam, A., Hemmati, S., Jafarabadi, M. A., & Gargari, B. P. (2016). Pomegranate (*Punica granatum* L.) peel hydro alcoholic extract ameliorates cardiovascular risk factors in obese women with dyslipidemia: A double blind, randomized, placebo controlled pilot study. *European Journal of Integrative Medicine, 8*(5), 676–682.

Hashemipour, M., Kargar, M., Ghannadi, A., & Kelishadi, R. (2016). The effect of Citrus Aurantifolia (Lemon) peels on cardiometabolic risk factors and markers of endothelial function in adolescents with excess weight: A triple-masked randomized controlled trial. *Medical Journal of the Islamic Republic of Iran, 30*, 429.

He, W., Li, X., Peng, Y., He, X., & Pan, S. (2019). Anti-oxidant and anti-melanogenic properties of essential oil from peel of Pomelo cv. Guan Xi. *Molecules, 24*(2), 242.

Hong, H. J., Jin, J. Y., Yang, H., Kang, W. Y., Kim, D. G., Lee, S., Choi, Y., Kim, J. H., Han, C. H., & Lee, Y. J. (2010). Dangyuja (Citrus grandis Osbeck) peel improves lipid profiles and alleviates hypertension in rats fed a high-fat diet. *Laboratory Animal Research, 26*(4), 361–367.

Hosni, K., Zahed, N., Chrif, R., Abid, I., Medfei, W., Kallel, M., Brahim, N. B., & Sebei, H. (2010). Composition of peel essential oils from four selected Tunisian Citrus species: Evidence for the genotypic influence. *Food Chemistry, 123*(4), 1098–1104.

Huang, R., Cao, M., Guo, H., Qi, W., Su, R., & He, Z. (2014). Enhanced ethanol production from pomelo peel waste by integrated hydrothermal treatment, multienzyme formulation, and fed-batch operation. *Journal of Agricultural and Food Chemistry, 62*(20), 4643–4651.

Ibrahim, M., Amin, M. N., Millat, M. S., Raju, J. A., Hussain, M. S., Sultana, F., Islam, M. M., & Hasan, M. M. (2018). Methanolic extract of peel of citrus maxima fruits exhibit analgesic, CNS depressant and anti-inflammatory activities in Swiss albino mice. *Biology, Engineering, Medicine and Science Reports, 4*(1), 07–11.

Iyayi, E. A., Kluth, H., & Rodehutscord, M. (2008). Effect of heat treatment on antinutrients and precaecal crude protein digestibility in broilers of four tropical crop seeds. *International Journal of Food Science & Technology, 43*(4), 610–616.

Jalani, N. S., Michell, W., Lin, W. E., Hanani, S. Z., Hashim, U., & Abdullah, R. (2018). Biosynthesis of silver nanoparticles using Citrus grandis peel extract. *Malaysian Journal of Analytical Sciences, 22*(4), 676–683.

John, S., Weiss, J. N., & Ribalet, B. (2011). Subcellular localization of hexokinases I and II directs the metabolic fate of glucose. *PloS one, 6*(3), e17674.

Katheresan, V., Kansedo, J., & Lau, S. Y. (2018). Efficiency of various recent wastewater dye removal methods: A review. *Journal of Environmental Chemical Engineering, 6*(4), 4676–4697.

Koleckar, V., Kubikova, K., Rehakova, Z., Kuca, K., Jun, D., Jahodar, L., & Opletal, L. (2008). Condensed and hydrolysable tannins as antioxidants influencing the health. *Mini Reviews in Medicinal Chemistry, 8*(5), 436–447.

Kuo, P.-C., Liao, Y.-R., Hung, H.-Y., Chuang, C.-W., Hwang, T.-L., Huang, S.-C., Shiao, Y.-J., Kuo, D.-H., & Wu, T.-S. (2017). Anti-inflammatory and neuroprotective constituents from the peels of Citrus grandis. *Molecules, 22*(6), 967.

Ladeji, O., Akin, C., & Umaru, H. (2004). Level of antinutritional factors in vegetables commonly eaten in Nigeria. *African Journal of Natural Sciences, 7*, 71–73.

Lellis, B., Fávaro-Polonio, C. Z., Pamphile, J. A., & Polonio, J. C. (2019). Effects of textile dyes on health and the environment and bioremediation potential of living organisms. *Biotechnology Research and Innovation, 3*(2), 275–290.

Lewis, A., & Segal, A. (2010). Hyperlipidemia and primary prevention of stroke: Does risk factor identification and reduction really work? *Current Atherosclerosis Reports, 12*(4), 225–229.

Li, H., Sun, Z., Zhang, L., Tian, Y., Cui, G., & Yan, S. (2016). A cost-effective porous carbon derived from pomelo peel for the removal of methyl orange from aqueous solution. *Colloids and Surfaces A: Physicochemical and Engineering Aspects, 489*, 191–199.

Li, H., Zhang, L., Sun, Z., Liu, Y., Yang, B., & Yan, S. (2015). One-step synthesis of magnetic 1, 6-hexanediamine-functionalized reduced graphene oxide–zinc ferrite for fast adsorption of Cr (vi). *RSC Advances*, *5*(40), 31787–31797.

Li, P.-l., Liu, M.-h., Hu, J.-h., & Su, W.-w. (2014). Systematic chemical profiling of Citrus grandis 'Tomentosa' by ultra-fast liquid chromatography/diode-array detector/quadrupole time-of-flight tandem mass spectrometry. *Journal of Pharmaceutical and Biomedical Analysis*, *90*, 167–179.

Liang, N., & Kitts, D. D. (2016). Role of chlorogenic acids in controlling oxidative and inflammatory stress conditions. *Nutrients*, *8*(1), 16.

Liew, S. Q., Ngoh, G. C., Yusoff, R., & Teoh, W. H. (2016). Sequential ultrasound-microwave assisted acid extraction (UMAE) of pectin from pomelo peels. *International Journal of Biological Macromolecules*, *93*, 426–435.

Liew, S. Q., Teoh, W. H., Yusoff, R., & Ngoh, G. C. (2019). Comparisons of process intensifying methods in the extraction of pectin from pomelo peel. *Chemical Engineering and Processing-Process Intensification*, *143*, 107586.

Lim, S., & Loh, S. (2016). In vitro antioxidant capacities and antidiabetic properties of phenolic extracts from selected citrus peels. *International Food Research Journal*, *23*(1), 211–219.

Liu, Y., Liu, A., Ibrahim, S. A., Yang, H., & Huang, W. (2018). Isolation and characterization of microcrystalline cellulose from pomelo peel. *International Journal of Biological Macromolecules*, *111*, 717–721.

Liu, Z., Zu, Y., & Yang, L. (2017). A process to preserve valuable compounds and acquire essential oils from pomelo flavedo using a microwave irradiation treatment. *Food Chemistry*, *224*, 172–180.

Lü, Z., Zhang, Z., Wu, H., Zhou, Z., & Yu, J. (2016). Phenolic composition and antioxidant capacities of Chinese local pummelo cultivars' peel. *Horticultural Plant Journal*, *2*(3), 133–140.

Malleshappa, P., Kumaran, R. C., Venkatarangaiah, K., & Parveen, S. (2018). Peels of citrus fruits: A potential source of anti-inflammatory and anti-nociceptive agents. *Pharmacognosy Journal*, *10*(6s), s172–s178.

Mancuso, M., Catalfamo, M., Laganà, P., Rappazzo, A. C., Raymo, V., Zampino, D., & Zaccone, R. (2019). Screening of antimicrobial activity of citrus essential oils against pathogenic bacteria and Candida strains. *Flavour and Fragrance Journal*, *34*(3), 187–200.

Maran, J. P., Swathi, K., Jeevitha, P., Jayalakshmi, J., & Ashvini, G. (2015). Microwave-assisted extraction of pectic polysaccharide from waste mango peel. *Carbohydrate Polymers*, *123*, 67–71.

Marín, F. R., Soler-Rivas, C., Benavente-García, O., Castillo, J., & Pérez-Alvarez, J. A. (2007). By-products from different citrus processes as a source of customized functional fibres. *Food Chemistry*, *100*(2), 736–741.

McDougall, G. J., & Stewart, D. (2005). The inhibitory effects of berry polyphenols on digestive enzymes. *BioFactors*, *23*(4), 189–195.

Methacanon, P., Krongsin, J., & Gamonpilas, C. (2014). Pomelo (Citrus maxima) pectin: Effects of extraction parameters and its properties. *Food Hydrocolloids*, *35*, 383–391.

Nasir, M., Akhtar, S., & Sharif, M. K. (2004). Effect of moisture and packaging on the Shelf life of wheat flour. *Internet Journal of Food Safety V*, *4*, 1–6.

Njoroge, S. M., Koaze, H., Karanja, P. N., & Sawamura, M. (2005). Volatile constituents of redblush grapefruit (Citrus paradisi) and pummelo (Citrus grandis) peel essential oils from Kenya. *Journal of Agricultural and Food Chemistry*, *53*(25), 9790–9794.

Noreen, A., Akram, J., Rasul, I., Mansha, A., Yaqoob, N., Iqbal, R., Tabasum, S., Zuber, M., & Zia, K. M. (2017). Pectins functionalized biomaterials; a new viable approach for biomedical applications: A review. *International Journal of Biological Macromolecules*, *101*, 254–272.

Nowicki, P., Kazmierczak-Razna, J., & Pietrzak, R. (2016). Physicochemical and adsorption properties of carbonaceous sorbents prepared by activation of tropical fruit skins with potassium carbonate. *Materials & Design*, *90*, 579–585.

Oboh, G., & Ademosun, A. O. (2011). Shaddock peels (Citrus maxima) phenolic extracts inhibit α-amylase, α-glucosidase and angiotensin I-converting enzyme activities: A nutraceutical approach to diabetes management. *Diabetes & Metabolic Syndrome: Clinical Research & Reviews*, *5*(3), 148–152.

Okwu, D., & Emenike, I. (2007). Nutritive value and mineral content of different varieties of citrus fruits. *Journal of Food Technology*, *5*(2), 105–108.

Ortuño, A., Garcia-Puig, D., Fuster, M., Perez, M., Sabater, F., Porras, I., Garcia-Lidon, A., & Del Rio, J. (1995). Flavanone and nootkatone levels in different varieties of grapefruit and pummelo. *Journal of Agricultural and Food Chemistry*, *43*(1), 1–5.

Ou, M.-C., Liu, Y.-H., Sun, Y.-W., & Chan, C.-F. (2015). The composition, antioxidant and antibacterial activities of cold-pressed and distilled essential oils of Citrus paradisi and Citrus grandis (L.) Osbeck. *Evidence-Based Complementary and Alternative Medicine*, *2015*, 1–9.

Palazzolo, E., Laudicina, V. A., & Germanà, M. A. (2013). Current and potential use of citrus essential oils. *Current Organic Chemistry*, *17*(24), 3042–3049.

Patel, P. (2017). Extraction of pectin from mango peel and application of pectin. *Journal of Natural Products and Resources*, *3*(1), 102–103.

Peng, C., Lang, J., Xu, S., & Wang, X. (2014). Oxygen-enriched activated carbons from pomelo peel in high energy density supercapacitors. *RSC Advances*, *4*(97), 54662–54667.

Perussello, C. A., Zhang, Z., Marzocchella, A., & Tiwari, B. K. (2017). Valorization of apple pomace by extraction of valuable compounds. *Comprehensive Reviews in Food Science and Food Safety*, *16*(5), 776–796.

Poli, A., Anzelmo, G., Fiorentino, G., Nicolaus, B., Tommonaro, G., & Di Donato, P. (2011). Polysaccharides from wastes of vegetable industrial processing: New opportunities for their eco-friendly re-use. In M. Elnashar (Ed.), *Biotechnology of biopolymers* (pp. 33–56). London: IntechOpen.

Puglisi, I., De Patrizio, A., Schena, L., Jung, T., Evoli, M., Pane, A., Van Hoa, N., Van Tri, M., Wright, S., & Ramstedt, M. (2017). Two previously unknown Phytophthora species associated with brown rot of Pomelo (Citrus grandis) fruits in Vietnam. *PLoS One*, *12*(2), e0172085.

Quoc, L., Huyen, V., Hue, L., Hue, N., Thuan, N., Tam, N., Thuan, N., & Duy, T. (2015). Extraction of pectin from pomelo (Citrus maxima) peels with the assistance of microwave and tartaric acid. *International Food Research Journal*, *22*(4), 1637.

Quynh, N. T. T., Phong, V. C., & Huong, P. T. (2010). Effect of pomelo (citrus grandis (l). osbeck) peel extract on lipid-carbohydrate metabolic enzymes and blood lipid, glucose parameters in experimental obese and diabetic mice. *VNU Journal of Science: Natural Sciences and Technology*, *26*(4), 244–232.

Rosnah, S., Nur Farhana, A., Amin, I., & Nik Suhaila, N. (2015). Effect of oven-drying on the colour of pomelo fruit waste. In *8th Asia-Pacific Drying Conference (ADC 2015)* (Kuala Lumpur, Malaysia).

Sarkic, A., & Stappen, I. (2018). Essential oils and their single compounds in cosmetics—A critical review. *Cosmetics*, *5*(1), 11.

Sarvamangala, D., Kondala, K., Murthy, U., Rao, B. N., Sharma, G., & Satyanarayana, R. (2013). Biogenic synthesis of AGNP's using Pomelo fruit—characterization and antimicrobial activity against Gram+ Ve and Gram− Ve bacteria. *International Journal of Pharmaceutical Sciences Review and Research*, *19*(2), 30–35.

Schrauzer, G. N. (2000). Selenomethionine: A review of its nutritional significance, metabolism and toxicity. *The Journal of Nutrition*, *130*(7), 1653–1656.

Shan, Y. (2016). Chapter 1 - Functional components of citrus peel. In Y. Shan (Ed.), *Comprehensive utilization of citrus by-products* (pp. 1–13). Academic Press.

Stone, N. J., Robinson, J. G., Lichtenstein, A. H., Bairey Merz, C. N., Blum, C. B., Eckel, R. H., Goldberg, A. C., Gordon, D., Levy, D., & Lloyd-Jones, D. M. (2014). 2013 ACC/AHA guideline on the treatment of blood cholesterol to reduce atherosclerotic cardiovascular risk in adults: A report of the American College of Cardiology/American Heart Association Task Force on Practice Guidelines. *Journal of the American College of Cardiology*, *63*(25 Part B), 2889–2934.

Sun, H., Ni, H., Yang, Y., Chen, F., Cai, H., & Xiao, A. (2014). Sensory evaluation and gas chromatography–mass spectrometry (GC-MS) analysis of the volatile extracts of pummelo (Citrus maxima) peel. *Flavour and Fragrance Journal*, *29*(5), 305–312.

Tao, N., Gao, Y., Liu, Y., & Ge, F. (2010). Carotenoids from the peel of Shatian pummelo (Citrus grandis Osbeck) and its antimicrobial activity. *American-Eurasian Journal of Agricultural and Environmental Science*, *7*(1), 110–115.

Teng, T.-S., Ji, A.-l., Ji, X.-Y., & Li, Y.-Z. (2017). Neutrophils and immunity: From bactericidal action to being conquered. *Journal of Immunology Research*, *2017*, 1–14.

Thakur, B. R., Singh, R. K., Handa, A. K., & Rao, M. (1997). Chemistry and uses of pectin—A review. *Critical Reviews in Food Science & Nutrition*, *37*(1), 47–73.

Tocmo, R., Pena-Fronteras, J., Calumba, K. F., Mendoza, M., & Johnson, J. J. (2020). Valorization of pomelo (Citrus grandis Osbeck) peel: A review of current utilization, phytochemistry, bioactivities, and mechanisms of action. *Comprehensive Reviews in Food Science and Food Safety*, *19*(4), 1969–2012.

Toh, J., Khoo, H., & Azrina, A. (2013). Comparison of antioxidant properties of pomelo [Citrus Grandis (L) Osbeck] varieties. *International Food Research Journal, 20*(4), 1661–1668.

Tonini, D., Albizzati, P. F., & Astrup, T. F. (2018). Environmental impacts of food waste: Learnings and challenges from a case study on UK. *Waste Management, 76*, 744–766.

Tundis, R., Loizzo, M., & Menichini, F. (2010). Natural products as α-amylase and α-glucosidase inhibitors and their hypoglycaemic potential in the treatment of diabetes: An update. *Mini Reviews in Medicinal Chemistry, 10*(4), 315–331.

Uraku, A. (2015). Nutritional potential of Citrus sinensis and Vitis vinifera peels. *Journal of Advancement in Medical and Life Sciences, 3*(4), 2348–2294X.

Walker, R., Janda, E., & Mollace, V. (2014). The use of bergamot-derived polyphenol fraction in cardiometabolic risk prevention and its possible mechanisms of action. In R. R. Watson, V. R. Preedy, & S. Zibadi (Eds.), *Polyphenols in human health and disease* (pp. 1087–1105). Elsevier.

Wandee, Y., Uttapap, D., & Mischnick, P. (2019). Yield and structural composition of pomelo peel pectins extracted under acidic and alkaline conditions. *Food Hydrocolloids, 87*, 237–244.

Wang, Q., Zhou, C., Kuang, Y.-j., Jiang, Z.-h., & Yang, M. (2020). Removal of hexavalent chromium in aquatic solutions by pomelo peel. *Water Science and Engineering, 13*(1), 65–73.

Wang, Y.-C., Chuang, Y.-C., & Hsu, H.-W. (2008). The flavonoid, carotenoid and pectin content in peels of citrus cultivated in Taiwan. *Food Chemistry, 106*(1), 277–284.

WHO/FAO. (2001). *Human vitamin and mineral requirements.* Report of a joint FAO/WHO expert consultation, Bangkok, Thailand. Food and Nutrition Division, FAO, Rome, 235–247.

Wu, Y., Cha, L., Fan, Y., Fang, P., Ming, Z., & Sha, H. (2017). Activated biochar prepared by pomelo peel using H 3 PO 4 for the adsorption of hexavalent chromium: Performance and mechanism. *Water, Air, & Soil Pollution, 228*(10), 1–13.

Xu, C.-J., Fraser, P. D., Wang, W.-J., & Bramley, P. M. (2006). Differences in the carotenoid content of ordinary citrus and lycopene-accumulating mutants. *Journal of Agricultural and Food Chemistry, 54*(15), 5474–5481.

Yu, E. A., Kim, G. S., Lee, J. E., Park, S., Yi, S., Lee, S. J., Kim, J. H., Jin, J. S., Abd El-Aty, A., & Shim, J. H. (2015). Flavonoid profiles of immature and mature fruit tissues of Citrus grandis Osbeck (Dangyuja) and overall contribution to the antioxidant effect. *Biomedical Chromatography, 29*(4), 590–594.

Yu, J., Ji, H., & Liu, A. (2018). Preliminary structural characteristics of polysaccharides from pomelo peels and their antitumor mechanism on S180 tumor-bearing mice. *Polymers, 10*(4), 419.

Zahra, N., Alim-un-Nisa, K., Saeed, M. K., Ahmad, I., & Hina, S. (2017). Nutritional evaluation and antioxidant activity of zest obtained from orange (Citrus sinensis) peels. *International Journal of Theoretical and Applied Science, 9*(1), 07–10.

Zain, M., Fazelin, N., Yusop, M., & Ahmad, I. (2015). Preparation and characterization of cellulose and nanocellulose from pomelo (Citrus grandis) albedo. *Nutrition & Food Sciences, 5*(1), 334.

Zhu, J., Liu, Q., Li, Z., Liu, J., Zhang, H., Li, R., Wang, J., & Emelchenko, G. (2017). Recovery of uranium (VI) from aqueous solutions using a modified honeycomb-like porous carbon material. *Dalton Transactions, 46*(2), 420–429.

CHAPTER 3

Avocado Wastes
Chemistry, Processing, and Utilization

Vildan Eyiz and Ismail Tontul

CONTENTS

3.1 INTRODUCTION

The avocado, commonly known as the avocado pear or alligator pear, is an evergreen tropical tree (*Persea americana* Mill.) native to Mexico, Central America and South America, but which is now widely grown worldwide. Avocado belongs to the laurel family (the Lauraceae) (Zafar and Sidhu, 2011).

The avocado fruit is an oval-shaped, pear-like fruit with a variable length of neck; the fruit can reach between 7.5 and 33 cm in length and 15 cm in width. The fruit's skin is green to purple and can have a glossy or matte finish, with a smooth or lumpy texture. The avocado seed consists of a very thin shell called the endocarp surrounding the core (Araújo et al., 2018).

According to the Food and Agriculture Organization (FAO), global avocado production was 7.18 million tons in 2019. Mexico is the main producer of avocados, accounting for 32% of world production. There are 64 avocado-producing countries worldwide. Avocado consumption has risen steadily globally over the past two decades. They are no longer regarded as an exotic fruit but as a part of many countries' daily diet. In addition to its oil, there has been increasing demand for fresh avocado and even processed foods derived from this berry, such as guacamole, frozen goods and avocado paste, with avocados also being used in the cosmetic, soap and shampoo industries.

Avocado is a fruit with a very high nutritional value. It is rich in the fat-soluble vitamins A, B, D and E and minerals (potassium, phosphorus, iron, magnesium, sulfur and copper). It also has a high content of unsaturated oils, which lowers "bad cholesterol" (low-density lipoprotein) in the blood,

DOI: 10.1201/9781003164463-3

and contains dietary fiber, carotenoids and phenolics (Mfonobong et al., 2013). Avocado can be used in meals, salads, desserts, ice creams and pastries (Jayanegara et al., 2011).

There have been studies showing that avocado has anticancer, anti-inflammatory and life-extending properties because of its antioxidant potential and content of vitamins, mono-unsaturated fatty acids, fiber and potassium (Kim and Uhl, 2011, Talabi et al., 2016) Avocado is primarily eaten as fresh pulp. Its industrialization started with the production of pulp (avocado purée or guacamole, packaged slices and pieces, dehydrated or dried pulp) and the extraction of oil (Rotta et al., 2016).

Direct or industrial avocado usage results in large volumes of by-products, such as leaves, peel and seed. In ancient cultures or folk medicine, avocado fruit and its by-products have commonly been used as infusions to achieve medicinal effects. Avocado seed and peel constitute 16% and 11% of the fruit's total weight, respectively (Jiménez-Velázquez et al., 2020). This means that approximately 1.9 million tons of avocado by-products is discarded annually (Salazar-López et al., 2020). As part of the circular economy, agricultural and food processing wastes and by-products, like those of the avocado industry, should be re-used or recycled (Colombo and Papetti, 2019).

3.2 CHEMISTRY OF AVOCADO WASTE STREAMS

3.2.1 Avocado Leaves

The chemical composition of avocado leaves has been reported to be 5.3% moisture, 4.0% fat, 25.5% protein, 7.3% carbohydrate, 38.4% fiber and 19.4% ash (Arukwe et al., 2012). The leaves also contain high concentrations of potassium, magnesium, sodium, calcium and phosphorus (Arukwe et al., 2012). Many studies have reported the presence of phenolic acids and flavonoids, such as procyanidins, in avocado leaves (De Almeida et al., 1998, Adeyemi, Okpo and Ogunti, 2002). Park et al. (2019) isolated ten flavonoids, four megastigmane glycosides and two lignans, along with a novel flavonol glycoside, from avocado leaves. Kose et al. (2020) studied the secondary metabolite content of ethanolic and aqueous extracts of avocado leaves. The main secondary metabolites of the ethanolic extract were determined to be chlorogenic acid (852.81 mg/kg), kaempferol (663.54 mg/kg), quercetin-3-O-arabinoside (253.18 mg/kg) and fumaric acid (214.32 mg/kg) On the other hand, pyrogallol (122.25 mg/kg) was the main secondary metabolite in the aqueous extract, followed by fumaric acid (59.18 mg/kg), kaempferol (50.15 mg/kg) and chlorogenic acid (28.83 mg/kg). The major constituents of the essential oil from the avocado leaves were β-pinene, p-cymene, caryophyllene, farnesene, humulene, estragole, cadinene and anethole (Bergh, Scora and Storey, 1973).

There is a growing interest in studying avocado leaves due to their health-promoting properties and high nutritional value. For example, avocado leaves are used in Brazil to treat various ailments such as urinary tract infections, bronchitis and rheumatism. Moreover, Brai, Odetola and Agomo (2007) reported that avocado leaves have diuretic, anti-inflammatory, hypertensive and hypoglycemic properties, and have been used to treat diarrhea, sore throat and bleeding. In a recent study, an aqueous extract of avocado leaves was reported to exhibit hypoglycemic activity (Antia, Okokon and Okon, 2005).

Kose et al. (2020) studied the antioxidant activity and anticholinergic activity of aqueous and ethanolic extracts of avocado leaves. They reported that both extracts exhibited lower antioxidant activity than standard synthetic and natural antioxidants (BHA (butylated hydroxyanisole) and BHT (butylated hydroxytoluene), trolox (6-hydroxy-2,5,7,8-tetramethylchroman-2-carboxylic acid) or α-tocopherol), whereas both extracts had higher anticholinergic activity than tacrine. Therefore, avocado leaf extracts were shown to have powerful anti-acetylcholinesterase and anti-butyrylcholinesterase effects.

In another study, the effects of aqueous or methanolic extracts of avocado leaves on rat bodyweight and liver lipids were examined. Rats treated with aqueous or methanolic extract exhibited a 14% or 25% decrease in bodyweight gain, respectively. This phenomenon was explained by increased catabolism in the treated rats (Brai, Odetola and Agomo, 2007).

In Indonesia, avocado leaves have been used in traditional medicine as a diuretic to treat urolithiasis. In a study, the calcium concentration in the kidneys of rats treated with an extract of avocado leaves was significantly lower ($P < 0.05$) than in control untreated rats (Wientarsih et al., 2012). This study also showed that the leaf extract could be recommended for preventative treatment of urolithiasis.

Plazola-Jacinto et al. (2019) used a different approach by microencapsulating biologically active compounds from avocado leaves in vegetable oils as a result of spray drying. These researchers reported the encapsulation efficiency of the leaf extract to be 50%.

3.2.2 Avocado Fruit Peel

The average weight of the peel of avocado fruit is 11% of the total weight (Salazar-López et al., 2020). The approximate composition of avocado peel was reported to be 2.9–11.0% total lipids, 1.5–6.3% protein, the ash content of 0.75–1.6% ash and 6.9–56.9% fiber (Jimenez et al., 2020). Figueroa et al. (2018a) identified 61 phenolic and polar compounds in avocado peel. These researchers reported that the most common groups were procyanidins, flavonols, hydroxybenzoic acids and hydroxycinnamic acids. The total phenolic content of avocado peel from the cultivar Hass was reported to be 25.3 mg (+)-catechin equivalent (CE)/g dry weight (dw), whereas it was 15.6 mg CE/g dw for cultivar Shepard (Kosińska et al., 2012). Catechin, procyanidin dimer B (I), 5-*O*-caffeoylquinic acid and quercetin-3-*O*-arabinosyl-glucoside were the main phenolic compounds present in the peel of 'Hass' fruit. On the other hand, quercetin derivatives were the dominant phenolics in the peel of 'Shepard' fruit (Kosińska et al., 2012). The phenolic composition of avocado peel, as determined by López-Cobo et al. (2016), is shown in Table 3.1.

In a recent study, the peel of 20 different fruits were compared in terms of total phenolic, total flavonoid and total tannin concentrations, as well as antioxidant activity (Suleria, Barrow and Dunshea, 2020). Avocado peel was ranked fifth in terms of total phenolic concentration, first in total tannin concentration and second in antioxidant activity.

The antioxidant-active flavonoids in avocado fruit are mainly found in the fruit peel (Rodríguez-Carpena et al., 2011), which is also rich in mineral compounds (Mariane et al., 2016). Therefore, the peel can be consumed as a source of phytomedicinals and phytonutrients (Marques, Hofman and Wearing, 2001).

Mariane et al. (2016) produced an aqueous avocado peel tea in their study. They reported the tea to be a good source of phenolic substances, flavonoids and antioxidants (Mariane et al., 2016).

Ramos-Aguilar et al. (2020) evaluated the phytochemical composition and biological properties of avocado peel. They determined that avocado peel contains high concentrations of polar and

Table 3.1 Phenolic Concentrations in Avocado Peel from
Avocados at Different Development Stages

Compounds	Ripe		Over-Ripe	
	Mean	SD	Mean	SD
Quinic acid	0.43	0.01	0.12	0.02
Chlorogenic acid	189.89	1.07	159.24	0.82
Quercetin-3,4'-diglucoside	270.05	2.19	29.50	3.73
Quercetin-3-*O*-arabinosyl-glucoside	19.76	0.22	60.78	0.30
Rutin	35.31	0.61	46.96	2.09
Total	515.45	0.00	296.61	2.22

Source: Adapted from López-Cobo et al. (2016)
SD: standard deviation

non-polar phytochemicals such as phenolic compounds, chlorophyll and lutein. They also reported novel bioactive compounds from avocado peel, such as some carotenoids, acids, sterols, and volemitol which were reported in avocado peel for the first time (Ramos-Aguilar et al., 2020).

3.2.3 Avocado Wastewater

Avocado wastewater is obtained during cold pressing of fruit to obtain avocado oil. As a result of processing 1000 kg of avocado fruit, 80 kg of avocado oil are obtained, and 500 kg of wastewater are generated (Figure 3.1). This wastewater causes problems for producers and processors because of its volume and organic material content.

The composition of avocado wastewater was determined to be 88.3% moisture, 2.6% dietary fiber, 1.1% protein, 2.1% ash and 6.3% lipid (Permal et al., 2020b).

The antioxidant activity and total phenolic content of different spray-dried preparations of avocado wastewater are shown in Table 3.2.

Permal et al. (2020b) suggested that spray-drying can turn avocado wastewater into a valuable product. They revealed that a higher spray-drying temperature (160°C) achieved higher antioxidant activity and total phenolic content in the resultant powder compared to other temperatures (110–160°C). Additionally, the highest drying yield was observed at a 160°C inlet temperature. In a subsequent study, the best carrier agent in terms of yield was determined to be 5% whey protein concentrate (Permal et al., 2020a). The avocado wastewater obtained was used in a cooked pork fat product as an antioxidant and compared with standard synthetic and natural antioxidants (BHT, BHA and α-tocopherol). When supplemented with avocado wastewater, thiobarbituric acid-reactive substances values of the pork product were found to be similar to the values present when supplemented with BHA or BHT and lower than those of the control or the α-tocopherol-supplemented sample.

3.2.4 Avocado Seed

The fact that avocado has a much larger seed than other fruits results in considerable volumes of avocado seed waste. This observation underlines the necessity of analyzing avocado seeds to identify the various compounds present. Although avocado seeds represent such a significant proportion of the whole fruit, there have been few scientific studies on the composition of avocado seeds.

Avocado seeds contain 43–81% carbohydrates, 3–15% lipids, 0.1–9% proteins, 2–4.2% fiber and 1.3–4.3% minerals (Araújo et al., 2018). López-Cobo et al. (2016) identified 11 different minor compounds in avocado seed, among which tyrosyl glucoside and 1-caffeoylquinic acid were the main compounds. The phenolic composition of avocado seeds as determined by López-Cobo et al. (2016) is given in Table 3.3.

Currently, the seed is a waste problem for avocado processors but represents an underused resource. Avocado seeds are rich in phenolic compounds, which have significant effects on human health (Dabas et al., 2013). Especially in South American countries, there is an ethnopharmacological use of avocado seeds for health-related treatments. Current research has shown that avocado seeds can improve hypercholesterolemia and help to treat hypertension, inflammatory conditions and diabetes. The seeds have also been reported to exhibit insecticidal, fungicidal and antimicrobial activities.

Recent studies on avocado seeds have generally focused on extracting and identifying bioactive compounds. For example, Segovia, Corral-Pérez and Almajano (2016) released polyphenol compounds from avocado seed by ultrasound-assisted extraction. The authors modeled the extraction process and reported avocado seeds to be a sustainable and cheap polyphenol source for industry. Similarly, Segovia et al. (2018) demonstrated that avocado seeds can be used to obtain extracts with high antioxidant power (Segovia et al., 2018).

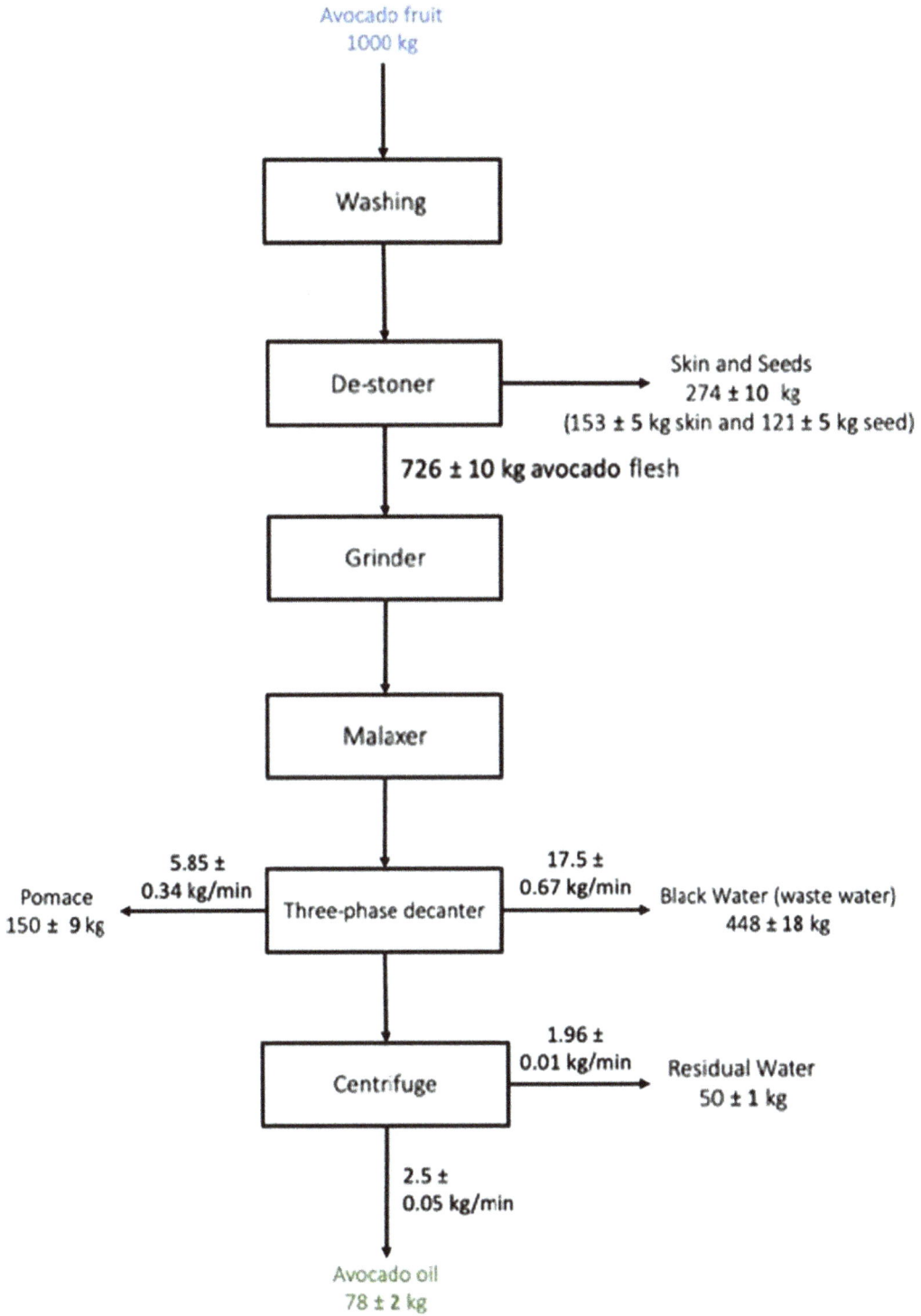

Figure 3.1   Cold pressed avocado oil production process. [Reprinted from Permal et al. (2020b) with permission from Elsevier.]

Table 3.2 Mean ± Standard Deviation Antioxidant Activity (CUPRAC, FRAP, Phosphomolybdenum and TPC) in Avocado Wastewater Powder at Different Spray-Drying Temperatures

Samples	CUPRAC (g TE/100 g powder)	FRAP (g TE/100 g powder)	Phosphomolybdenum (mg TE/100 g powder)	TPC (g GAE/100 g powder)
110°C	8.8 ± 0.87	3.1 ± 0.13	5.0 ± 1.18	3.9 ± 0.24
120°C	8.0 ± 0.84	3.6 ± 3.60	6.5 ± 0.74	4.4 ± 0.38
130°C	11.3 ± 1.36	4.1 ± 4.07	7.3 ± 0.31	5.1 ± 0.58
140°C	11.1 ± 0.70	4.0 ± 0.33	7.2 ± 0.41	4.7 ± 0.86
150°C	12.6 ± 0.52	4.2 ± 4.18	7.4 ± 0.38	5.0 ± 0.44
160°C	11.1 ± 0.96	5.4 ± 0.56	7.6 ± 0.47	5.1 ± 0.30

Source: Adapted from Permal et al. (2020b)
CUPRAC: cupric ion reducing antioxidant capacity, FRAP: ferric reducing antioxidant power, GAE: gallic acid equivalent, TE: Trolox equivalent, TPC: total phenolic concentration

Table 3.3 Phenolic Concentrations (mg/100 g Dry Matter) in Seeds from Avocados at Different Development Stages

Compounds	Ripe		Over-Ripe	
	Mean	SD	Mean	SD
Quinic acid	0.08	0.01	0.08	0.01
Citric acid	4.63	0.14	12.39	0.30
Hydroxytyrosol glucoside	38.95	0.61	25.22	1.32
L-Caffeoylquinic acid	112.29	0.41	243.78	9.52
Tyrosol glucoside	223.66	1.33	339.14	11.85
3-*O*-*p*-coumaroylquinic acid	7.01	0.05	37.56	1.73
4-Caffeoylquinic acid	6.69	0.05	10.39	0.62
Vanillic acid glucoside	6.74	0.27	7.86	0.01
Total	400.05	0.00	676.43	24.74

Source: Reprinted from López-Cobo et al. (2016)

Figueroa et al. (2018b) characterized the profiles of phenolics and other polar compounds from avocado seeds and seed coats, using accelerated solvent extraction and liquid chromatography in their study. The researchers identified 84 compounds in the samples, with 45 of these compounds being determined for the first time in these tissues. Both samples were determined to be rich in condensed tannins, phenolics acids and flavonoids, although the seed was richer in hydroxycinnamic acids, phenolic alcohols, catechins and condensed tannins than the seed coat. As a result of the study, the authors reported that avocado seeds and seed coats constitute valuable and inexpensive sources of bioactive ingredients for the food, cosmetic or pharmaceutical industries. Table 3.2 shows the polyphenolic compounds found in avocado peel and seed tissues.

The effect of solid-state fermentation of avocado seed by the fungus *Aspergillus niger* on the release of bioactive compounds with antioxidant effect was investigated (Yepes-Betancur et al., 2020). The growth conditions of *A. niger* were optimized and the study revealed the ability of *A. niger* to degrade compounds present in the avocado seed and improve the antioxidant capacity. As a result, it has been demonstrated that avocado seed can be used as a potential substrate for solid-state fermentation (Yepes-Betancur et al., 2020).

In another study, the avocado seeds were used as a raw material to produce activated carbon by traditional pyrolysis. The result concluded that avocado seed could be used as a raw material to produce activated carbon with a high surface area which could be used to treat water contaminated with organic pollutants (Leite et al., 2018).

In a research study, the recovery of oil and bioactive compounds from avocado seeds and peel with Soxhlet (using hexane, ethanol and ethyl acetate as solvents) or supercritical carbon dioxide (with ethanol and ethyl acetate as co-solvents) was examined, as was the fatty acid profile, total phenolic content and antioxidant activity of samples. The highest extraction yields were achieved using ethanol as solvent in Soxhlet (14 wt%) or as a co-solvent in supercritical carbon dioxide (6.9 wt%) (Páramos et al., 2020). In another study on avocado seed oil, Tambun, Tambun and Tarigan (2020) produced fatty acids directly from avocado seeds by activating the lipase enzyme found in avocado seeds.

3.3 POTENTIAL USES OF AVOCADO WASTE

Avocado by-products are rich in nutrients and can be used in different industries because of the presence of bioactive substances (Figure 3.2) and because avocado by-products are natural products that do not pollute the environment (Colombo and Papetti, 2019).

Rodríguez-Carpena et al. (2011) investigated the effects of residues from avocado fruit processing on lipid and protein oxidation, such as discoloration in raw pork meatballs stored at 4°C for 15 days. As a result of their study, the researchers revealed that the avocado by-products prevented discoloration and oxidation of meatballs. In a similar study, avocado peel extract was used directly or after microencapsulation to preserve ground meat. The use of the extract resulted in less oxidation and lower bacterial growth during storage. Moreover, vacuum packaging provided a synergistic effect with avocado peel extract on samples (Calderon-Oliver, Escalona-Buendia and Ponce-Alquicira, 2020).

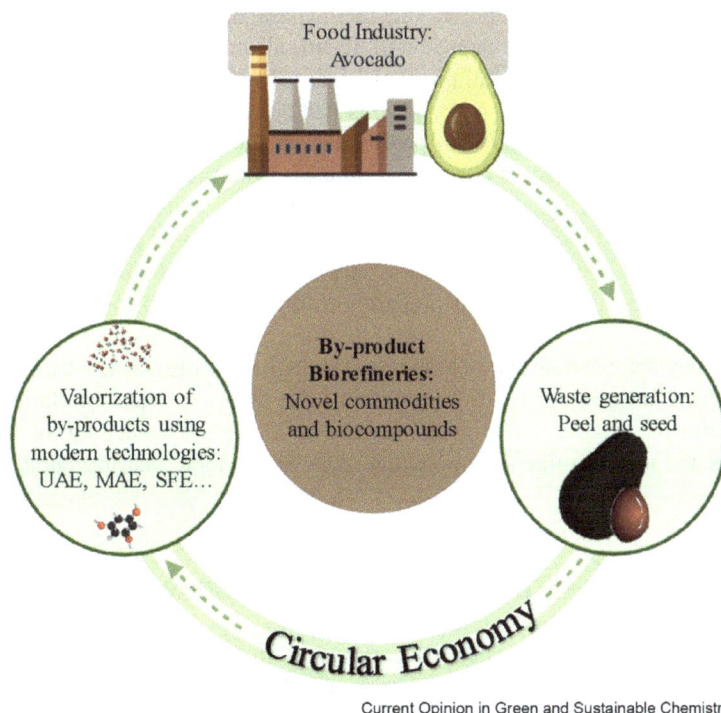

Current Opinion in Green and Sustainable Chemistry

Figure 3.2 Avocado by-products. [Adapted from Del Castillo-Llamosas et al. (2021).]

Cerda-Opazo et al. (2021) produced a nano-emulsion loaded with an extract from avocado peel. The emulsion proved to be a valuable resource for the enrichment of various food products with anti-inflammatory and anticancer properties (Cerda-Opazo et al., 2021).

Chel-Guerrero et al. (2016) revealed in their studies that avocado seeds are a starch source. The authors employed two different techniques based on wet fractionation and reported that both methods achieved similar starch yields (~20%). The starch had an amylose content of 15–16%. The authors reported that avocado seed starch could be used as a thickening and gelling agent in food systems because of its high thermal process's stability (Kringel et al., 2020). In another study, bioplastics were produced from avocado seed starch. It has been reported that bioplastic from avocado seeds will reduce environmental problems caused by overuse of traditional synthetic plastics (Ginting and Mora Sartika, 2018).

A study on animal nutrition determined that, when avocado by-products were included in multi-nutrient blocks for dairy goats, the fatty acid profile of the milk was improved (de Evan et al., 2020).

Avocado seeds are obtained during pulp production, but they are not yet used industrially. A study has evaluated the potential usage of avocado seed flour. After processing, the flour yield was determined to be 46.28%. The functional properties of the avocado flour have been associated with strong molecular interactions with apparently unique amylose structures; therefore, it has been determined that avocado seed flour may have potential for use in particular food applications (Rivera-González, Amaya-Guerra and de la Rosa-Millán, 2019).

The residue paste of the avocado oil industry is an important source of amino acids. Wang et al. (2019) obtained an edible protein from avocado processing residues that absorbs water and oil well, removes radicals and has good emulsification properties.

Fidelis et al. (2015) developed thermoplastic starch films and poly(butylene adipate-co-terephthalate) films with added avocado peel extract. Addition of avocado peel extract reduced the water vapor permeability of the films and improved their mechanical properties and antioxidant capacity. For these reasons, researchers have reported that the material developed can be used in active packaging. In another study, Dalle Mulle Santos et al. (2016) also made use of avocado by-products in films, and good thermal and mechanical properties were observed in the films developed.

Araújo et al. (2018) reported that many patents had been awarded related to the utilization of avocado by-products obtained during fruit processing. Of these, ten were food-related, including using the avocado seeds in the tea and beverage industry, and four were related to the cosmetics industry, including the use of facial cleansing products. Other patents included the use of avocado lipids to treat leukemia and the use of avocado extracts as modulators of pp2a methylation and as promoters of antioxidant and anti-inflammatory activities, while four health-related patents have been proposed using avocado by-products preparations as sunscreen, to control mosquitoes and for hair repair.

All these studies demonstrate the potential and diverse uses of avocado fruit residues. When avocado by-products are evaluated in different industries, they will reduce pollution and improve nutrition and health and achieve economic gains. Further studies are needed for the safe assessment of these residues and to determine their potential usage in the food, pharmaceutical and cosmetic industries.

3.4 HEALTH BENEFITS OF AVOCADO WASTES

The type and concentration of phenolic compounds in the different parts of avocado vary depending on the cultivar (Figure 3.3). The primary emphasis in avocado by-products exploitation is definitely on the recovery of bioactive compounds from various by-products with which to use phytochemicals as food or cosmetic ingredients (Valdivia et al., 2002). Avocado fruit by-products (peel and seeds) obtained during the processing of avocados have various beneficial health effects (Figure 3.4). Bioactive compounds found in avocado by-products have been reported to exhibit

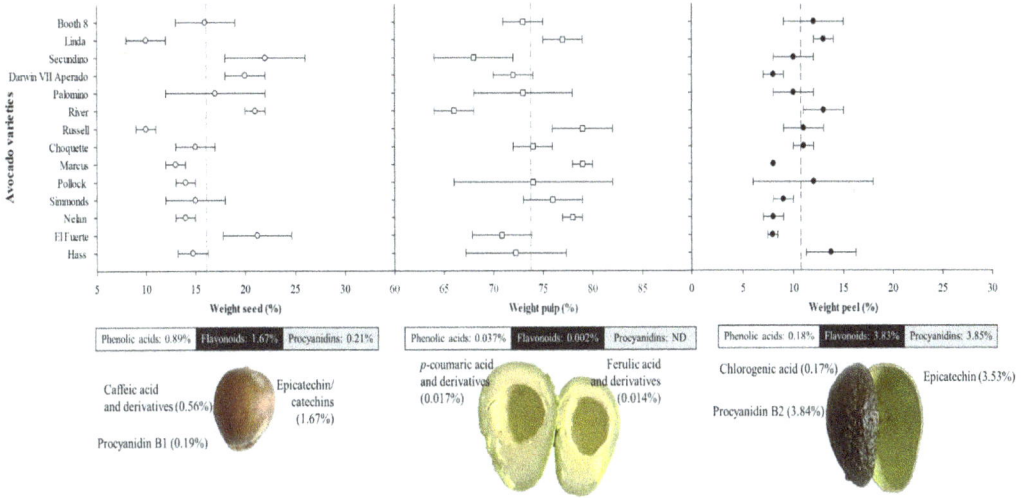

Figure 3.3 Phenolic compounds of avocado parts depending on the cultivar. [Adapted from Salazar-López et al. (2020).]

Leaves
Antioxidant
Antidiabetic
Hypotensive
Cardioprotective

Seed
Hypolipidemic
Hypotensive
Antidiabetic
Antioxidant
Cardioprotective

Peels
Antioxidant
Antidiabetic

Figure 3.4 Health benefits of different waste streams obtained from avocado.

various biological activities. The activity of bioactive components in the avocado pulp, peel, seed, and leaves has been demonstrated in *in-vitro* and *in-vivo* models (Table 3.4).

3.5 CONCLUSIONS

By-products (particularly seeds and peel) from avocado fruit processing are rich sources of nutrients and phytochemicals. These residues contain various phenolic compounds, carotenoids, phytosterols, and tocopherols. Various studies have shown that by-products from avocado fruit

Table 3.4   Some Published Values of Biological Activities of Avocado By-Products (Peel, Seed and Leaves)

Avocado Tissue	Biological Activities: Model	Main Findings	Reference
Peel and seed	Antioxidant activity, antimicrobial activity	Peels had higher antioxidant and antibacterial effects than seeds.	Melgar et al. (2018)
Peel and seed	Antioxidant activity, anti-inflammatory activity, cytotoxic activity	Peel extracts had higher antioxidant and antibacterial effects than the seed extract. Peel extract of 'Fuerte' had higher antioxidant activity than the peel extract of 'Hass'. On the other hand, the antioxidant activity of seeds of 'Hass' was higher than that of 'Fuerte'. The inflammatory activity of 'Fuerte peel' was found to be higher because of the high concentration of phenolics.	Tremocoldi et al. (2018)
Pulp and seed	Antioxidant activity, anti-inflammatory activity, anticancer activity	Both the pulp and seed extracts showed anti-inflammatory and anticancer activities against cell lines of colon cancer or liver cancer, depending on the dose. Seed extract was more powerful than pulp extract.	Alkhalaf et al. (2019)
Seed	Cytotoxic activity	IC_{50} values of aqueous extract was 560.2 µg/mL whereas it was only 107.15 µg/mL for ethanolic extract.	Kristanty, Suriawati and Sulistiyo (2014)
Seed	Cytotoxic activity	An isolated triterpenoid compound from avocado seed ethanolic extract showed significant cytotoxic activity against MCF-7 and HepG2 cell lines.	Abubakar, Achmadi and Suparto (2017)
Pulp, seed and leaves	Pro-apoptotic effect	Pulp, seed and leaf extracts promoted apoptosis by different mechanisms.	Bonilla-Porras et al. (2014)
Seed	Anti-inflammatory activity, cytotoxic activity	200 mg/mL avocado seed extract achieved inhibition of the proliferation of human lung and gastric cancer cells.	Vo, Le and Ngo (2019)
Seed	Hypolipidemic activity	The oral LD_{50} for avocado seed flour was determined as 1767 mg/kg bodyweight. It significantly reduced the levels of total cholesterol, LDL-C, and atherogenic index in a hypercholesterolemic mice model.	Pahua-Ramos et al. (2012)
Seed	Anti-hyperglycemic activity, anti-hypercholesteremia activity	Avocado seed reduced glucose and cholesterol levels in rat blood and enhanced liver glycogen storage.	Uchenna, Shori and Baba (2017)
Seed	Anti-diabetic activity	600 mg/kg seed extract reduced glucose levels by >70%. Seed extract showed restorative effects on pancreatic islet cells. Seed extract can be useful in diabetes treatments.	Edem, Ekanem and Ebong (2009)
Leaf and seed	Anticholinesterase activity, antioxidant activity	Seed extract had higher anticholinesterase effects compared to the leaf extract. Antioxidant activity of leaf extract was higher than that of seed extract.	Oboh et al. (2016)
Seed	Genotoxic activity	No genotoxic activity was observed at <250 mg/kg concentration in mice.	Padilla-Camberos et al. (2013)
Seed	Anthelmintic activity	Ethanolic extract provided higher anthelmintic activity compared to hot water extract. Dried seeds are a better option than fresh seeds based on anthelmintic activity.	Soldera-Silva et al. (2018)

(Continued)

Table 3.4 (Continued)   Some Published Values of Biological Activities of Avocado By-Products (Peel, Seed and Leaves)

Avocado Tissue	Biological Activities: Model	Main Findings	Reference
Peel and seed	Antioxidant and antimicrobial activity	Peel and seed extract mixture showed a synergistic effect with nisin. Avocado peel extract had higher antioxidant activity than avocado seed extract.	Calderón-Oliver et al. (2016)
Peel	Larvicidal activity	Methanolic avocado peel extract showed promising larvicidal activity.	Louis et al. (2020)

processing inhibit membrane lipid peroxidation because of their antioxidant properties. Different extraction techniques need to be developed to allow application of avocado by-products on a commercial scale. Bioactive compounds extracted from avocado residues could be used to preserve and thicken foods and improve the quality and stability of meat products. In addition, more advanced sensory and nutritional studies need to be performed to demonstrate the potential value of avocado processing residues in food products.

REFERENCES

Abubakar, Andi Nur Fitriani, Suminar Setiati Achmadi, and Irma Herawati Suparto. 2017. "Triterpenoid of avocado (*Persea americana*) seed and its cytotoxic activity toward breast MCF-7 and liver HepG2 cancer cells." *Asian Pacific Journal of Tropical Biomedicine* 7 (5):397–400. https://doi.org/10.1016/j.apjtb .2017.01.010.

Adeyemi, O. O., S. O. Okpo, and O. O. Ogunti. 2002. "Analgesic and anti-inflammatory effects of the aqueous extract of leaves of *Persea americana* Mill. (Lauraceae)." *Fitoterapia* 73 (5):375–380.

Alkhalaf, Maha I., Wafa S. Alansari, Eman Ahmed Ibrahim, and Manal E. A. Elhalwagy. 2019. "Antioxidant, anti-inflammatory and anti-cancer activities of avocado (*Persea americana*) fruit and seed extract." *Journal of King Saud University - Science* 31 (4):1358–1362. https://doi.org/10.1016/j.jksus .2018.10.010.

Antia, B. S., J. E. Okokon, and P. A. Okon. 2005. "Hypoglycemic activity of aqueous leaf extract of *Persea americana* Mill." *Indian Journal of Pharmacology* 37 (5):325.

Araújo, Rafael G., Rosa M. Rodriguez-Jasso, Héctor A. Ruiz, Maria Manuela E. Pintado, and Cristóbal Noé Aguilar. 2018. "Avocado by-products: Nutritional and functional properties." *Trends in Food Science & Technology* 80:51–60.

Arukwe, U., B. A. Amadi, M. K. C. Duru, E. N. Agomuo, E. A. Adindu, P. C. Odika, K. C. Lele, L. Egejuru, and J. Anudike. 2012. "Chemical composition of *Persea americana* leaf, fruit and seed." *International Journal of Research and Reviews in Applied Sciences* 11 (May):346–349.

Bergh, B. O., R. W. Scora, and W. B. Storey. 1973. "A comparison of leaf terpenes in Persea subgenus *Persea*." *Botanical Gazette* 134 (2):130–134.

Bonilla-Porras, Angelica R., Andrea Salazar-Ospina, Marlene Jimenez-Del-Rio, Andres Pereañez-Jimenez, and Carlos Velez-Pardo. 2014. "Pro-apoptotic effect of *Persea americana* var. Hass (avocado) on Jurkat lymphoblastic leukemia cells." *Pharmaceutical Biology* 52 (4):458–465. https://doi.org/10.3109 /13880209.2013.842599.

Brai, B. I. C., A. A. Odetola, and P. U. Agomo. 2007. "Effects of *Persea americana* leaf extracts on body weight and liver lipids in rats fed hyperlipidaemic diet." *African Journal of Biotechnology* 6 (8):1007–1011.

Calderon-Oliver, M., H. B. Escalona-Buendia, and E. Ponce-Alquicira. 2020. "Effect of the addition of microcapsules with avocado peel extract and nisin on the quality of ground beef." *Food Science & Nutrition* 8 (3):1325–1334. https://doi.org/10.1002/fsn3.1359.

Calderón-Oliver, Mariel, Héctor B. Escalona-Buendía, Omar N. Medina-Campos, José Pedraza-Chaverri, Ruth Pedroza-Islas, and Edith Ponce-Alquicira. 2016. "Optimization of the antioxidant and antimicrobial response of the combined effect of nisin and avocado byproducts." *LWT - Food Science and Technology* 65:46–52. https://doi.org/10.1016/j.lwt.2015.07.048.

Cerda-Opazo, Paulina, Martin Gotteland, Felipe A. Oyarzun-Ampuero, and Lorena Garcia. 2021. "Design, development and evaluation of nanoemulsion containing avocado peel extract with anticancer potential: A novel biological active ingredient to enrich food." *Food Hydrocolloids* 111:106370.

Chel-Guerrero, Luis, Enrique Barbosa-Martín, Agustino Martínez-Antonio, Edith González-Mondragón, and David Betancur-Ancona. 2016. "Some physicochemical and rheological properties of starch isolated from avocado seeds." *International Journal of Biological Macromolecules* 86:302–308.

Colombo, Raffaella, and Adele Papetti. 2019. "Avocado (*Persea americana* Mill.) by-products and their impact: From bioactive compounds to biomass energy and sorbent material for removing contaminants. A review." *International Journal of Food Science & Technology* 54 (4):943–951.

Dabas, Deepti, Rachel M. Shegog, Gregory R. Ziegler, and Joshua D. Lambert. 2013. "Avocado (*Persea americana*) seed as a source of bioactive phytochemicals." *Current Pharmaceutical Design* 19 (34):6133–6140.

Dalle Mulle Santos, Cassandra, Carlos Henrique Pagno, Tania Maria Haas Costa, Débora Jung Luvizetto Faccin, Simone Hickmann Flôres, and Nilo Sergio Medeiros Cardozo. 2016. "Biobased polymer films from avocado oil extraction residue: Production and characterization." *Journal of Applied Polymer Science* 133 (37):43957.

De Almeida, A. P., M. M. F. S. Miranda, I. C. Simoni, M. D. Wigg, M. H. C. Lagrota, and S. S. Costa. 1998. "Flavonol monoglycosides isolated from the antiviral fractions of *Persea americana* (Lauraceae) leaf infusion." *Phytotherapy Research: An International Journal Devoted to Pharmacological and Toxicological Evaluation of Natural Product Derivatives* 12 (8):562–567.

de Evan, Trinidad, María Dolores Carro, Julia Eugenia Fernández Yepes, Ana Haro, Lesly Arbesú, Manuel Romero-Huelva, and Eduarda Molina-Alcaide. 2020. "Effects of feeding multinutrient blocks including avocado pulp and peels to dairy goats on feed intake and milk yield and composition." *Animals* 10 (2):194.

Del Castillo-Llamosas, Alexandra, Pablo G. del Río, Alba Pérez-Pérez, Remedios Yáñez, Gil Garrote, and Beatriz Gullón. 2021. "Recent advances to recover value-added compounds from avocado by-products following a biorefinery approach." *Current Opinion in Green and Sustainable Chemistry* 28:100433. https://doi.org/10.1016/j.cogsc.2020.100433.

Edem, D., I. Ekanem, and P. Ebong. 2009. "Effect of aqueous extracts of alligator pear seed (*Persea americana* Mill) on blood glucose and histopathology of pancreas in alloxan-induced diabetic rats." *Pakistan Journal of Pharmaceutical Science* 22 (3):272–276.

Fidelis, J. C. F., A. R. G. Monteiro, M. R. S. Scapim, C. C. F. Monteiro, D. R. Morais, T. Claus, J. V. Visentainer, and F. Yamashita. 2015. "Development of an active biodegradable film containing tocopherol and avocado peel extract." *Italian Journal of Food Science* 27 (4):468–475.

Figueroa, J. G., I. Borras-Linares, J. Lozano-Sanchez, and A. Segura-Carretero. 2018a. "Comprehensive identification of bioactive compounds of avocado peel by liquid chromatography coupled to ultra-high-definition accurate-mass QTOF." *Food Chemistry* 245:707–716. https://doi.org/10.1016/j.foodchem.2017.12.011.

Figueroa, Jorge G., Isabel Borrás-Linares, Jesús Lozano-Sánchez, and Antonio Segura-Carretero. 2018b. "Comprehensive characterization of phenolic and other polar compounds in the seed and seed coat of avocado by HPLC-DAD-ESI-QTOF-MS." *Food Research International* 105:752–763.

Ginting, Hendra S., and Hidayatul Azmi Mora Sartika. 2018. "Production of bioplastic from avocado seed starch reinforced with microcrystalline cellulose from sugar palm fibers." *Journal of Engineering Science and Technology* 13 (2):381–393.

Jayanegara, A., E. Wina, C. R. Soliva, S. Marquardt, M. Kreuzer, and F. Leiber. 2011. "Dependence of forage quality and methanogenic potential of tropical plants on their phenolic fractions as determined by principal component analysis." *Animal Feed Science and Technology* 163 (2–4):231–243.

Jimenez, Paula, Paula Garcia, Vilma Quitral, Karla Vasquez, Claudia Parra-Ruiz, Marjorie Reyes-Farias, Diego F. Garcia-Diaz, Paz Robert, Cristian Encina, and Jessica Soto-Covasich. 2020. "Pulp, leaf, peel and seed of avocado fruit: A review of bioactive compounds and healthy benefits." *Food Reviews International*:1–37. https://doi.org/10.1080/87559129.2020.1717520.

Jiménez-Velázquez, Perla, Salvador Valle-Guadarrama, Iran Alia-Tejacal, Yolanda Salinas-Moreno, Leticia García-Cruz, Artemio Pérez-López, and Diana Guerra-Ramírez. 2020. "Separation of bioactive compounds from epicarp of 'Hass' avocado fruit through aqueous two-phase systems." *Food and Bioproducts Processing* 123:238–250.

Kim, M.-J., and K. Uhl. 2011. "Sex and lifestyle drugs: The pursuit of the fountain of youth." *Clinical Pharmacology & Therapeutics* 89 (1):3–9.

Kose, L. P., Z. Bingol, R. Kaya, A. C. Goren, H. Akincioglu, L. Durmaz, E. Koksal, S. H. Alwasel, and I. Gulcin. 2020. "Anticholinergic and antioxidant activities of avocado (*Folium perseae*) leaves – phytochemical content by LC-MS/MS analysis." *International Journal of Food Properties* 23 (1):878–893. https://doi .org/10.1080/10942912.2020.1761829.

Kosińska, Agnieszka, Magdalena Karamać, Isabel Estrella, Teresa Hernández, Begoña Bartolomé, and Gary A. Dykes. 2012. "Phenolic compound profiles and antioxidant capacity of *Persea americana* Mill. Peels and seeds of two varieties." *Journal of Agricultural and Food Chemistry* 60 (18):4613–4619. https://doi .org/10.1021/jf300090p.

Kringel, D. H., A. R. G. Dias, E. D. Zavareze, and E. A. Gandra. 2020. "Fruit wastes as promising sources of starch: Extraction, properties, and applications." *Starch-Starke* 72 (3–4):9. https://doi.org/10.1002/star .201900200.

Kristanty, Ruth Elenora, Junie Suriawati, and Joko Sulistiyo. 2014. "Cytotoxic activity of avocado seeds extracts (*Persea americana* Mill.) on t47d cell lines." *International Research Journal of Pharmacy* 5 (7):557–559.

Leite, Anderson B., Caroline Saucier, Eder C. Lima, Glaydson S. dos Reis, Cibele S. Umpierres, Beatris L. Mello, Mohammad Shirmardi, Silvio L. P. Dias, and Carlos H. Sampaio. 2018. "Activated carbons from avocado seed: Optimisation and application for removal of several emerging organic compounds." *Environmental Science and Pollution Research* 25 (8):7647–7661.

López-Cobo, Ana, Ana María Gómez-Caravaca, Federica Pasini, María Fiorenza Caboni, Antonio Segura-Carretero, and Alberto Fernández-Gutiérrez. 2016. "HPLC-DAD-ESI-QTOF-MS and HPLC-FLD-MS as valuable tools for the determination of phenolic and other polar compounds in the edible part and by-products of avocado." *LWT* 73:505–513.

Louis, M. R. Lima Mirabel, V. Pushpa, K. Balakrishna, and P. Ganesan. 2020. "Mosquito larvicidal activity of Avocado (*Persea americana* Mill.) unripe fruit peel methanolic extract against *Aedes aegypti*, *Culex quinquefasciatus* and *Anopheles stephensi*." *South African Journal of Botany* 133:1–4. https://doi.org /10.1016/j.sajb.2020.06.020.

Mariane, Eliza, Damila Rodrigues, Franca Biondo, Polyana Batoqui, Vanessa Jorge, and Jesui Vergilio. 2016. "Use of avocado peel (*Persea americana*) in tea formulation: A functional product containing phenolic compounds with antioxidant activity." *Acta Scientiarum-Technology* 38:23–29.

Marques, J., P. J. Hofman, and A. H. Wearing. 2001. "'Hass' Avocado fruit quality and minerals as affected by rootstocks." *Journal of Horticultural Science and Biotechnology* 78 (5):673–679.

Melgar, Bruno, Maria Inês Dias, Ana Ciric, Marina Sokovic, Esperanza M. Garcia-Castello, Antonio D. Rodriguez-Lopez, Lillian Barros, and Isabel C. R. F. Ferreira. 2018. "Bioactive characterization of *Persea americana* Mill. by-products: A rich source of inherent antioxidants." *Industrial Crops and Products* 111:212–218. https://doi.org/10.1016/j.indcrop.2017.10.024.

Mfonobong, Afahakan, Umar Ismail, Inuwa Hajiya Mairo, Zubairu Maimuna, and Dawud Fatima. 2013. "Hypolipidemic and antioxidant effects of petroleum ether and methanolic fractions of *Persea americana* Mill seeds in Wistar rats fed a high-fat high-cholesterol diet." *International Journal of Pharmaceutical Sciences* 3 (12):1–10.

Oboh, Ganiyu, Veronica O. Odubanjo, Fatai Bello, Ayokunle O. Ademosun, Sunday I. Oyeleye, Emem E. Nwanna, and Adedayo O. Ademiluyi. 2016. "Aqueous extracts of avocado pear (*Persea americana* Mill.) leaves and seeds exhibit anti-cholinesterases and antioxidant activities in vitro." *Journal of Basic and Clinical Physiology and Pharmacology* 27 (2):131–140. https://doi.org/10.1515/jbcpp-2015-0049.

Padilla-Camberos, Eduardo, Moisés Martínez-Velázquez, José Miguel Flores-Fernández, and Socorro Villanueva-Rodríguez. 2013. "Acute toxicity and genotoxic activity of avocado seed extract (*Persea americana* Mill., c.v. Hass)." *The Scientific World Journal* 2013:245828. https://doi.org/10.1155/2013 /245828.

Pahua-Ramos, María Elena, Alicia Ortiz-Moreno, Germán Chamorro-Cevallos, María Dolores Hernández-Navarro, Leticia Garduño-Siciliano, Hugo Necoechea-Mondragón, and Marcela Hernández-Ortega. 2012. "Hypolipidemic effect of avocado (*Persea americana* Mill) seed in a hypercholesterolemic mouse model." *Plant Foods for Human Nutrition* 67 (1):10–16. https://doi.org/10.1007/s11130-012-0280-6.

Páramos, Patrícia R. S., José F. O. Granjo, Marcos L. Corazza, and Henrique A. Matos. 2020. "Extraction of high value products from avocado waste biomass." *The Journal of Supercritical Fluids* 165:104988. https://doi.org/10.1016/j.supflu.2020.104988.

Park, SeonJu, Youn Hee Nam, Isabel Rodriguez, Jun Hyung Park, Hee Jae Kwak, Youngse Oh, Mira Oh, Min Seon Park, Kye Wan Lee, and Jung Suk Lee. 2019. "Chemical constituents of leaves of *Persea americana* (avocado) and their protective effects against neomycin-induced hair cell damage." *Revista Brasileira de Farmacognosia* 29 (6):739–743.

Permal, R., W. L. Chang, T. Chen, B. Seale, N. Hamid, and R. Kam. 2020a. "Optimising the spray drying of avocado wastewater and use of the powder as a food preservative for preventing lipid peroxidation." *Foods* 9 (9). https://doi.org/10.3390/foods9091187.

Permal, Rahul, Wee Leong Chang, Brent Seale, Nazimah Hamid, and Rothman Kam. 2020b. "Converting industrial organic waste from the cold-pressed avocado oil production line into a potential food preservative." *Food Chemistry* 306:125635. https://doi.org/10.1016/j.foodchem.2019.125635.

Plazola-Jacinto, C. P., Pérez-Pérez, V., Pereyra-Castro, S. C., Alamilla-Beltrán, L., and Ortiz-Moreno, A. (2019). "Microencapsulation of biocompounds from avocado leaves oily extracts." *Revista Mexicana de Ingeniería Química* 18 (3):1261–1276.

Ramos-Aguilar, Ana L., Juan Ornelas-Paz, Luis M. Tapia-Vargas, Alfonso A. Gardea-Béjar, Elhadi M. Yahia, José de Jesús Ornelas-Paz, Saúl Ruiz-Cruz, Claudio Rios-Velasco, and Pilar Escalante-Minakata. 2020. "Effect of cultivar on the content of selected phytochemicals in avocado peels." *Food Research International* 140:110024.

Rivera–González, Gerardo, Carlos Abel Amaya–Guerra, and Julián de la Rosa–Millán. 2019. "Physicochemical characterisation and in vitro Starch digestion of Avocado Seed Flour (*Persea americana* V. Hass) and its starch and fibrous fractions." *International Journal of Food Science & Technology* 54 (7):2447–2457.

Rodríguez-Carpena, Javier-Germán, David Morcuende, María-Jesús Andrade, Petri Kylli, and Mario Estévez. 2011. "Avocado (*Persea americana* Mill.) phenolics, in vitro antioxidant and antimicrobial activities, and inhibition of lipid and protein oxidation in porcine patties." *Journal of Agricultural and Food Chemistry* 59 (10):5625–5635.

Rotta, E. M., D. R. de Morais, P. B. F. Biondo, V. J. dos Santos, M. Matsushita, and J. V. Visentainer. 2016. "Use of avocado peel (*Persea americana*) in tea formulation: A functional product containing phenolic compounds with antioxidant activity." *Acta Scientiarum-Technology* 38 (1):23–29. https://doi.org/10.4025/actascitechnol.v38i1.27397.

Salazar-López, Norma Julieta, J. Abraham Domínguez-Avila, Elhadi M. Yahia, Beatriz Haydee Belmonte-Herrera, Abraham Wall-Medrano, Efigenia Montalvo-González, and G. A. González-Aguilar. 2020. "Avocado fruit and by-products as potential sources of bioactive compounds." *Food Research International* 138:109774. https://doi.org/10.1016/j.foodres.2020.109774.

Segovia, Francisco J., Juan J. Corral-Pérez, and María P. Almajano. 2016. "Avocado seed: Modeling extraction of bioactive compounds." *Industrial Crops and Products* 85:213–220.

Segovia, Francisco J., Gádor Indra Hidalgo, Juliana Villasante, Xavier Ramis, and María Pilar Almajano. 2018. "Avocado seed: A comparative study of antioxidant content and capacity in protecting oil models from oxidation." *Molecules* 23 (10):2421.

Soldera-Silva, Andressa, Melina Seyfried, Luciano Henrique Campestrini, Selma Faria Zawadzki-Baggio, Alessandro Pelegrine Minho, Marcelo Beltrão Molento, and Juliana Bello Baron Maurer. 2018. "Assessment of anthelmintic activity and bio-guided chemical analysis of *Persea americana* seed extracts." *Veterinary Parasitology* 251:34–43. https://doi.org/10.1016/j.vetpar.2017.12.019.

Suleria, H. A. R., C. J. Barrow, and F. R. Dunshea. 2020. "Screening and characterization of phenolic compounds and their antioxidant capacity in different fruit peels." *Foods* 9 (9):26. https://doi.org/10.3390/foods9091206.

Talabi, Justina Y., Olukemi A. Osukoya, O. O. Ajayi, and G. O. Adegoke. 2016. "Nutritional and antinutritional compositions of processed Avocado (*Persea americana* Mill) seeds." *Asian Journal of Plant Science and Research* 6 (2):6–12.

Tambun, R., J. O. A. Tambun, and I. A. A. Tarigan. 2020. "Fatty acid production from avocado seed by activating lipase enzyme in the seed." In *IOP Conference Series: Materials Science and Engineering*, Nommensen HKBP University, Indonesia.

Tremocoldi, Maria Augusta, Pedro Luiz Rosalen, Marcelo Franchin, Adna Prado Massarioli, Carina Denny, Érica Regina Daiuto, Jonas Augusto Rizzato Paschoal, Priscilla Siqueira Melo, and Severino Matias de Alencar. 2018. "Exploration of avocado by-products as natural sources of bioactive compounds." *PLOS ONE* 13:1–12. https://doi.org/10.1371/journal.pone.0192577.

Uchenna, Uzukwu Emmanuel, Amal Bakr Shori, and Ahmad Salihin Baba. 2017. "Inclusion of avocado (*Persea americana*) seeds in the diet to improve carbohydrate and lipid metabolism in rats." *Revista Argentina de Endocrinología y Metabolismo* 54 (3):140–148. https://doi.org/10.1016/j.raem.2017.07.005.

Valdivia, Ma Ángeles, Ma Emilia Bustos, Javier Ruiz, and Luisa F. Ruiz. 2002. "The effect of irradiation in the quality of the avocado frozen pulp." *Radiation Physics and Chemistry* 63 (3–6):379–382.

Vo, Thanh, Phuong Le, and Dai Ngo. 2019. "Free radical scavenging and anti-proliferative activities of avocado (*Persea americana* Mill.) seed extract." *Asian Pacific Journal of Tropical Biomedicine* 9 (3):91–97. https://doi.org/10.4103/2221-1691.254602.

Wang, Lu, Xue Lin, Jiachao Zhang, Weimin Zhang, Xiaoping Hu, Wu Li, Congfa Li, and Sixin Liu. 2019. "Extraction methods for the releasing of bound phenolics from *Rubus idaeus* L. leaves and seeds." *Industrial Crops and Products* 135:1–9.

Wientarsih, Ietje, Rini Madyastuti, Bayu Febram Prasetyo, and Anggara Aldobrata. 2012. "Anti lithiasis activity of Avocado (*Persea americana* Mill) leaves extract in white male rats." *HAYATI Journal of Biosciences* 19 (1):49–52.

Yepes-Betancur, Diana Paola, Carlos Julio Márquez-Cardozo, Edith Marleny Cadena-Chamorro, Jaison Martinez-Saldarriaga, Cristian Torres-León, Alberto Ascacio-Valdes, and Cristobal N. Aguilar. 2020. "Solid-state fermentation – assisted extraction of bioactive compounds from Hass avocado seeds." *Food and Bioproducts Processing*. https://doi.org/10.1016/j.fbp.2020.10.012.

Zafar, Tasleem, and Jiwan S. Sidhu. 2011. "Avocado: Production, quality, and major processed products." In *Handbook of Vegetables and Vegetable Processing*, edited by Nirmal K Sinha, 525–543. Boca Raton: Wiley Online Library.

CHAPTER 4

Date Wastes and By-Products
Chromistry, Processing, and Utilization

Bibi Marzieh Razavizadeh and Razieh Niazmand

CONTENTS

4.1 INTRODUCTION

The date palm (*Phoenix dactylifera*) is one of the earliest domesticated trees, having been culti-vated since ancient times, and it has a sacred status among various tribes and nations (Ragava and Loganathan 2016). The date palm is cultivated in semiarid and dry areas in mainly the Middle East and North Africa. The fruits are known as dates, which are a high-energy food source because of the high sugar concentration at maturity of 72% to 88% (Chao and Krueger 2007). Dates are oval-cylindrical in form, 3 to 7 cm in length, with a diameter of 2 to 3 cm, and, when unripe, the color varies from bright red to bright yellow, depending on variety (Ragava and Loganathan 2016). The date fruit is a multiplex of the layers pericarp, mesocarp, and endocarp, which contains a single seed about 2–2.5 cm long and 6–8 mm thick (Ghnimi et al. 2017). The growth period of date fruit is about 200 days and, during this period, the fruit goes through four distinct maturity stages, namely Khimri, Khalal, Rutab, and Tamar (Zaid 2002). The date fruits may be consumed during

DOI: 10.1201/9781003164463-4

the Khalal stage, when the fruit is still very astringent due to its high tannin content (Awad 2007). However, dates are mainly harvested during the Rutab and Tamar stages when they are fully ripe, the sugar concentration is high, and the moisture and tannin contents are low. About 80% to 85% of the sugar is sucrose in the Khalal stage. During the ripening of the fruit, the sucrose is hydrolyzed into reduced sugars such as glucose and fructose (Chao and Krueger 2007).

The annual global production of date fruit is about 7.5 million tons in an area of 1.1 million hectares of groves with an average yield of 6,834 kg/ha. In Iran, more than 1.2 million tons of dates are produced annually on 203,763 hectares. Egypt, Iran, Saudi Arabia, UAE, Pakistan, Algeria, Iraq, Sudan, Oman, and Libya are the top ten date producers in the world, with Egypt having the highest total yield and Algeria having the highest area under cultivation (FAO 2018; National Date Association of Iran 2018).

The average *per capita* global consumption of dates is estimated at more than one kilogram, although this index is much higher in Saudi Arabia at about 34 kilograms (FAO 2018), while the *per capita* consumption of dates in Iran in 2018 was 8.6 kg.

This product is produced in countries with hot and semi-arid climates having suitable production temperatures. At present, countries such as Egypt, Iran, Iraq, Saudi Arabia, Tunisia, UAE, and so on can be mentioned to have the necessary conditions for cultivation of this fruit and are among the major producers of the product (National Date Association of Iran 2018).

Date palm includes more than 3,000 varieties from all around the world (FAOSTAT 2016, Ghnimi et al. 2017). The three principal cultivar groups of date fruit are soft, semi-dry, and dry dates, depending upon the water content (FAOSTAT 2016, Ghnimi et al. 2017). The main difference between the various types of dates is in the type of sugar and also the amount of water in them. The total concentration of sugar in different cultivars of dates varies, but the difference is not large, being only a few percentages. The sugars in dates are the reducing sugars fructose and glucose, as well as sucrose. Dried dates have a higher sucrose concentration, the semi-dried dates contain identical amounts of inverted sugars (glucose and fructose) and sucrose, whereas soft dates contain more inverted sugars than sucrose (Chandrasekaran and Bahkali 2013). In general, soft dates have less carbohydrate than dried or semi-dried dates (FAOSTAT 2016, Ghnimi et al. 2017).

As mentioned above, there are many date cultivars, but the most important cultivars have a considerable global consumer market, including Deglet Noor, Medjool, and Khalas (FAOSTAT 2016). The first two dates named are the best commercial cultivars in the world and are of greatest importance. Figure 4.1 shows pictures of some common date cultivars. The Medjool date is the most expensive date in the world and is native to Morocco, has brown skin and is wrinkled and firm in appearance, but its texture is fleshy, moist, and soft. The Deglet Noor date is one of the most commercially important dates, and is mostly consumed among the Arab communities. This type of date is native to Iraq. The most important date cultivars in Egypt are Hayany or Hayani, Khazrawi, and Zaghloul, those in Saudi Arabia, are Ajwah, Al-Khunaizi, Barhee or Barhi, and Khazrawi, and the most important cultivars in the United Arab Emirates are Dabbas, Khenaizi, and Lulu (Djouab et al. 2016, Aromanadates 2018, IOTCO 2018). The date cultivars with the highest area under cultivation in Iran include Mazafati, Kabkab, Sayer, Rabbi, Shahani, and Piaroom (National Date Association of Iran 2018).

4.2 WASTES AND BY-PRODUCTS OF DATE FRUIT

Date fruit can be eaten fresh, dried, or processed in various forms. Although date fruit can be used in various forms, as whole date, unpitted date, sliced, or extruded fragments, it may also be used for processing, generating products including juice concentrates and fermented date products. However, date waste and date by-products, including date seeds and date press cake, can be obtained during date processing (Chao and Krueger 2007, FAOSTAT 2016).

Khalas

Medjool

Ajwah

Sayer

Zaghloul

Deglet noor

Kabkab

Mazafati

Piaroom

Khazravi

Rabi

Barhi

Figure 4.1 Important date cultivars found in the world.

4.2.1 Date Seeds

Date seeds (also named kernels, stones, or pits) are a low-cost waste stream generated from many date products, such as date powders, date confectionery, date syrup, pitted dates, and date juice (Rahman et al. 2007). In the case of juice extraction, seeds may be mixed with the remaining press cake or they have been sieved out in the process (FAOSTAT 2016). They are commonly discarded or used for animal feed, although they possess high value-added compounds, which makes them suitable for being used in food and pharmaceutical industries (El Hadrami and Al-Khayri 2012). For example, in Tunisia, on average about 1,000 tons of date seed oil is extracted from 100,000 tons of date fruits each year (Chandrasekaran and Bahkali 2013).

The individual weight of a date seed is in the range of 0.5–4 g (Al-Farsi and Lee 2011) and they constitute 10–18% of the date fruit weight (Afiq et al. 2013). Date seed has an abdominal side characterized by a depth and width track along its length. The back side of the seed is convex with a small, shallow hole (micropyle), beneath which lies the embryo.

The main date-producing countries worldwide are Egypt, Saudi Arabia, and Iran with more than 8.5 million tons of date fruits produced in 2018 (FAO 2018). It is estimated that 93,500–1.53 million tons of date seeds are produced globally each year, most of which are discarded (FAO 2018). Today, however, due to the remarkable properties of date seeds, industries tend to use them in the generation of value-added products. Due to its lack of caffeine, date seed coffee, for example, has been widely used (Hossain et al. 2014).

Some other products which have been derived from date seeds include oil, activated carbon, and animal feed (Oladipupo Kareem et al. 2019).

4.2.2 Date Press Cake

Date Press Cake (DPC) is an unavoidable by-product of date processing. It is the solids remaining after pressing the dates to extract the juice (Ashraf and Hamidi-Esfahani 2011). DPC contains date flesh with some remaining sugar. Depending on the extraction type, DPC may also include the seeds (FAOSTAT 2016). On average, juicing of date results in 17–28% date press cake (Majzoobi et al. 2019).

4.3 CHEMISTRY OF DATE WASTE AND BY-PRODUCTS

From a nutritional point of view, dates are valuable fruits. They are rich in reducing sugars, mainly fructose and glucose, dietary fiber, vitamins, and minerals. They also contain huge amounts of phytochemicals, such as carotenoids, polyphenols, and flavonoids, which, as natural antioxidants, are responsible for several human health benefits (El Hilaly et al. 2018). Therefore, date waste and by-products can exhibit these nutritional and health-promoting properties to some extent, depending on the process history. The properties of these date residues vary according to the date fruit variety, climate, and process.

4.3.1 Date Seed

There are many nutrients in date seeds (carbohydrates, fiber, oils, minerals, and protein) that are also present in the date seed powder so that its nutritional value is high (Chandrasekaran and Bahkali 2013, Wahini 2016). It has been reported that the drying method affects the nutritional value of the flour obtained, with the concentrations of protein and carbohydrate in date seed flour dried in the sun being greater than in the date seed flour dried by oven heating (26.54% and 31.54% *versus* 5.03% and 25.64%, respectively) (Wahini 2016).

From the literature, the chemical composition of date seeds is shown in Table 4.1 (Deng et al. 2012, Shina et al. 2013, Parvin et al. 2015, FAOSTAT 2016, Ghnimi et al. 2017, Ahmad and Imtiaz 2019).

Compared with the date flesh, which contains 50–80% carbohydrates (mostly glucose, fructose, and minor quantities of sucrose, as well as cellulose, starch, and other polysaccharides), date seeds contain about 83.0% carbohydrates, predominantly insoluble fiber type carbohydrates (cellulose 42%, hemicellulose 8–17.5%), 25% total sugars, and other components (FAOSTAT 2016) (Ahmad and Imtiaz 2019, Metoui et al. 2018). An average concentration (2–4 g/kg) of soluble sugars includes glucose, fructose, raffinose, stachyose, sucrose, and galactose in date seeds (Al Juhaimi

Table 4.1   Date Seed Chemical Composition

Constituent	Quantity (w/w, %)
Moisture	5–12
Protein (N% × 6.25)	5–7
Lipid	5.7–10
Crude fiber	10–24
Carbohydrate	55–70
Ash	0.8–2

et al. 2018). Furthermore, carbohydrates are stored in the date seed endosperm, mostly in the form of $(1 \rightarrow 4)$ β-D mannan. It has been reported that other hemicellulose components include gluco- and galacto-mannans and alkali-soluble heteroxylans have been identified in date seeds (Ghnimi et al. 2017, Ahmad and Imtiaz 2019), with the heteroxylans mostly containing xylose (82%) and 4-O-methylglucuronic acid (17%) and very small amounts of arabinoses, galactose, glucose, and mannose (Assirey 2015).

Date seeds have a high dietary fiber content. Al-Farsi and Lee (2008) studied the optimization of the extraction of dietary fiber from date seeds. They evaluated the effects of some parameters including solvent/sample ratio, temperature, the duration of extraction, extraction numbers, and the type of solvent on phenolic compounds. They found that the total dietary fiber concentration after extraction of date seeds in water was 83.5 g/100 g, compared with acetone (82.2 g/100 g), whereas Hamada, Hashim, and Sharif (2002) determined the total dietary fiber for different types of date seeds to be in the range 64.5–68.8 g/ 100 g of date seeds.

Date seeds contain considerable amounts of protein. Date seeds contain albumins, globulins, prolamins, and glutelins as soluble proteins. Also, methionine and cystine are the dominant amino acids in date seeds which are both essential and contain sulfur (Zhang et al. 2013). Moreover, it was reported that the date seeds of the two varieties Ruzeiz and Sifri contain glutamic acid, aspartic acid, and arginine which make approximately half of the total amino acids in the date seed, whereas lysine and tryptophan are present in lower concentrations (Al-Farsi and Lee 2011, Akasha, Campbell, and Euston 2012).

Phenolic and antioxidant compounds have been identified in high concentrations in date seeds, representing 3,100–4,400 gallic acid equivalents/100 g (or 580–930 µM trolox equivalents) (Ghnimi et al. 2017, Metoui et al. 2018). The most abundant phenolic constituents in date seeds include gallic acid, protocatechuic acid, vanillic acid, ferulic acid, caffeic acid, p-hydroxybenzoic acid, p-coumaric acid, m-coumaric acid and o-coumaric acid. Of these phenolics, hydroxybenzoic acid, protocatechuic acid, and m-coumaric acid, with concentrations of 9.89, 8.84, and 8.42 mg, respectively, showed antioxidant potential in date seed (Al-Farsi and Lee 2011).

Date seed also contains flavonoids, the concentrations of which differ according to variety and environmental situation. For example, in 'Ajwa' date seed, the total flavonoid concentration was 1.35–3.67 mg/ 100 g of quercetin equivalent (Ahmad and Imtiaz 2019).

Date seeds have been shown to be rich in some minerals such as potassium (K), sodium (Na), calcium (Ca), magnesium (Mg), cadmium (Cd), aluminum (Al), phosphorus (P), sulfur (S), iron (Fe), zinc (Zn), copper (Cu), and lead (Pb) (Ahmad and Imtiaz 2019); the concentrations of potassium, phosphorus, magnesium, and calcium (350–400, 200, 70, and 40 mg/100 g, respectively) were found to be high (Al Juhaimi et al. 2018).

Date seeds are considered to be a rich source of lipids, although the date fruit contains only a low concentration of lipids (Ahmad and Imtiaz 2019). More details of lipid extracted from date seeds are described in Subsection 4.3.1.1 under "Date Seed Oil".

It has been claimed that some varieties of date seeds of some varieties contain estrone, such as 'Thamani' and 'Sukkari' date seeds with concentrations of 1.4 and 3.3 mg/100 g, respectively

(Ghnimi et al. 2017). Therefore, using date seeds as an ingredient in some foods (for example, bread) in an appropriate concentration may be beneficial as a hormone replacement therapy for postmenopausal women.

4.3.1.1 Date Seed Oil

Date seed oil is considered to be a safe and valuable product for human usage. Date seed oil is a yellow liquid at room temperature. According to Al Juhaimi et al. (2018), the seed oil concentration of different date varieties was between 5.77% and 10.71%, and the general properties of the oil were as follows: peroxide value of <30 meq O_2/kg, iodine value of 67.18–71.23 g I_2/100 g, saponification value of 203.27–213.18 mg KOH/g, and an acidic value of 1.09–1.44 mg KOH/g (Al Juhaimi et al. 2018).

Date seed oil is composed of 50% monosaturated fatty acids, 40% saturated acids, and 10% polyunsaturated acids, of which the most abundant fatty acids in each category are oleic acid (C18:1), lauric acid (C12:0), and linoleic acid (C18:2), respectively (Brouk and Fishman 2016).

In other research studies, the properties of date seed oil extracted from 18 date varieties (Khalas, Barhe, Lulu, Shikat Alkahlas, Sokkery, Bomaan, Sagay, Shishi, Maghool, Sultana, Fard, Maktoomi, Naptit Saif, Jabri, Khodary, Dabbas, Raziz, and Shabebe) were compared. The average fatty acid concentration of them were as follows: myristic acid (C14:0) at 14.52%, palmitic acid (C16:0) at 12.41%, margaric acid (C17:0) at 0.09%, stearic acid (C18:0) at 3.36%, arachidic acid (C20:0) at 0.34%, palmitoleic acid (C16:1) at 0.20%, oleic acid at 47.47%, linoleic acid at 10.23%, and linolenic acid (C18:3) at 0.16%. From a health viewpoint, the fatty acid profile of date seed oil is characterized by high oleic acid content. Therefore, date seed oil is considered to be a safe and valuable product for human usage due to its high oleic acid content. The oleic acid in date seed oil possesses positive health aspects, such as minimum levels of the *trans*-isomer, its potential to diminish LDL blood cholesterol, and its considerable oxidative stability (Habib et al. 2013).

The main unsaponifiable compounds in date seed oil are sterols at concentrations of 300–350 mg/100 g, consisting largely of β-sitosterol (75%), campesterol (10%), and Δ5-avenasterol (10%) (Nehdi et al. 2010).

The average concentrations of tocopherols, polyphenols, and carotenoids date seed oil are shown in Table 4.2. It can be seen that α-tocotrienol, followed by γ-tocopherol, constitute the main tocopherols in date seed oil. Natural vitamin E exists in chemical forms of tocopherols and tocotrienols. Vitamin E is an antioxidant acting as a radical scavenger to protect unsaturated fatty acid from oxidation, leading to oil stability. It also has an important function in health and decreases the risk of atherosclerosis (Habib et al. 2013).

The yellow color of the date seed oil is because of the presence of high concentrations of carotenoids (5.51 mg/100 g) (Nehdi et al. 2010). Various carotenoids are found in date seed oil, including lutein, cryptoxanthine, echinenone, lycopene, α-carotene, β-carotene, and γ-carotene (Table 4.2). Beta-carotene constitutes the highest concentration of a carotenoid whereas lycopene is the lowest concentration carotenoid detected in date seed oil (Habib et al. 2013).

The main polyphenols detected in date seed oil are gallic acid, syringic acid, protocatechuic acid, rutin, catechin, and caffeic acid (Table 4.2), most of which is caffeic acid followed by gallic acid (Al Juhaimi et al. 2018). Recently, the phenolic compounds have received increasing attention for their health benefits, such as preventing cardiovascular disease, diabetes, and cancer (Brouk and Fishman 2016).

4.3.1.2 Date Seed Coffee

The characteristics of seeds from date varieties are influenced by environmental and agricultural conditions such as irrigation, diurnal temperatures, growth duration, and post-harvest treatments

Table 4.2 The Tocopherol, Polyphenol, and Carotenoid Concentration of Date Seed Oil

Compounds	Concentration	
	According to Habib et al. (2013)	According to Al Juhaimi et al. (2018)
Tocopherols (mg/100 g)	1.01–1.86	0.54–1.17
α-Tocopherol	-	0.69–1.33
β-Tocopherol	0.40–0.70	7.61–11.84
γ-Tocopherol	-	31.76–37.41
α-Tocotrienol	-	4.47–8.47
γ-Tocotrienol	-	1.13–2.81
δ-Tocopherol	-	2.43–6.91
Polyphenols (mg/kg)	-	1.28–4.86
Gallic acid	-	0.031–0.089
Syringic acid	-	0.86–1.23
Protocatechuic acid	-	2.86–7.23
Rutin	-	1.13–3.43
Catechin	0.06–0.27	-
Caffeic acid	0.03–0.15	-
Carotenoids (mg/100 g)	0.06–0.19	-
Lutein	0.01–0.03	-
Cryptoxanthin	0.02–0.08	-
Echinenone	1.18–2.68	-
Lycopene	0.03–0.49	-
α-Carotene		
β-Carotene		
γ-Carotene		

(Habib and Ibrahim 2009). Therefore, the characteristics of the coffee obtained from date seeds are also influenced by these factors as well as the processing conditions. For example, moisture content, water activity, and water extract in roasted date seed powder were determined to be 1.63%, 6.57%, and 0.09, respectively (Rahman et al. 2007), while the concentrations of carbohydrates, protein, fat, crude fiber, and ash remained within the range of the concentrations reported in unroasted date seeds (Rahman et al. 2007).

According to the results of Warnasih et al. (2019), date seed coffee powder contains 4.42% moisture, 1.17% ash, 8.55% protein, 7.34% fat, 16.39% sugar, 78.52% carbohydrate, 340.65 mg GAE/100 g total phenol, and 23.81 µg/mL(IC_{50}) antioxidant activity. It has been reported that potassium is the mineral present at the highest concentration in date seed coffee powder, followed by phosphorus, magnesium, calcium, and sodium concentrations, namely 375.87, 125.58, 77.55, 18.73 and 15.23 mg/100 g, respectively (Niazi et al. 2017).

The process of roasting date seeds, along with their variety and pollination, affect the concentrations of sugars and fatty acids in date seeds. However, it has been reported that these variations are minor. The analysis of fatty acid profiles indicated that oleic, linoleic, lauric, and palmitic acids were the principal fatty acids in date seeds over a number of varieties (Rahman et al. 2007).

In comparison with the high levels of caffeine in normal coffee beans (20–40%,), the caffeine concentration of date seed coffee is zero (Warnasih et al. 2019). This is a beneficial characteristic of date seed coffee from a health point of view due to caffeine's negative effects (Diego et al. 2008). In terms of sensory traits, however, the date seed coffee drink is comparable with conventional coffee. According to the results, drinks prepared with 9% date seed coffee obtained similar scores for color (8.5 of 10), taste (9 of 10), odor (7.7 of 10), flavor (8.3 of 10), and overall acceptability (8.4 of 10) as control coffee samples, although drinks containing lower concentrations of date seed coffee showed undesirable scores (Venkatachalam and Sengottian 2016).

The total phenolic concentration and radical scavenging capacity (ORAC) of Espressodate (as commercial hot date-seed-based beverages), Espressodate mixed with 10% unroasted Majul date seeds, and normal coffee prepared by various procedures were compared by Brouk and Fishman (2016). These authors revealed that the roasting process decreases the content of phenolic compounds and also the antioxidant activity. Therefore, the date seed coffee presented lower amounts of phenolic compounds and antioxidant activity than did normal coffee. However, mixing Expressodate with unroasted date seeds improved these properties.

4.3.2 Date Press Cake (DPC)

Depending on the method of date extraction, the chemical composition of dried DPC varies. Boudechiche et al. (2010) reported that fresh DPC possessed nearly 70% water, 9–22% crude fiber, 5–8% protein, and 15% sugar content on average (Boudechiche, Araba, and Ouzrout 2010). The seedless DPC of several date varieties from Oman was reported to contain an average of 4.39% protein, 1.72% fat, 1.95% ash, 82.36% carbohydrate, and 28.57% dietary fiber, as well as high concentrations of phenolic compounds and antioxidants (Al-Farsi et al. 2007), whereas another study determined the chemical composition of three types of DPC on average as follows: 91.9% dry matter, 5.93% protein, 2.20% fat, 2.70% ash, and 14.17% dietary fiber (FAOSTAT 2016). Comparing these reports, it is deduced that the seedless DPC contains lower fat than the DPC with seeds. Majzoobi et al. (2019) studied DPC obtained from 'Shahani' dates in a ground form as having different particle sizes, namely coarse (355 μm) and fine particles (167 μm). They reported that the chemical composition of 'Shahani' DPC contained 13.37% moisture, 79.06% carbohydrate, 11.74% crude fiber, 6.35% protein, and 4.92% fat on average. The mineral analysis of the sample detected Fe, Zn, Cu, P, Mn, Mg, Ca, Na, and K, with the maximum concentration for Mg followed by P, while Na had the lowest concentration in the DPC samples. The fatty acid profile revealed that oleic acid (C18:1) was the main fatty acid in DPC, while the major unsaturated fatty acid was myristic acid and the minor unsaturated fatty acids included behenic, luric, and capric acids.

According to the findings of this research group, the contents of moisture, ash, carbohydrate and protein were similar for the coarse and the fine particles of date press cake.

However, functional properties such as antioxidant activity, and the concentrations of oleic acid, total phenolic and flavonoid were affected by the particle size of DPC so that the amount of these changed as follows: antioxidant activity (9.12 and 9.53 mg Vit C/g), and the concentrations of oleic acid (42.25 and 39.89%), total phenolic (16.18 and 18.78 mg/g) and flavonoid (1.84 and 1.95 mg/g) for the coarse and the fine particles, respectively. With decreasing particle size of the DPC, the properties of consistency, firmness, stickiness, cohesiveness, and viscosity increased. However, the DPC particle size did not significantly affect ($P > 0.05$) the pH or crust moisture content (Majzoobi et al. 2019, 2020).

4.4 PROCESSING AND UTILIZATION OF DATE WASTES AND BY-PRODUCTS

4.4.1 Date Seed

To use date seeds, they have to be converted into powder or flour (FAOSTAT 2016). Powder making from date seeds is performed in several steps, including washing to remove excess fruit residue, soaking in hot water for 1 h to soften the texture of the date kernels, draining, drying (in the sun or in the oven at 50–60°C for 24 h), roasting the kernels (temperature range of 160–200°C for 10–30 min), cooling at room temperature, grinding (using a hammer and then a mill), and sieving the date seeds (to achieve a smooth powder) (Hossain et al. 2014, Wahini 2016, Fikry et al. 2019). The scheme of processing date seeds and the products is shown in Figure 4.2.

Date seed valorized a surplus or waste product of date fruit processing, and its utilization would provide an economic advantage. Date seeds have plenty of potential for producing biomass, for water treatment (as active carbon), as an ingredient with health benefits in food formulations or for animal and poultry feeds. On the other hand, because of the low economic benefits, their application in water treatment and biomass production is of low priority (Hossain et al. 2014).

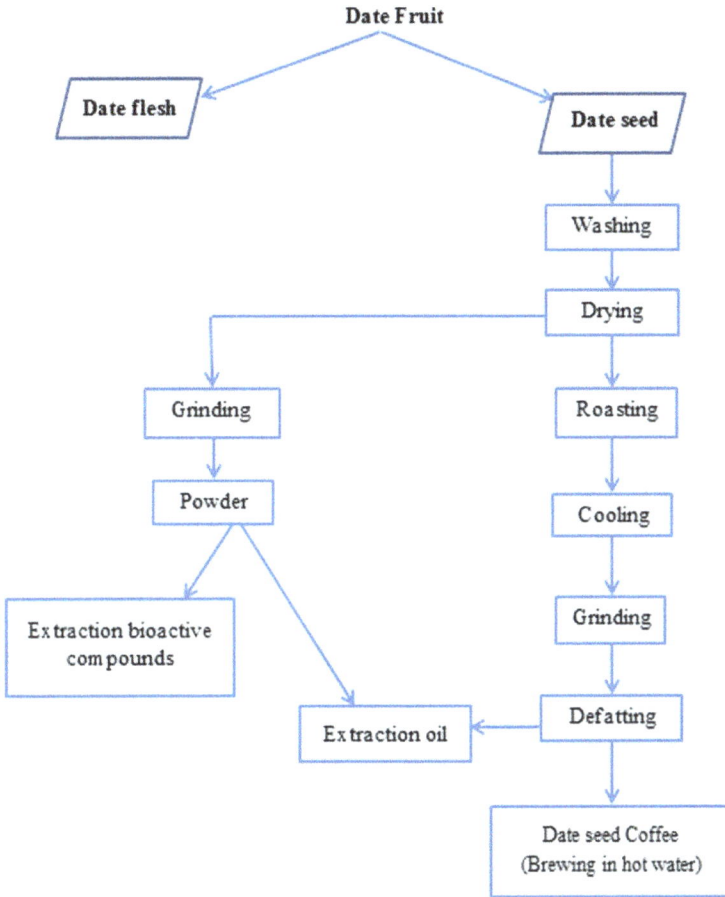

Figure 4.2 Flow diagram for processing of date seed (adapted from Hossain et al. 2014, Fikry et al. 2019).

Date seeds contain high levels of phenolic acids and flavonoids that possess many beneficial effects. These effects include the reduction of cardiovascular diseases, as well as anti-inflammatory, anticarcinogenic, antimicrobial, and antioxidant properties (Al-Farsi 2014). These properties can help to control blood sugar and to prevent DNA structure damage (in the form of antioxidants), prevent kidney stones, bladder diseases, and infectious diseases (Brouk and Fishman 2016). Due to the high antioxidant activity and the subsequent health benefits, date seeds were used in feed for male rats as a basal diet including 0, 70 or 140 g/kg date seeds over 30 days. The results indicated that date seeds exhibited a protective effect against *in-vivo* oxidative damage. These properties were probably related to the functionality of bioactive antioxidants (Habib and Ibrahim 2011).

Salama, Ismael, and Megeed (2019) studied the *in-vivo* anti-inflammatory properties of date seed powder nanoparticles infused into rats which had previously been fed a high-fat diet (HFD). According to their findings, HFD increased cholesterol and inflammatory markers significantly compared to the controls, and the authors concluded that applying polyvinyl alcohol (PVA) nanoparticles of date seed powder (DSP) in a 2:3 (w/w) ratio of PVA:DSP resulted in an acceptable particle size and desirable physicochemical properties.

Date seeds in the form of powder (flour) have been used in bread. The prepared breads, comprising 0, 5, 10, or 15% date seed flour, had slightly lower protein and higher fat than the control bread, but their total and soluble dietary fiber was considerably higher than that of the control bread.

According to the research findings, the inclusion of 10% of wheat flour with date seed powder in the bread formulation was shown to enhance total dietary fiber content (Brouk and Fishman 2016). Because of the high dietary fiber content of date seeds, they have nutritional value and thus can be used for preparing some foods or dietary supplements that are fiber based, such as bread, biscuits, and cakes (Hossain et al. 2014).

Date seeds were converted to protein powder by simple extraction in acidic or alkaline medium, precipitation of phenol/trichloroacetic acid (Ph/TCA) with acetone, and with or without ultrafiltration methods. The powder with the highest protein content (68%) was produced by Ph/TCA method, whereas the lowest content was 8% (Akasha, Campbell, and Euston 2012).

Several studies were performed on the decontaminating use of date seed powder for the removal of pollutants and chemical compounds; for example, 90% of phenol and p-nitrophenol could be removed from wastewater by date seed powder (Ahmed and Theydan 2012) or dye could be removed (Belala et al. 2011).

One of the most widely applied adsorbents is activated carbon. Activated carbon was prepared from date seeds at different burn-off rates. The results indicated that the activated carbon obtained had a higher adsorption capacity than commonly available activated carbons (Abdulkarim et al. 2002). For example, activated carbon prepared from date seeds was used to remove iodine, phenol, and methylene blue (Girgis and El-Hendawy 2002), to eliminate heavy metals (such as Cd, Pb, Cu, and Zn), dyes, phenolic compounds, and pesticides (Ahmed and Theydan 2012, Aldawsari et al. 2017), and to remove boron and phenol from drinking water (Al-Ithari et al. 2011).

Date seeds were also found to be effective at increasing the yield of citric acid in the fermentation process due to the presence of some elements, including Mg, Fe, Ca, Mn, Zn and Ni (Hossain et al. 2014). Several studies have been published on the microbial conversion of date seeds, such as utilization of the date seed hydrolysate and whey for citric acid production by *Candida lipolytica* (Abou-Zeid and Khoja 1993), and the single-cell protein produced from date seeds using the fungi *Aspergillus oryzae*, and *Candida utilis* (Hossain et al. 2014, FAOSTAT 2016). However, from a practical point of view, these studies have not yet been followed up.

4.4.2 Date Seed Oil

Despite the benefits, the quantity of extracted oil from date seeds is small and the processing is energy intensive, which limits the utilization of this oil. Therefore, recent research has focused on improving the extraction methods in order to increase the yield and reduce the energy consumption (Ben-Youssef et al. 2017).

There are conventional methods for oil extraction, namely solvent maceration and Soxhlet extraction, with the former method being used industrially. Despite high-yield extraction by the Soxhlet method, it takes a long time and requires large volumes of solvent (Virot et al. 2007). The effects of different extraction parameters (date seed powder particle size, drying method, solvent type, the feed: solvent ratio) were investigated by some researchers (Ali, Al-Hattab, and Al-Hydary 2015, Oladipupo Kareem et al. 2019). The conventional methods of oil extraction are solvent maceration and Soxhlet extraction, with the former being the more common method in use industrially. Nowadays, some advanced technologies, like microwave or ultrasound, are also used for assisting the extraction.

Soxhlet extraction is a type of solvent extraction that is repeated in several cycles. Generally, a fine powder of date seeds is positioned in a spongy cartridge, and then put in the extractor. The solvent is transferred from the flask to the upper part of the system. The solvent condenses through the action of cooling water at the top of the device and accumulates around and in the cartridge. The mixture of solvent and extracted oil adjacent to the lid/ mouth of the siphon is returned to the flask and is evaporated again, and this cycle repeats several times (Jensen 2007). The operation time in this method can be optimized, although the extraction time is fixed at 8 h, according to some standards (ISO-659-1988 1988). Despite the Soxhlet extraction achieving the highest extraction yield in comparison with other techniques, it is time consuming, uses large volumes of solvent, needs solvent evaporation, and the properties of the oil may be modified during extraction due to the long extraction period, and the prolonged exposure to oxygen (Virot et al. 2007).

In the maceration technique, milled date seeds are incubated with a suitable solvent (mostly hexane and petroleum ether) and continuously stirred at a moderate temperature (usually less than 60°C) for several hours to extract the oil. Then, the mixture is filtrated and the solvent is evaporated to recover the extracted oil (Organization et al. 2008, Fine et al. 2013). A microwave oven may be used as an assisted technique, accompanied by solvent extraction, to enhance the yield of oil extraction. In this technique, the microwave rays cause internal heating and consequently expand the plant cells at atmospheric pressure leading to greater release of oil. A cooling apparatus placed out of the microwave oven condenses the solvent (Fakhfakh et al. 2019).

Ultrasonic operation at an intensity of 20–100 kHz is also used as a solvent-assisted technique to increase the yield at low temperature. It causes the plant cell destruction to achieve more oil extraction (Chemat and Khan 2011). Nowadays, there is a move toward green extraction strategies, such as supercritical carbon dioxide (CO_2) extraction, thanks to its great advantages, such as low temperature (31.1°C), safe, non-toxic, non-flammable, and inexpensive solvent, and short extraction time. When the temperature and pressure exceed their critical points (temperature of 31.1°C and pressure of 7.38 MPa), CO_2 enters its supercritical state. However, the utilization of this technique is limited because of the need for expensive equipment (Fakhfakh et al. 2019).

Oladipupo Kareem et al. (2019) investigated the effect of different extraction parameters (particle size of date seed powder, drying method, solvent type, feed: solvent ratio, and pretreatments) on the yield of oil extracted from the waste date seeds ('Sukkary' variety) by the Soxhlet method. Their results revealed that the highest oil extraction yield of 10.34% was reached at a particle size of smaller than 200 μm, oven drying at a temperature of 70°C, using an n-hexane solvent: feed ratio of 1:40, and hydrochloric acid pretreatment. The effect on oil yield from date seeds by the Soxhlet method of different parameters, such as particle sizes (0.425 mm, 1 mm, or 2 mm), extraction times (1, 2, 4, or 6 hours), and solvent type (methanol, 2-propanol, chloroform, n-hexane, or toluene), were examined on yield of oil extraction. The findings revealed that the highest yield of 8.5% (w/w) was achieved by the 120-min extraction with n-hexane from a particle size of 0.425 mm. The physical properties of the oil obtained were as follows: viscosity of 29 cP (centipoise), density of 0.925 g cm^{-2} and a refractive index of 1.444 (Ali, Al-Hattab, and Al-Hydary 2015).

Although date seed oil is edible, it is rarely used in the food industry due to the low oil content of the seeds. However, in some food research, date seed oil has been used, resulting in desirable properties. Sensory characteristics of mayonnaise were improved by using date seed oil in comparison with corn oil (Basuny and Al-Marzooq 2011). It was reported that date seed oil can be used in frying or cooking processes due to its high thermal and oxidative stability (Besbes et al. 2005). Furthermore, date seed oil has the capacity to be applied in the fields of cosmetics (creams, soaps, and shampoos), pharmaceuticals (liniment for indolent (slow-growing) tumors, or for protection against UV-A and UV-B), and medicine (to prevent atherosclerosis, to increase motility, viability, and count of sperm) (Hossain et al. 2014).

It was found that date seed extract was able to return the function of a poisoned liver to the normal status, and to guard it from subsequent carbon tetrachloride hepatotoxicity in the rats' liver (Al-Qarawi et al. 2009).

Extracts of date seeds in water or ethanol exhibited antibacterial activity toward *Klebsiella pneumoniae*, *Proteus vulgaris*, *Staphylococcus aureus*, *Escherichia coli*, and *Bacillus subtilis*. The inhibitory effect of date seed extracts toward bacteria was more than the effect of antibiotics (Saddiq and Bawazir 2010). On the other hand, date seeds produced different types of extracts in several solvents (water, ethanol, methanol, or acetone), with significant antidiabetic effects, especially the aqueous extract. Hence, date seed extract decreases blood sugar levels and may be used as an antidiabetic medicine (Hafez El-Far et al. 2016).

Metoui et al. (2018) reported antimicrobial activity of extracts of seeds from 11 common date cultivars from Tunisia against *Escherichia coli*, *Salmonella typhimurium*, *Staphylococcus aureus*, and *Staphylococcus epidermis*. Their results indicated antibacterial activities for all the date seed extracts except for *Enterococcus faecalis*. They concluded that the extracted date seed oil can be used for medicinal purposes. On the other hand, Al-rajhi et al. (2019) reported the antibacterial activity of the date seed powder extract, with date seed cake having potential antibacterial activity against a wide spectrum of bacteria.

The use of date seed oil in biodiesel production has also been reported due its low fatty acid content (Golshan Tafti, Solaimani Dahdivan, and Yasini Ardakani 2017).

4.4.3 Date Seed Coffee

The usual procedure to produce date seed coffee is as follows. The date seeds are washed in order to be separated from any adhering date flesh. Then, the seeds are dried and roasted for a given time to reach a light brown color. The roasted seeds are ground and sieved to achieve fine roasted seed powder usually of less than 60 mesh (Figure 4.2) (Niazi et al. 2017, Warnasih et al. 2019).

Date seeds might be soaked in water and dried before roasting or defatted after that. The temperature and duration of roasting might be varied, depending on the hardness of the date seeds. Most studies suggested roasting at 125°C for 30 minutes (Venkatachalam and Sengottian 2016, Warnasih et al. 2019).

However, Fikry et al. (2019) examined the effect of roasting temperature (160, 180, or 200°C) and duration (10, 20, or 30 min) on the physicochemical and sensory properties of date seeds. According to their results, the optimum conditions were roasting at 200°C for 20 min.

4.4.4 Date Feed

Date wastes may include immature dates, cull dates, drier dates, or second-grade dates with less commercial value, which are not considered suitable for the main uses. Cull and drier dates, especially, are used as an animal feed source. Since the date without the pit is not a balanced feedstuff (high in carbohydrates but low in protein and fat) (FAOSTAT 2016, Ghnimi et al. 2017), it must be mixed with dry material, such as barley, maize, or soybean meal, as part of the feed (Hossain et al. 2014). Generally, the feeds were made by adding different amounts of date pits (namely 5 to 27%) with date flesh (8 to 43%), and other ingredients (Hossain et al. 2014). It was found that the incorporation of dates at up to 20% in a mixture with barley gave the best results in the diet of Awassi lambs (Hassan, Al-Baiati, and Almosawy 2013). Al-Suwaiegh (2016) used date seeds in four concentrations (0, 10, 15, or 20%) as replacement for concentrate feed in the diet of Ardi goats. The results suggested that the partial inclusion of date seeds in concentrate feeds could increase their productive performance by up to 20%.

Normally, the date feed for animals is prepared in three stages, namely mashing, screening, and mixing. The date mash is obtained by adding water (50%) and removing the seeds from the pulp *via* a rotary screen, followed by mixing it with flaked barley (1:1) and drying it in a rotary dryer to be used as a feed component (FAOSTAT 2016). To produce storable compound mixes or pellets, the mash moisture content is a limiting factor. In this case, the moisture content in the final mix should not be more than 12–15% (FAOSTAT 2016).

Due to the content of carbohydrate in date seed, it could be used in animal diets. Several studies have investigated the usage of date seeds or date press cake as feed for animals such as cows, sheep, poultry, camels, and fish (Ali, Bashir, and Alhadrami 1999, Al Dhaheri et al. 2004, Hossain et al. 2014, FAOSTAT 2016). The incorporation of date seed into the diet of farmed animals has some benefits like increasing bodyweight gain, enhancing the plasma level of estrogens or testosterone, increasing the feed efficiency, and improving the palatability of meat (Hossain et al. 2014, FAOSTAT 2016).

4.4.5 Date Press Cake

Date Press Cake (DPC) is usually used as animal feed and as substrate for microbial conversion. However, it may be discarded into drains and onto open lands, causing great economic loss and environmental problems because of its bulky nature (30% of date weight), high moisture content (70%), and the risk of spoilage (FAOSTAT 2016).

Some researchers produced activated carbon from DPC and evaluated its efficiency for adsorption of different pollutants, such as Pb(II), Cr(VI), and methylene blue from aqueous solutions (Norouzi et al. 2018, Heidarinejad et al. 2018, Heidarinejad, Rahmanian, and Heidari 2019). Recently, a study showed that DPC has a great potential to be utilized in value-added products (Majzoobi et al. 2020).

DPC has been used as the corn grain replacement in ruminant feeding. Fahmy et al. (2019) compared DPC and corn grain for evaluating the effect of ensiling and/or exogenous fibrolytic enzyme used in the *in-vitro* batch culture technique. They concluded that the rumen activity did not change significantly when either DPC or corn grains was used. On the other hand, Morsy et al. (2020) evaluated substituting corn grain with different amounts of DPC (25, 50, 75, or 100%) in ruminant feeding with or without the fibrolytic enzyme by the *in-vitro* batch culture method. They concluded that partial replacing of the corn grains with 25% DPC achieved the best results.

Heidarinejad, Rahmanian, and Heidari (2019) studied the efficiency of Pb(II) adsorption by KOH-activated carbon prepared from date press cake (DPC). They found that the ratio of 4:1 KOH: activated carbon was best able to adsorb Pb(II) from either aqueous or saline solutions.

4.5 CONCLUSIONS

Dates wastes and by-products are valuable fruit materials due to being rich in carbohydrates, fiber, minerals, vitamins, and phytocompounds. These date processing residues have potential health benefits, and a number of value-added products can be derived from them with economic benefits. They can also be used in the pharmaceutical and food industries as supplements or ingredients. Therefore, industrialization and exploitation of date wastes and byproducts are increasing day by day. Further studies need to be carried out with respect to industrially producing or deriving food ingredients or biocomponents from these wastes and by-products. An advantageous side effect is that proper management of date residues will minimize their impact on the environment.

REFERENCES

Abdulkarim, M. A., N. A. Darwish, Y. M. Magdy, and A. Dwaidar. 2002. "Adsorption of Phenolic Compounds and Methylene Blue onto Activated Carbon Prepared from Date Fruit Pits." *Engineering in Life Sciences* 2 (6):161–165. https://doi.org/10.1002/1618-2863(200206)2:6.

Abou-Zeid, A. A., and Samir M. Khoja. 1993. "Utilization of Dates in the Fermentative Formation of Citric Acid by Yarrowia Lipolytica." *Zentralblatt für Mikrobiologie* 148 (3):213–221. https://doi.org/10.1016/S0232-4393(11)80093-4.

Afiq, M. J. Abdul, R. Abdul Rahman, Y. B. Che Man, H. A. Al-Kahtani, and T. S. T. Mansor. 2013. "Date Seed and Date Seed Oil." *International Food Research Journal* 20 (5):2035.

Ahmad, A., and Imtiaz, H. 2019. "Chemical Composition of Date Pits: Potential to Extract and Characterize the Lipid Fraction." In *Sustainable Agriculture Reviews*, edited by M. Naushad and E. Lichtfouse, 55–77. Cham: Springer.

Ahmed, Muthanna J., and Samar K. Theydan. 2012. "Equilibrium Isotherms, Kinetics and Thermodynamics Studies of Phenolic Compounds Adsorption on Palm-Tree Fruit Stones." *Ecotoxicology and Environmental Safety* 84:39–45. https://doi.org/10.1016/j.ecoenv.2012.06.019.

Akasha, Ibrahim, Lydia Campbell, and Stephen Euston. 2012. "Extraction and Characterisation of Protein Fraction from Date Palm Fruit Seeds." *World Academy of Science, Engineering and Technology* 70:70.

Al-Farsi, Mohamed Ali. 2014. "Enrichment of Date Paste." *Journal of Human Nutrition & Food Science* 2:1032.

Al-Farsi, Mohamed Ali, Cesarettin Alasalvar, Mohammed Al-Abid, Khalid Al-Shoaily, Mansorah Al-Amry, and Fawziah Al-Rawahy. 2007. "Compositional and Functional Characteristics of Dates, Syrups, and Their By-Products." *Food Chemistry* 104 (3):943–947.

Al-Farsi, Mohamed Ali, and Chang Young Lee. 2011. "Usage of Date (*Phoenix dactylifera* L.) Seeds in Human Health and Animal Feed." In *Nuts and Seeds in Health and Disease Prevention*, edited by V. R. Preedy, R. R. Watson, and V. B. Patel, 447–452. London: Elsevier.

Al-Farsi, Mohamed Ali, and Chang Lee. 2008. "Nutritional and Functional Properties of Dates: A Review." *Critical Reviews in Food Science and Nutrition* 48:877–887. https://doi.org/10.1080/10408390701724264.

Al-Ithari, Afrah J., Arumugam Sathasivan, Roxanne Ahmed, Hari B. Vuthaluru, Weixi Zhan, and Mushtaque Ahmed. 2011. "Superiority of Date Seed Ash as an Adsorbent over Other Ashes and Ferric Chloride in Removing Boron from Seawater." *Desalination and Water Treatment* 32 (1–3):324–328. https://doi.org/10.5004/dwt.2011.2717.

Al-Qarawi, A. A., Hassan Mousa, Badreldin Ali, Hassan Abdel-Rahman, and S. A. El-Mougy. 2009. "Protective Effect of Extracts from Dates (*Phoenix dactylifera* L.) on Carbon Tetrachloride-Induced Hepatotoxicity in Rats." *International Journal of Applied Research in Veterinary Medicine* 2:176–180.

Al-Suwaiegh, S. B. 2016. "Effect of Feeding Date Pits on Milk Production, Composition and Blood Parameters of Lactating Ardi Goats." *Asian-Australasian Journal of Animal Sciences* 29 (4):509–515. https://doi.org/10.5713/ajas.15.0012.

Al Dhaheri, Ayesha, Ghaleb Alhadrami, N. Aboalnaga, Ibrahim Wasfi, and Mamdouh El-Ridi. 2004. "Chemical Composition of Date Pits and Reproductive Hormonal Status of Rats Fed Date Pits." *Food Chemistry* 86:93–97. https://doi.org/10.1016/j.foodchem.2003.08.022.

Al Juhaimi, Fahad, Mehmet Musa Özcan, Oladipupu Q. Adiamo, Omer N. Alsawmahi, Kashif Ghafoor, and Elfadil E. Babiker. 2018. "Effect of Date Varieties on Physico-Chemical Properties, Fatty Acid Composition, Tocopherol Contents, and Phenolic Compounds of Some Date Seed and Oils." *Journal of Food Processing and Preservation* 42 (4):e13584.

Aldawsari, A., M. A. Khan, B. H. Hameed, A. A. Alqadami, M. R. Siddiqui, Z. A. Alothman, and Aybh Ahmed. 2017. "Mercerized Mesoporous Date Pit Activated Carbon-A Novel Adsorbent to Sequester Potentially Toxic Divalent Heavy Metals from Water." *PLoS One* 12 (9):e0184493. https://doi.org/10.1371/journal.pone.0184493.

Ali, B. H., A. K. Bashir, and G. Alhadrami. 1999. "Reproductive Hormonal Status of Rats Treated with Date Pits." *Food Chemistry* 66 (4):437–441. https://doi.org/10.1016/S0308-8146(98)00060-0.

Ali, Mortadha A., Tahseen A. Al-Hattab, and Imad A. Al-Hydary. 2015. "Extraction of Date Palm Seed Oil (*Phoenix dactylifera*) by Soxhlet Apparatus." *International Journal of Advances in Engineering & Technology* 8 (3):261.

Alrajhi, Maha, Mabrouk Al-Rasheedi, Salah Elnaeem M. Eltom, Yasir Alhazmi, Mustafa Mohammed Mustafa, and ALreshidi Mateq Ali. 2019. "Antibacterial Activity of Date Palm Cake Extracts (*Phoenix dactylifera*)." *Cogent Food & Agriculture* 5 (1):1625479. https://doi.org/10.1080/23311932.2019.1625479.

Aromanadates. 2018. Avilable at: https://www.aromanadates.com/dates-and-associated-countries.

Ashraf, Zahra, and Zohreh Hamidi-Esfahani. 2011. "Date and Date Processing: A Review." *Food Reviews International* 27 (2):101–133. https://doi.org/10.1080/87559129.2010.535231.

Assirey, Eman Abdul Rahman. 2015. "Nutritional Composition of Fruit of 10 Date Palm (*Phoenix dactylifera* L.) Cultivars Grown in Saudi Arabia." *Journal of Taibah University for Science* 9 (1):75–79. https://doi.org/10.1016/j.jtusci.2014.07.002.

Awad, M. A. 2007. "Increasing the Rate of Ripening of Date Palm Fruit (*Phoenix dactylifera* L.) cv. Helali by Preharvest and Postharvest Treatments." *Postharvest Biology and Technology* 43 121–127.

Basuny, Amany Mohamed Mohamed, and Maliha Ali Al-Marzooq. 2011. "Production of Mayonnaise from Date Pit Oil." *Food and Nutrition Science* 2 (9):938–943.

Belala, Zohra, Mejdi Jeguirim, Meriem Belhachemi, Fatima Addoun, and Gwenaëlle Trouvé. 2011. "Biosorption of Basic Dye from Aqueous Solutions by Date Stones and Palm-Trees Waste: Kinetic, Equilibrium and Thermodynamic Studies." *Desalination* 271 (1):80–87. https://doi.org/10.1016/j.desal.2010.12.009.

Ben-Youssef, Sahar, Jawhar Fakhfakh, Cassandra Breil, Maryline Abert-Vian, Farid Chemat, and Noureddine Allouche. 2017. "Green Extraction Procedures of Lipids from Tunisian Date Palm Seeds." *Industrial Crops and Products* 108:520–525.

Besbes, Souhail, Christophe Blecker, Claude Deroanne, Georges Lognay, Nour-Eddine Drira, and Hamadi Attia. 2005. "Heating Effects on Some Quality Characteristics of Date Seed Oil." *Food Chemistry* 91 (3):469–476.

Boudechiche, Lyes, Abdelilah Araba, and R. Ouzrout. 2010. "Influence of Date Waste Supplementation of Ewes in Late Gestation on the Performance During Lactation." *Livestock Research for Rural Development* 22 (3):2010.

Brouk, Moran, and Ayelet Fishman. 2016. "Antioxidant Properties and Health Benefits of Date Seeds." In *Functional Properties of Traditional Foods*, edited by K. Kristbergsson and S. Otles, 233–240. New York: Springer.

Chandrasekaran, M., and Ali H. Bahkali. 2013. "Valorization of Date Palm (*Phoenix dactylifera*) Fruit Processing By-Products and Wastes Using Bioprocess Technology – Review." *Saudi Journal of Biological Sciences* 20 (2):105–120. https://doi.org/10.1016/j.sjbs.2012.12.004.

Chao, C. C. T., and R. R. Krueger. 2007. "The Date Palm (*Phoenix dactylifera* L.): Overview of Biology, Uses, and Cultivation." *HortScience* 42 (5):1077–1082. https://doi.org/10.21273/HORTSCI.42.5.1077.

Chemat, Farid, and Muhammed Kamran Khan. 2011. "Applications of Ultrasound in Food Technology: Processing, Preservation and Extraction." *Ultrasonics Sonochemistry* 18 (4):813–835.

Deng, Gui-Fang, Chen Shen, Xiang-Rong Xu, Ru-Dan Kuang, Ya-Jun Guo, Li-Shan Zeng, Li-Li Gao, Lin xi, Jie-Feng Xie, En-Qin Xia, Sha Li, Shan Wu, Feng Chen, Wen-Hua Ling, and Hua-Bin Li. 2012. "Potential of Fruit Wastes as Natural Resources of Bioactive Compounds." *International Journal of Molecular Sciences* 13:8308–8323. https://doi.org/10.3390/ijms13078308.

Diego, Miguel, Tiffany Field, Maria Hernandez-Reif, Yanexy Vera, Karla Gil, and Adolfo Gonzalez-Garcia. 2008. "Caffeine Use Affects Pregnancy Outcome." *Journal of Child & Adolescent Substance Abuse* 17 (2):41–49. https://doi.org/10.1300/J029v17n02_03.

Djouab, Amrane, Salem Benamara, Hassina Gougam, Hayet Amellal, and Karima Hidous. 2016. "Physical and Antioxidant Properties of Two Algerian Date Fruit Species (*Phoenix dactylifera* L. and *Phoenix canariensis* L.)." *Emirates Journal of Food and Agriculture*:1. https://doi.org/10.9755/ejfa.2015-12-1056.

El Hadrami, Abdelbasset, and Jameel M. Al-Khayri. 2012. "Socioeconomic and Traditional Importance of Date Palm." *Emirates Journal of Food and Agriculture* 24 (5):371.

El Hilaly, Jaouad, Jamal Ennassir, Mohamed Benlyas, Chakib Alem, Mohamed-Yassine Amarouch, and Younes Filali-Zegzouti. 2018. "Anti-Inflammatory Properties and Phenolic Profile of Six Moroccan Date Fruit (*Phoenix dactylifera* L.) Varieties." *Journal of King Saud University-Science* 30 (4):519–526.

Fahmy, M., Tarek Morsy, Hany Gado, O. Matloup, S. Kholif, and Nasr El-Bordeny. 2019. "In Vitro Evaluation of Ensiling and/or Exogenous Fibrolytic Enzyme Supp Date Press Cake." *Arab Universities Journal of Agricultural Sciences* 27:347–355. https://doi.org/10.21608/ajs.2019.43546.

Fakhfakh, Jawhar, Sahar Ben-Youssef, Mu Naushad, and Noureddine Allouche. 2019. "Different Extraction Methods, Physical Properties and Chemical Composition of Date Seed Oil." In *Sustainable Agriculture Reviews 34*, edited by M. Naushad and E. Lichtfouse, 125–153. Cham: Springer.

FAO. 2018. *Statistical Databases.* http://faostat.fao.org (accessed 23.11.20).

FAOSTAT. 2016. *Food and Agriculture Organization of the United Nations.* Rome, Italy.

Fikry, Mohammad, Yus Aniza Yusof, Alhussein M. Al-Awaadh, Russly Abdul Rahman, Nyuk Ling Chin, Esraa Mousa, and Lee Sin Chang. 2019. "Effect of the Roasting Conditions on the Physicochemical, Quality and Sensory Attributes of Coffee-Like Powder and Brew from Defatted Palm Date Seeds." *Foods* 8 (2):61.

Fine, Frederic, Maryline Abert Vian, Anne-Sylvie Fabiano Tixier, Patrick Carre, Xavier Pages, and Farid Chemat. 2013. "Les agro-solvants pour l'extraction des huiles végétales issues de graines oléagineuses." *OCL* 20 (5):A502.

Ghnimi, Sami, Syed Umer, Azharul Karim, and Afaf Kamal-Eldin. 2017. "Date Fruit (*Phoenix dactylifera* L.): An Underutilized Food Seeking Industrial Valorization." *NFS Journal* 6:1–10. https://doi.org/10.1016/j.nfs.2016.12.001.

Girgis, Badie S., and Abdel-Nasser A. El-Hendawy. 2002. "Porosity Development in Activated Carbons Obtained from Date Pits under Chemical Activation with Phosphoric Acid." *Microporous and Mesoporous Materials* 52 (2):105–117. https://doi.org/10.1016/S1387-1811(01)00481-4.

Golshan Tafti, A., N. Solaimani Dahdivan, and S. A. Yasini Ardakani. 2017. "Physicochemical Properties and Applications of Date Seed and its Oil." *International Food Research Journal* 24 (4):1399–1406.

Habib, Hosam M., and Wissam H. Ibrahim. 2009. "Nutritional Quality Evaluation of Eighteen Date Pit Varieties." *International Journal of Food Sciences and Nutrition* 60 (sup1):99–111.

Habib, Hosam M., and Wissam H. Ibrahim. 2011. "Effect of Date Seeds on Oxidative Damage and Antioxidant Status In Vivo." *Journal of the Science of Food and Agriculture* 91 (9):1674–1679. https://doi.org/10 .1002/jsfa.4368.

Habib, Hosam M., Hina Kamal, Wissam H. Ibrahim, and Ayesha S. Al Dhaheri. 2013. "Carotenoids, Fat Soluble Vitamins and Fatty Acid Profiles of 18 Varieties of Date Seed Oil." *Industrial Crops and Products* 42:567–572.

Hafez El-Far, J. A., H. M. Shaheen, M. Abd El-Daim, S. K. Al Jaouni, and J. S. A. Mousa. 2016. "Date Palm (*Phoenix dactylifera*): Protection and Remedy Food." *Current Trends in Nutraceuticals* 1 (2):1–10.

Hamada, J. S., I. B. Hashim, and F. A. Sharif. 2002. "Preliminary Analysis and Potential Uses of Date Pits in Foods." *Food Chemistry* 76 (2):135–137. https://doi.org/10.1016/S0308-8146(01)00253-9.

Hassan, S. A., H. Y. Al-Baiati, and J. E. Almosawy. 2013. "Effect of Substitution of Barley by Whole Dates on Performance and Digestion of Awassi Lambs." *KSÜ Doğa Bilimleri Dergisi* 16 (3):12–15.

Heidarinejad, Zoha, Omid Rahmanian, Mehdi Fazlzadeh, and Mohsen Heidari. 2018. "Enhancement of Methylene Blue Adsorption onto Activated Carbon Prepared from Date Press Cake by Low Frequency Ultrasound." *Journal of Molecular Liquids* 264:591–599. https://doi.org/10.1016/j.molliq.2018.05.100.

Heidarinejad, Zoha, Omid Rahmanian, and Mohsen Heidari. 2019. "Production of KOH-Activated Carbon from Date Press Cake: Effect of the Activating Agent on its Properties and Pb(II) Adsorption Potential." *Desalination and Water Treatment* 165:232–243. https://doi.org/10.5004/dwt.2019.24501.

Hossain, Mohammad Zakir, Mostafa I. Waly, Vandita Singh, Venitia Sequeira, and Mohammad Shafiur Rahman. 2014. "Chemical Compositions of Date-Pits and Its Potential for Developing Value-Added Product - A Review." *Polish Journal of Food and Nutrition Sciences* 64 (4):215–226. https://doi.org/10 .2478/pjfns-2013-0018.

IOTCO. 2018. "Dates of Saudi Arabia." Available at: http://saudiarabiadates.com/dates.htm.

ISO-659-1988. 1988. "International Organization for Standardization (ISO)." Geneva: ISO.

Jensen, William B. 2007. "The Origin of the Soxhlet Extractor." *Journal of Chemical Education* 84 (12):1913.

Majzoobi, M., G. Karambakhsh, M. T. Golmakani, G. Mesbahi, and A. Farahnaky. 2020. "Effects of Level and Particle Size of Date Fruit Press Cake on Batter Rheological Properties and Physical and Nutritional Properties of Cake." *Journal of Agricultural Science and Technology (JAST)* 22 (1):121–133.

Majzoobi, Mahsa, G. Karambakhsh, G. Mesbahi, and A. Farahnaky. 2019. "Chemical Composition and Functional Properties of Date Press Cake, an Agro-Industrial Waste." *Journal of Agricultural Science and Technology* 21:1807–1817.

Metoui, Mounira, Awatef Essid, Amira Bouzoumita, and Ali Ferchichi. 2018. "Chemical Composition, Antioxidant and Antibacterial Activity of Tunisian Date Palm Seed." *Polish Journal of Environmental Studies* 28. https://doi.org/10.15244/pjoes/84918.

Morsy, Tarek A., Osama H. Matloup, Hany M. Gado, Nasr E. EL-Bordeny, Sobhy M. Kholif, and Mahmoud Fahmy. 2020. "Influence of Replacing Corn with Levels of Treated Date Press Cake on In Vitro Ruminal Fermentation, Degradability and Gas Production." *International Journal of Dairy Science* 15:72–79.

National Date Association of Iran. 2018. Available in Electronic Portal. http://www.naid.ir/article/.

Nehdi, I., S. Omri, M. I. Khalil, and S. I. Al-Resayes. 2010. "Characteristics and Chemical Composition of Date Palm (*Phoenix canariensis*) Seeds and Seed Oil." *Industrial Crops and Products* 32 (3):360–365.

Niazi, Sobia, Seemab Rasheed, Imran Mahmood Khan, Ayesha Safdar, and Fozia Niazi. 2017. "Chemical, Mineral and Phytochemical Screening Assay of Date Palm Seeds for the Development of Date Palm Seed Coffee." *International Journal of Bioengineering & Biotechnology* 2 (2):6.

Norouzi, Samira, Mohsen Heidari, Vali Alipour, Omid Rahmanian, Mehdi Fazlzadeh, Fazel Mohammadi-moghadam, Heshmatollah Nourmoradi, Babak Goudarzi, and Kavoos Dindarloo. 2018. "Preparation, Characterization and Cr(VI) Adsorption Evaluation of NaOH-Activated Carbon Produced from Date Press Cake; An Agro-Industrial Waste." *Bioresource Technology* 258:48–56. https://doi.org/10.1016/j .biortech.2018.02.106.

Oladipupo Kareem, Mujeeb, Anjali Achazhiyath Edathil, K. Rambabu, G. Bharath, Fawzi Banat, G. S. Nirmala, and K. Sathiyanarayanan. 2019. "Extraction, Characterization and Optimization of High Quality Bio-Oil Derived from Waste Date Seeds." *Chemical Engineering Communications*:1–11. https://doi.org/10.1080/00986445.2019.1650034.

Parvin, Sultana, Dilruba Easmin, Afzal Sheikh, Mrityunjoy Biswas, Subed Sharma, Dev Sharma, Md Rabbani, Sarowar Jahan, Amirul Islam, Narayan Roy, and Shovon Shariar. 2015. "Nutritional Analysis of Date Fruits (*Phoenix dactylifera* L.) in Perspective of Bangladesh." *American Journal of Life Sciences* 3:274–278. https://doi.org/10.11648/j.ajls.20150304.14.

Ragava, Sc., and M. Loganathan. 2016. "Microbial Evaluation and Control of Microbes in Commercially Available Date (*Phoenix dactylifera* Lynn.) Fruits." *Journal of Food Processing & Technology* 7:598. https://doi.org/10.4172/2157-7110.1000598.

Rahman, M. S., S. Kasapis, N. S. Z. Al-Kharusi, I. M. Al-Marhubi, and A. J. Khan. 2007. "Composition Characterisation and Thermal Transition of Date Pits Powders." *Journal of Food Engineering* 80 (1):1–10.

Saddiq, A., and A. E. Bawazir. 2010. "Antimicrobial Activity of Date Palm (*Phoenix dactylifera*) Pits Extracts and Its Role in Reducing the Side Effect of Methyl Prednisolone on Some Neurotransmitter Content in the Brain, Hormone Testosterone in Adulthood." *Journal of the American Society for Horticultural Science. American Society for Horticultural Science* 882:665–690. https://doi.org/10.17660/ActaHortic .2010.882.74.

Salama, A. A., N. M. Ismael, and M. M. Megeed. 2019. "Using Date Seed Powder Nanoparticles and Infusion as a Sustainable Source of Nutraceuticals." *Journal of Food and Nutrition Sciences* 7 (3):39–48. https:// doi.org/10.11648/j.jfns.20190703.11.

Shina, Sadiq, Thompson Izuagie, M. Shuaibu, A. I. Dogoyaro, A. Garba, and S. Abubakar. 2013. "The Nutritional Evaluation and Medicinal Value of Date Palm (*Phoenix dactylifera*)." International Journal of Modern Chemistry 4:147–154.

United Nations Industrial Development Organization, Suckdev Swami Handa, Suman Preet Singh Khanuja, Gennaro Longo, and Dev Dutt Rakesh. 2008. *Extraction Technologies for Medicinal and Aromatic Plants*. Padriciano: Earth, Environmental and Marine Sciences and Technologies.

Venkatachalam, Chitra Devi, and Mothil Sengottian. 2016. "Study on Roasted Date Seed Non-Caffeinated Coffee Powder as a Promising Alternative." *Asian Journal of Research in Social Sciences and Humanities* 6 (6):1387–1394.

Virot, Matthieu, Valérie Tomao, Giulio Colnagui, Franco Visinoni, and Farid Chemat. 2007. "New Microwave-Integrated Soxhlet Extraction: An Advantageous Tool for the Extraction of Lipids from Food Products." *Journal of Chromatography A* 1174 (1–2):138–144.

Wahini, M. 2016. "Exploration of Making Date Seed's Flour and its Nutritional Contents Analysis." *MS&E* 128 (1):012031.

Warnasih, Siti, Ade Heri Mulyati, Diana Widiastuti, Zuniar Subastian, Laksmi Ambarsari, and Purwantiningsih Sugita. 2019. "Chemical Characteristics, Antioxidant Activity, Total Phenol, and Caffeine Contents in Coffee of Date Seeds (*Phoenix dactylifera* L.) of Red Sayer Variety." *The Journal of Pure and Applied Chemistry Research* 8 (2):179–184.

Zaid, A. 2002. *Date Palm Cultivation*. Edited by FAO report. Rome.

Zhang, Chuan-Rui, Saleh A. Aldosari, Polana S. P. V. Vidyasagar, Karun M. Nair, and Muraleedharan G. Nair. 2013. "Antioxidant and Anti-Inflammatory Assays Confirm Bioactive Compounds in Ajwa Date Fruit." *Journal of Agricultural and Food Chemistry* 61 (24):5834–5840. https://doi.org/10.1021/jf401371v.

CHAPTER 5

Apple Wastes and By-Products
Chemistry, Processing, and Utilization

Naseer Ahmed, Krishan Kumar, Jaspreet Kaur, Qurat-Ul-Eain Hyder Rizvi, Sumaira Jan, Divya Chauhan, Priyanka Thakur, Tajendra Pal Singh, Chhaya and Shiv Kumar

CONTENTS

5.1 INTRODUCTION

Apple (*Malus* × *domestica* L. Borkh.) is the favorite fruit of millions and is widely grown in temperate regions (Kaushal et al., 2002; Kaushal and Joshi, 1995; Agrahari and Khurdiya, 2003). Apples have long been grown by humans and were planted by the ancient Romans and were valued for their fruits. *Malus pumila*, the ancestor of today's cultivated apple, originated in the region between the Caspian and Black Seas in Southwest Asia. Apples were a staple of the diet of Central European Stone Age lake dwellers, who dried them in the sun after breaking them up. The apple was introduced to the USA by settlers from Europe and was grown mostly by sowing seeds. The apple was transported to the West of the USA by John Chapman. At the time of the introduction of apple to the USA, most of the apple harvest was used for the preparation of cider (Upshall, 1970).

In the 2019/2020 harvest year, China was the world's largest apple producer. At this time, China's apple production was around 41 million tons (Ahmad et al., 2021). The European Union

DOI: 10.1201/9781003164463-5

was in second place with around 11.48 million tons of apples. In 2020/21, the world produced 89.58 million tonnes of apples. Over the past two decades, the worldwide output of apples has significantly expanded. There was an increase in fruit production globally from 29.4 million tonnes to 53.63 million tonnes between 2010 and 2017. China produced 44.45 million tonnes of fruit, followed by the United States with 4.46 million tonnes. India ranked fifth with 2.87 million tonnes of apples (Ahmad et al., 2021). There are 2.87 million tonnes of apples produced annually in India on 0.25 million hectares, making it the fifth biggest apple grower in the world and accounting for one-third of worldwide apple output (APEDA, 2018). Roughly 71% of apples are consumed fresh, while about 20% are processed into value-added goods. Products that include ready-to-serve (RTS), natural ingredients, RTS-enriched apple juice, apple juice, wine, apple cider and vermouth are all included in the value-added products (Kaushal et al., 2002). In India, Jammu, Kashmir, Uttaranchal and Himachal Pradesh are the main apple-producing states. It is an important horticultural product and the basis of the rural economies of these states (Agrahari and Khurdiya, 2003). Today, food is given not only to satisfy appetite, but also to supply necessary nutrients, prevent food-borne illnesses and promote the physical and mental health of its consumers. There are several functional apple-derived additives that may help improve the quality of a food product, such as micronutrients, phytosterols and omega-3 fatty acids, which can be used in the formulation of functional foods (Betoret et al., 2011).

5.2 CHEMISTRY OF APPLE WASTES AND BY-PRODUCTS

5.2.1 Apple Pomace

Apple pomace is a by-product of the apple juice processing industry and may account for 25–35% of the raw fruit (Schieber et al., 2002). Apple pomace is a rich source of polyphenols, minerals and dietary fiber (Sudha et al., 2007). Traditionally, pectin has been processed from apple pomace for use in animal feed, making it a cost-effective and environmentally protective use. Acid is used to extract and precipitate pectin from apple pomace, which accounts for 10–15% pomace dry weight. For the best of both worlds, try using pectin made from apple pomace instead of citrus pectin. When the pomace by-product is disposed of in landfills, it can lead to serious environmental problems because of groundwater contamination (Bhushan et al., 2008; Shalini and Gupta, 2010; Perussello et al., 2017; Singha and Muthukumarappan, 2018). Because of its high water content (more than 70%) and high degradable organic load, apple pomace is a good candidate for composting (chemical and biochemical oxygen demand). As a consequence, the pomace used as by-products is more vulnerable to uncontrolled fermentation, whereas the landfill pomace leads to environmental degradation and public health concerns. If you're going to process apple pomace, you need to know how to achieve it in a method that is both safe and effective. There is a huge quantity of pomace produced throughout the apple juice extracting and processing processes, and commercial applications for it might have a significant economic effect (Shalini and Gupta, 2010). Apple pomace is a heterogeneous mixture composed mostly of skin and flesh (95%), with a trace of seeds (2–4%) and stems (1%) (Bhushan et al., 2008; Grigelmo-Miguel and Martín-Belloso, 1999). Numerous studies have been conducted to determine the nutritional composition of the apple pomace by-product (Table 5.1). Apple juice manufacturing uses a number of processing methods, and Figure 5.1 illustrates how each one affects the final product (Vendruscolo et al., 2008).

Apple pomace comprises large quantities of phytochemicals and starch, as well as trace amounts of calcium and other minerals, as well as vitamins (Bhushan et al., 2008). Glucose, fructose and galactose are all present in significant quantities in apple pomace, but the majority of the carbohydrates are in the form of insoluble components, like cellulose (127.9 g carbohydrates per kg pomace dry weight), hemicelluloses (7.2 to 43.6 g per kg), and lignin (15.3 to 23.5 g per kg)

Table 5.1   Nutritional Value of Apple Pomace (Dry Weight Basis)

Component	Concentration Range (%)	Component	Concentration Range (%)
Moisture	4.4–10.5	Total dietary fiber	14.5–42.5
Protein	1.2–4.7	Ash	1.7–2.5
Lipids	0.6–4.2	Carbohydrates	45.1–83.1

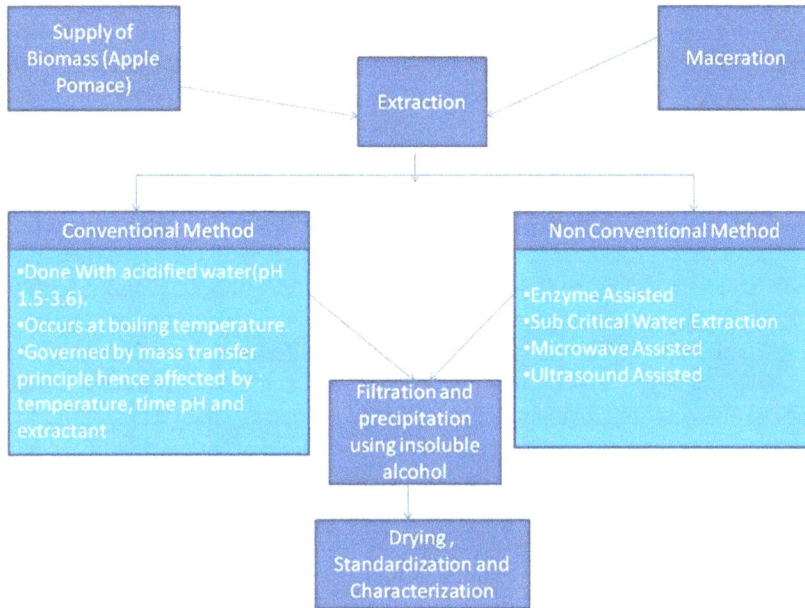

Figure 5.1   Extraction of pectin from apple pomace.

(Dhillon et al., 2013). Along with carbohydrates, minerals such as phosphorus (0.07–0.076%), calcium (0.06–0.1%), magnesium (0.02–0.36%) and iron (31.8–38.3 mg/kg), were estimated on a DW basis in apple pomace by-product and waste. Additionally, apple pomace contains a high concentration of polyphenols (31% to 51% DW), especially cinnamate esters, dihydrochalcones and flavonols (Will et al., 2006).

5.2.2 Apple Seeds

Worldwide, a significant weight of apple seeds is produced during juice production processes. Around 20,000 tonnes of seed are produced yearly in New Zealand and are primarily used as livestock feed. Apple seeds have been the subject of several studies, the majority of which have focused on low-value products (Kennedy et al., 1999). The phenolic compounds in apple seeds have attracted considerable interest for their antioxidant activity and possible health benefits (Rice-Evans et al., 1996). Apple seeds are a rich source of protein and lipid feed ingredients (Kamel et al., 1985). Seeds are a natural component of apple pomace, accounting for 2–3% of the dry material (Carson et al., 1994). You will find protein, fiber and lipids mostly in the endosperm of apple seeds, but you can also find them in other parts of the seeds (Yu et al., 2007). Seeds can be used to make oil because they contain a high concentration of lipids (about 29%). Linoleic acid, which accounts for around 44% of the fatty acids in apple seed oil, is one of the most abundant unsaturated fatty acids (Arain et al., 2012; Walia et al., 2014). There is still a lot of protein and fiber in seed residues after pressing to extract oil, that could be used for other roles. However, apple seeds also have a lot of bad things in them, like amygdalin (1.0–3.9 mg/g seeds), which is a mandelonitrile gentiobioside-based metabolite that is categorized as a cyanogenic glycoside (CG). CGs are plant secondary metabolites that can be hydrolyzed in mammals to make hydrogen cyanide, which is very toxic (Haque and Bradbury, 2002). Inhibition occurs when cyanide binds to the mitochondrial respiratory pathway's last terminal cytochrome oxidase and blocks its action, resulting in a harmful effect (at a minimum dosage of 0.5–3.5 mg/kg body weight). Oxygen can no longer be used by cyanide-treated cells as a result (Zagrobelny et al., 2004). To put it another way, the extended use of amygdalin can have a negative impact on an individual's bodily systems. However, amygdalin has also been shown to have positive benefits on animals. When amygdalin was administered intraperitoneally into rats with chronic kidney disease, it reduced renal fibrosis (Guo et al., 2013). Apple pomace is used to extract apple seed oil. Pomace (25–30% of the entire processed fruit) has four physiological components: flesh (60–65%), peel (30–35%), seed (3–4.5%) and stem (0.5–1%). Apple seeds are an important by-product of the fruit. In certain types of apple, the seeds may make up as much as 0.7% of apple fresh weight. In general, the seed oil content ranges from 15% to 22%, although it has been reported to reach as high as 29.54% (Fromm et al. 2012). Apple seed lipids may be affected by environmental variables under which the apple trees are grown, such as temperature, water availability, exposure, and horticultural techniques (Fromm et al., 2012). Apple seed oil can be extracted using a number of different methods. The higher amount of unsaturated fatty acids and the presence of natural antioxidants makes the apple seed oil more valuable (Walia et al., 2014). Phloridzin and quercetin-3-galactoside are the most abundant polyphenols in apple pomace, which includes the seeds (Lu and Foo, 1997). Potential antioxidant, antibacterial, and anti-angiogenic properties of the oil from apple seeds have been reported (Walia et al., 2014) but an evaluation of apple seed oil as a food additive will need further investigation (Walia et al., 2014; Fotschki et al. 2015), although it is already used in a range of cosmetic items, including body lotion, skin cream, toothpaste and natural moisturizer. Apple seed oil is also used in lip balms and skin butter. Lipid-soluble antioxidant molecules present in apple seed oil include tocopherols (having a saturated side chain) or tocotrienols (containing three unsaturated side chains). Tocopherols and tocotrienols each have four isomers, which represent vitamin E, although alpha-tocopherol is better known as vitamin E. Apple seed oil has been shown to contain tocopherols (Arain et al., 2012), with tocotrienols making up less than 2% of the vitamin E molecules. Tocopherol-rich preparations of apple seed oil might be used as a natural source of vitamin E, which plays a critical role in human nutrition and health (Pieszka et al., 2015).

5.2.3 Apple Peel

Waste management is a major problem in both developed and developing nations. An enormous quantity of waste is generated by the food producing industry each year and environmental damage and contamination can occur as a result of the inappropriate disposal of these waste products, which mostly consist of biodegradable organic matter (Shyamala and Jamuna, 2010). By-products from the fruit and vegetable sectors, in particular, are a problem because of their low cost and abundance. Dietary fiber may be isolated from a variety of industrial waste items, including apple waste (Figuerola et al., 2005). Apple peels are a by-product of apple, dried apple, apple sauce, apple juice and apple pie processing, and have a nutritional value similar to that of apple pomace. In 2003, the apple manufacturing industry in Nova Scotia, Canada generated 2–3 million kg of apple peel (Rupasinghe et al., 2008). The average dietary fiber content of dried apple peel varies between 21.36% and 39.75%, depending on the variety (Rupasinghe et al., 2008), and contains a ratio of 1:13.7 of soluble to insoluble dietary fiber, although a previous analysis had reported that apple peels had a soluble/insoluble dietary fiber ratio of around 1:2 (Figuerola et al., 2005). Therefore, apple peel is a better source of dietary fiber than apple pomace and apple pulp for the development of functional foods. Dietary fiber composition and physicochemical properties are determined by both the raw material properties and the processing steps (Chau et al., 2004). The presence of bioactive molecules in such agro-industrial waste, like fatty acids and phenolic compounds, increases the demand for such fruit waste and by-products (Socaci et al., 2017). Bioactive compounds found in fruit waste and by-products confer additional health benefits (Cătoi et al., 2006, 2013; Cătoi et al., 2013). Apple peel is a common agro-industrial waste and contains about 80% of the polyphenols in an apple (Leccese et al., 2009). It has a gross antioxidant potential five to six times greater than that of the apple flesh. Flavonols, anthocyanins, flavan-3-ols, phenolic acids and dihydrochalcones were all found in apple peel, and have been associated with a variety of health advantages, including improved cardiovascular health (Boyer and Liu, 2004). In comparison with the flesh, the peel produces high concentrations of flavonoids, including quercetin glycosides (Rupasinghe et al., 2008). The apple peel phytochemicals possessed antioxidant and inhibitory properties to liver and colon cancer. Several chronic illnesses, including cancers, cardiovascular disease, asthma, and type 2 diabetes, may be avoided by the uptake of antioxidants, which neutralize free radicals (He and Liu, 2008). Moreover, Liu et al. (2000) compared the antioxidant activity of peeled and unpeeled apples and discovered that both had a high antioxidant activity and the ability to prevent the development of human cancer cells in *in-vitro* experiments. Unpeeled apples had greater antioxidant and anti-proliferative properties than peeled apples, implying that apple peels could be an important functional food component (Liu, Eberhardt and Lee, 2000). Apple peels have been proposed as a source of dietary fiber and phytochemicals that could be used in value-added goods (Huber and Rupasinghe, 2009; Wolfe et al., 2003). The phenolic content of apple peel exhibited an antioxidant potential almost twice that of apple pulp. Mainly, flavanols and procyanidins are found in apple peel, both of which have strong antioxidant properties (Chinnici et al., 2004).

5.3 PROCESSING OF APPLE WASTES AND BY-PRODUCTS

5.3.1 Pectin Extraction from Apple Pomace

In the cell walls of terrestrial plants such as major fruit crops like apples, oranges and apricots, pectin is a structural heteropolysaccharide. The non-woody parts of the plant have highest concentrations of pectin. Natural sugars like galactose may be found in pectin's side chains as well as in the galacturonic acid (GalA) residues that make up the bulk of pectin. Homogalacturonan, rhamnogalacturonan-II and rhamnogalacturonan-I are the main polysaccharides in pectin. Homogalacturonic areas are "smooth" and "hairy" regions that carry the bulk of the neutral sugars in pectins. During the processing of apples, pectin has been shown to interact with and bind to polyphenols such as procyanidin that are produced in cells (mainly in the vacuoles) and released, for example, by pressing. A poor pectin extraction yield and a low degree of methylation are both consequences of interaction between pectin and acid. As polyphenols oxidize, their extractability and degree of methylation will continue to decline (Fernandes et al., 2019). Pectin is a gelling agent, emulsifier and carrier polymer commonly used in the food sector for encapsulating food products and assisting in the protection and regulated release of macromolecules. Additional uses include edible coatings and biodegradable food packaging materials. Pectin aids in the release of medicines and genes, wound healing and tissue engineering, where it promotes bone tissue regeneration (Munarin et al., 2011). Apple pomace and citrus peel are the principal sources of separated pectin, according to Willats et al. (2006). Figure 5.1 depicts the process of pectin production from apple pomace and the advantages and disadvantages of pectin with respect to extraction methods (Table 5.2).

5.3.2 Extraction of Oil from Apple Seed

Plants contain useful substances that may be extracted for identification and utilization (Stévigny et al., 2007). The best extraction method for maximizing yield and purity varied, depending on the nature of the target chemical. Various chemical and mechanical methods, such

Table 5.2 Pectin Extraction Methods

S. No.	Extraction Method	Advantage	Disadvantage
1.	Conventional or hot acidic solution extraction	a. Comprehensive understanding of the fundamental processes and how operating conditions influence the yield, duration and efficiency of pectin extraction. b. Optimizing the process parameters results in satisfactory extraction rates and yields.	Contrary to more modern extraction methods, the time required for extraction is much longer. Thermal degradation of pectin starts to reduce extraction yield at temperatures above 80°C. Machinery damage and the necessity for effluent treatment as a consequence of excessive solvent usage. Concerns concerning the toxicity and flammability of industrial chemicals. A lack of ability to discriminate between conventional and hot acid solution.
2.	Enzyme-assisted extraction	a. Excessive solvent and extraction time may be reduced thanks to the specificity of enzymes. b. It is possible to convert low methoxy pectin to high methoxy pectin by the use of enzymes. Reduces machine deterioration and sewage treatment expenses by allowing the use of moderate temperatures and pH. Because the amount of solvent used is minimized, this process is more environmentally friendly. c. Pectin with a greater molecular weight and degree of esterification may be obtained by tailoring the extraction process. d. The resultant polysaccharides are functional and structurally intact to the fullest degree feasible.	Enzyme expenditure, which is somewhat compensated for by the cost reductions associated with acid and the maintenance/control of increased temperatures. Scalability is difficult.
3.	Supercritical water extraction	a. Extracting pectin at a faster pace will enhance the solubility of pectin and hence decrease extraction periods. b. Pectin yields rise when pressure is applied. Continuous or semi-continuous extraction is an option. Pectin solubility rises to 160°C, boosting extraction rate, but thermal degradation happens as early as 80°C when the standard method is used. c. Pectin extraction quality has been improved. There is less need for solvents, which reduces the costs involved with cleaning solvents and waste management. d. For the most ecologically friendly extraction method available, water is all that is needed. e. In terms of pollution and flammability, there is an increase in safety. This method protects the pectin's original composition.	Expensive solvents may be replaced with cheaper water, which reduces the need for further implementation costs. Scaling up is being tried as well.

(Continued)

Table 5.2 (Continued) Pectin Extraction Methods

S. No.	Extraction Method	Advantage	Disadvantage
4.	Microwave-assisted extraction	a. Extraction time is often the shortest of all extraction technologies and more than ten times quicker than classical extraction. b. Conventional extraction uses a lot of solvent, which is bad for the environment. The concentration gradients that are created by conventional heating are not desirable. Increased microwave power and irradiation time are required to enhance yield, polymerization degree and galacturonic acid concentration.	Due to decreasing water content in foods, extraction rates and yields are lowered when using microwave radiation. The acidic fluid will become viscous, preventing extraction, if the solid/solvent ratio is not changed. Machinery depreciation and effluent treatment might be exacerbated by reduced solvent needs.
5.	Ultrasound-assisted extraction	a. Generates an extract with improved colour without changing its chemical structure. b. Needs less time and effort to extract than the traditional approach, resulting in increased process performance and less negative environmental impacts. c. Requires lower operating temperatures, maintaining the material's composition. d. Occasional sonication, rather than constant sonication, can result in a higher yield	Matrix-dependent increases in yield and extraction rates may not be found in all plant cells. Large, solid particle concentration lowers ultrasonic intensity; hence the solvent/matrix ratio must be regulated to avoid this. With increased sonication power and duration, gels get softer.

Sources: Adapted from Adetunji et al. 2017; Bagherian et al. 2011; Guo et al. 2012; Yeoh et al., 2008; De Oliveira et al., 2015; Knorr and Angersbach, 1998

as steam distillation, are employed to extract chemicals from plants (Shirsath et al., 2012). Soxhlet, hydro-distillation and alcohol maceration are some of the current methods for extracting essential oils, fats, and oils from plant tissues (Wang and Weller, 2006). Traditional Soxhlet extraction methods are frequently rendered ineffective due to high mass transfer resistances caused by the system's multi-phase composition (Jadhav et al. 2009). Depending on the diffusion speeds of the solvents, this separation procedure might take a long time to achieve results. Conventional extraction methods, on the other hand, use a lot of energy (Puri et al., 2012). These techniques are manual processes, and reproducibility is a major challenge (Shen and Shao, 2005). Due to the use of high temperatures, heat-sensitive components are damaged and extraction yields are reduced. The pH, temperature and pressure conditions under which these active compounds are extracted may modify their properties. Extracting bioactive components, oils, lipids and essential oils has become increasingly important as a result of the rise in demand for these products, which has led to the development of new extraction methods that are both cost-effective and environmentally benign (Ibañez et al., 2012). It is for this reason that novel methods of extracting active natural substances, like supercritical fluid extraction (SFE), have been developed (Sajfrtová et al., 2010). SFE has attracted considerable interest in recent years as an alternative to established methods for separating numerous valuable compounds from natural sources (Gomes et al., 2007). To reduce solvent usage, SFE is carried out at lower temperatures and for shorter periods of time than typical extraction procedures. CO_2 extraction uses a non-toxic, inert solvent and is affordable, and CO_2 vaporizes quickly in the presence of oxygen (Herrero et al., 2010). In the case of extracting polar compounds, adding a small quantity of non-polar solvents to polar solvents as a co-solvent may significantly improve extraction efficiency. In the food and pharmaceutical industries, ether is the most widely utilized solvent because of its excellent flowability with CO_2 and non-toxicity, as well as its permissible use (Herrero et al., 2010). Selective extraction of target chemicals using super-critical CO_2 (SC-CO_2) has been proven to be safe and without the risk of heat degradation of the final product. SC-CO_2 extracts are also Generally Accepted as Safe (GRAS) in food items. It is possible to extract important chemicals from solid plant matrices and seed oil using SC-CO_2 extraction, a technology which has reached a rather advanced stage of development.

5.3.3 Extraction of Protein Isolates

Protein isolates are the most refined type of protein goods, having the highest concentration of protein but possessing no dietary fiber, in comparison to flour and concentrates. These isolates developed in the 1950s in the United States. They are readily digested and may be added to a variety of food items. Isolated proteins are now being credited with the production of a new kind of synthetic food. It is a wonderful, raw ingredient for beverages, infant and child milk meals, textured protein products and other specialty foods because of its high protein content and acceptable color, taste, and functional features. Soybean, peanut, canola, cashew nut, almonds, sesame, apple seed, and chili seed are some of the matrices that have been used from which to extract protein isolates.

5.3.3.1 Extraction Methods

5.3.3.1.1 Isoelectric Precipitation

The isoelectric point (pI) of a protein is the pH value at which the protein's net charge is zero. Because there are no electrostatic forces separating them at this stage, proteins aggregate and precipitate at the pI. Due to the fact that each protein has a distinct isoelectric point due to its unique amino-acid sequence, they may be separated by changing the pH of a solution. When the pH is changed to a certain protein's isoelectric point, that protein precipitates, leaving the other proteins in solution.

5.3.3.1.2 Alkaline Extraction

Alkaline reagents were shown to be more efficient in extracting protein from dietary beans. Many alterations may occur during alkali extraction, such as degradation of lysine and alanine formation and racemization, which lowers the protein's quality. A new method of producing protein isolates and concentrates with the highest possible yield and with the inclusion of little or no antinutritional components is needed.

5.3.3.1.3 Ultra-Filtration Method

Membrane processing may be used to extract protein from a solution by placing it in a cell with a semi-permeable membrane and applying pressure. It is possible for smaller molecules to pass through the membrane, whereas the larger ones stay suspended in the liquid. Membranes with cut-off molecular weight values ranging from 500 to 300,000 Da are commercially available.

5.4 UTILIZATION OF APPLE WASTES AND BY-PRODUCTS

5.4.1 Milk and Milk Products

Apple pomace has been used as a natural stabilizer and texturizer in set-type yoghurt (Wang et al., 2019). *Lactobacillus bulgaricus* and *Streptococcus thermophilus* were introduced to skimmed milk at 42°C and left to ferment for a given length of time. The addition of 1% apple pomace resulted in much greater pH start-up and shorter gelation time. After 28 days of storage, all pomace-fortified yoghurts had a better consistency and cohesion. Wang has recently explored the use of freeze-dried apple pomace powder as a dairy supplement. It was shown that 1% apple pomace powder improved gelation pH and reduced the time it took to ferment yoghurt, leading to a more viscoelastic, homogeneous and firmer yoghurt gel. An additional 3% apple pomace is included in the yoghurt, when the results showed that the syneresis of the matrix decreased significantly following 28 days of cold storage, as well as exhibiting an increase in viscosity, stiffness and cohesiveness. Roles of apple pomace as a natural stabilizer and texturizer and as a source of fiber and polyphenols in dairy products, such as yoghurt, were emphasized.

5.4.2 Meat and Meat Products

The use of apple pomace in meat products is now being researched as a means to boost the dietary fiber in meat products. Mutton items, such as nuggets and goshtaba, together with sausages, have been the subject of attempts at improvement (Huda et al., 2014; Rather et al., 2015; Verma, Sharma, and Banerjee, 2010). As a result, Younis and Ahmad (2018) developed buffalo meat patties in which the meat content was partially replaced by 2–8% of apple pomace, which they found to be acceptable to consumers (Younis and Ahmad, 2018). When it comes to assessing the level of replacement, fat, moisture and crude fiber content were shown to be the most reliable indicators. The thickness of patties, as well as the textural features of the patties, such as hardness and roughness, were linked to the cooking yield. When the replacement rate exceeded 6%, a decrease in cohesion and springiness was observed. It was found that a comparable degree of replacement was suitable for buffalo sausages. Softer chicken patties were discovered to be created by substituting 10% or 20% of the meat in the patties with apple pomace (Jung et al., 2015). According to Verma et al. (2010), low-fat chicken nuggets with 8% to 12% (w/w) pomace exhibited decreased hardness of the nuggets. It was found that all of the beef items examined had deeper and redder color changes, as well as an increased level of total dietary fiber. Future study should focus on how to incorporate apple pomace into pork products since there has been a lack of research on this topic.

5.4.3 Bakery Products

Apple wastes and by-products may be used in food manufacturing as a functional ingredient due to their nutritional value and nutraceutical potential. Apple wastes and by-products can be processed into a fine dry powder for development of food products in the bakery, meat, milk and other processing industries. Previously, apple pomace or apple peel powder had been incorporated into bakery production to improve the nutritional value as well as the rheological and sensory parameters (Rupasinghe et al., 2008; Sudha et al., 2007). Cake enriched with 25% dried apple pomace powder had a 13.73% increase in dietary fiber content and a 50% increase in phenolic content compared with the control. Compared with the control, cakes enriched with 15% dried apple pomace powder showed a minor increase in water absorption of 10.5%. Dried apple peel powder, on the other hand, speeded up the manufacturing process and lowered the batter temperature and viscosity. An enhancement in browning and a grittier texture may be achieved by adding apple pomace powder, which has a significant positive impact on sensory attributes. It was also found that increasing the quantity of apple pomace powder in the cake boosted its fruity flavor, according to a panel of experts (Sudha et al., 2007; Yadav et al., 2016). Moreover, Jung et al. (2015) also used apple peel powder as a dietary fiber-fortified component in cake baking and discovered that, when the powder was used at a concentration greater than 3 g apple peel powder per 100 g serving, the cake's characteristics deteriorated. As a result, the amount of apple by-products used in food production must also be investigated in terms of sensory qualities. Apple peel powder has also been applied to different bakery products. Rupasinghe et al. (2008) processed muffins with two forms of apple peel powder (from cultivar 'Ida Red' or 'Northern Spy'). Muffins baked with varied amounts of dried apple peel powder (0%, 4%, 8%, 16%, 24% or 32%) were tested for their nutritional content. In this study, it was shown that apple peel powder significantly enhanced the muffins' total dietary fiber, phenolic concentration and antioxidant potential. As a result of this research, apple skin might be employed as a functional ingredient in baked goods.

Table 5.3 summarizes the usage of apple pomace in various forms of plant-based products and the resulting impact. The studies looked at incorporating apple pomace flour into a variety of items, including cakes, muffins, cookies, bread, biscuits, crackers and extruded snacks.

Phosphoric compounds extracted and partially precipitating with pectin are the primary reason for the color of apple pomace and pectin. Apple pomace has been discovered to be a potent source of phytochemicals, which are mostly contained in the peel and are only partially absorbed into the juice of the apple fruit. The principal constituents include hydroxycinnamates, phloretin glycosides, quercetin glycosides, and procyanidins, among others.

5.5 CONCLUSIONS

An astonishing number of essential nutrients and phytochemicals may be found in the by-products of the apple juice business. This comprehensive study found that apple wastes and by-products may be employed as fortifying elements in plant food items, meat and dairy products, as well as in other food products. The high dietary fiber content of apple waste and by-products were translated to the treated end-product, enhancing its nutritional value. Due to apple waste and by-products, which are the primary polyphenolic source of the entire fruit, an increase in the antioxidant activity has been reported. In plants, fortification led to an undesired shift in hue to a reddish one, although this change was well received in processed meat by the sensory panel.

Table 5.3 Effect of Incorporation of Apple Pomace in Bakery

S. No.	Name of the Product	Condition	Effect	Reference
1.	Cake	Apple pomace up to 30% is added as replacement for wheat flour.	The acceptability and antioxidant properties and fruity flavor of the final product increased with a decrease in volume.	Sudha et al. (2007)
2.	Cookies and muffins	The wheat flour was replaced by 10–15% apple pomace.	The fiber content increased and the color of the muffins became more attractive.	Jung et al. (2015); De Toledo et al. (2017)
3.	Gluten-free bread	12.5% apple pomace was included in bread mixture.	The crumb hardness and acceptability increased while the cohesiveness decreased.	Rocha et al. (2015); Bchir et al. (2014)
4.	Biscuits	The wheat flour was replaced by 10–12% apple pomace.	The glycemic index decreased to the acceptable level with an increase in the sensory acceptability of the final product	Alongi et al. (2019)
5.	Gluten-free brown rice crackers. Sugar snap cookies	Dried apple pomace may be added to replace 3–9% wheat flour, and wheat flour can be fortified with 15–30% dried apple pomace.	The fiber, antioxidant potential and sensory acceptance increased along with texture and color.	Mir et al. (2017); Parra et al. (2019)

REFERENCES

Adetunji, Lanrewaju Ridwan, Ademola Adekunle, Valérie Orsat, & Vijaya Raghavan. (2017). Advances in the pectin production process using novel extraction techniques: A review. *Food Hydrocolloids* 62:239–250.

Agrahari, PR, & DS Khurdiya. (2003). Studies on preparation and storage of RTS beverage from pulp of culled apple pomace. *Indian Food Packer* 57 (2):56–61.

Ahmad, Rayees, Barkat Hussain, & Tariq Ahmad. (2021). Fresh and dry fruit production in Himalayan Kashmir, Sub-Himalayan Jammu and Trans-Himalayan Ladakh, India. *Heliyon* 7 (1):e05835.

Alongi, Marilisa, Sofia Melchior, & Monica Anese. (2019). Reducing the glycemic index of short dough biscuits by using apple pomace as a functional ingredient. *LWT* 100:300–305.

Apeda, A. (2018). *Processed Food Products Export Development Authority.* New Delhi, India: Ministry of Commerce & Industry, GOI.

Arain, S, STH Sherazi, MI Bhanger, N Memon, SA Mahesar, & MT Rajput. (2012). Prospects of fatty acid profile and bioactive composition from lipid seeds for the discrimination of apple varieties with the application of chemometrics. *Grasas y Aceites* 63 (2):175–183.

Bagherian, Homa, Farzin Zokaee Ashtiani, Amir Fouladitajar, & Mahdy Mohtashamy. (2011). Comparisons between conventional, microwave-and ultrasound-assisted methods for extraction of pectin from grapefruit. *Chemical Engineering and Processing: Process Intensification* 50 (11–12):1237–1243.

Bchir, Brahim, Holy Nadia Rabetafika, Michel Paquot, & Christophe Blecker. (2014). Effect of pear, apple and date fibres from cooked fruit by-products on dough performance and bread quality. *Food and Bioprocess Technology* 7 (4):1114–1127.

Betoret, E, Betoret, N, Vidal, D, & Fito, P. (2011). Functional foods development: Trends and technologies. *Trends in Food Science and Technology* 22 (9):498–508.

Bhushan, Shashi, Kalpana Kalia, Madhu Sharma, Bikram Singh, & Paramvir Singh Ahuja. (2008). Processing of apple pomace for bioactive molecules. *Critical Reviews in Biotechnology* 28 (4):285–296.

Boyer, Jeanelle, & Rui Hai Liu. (2004). Apple phytochemicals and their health benefits. *Nutrition Journal* 3 (1):1–15.

Carson, KJ, JL Collins, & MP Penfield. (1994). Unrefined, dried apple pomace as a potential food ingredient. *Journal of Food Science* 59 (6):1213–1215.

Cătoi, Adriana Florinela, Alina Pârvu, Romeo Florin Galea, Ioana Delia Pop, Adriana Mureşan, & Cornel Cătoi. (2013). Nitric oxide, oxidant status and antioxidant response in morbidly obese patients: The impact of 1-year surgical weight loss. *Obesity Surgery* 23 (11):1858–1863.

Catoi, AF, A Parvu, A Muresan, RF Galea, & C Catoi. (2006). Evaluation of nitric oxide synthesis in morbid obesity. Paper read at *Free Radical Research*.

Chau, Chi-Fai, Chien-Hung Chen, & Mao-Hsiang Lee. (2004). Comparison of the characteristics, functional properties, and in vitro hypoglycemic effects of various carrot insoluble fiber-rich fractions. *LWT-Food Science and Technology* 37 (2):155–160.

Chinnici, Fabio, Alessandra Bendini, Anna Gaiani, & Claudio Riponi. (2004). Radical scavenging activities of peels and pulps from cv. Golden Delicious apples as related to their phenolic composition. *Journal of Agricultural and Food Chemistry* 52 (15):4684–4689.

de Oliveira, Cibele Freitas, Diego Giordani, Poliana Deyse Gurak, Florencia Cladera-Olivera, & Ligia Damasceno Ferreira Marczak. (2015). Extraction of pectin from passion fruit peel using moderate electric field and conventional heating extraction methods. *Innovative Food Science & Emerging Technologies* 29:201–208.

de Toledo, Nataly Maria Viva, Larissa Peixoto Nunes, Paula Porrelli Moreira da Silva, Marta Helena Fillet Spoto, & Solange Guidolin Canniatti-Brazaca. (2017). Influence of pineapple, apple and melon by-products on cookies: Physicochemical and sensory aspects. *International Journal of Food Science & Technology* 52 (5):1185–1192.

Dhillon, Gurpreet Singh, Surinder Kaur, & Satinder Kaur Brar. (2013). Perspective of apple processing wastes as low-cost substrates for bioproduction of high value products: A review. *Renewable and Sustainable Energy Reviews* 27:789–805.

Fernandes, Pedro AR, Carine Le Bourvellec, & Catherine MGC Renard. (2019). Revisiting the chemistry of apple pomace polyphenols. *Food Chemistry* 294:9–18.

Figuerola, Fernando, María Luz Hurtado, Ana María Estévez, Italo Chiffelle, & Fernando Asenjo. (2005). Fibre concentrates from apple pomace and citrus peel as potential fibre sources for food enrichment. *Food Chemistry* 91 (3):395–401.

Fotschki, Bartosz, A Jurgonski, J Juskiewicz, & Z Zdunczyk. (2015). Metabolic effects of dietary apple seed oil in rats. *Żywnoś ć Nauka Technologia Jakoś ć* 1 (98):220–231.

Fromm, Matthias, Sandra Bayha, Reinhold Carle, & Dietmar R Kammerer. (2012). Comparison of fatty acid profiles and contents of seed oils recovered from dessert and cider apples and further Rosaceous plants. *European Food Research and Technology* 234 (6):1033–1041.

Gomes, Paula B, Vera G Mata, & Alírio E Rodrigues. (2007). Production of rose geranium oil using supercritical fluid extraction. *The Journal of Supercritical Fluids* 41 (1):50–60.

Grigelmo-Miguel, Nuria, & Olga Martín-Belloso. (1999). Comparison of dietary fibre from by-products of processing fruits and greens and from cereals. *LWT-Food Science and Technology* 32 (8):503–508.

Guo, Junqi, Weizheng Wu, Mingxiong Sheng, Shunliang Yang, & Jianming Tan. (2013). Amygdalin inhibits renal fibrosis in chronic kidney disease. *Molecular Medicine Reports* 7 (5):1453–1457.

Guo, Xingfeng, Dongmei Han, & Huping Xi. (2012). Extraction of pectin from navel orange peel assisted by ultra-high pressure, microwave or traditional heating: A comparison. *Carbohydrate Polymers* 88 (2):441–448.

Haque, M Rezaul, & J Howard Bradbury. (2002). Total cyanide determination of plants and foods using the picrate and acid hydrolysis methods. *Food Chemistry* 77 (1):107–114.

He, Xiangjiu, & Rui Hai Liu. (2008). Phytochemicals of apple peels: Isolation, structure elucidation, and their antiproliferative and antioxidant activities. *Journal of Agricultural and Food Chemistry* 56 (21):9905–9910.

Herrero, Miguel, Jose A Mendiola, Alejandro Cifuentes, & Elena Ibáñez. (2010). Supercritical fluid extraction: Recent advances and applications. *Journal of Chromatography A* 1217 (16):2495–2511.

Huber, GM, & HPV Rupasinghe. (2009). Phenolic profiles and antioxidant properties of apple skin extracts. *Journal of Food Science* 74 (9):C693–C700.

Huda, Aamina B, Shahnaz Parveen, Sajad A Rather, Rehana Akhter, & Massarat Hassan. (2014). Effect of incorporation of apple pomace on the physico-chemical, sensory and textural properties of mutton nuggets. *International Journal of Advanced Research* 2 (4):974–983.

Ibañez, Elena, Miguel Herrero, Jose A Mendiola, & María Castro-Puyana. (2012). Extraction and characterization of bioactive compounds with health benefits from marine resources: Macro and micro algae, cyanobacteria, and invertebrates. In M Hayes (Ed.), *Marine Bioactive Compounds* (pp. 55–98). Boston, MA: Springer.

Jadhav, Dnyaneshwar, BN Rekha, Parag R Gogate, & Virendra K Rathod. (2009). Extraction of vanillin from vanilla pods: A comparison study of conventional soxhlet and ultrasound assisted extraction. *Journal of Food Engineering* 93 (4):421–426.

Jung, Jooyeoun, George Cavender, & Yanyun Zhao. (2015). Impingement drying for preparing dried apple pomace flour and its fortification in bakery and meat products. *Journal of Food Science and Technology* 52 (9):5568–5578.

Kamel, Basil S, H Dawson, & Y Kakuda. (1985). Characteristics and composition of melon and grape seed oils and cakes. *Journal of the American Oil Chemists' Society* 62 (5):881–883.

Kaushal, NK, & VK Joshi. (1995). Preparation and evaluation of apple pomace based cookies. *Indian Food Packer* 49:17–24.

Kaushal, NK, VK Joshi, & RC Sharma. (2002). Effect of stage of apple pomace collection and the treatment on the physico-chemical and sensory qualities of pomace Papad (fruit cloth). *Journal of Food Science and Technology (Mysore)* 39 (4):388–393.

Kennedy, M, D List, & Y Lu, (1999). Apple pomace and products derived from apple pomace: Uses, composition and analysis. In HF Linskens & JF Jackson (Eds.), *Analysis of Plant Waste Materials*, vol. 20 (pp. 75–118). Berlin, Heidelberg: Springer.

Knorr, Dietrich, & Alexander Angersbach. (1998). Impact of high-intensity electric field pulses on plant membrane permeabilization. *Trends in Food Science & Technology* 9 (5):185–191.

Leccese, Annamaria, Susanna Bartolini, & Raffaella Viti. (2009). Antioxidant properties of peel and flesh in 'GoldRush'and 'Fiorina'scab-resistant apple (*Malus domestica*) cultivars. *New Zealand Journal of Crop and Horticultural Science* 37 (1):71–78.

Liu, RH, MV Eberhardt, & CY Lee. (2000). Nutrition: Antioxidant activity of fresh apples. *Nature* 405 (6789):903–904.

Lu, Yinrong, & L Yeap Foo. (1997). Identification and quantification of major polyphenols in apple pomace. *Food Chemistry* 59 (2):187–194.

Mir, Shabir Ahmad, Sowriappan John Don Bosco, Manzoor Ahmad Shah, Swaminathan Santhalakshmy, & Mohammad Maqbool Mir. (2017). Effect of apple pomace on quality characteristics of brown rice based cracker. *Journal of the Saudi Society of Agricultural Sciences* 16 (1):25–32.

Munarin, Fabiola, SG Guerreiro, & MA Grellier, (2011). Pectin-based injectable biomaterials for bone tissue engineering. *Biomacromolecules* 12 (3):568–577.

Parra, Andrés F Rocha, Marta Sahagún, Pablo D Ribotta, Cristina Ferrero, & Manuel Gómez. (2019). Particle size and hydration properties of dried apple pomace: Effect on dough viscoelasticity and quality of sugar-snap cookies. *Food and Bioprocess Technology* 12 (7):1083–1092.

Perussello, Camila A, Zhihang Zhang, Antonio Marzocchella, & Brijesh K Tiwari. (2017). Valorization of apple pomace by extraction of valuable compounds. *Comprehensive Reviews in Food Science and Food Safety* 16 (5):776–796.

Pieszka, Marek, Władysław Migdał, & Robert Gąsior, (2015). Native oils from apple, blackcurrant, raspberry, and strawberry seeds as a source of polyenoic fatty acids, tocochromanols, and phytosterols: A health implication. *Journal of Chemistry* 2015:1–8.

Puri, Munish, Deepika Sharma, & Colin J Barrow. (2012). Enzyme-assisted extraction of bioactives from plants. *Trends in Biotechnology* 30 (1):37–44.

Rather, Sajad A, Rehana Akhter, FA Masoodi, Adil Gani, & SM Wani. (2015). Utilization of apple pomace powder as a fat replacer in goshtaba: A traditional meat product of Jammu and Kashmir, India. *Journal of Food Measurement and Characterization* 9 (3):389–399.

Rice-Evans, Catherine A, Nicholas J Miller, & George Paganga. (1996). Structure-antioxidant activity relationships of flavonoids and phenolic acids. *Free Radical Biology and Medicine* 20 (7):933–956.

Rocha Parra, Andrés F, Pablo D Ribotta, & Cristina Ferrero. (2015). Apple pomace in gluten-free formulations: Effect on rheology and product quality. *International Journal of Food Science & Technology* 50 (3):682–690.

Rupasinghe, HP Vasantha, Laixin Wang, Gwendolyn M Huber, & Nancy L Pitts. (2008). Effect of baking on dietary fibre and phenolics of muffins incorporated with apple skin powder. *Food Chemistry* 107 (3):1217–1224.

Sajfrtová, Marie, Ivana Ličková, Martina Wimmerová, Helena Sovová, & Zdeněk Wimmer. (2010). β-Sitosterol: Supercritical carbon dioxide extraction from sea buckthorn (*Hippophae rhamnoides* L.) seeds. *International Journal of Molecular Sciences* 11 (4):1842–1850.

Schieber, Andreas, Petra Hilt, Jürgen Conrad, Uwe Beifuss, & Reinhold Carle. (2002). Elution order of quercetin glycosides from apple pomace extracts on a new HPLC stationary phase with hydrophilic endcapping. *Journal of Separation Science* 25 (5–6):361–364.

Shalini, Rachana, & DK Gupta. (2010). Utilization of pomace from apple processing industries: A review. *Journal of Food Science and Technology* 47 (4):365–371.

Shen, Jinchao, & Xueguang Shao. (2005). A comparison of accelerated solvent extraction, Soxhlet extraction, and ultrasonic-assisted extraction for analysis of terpenoids and sterols in tobacco. *Analytical and Bioanalytical Chemistry* 383 (6):1003–1008.

Shirsath, SR, SH Sonawane, & PR Gogate. (2012). Intensification of extraction of natural products using ultrasonic irradiations—A review of current status. *Chemical Engineering and Processing: Process Intensification* 53:10–23.

Shyamala, BN, & P Jamuna. (2010). Nutritional content and antioxidant properties of pulp waste from *Daucus carota* and *Beta vulgaris*. *Malaysian Journal of Nutrition* 16 (3):397–408.

Singha, Poonam, & Kasiviswanathan Muthukumarappan. (2018). Single screw extrusion of apple pomace-enriched blends: Extrudate characteristics and determination of optimum processing conditions. *Food Science and Technology International* 24 (5):447–462.

Socaci, Sonia A, AC Farcas, Dan C Vodnar, & M Tofana. (2017). Food wastes as valuable sources of bioactive molecules. In N Shiomi & V Waisundara (Eds.), *Superfood and Functional Food—The Development of Superfoods and Their Roles as Medicine*. Rijeka, Croatia: InTech:75–93.

Stévigny, Caroline, Luca Rolle, Nadia Valentini, & Giuseppe Zeppa. (2007). Optimization of extraction of phenolic content from hazelnut shell using response surface methodology. *Journal of the Science of Food and Agriculture* 87 (15):2817–2822.

Sudha, ML, V Baskaran, & K Leelavathi. (2007). Apple pomace as a source of dietary fiber and polyphenols and its effect on the rheological characteristics and cake making. *Food Chemistry* 104 (2):686–692.

Upshall, Wh. (1970). Northern spy. In D Carlson and P Larsen (Eds.), *North American Apples: Varieties, Rootstocks, Outlook*, vol. 115 (pp. 18–23). East Lansing, MI: Michigan State University Press.

Vendruscolo, Francielo, Patrícia M Albuquerque, Fernanda Streit, Elisa Esposito, & Jorge L Ninow. (2008). Apple pomace: A versatile substrate for biotechnological applications. *Critical Reviews in Biotechnology* 28 (1):1–12.

Verma, Arun K, BD Sharma, & Rituparna Banerjee. (2010). Effect of sodium chloride replacement and apple pulp inclusion on the physico-chemical, textural and sensory properties of low fat chicken nuggets. *LWT-Food Science and Technology* 43 (4):715–719.

Walia, Mayanka, Kiran Rawat, Shashi Bhushan, Yogendra S Padwad, & Bikram Singh. (2014). Fatty acid composition, physicochemical properties, antioxidant and cytotoxic activity of apple seed oil obtained from apple pomace. *Journal of the Science of Food and Agriculture* 94 (5):929–934.

Wang, Lijun, & Curtis L Weller. (2006). Recent advances in extraction of nutraceuticals from plants. *Trends in Food Science & Technology* 17 (6):300–312.

Wang, Xinya, Eleana Kristo, & Gisèle LaPointe. (2019). The effect of apple pomace on the texture, rheology and microstructure of set type yogurt. *Food Hydrocolloids* 91:83–91.

Will, Frank, Melanie Olk, Isabelle Hopf, & Helmut Dietrich. (2006). Characterization of polyphenol extracts from apple juice. *Deutsche Lebensmittel-Rundschau* 102 (7):297–302.

Wolfe, Kelly, Xianzhong Wu, & Rui Hai Liu. (2003). Antioxidant activity of apple peels. *Journal of Agricultural and Food Chemistry* 51 (3):609–614.

Yadav, Sanjay, Ashok Malik, Ashok Pathera, Rayees Ul Islam, & Diwakar Sharma. (2016). Development of dietary fibre enriched chicken sausages by incorporating corn bran, dried apple pomace and dried tomato pomace. *Nutrition & Food Science* 46 (1):16–29.

Yeoh, S, JTAG Shi, & TAG Langrish. (2008). Comparisons between different techniques for water-based extraction of pectin from orange peels. *Desalination* 218 (1–3):229–237.

Younis, Kaiser, & Saghir Ahmad. (2018). Quality evaluation of buffalo meat patties incorporated with apple pomace powder. *Buffalo Bulletin* 37 (3):389–401.

Yu, Xiuzhu, Frederick R Van De Voort, Zhixi Li, & Tianli Yue. (2007). Proximate composition of the apple seed and characterization of its oil. *International Journal of Food Engineering* 3 (5):21–27.

Zagrobelny, Mika, Søren Bak, Anne Vinther Rasmussen, Bodil Jørgensen, Clas M Naumann, & Birger Lindberg Møller. (2004). Cyanogenic glucosides and plant–insect interactions. *Phytochemistry* 65 (3):293–306.

CHAPTER 6

Orange Wastes and By-Products
Chemistry, Processing, and Utilization

Priyanka Suthar, Bababode Adesegun Kehinde, Shafiya Rafiq,
Nazmin Ansari, Barinderjit Singh and Harish Kumar

CONTENTS

6.1 INTRODUCTION

Sweet oranges (*Citrus × sinensis* L.) are non-climacteric fruit produced on an evergreen tree. The fruit originated mainly from southern regions of China, but today sweet oranges are grown and used commercially almost around the entire globe, making oranges one of the most cultivated fruit trees in the world. In 2014, the global orange production reached 72.9 million tons and production increased to 78.6 million tons by 2019 with 8.6% (approx.) contribution from Mediterranean countries (Ganatsios et al., 2021). This fruit crop held the position of 20th most important agricultural commodity in the world in 2008 (Rezzadori et al., 2012). Being non-climacteric crops, once oranges are harvested their quality cannot be improved, so that, under favourable climatic conditions, oranges should be kept on the tree itself rather than harvested and stored at low temperatures (Etebu and Nwauzoma, 2014). Blood oranges are different from the other types of sweet oranges based on the anthocyanin content present in the edible flesh of the fruit; in some cases, anthocyanins are present in the peel as well. 'Moror', 'Sanguinello' and 'Tarocco' are well-known commercial blood orange cultivars. As with other varieties of oranges, blood oranges are consumed fresh as they are rich in other bioactive ingredients, like flavonoids, hydroxycinnamic acids and ascorbic acid. As one of the bioactive components of blood orange, when consumed, anthocyanins are helpful in preventing various non-communicable diseases like diabetes, cancer and inflammation due their strong antioxidant and free radical scavenging activities. Other bioactive compounds present in blood oranges include vitamin C (ascorbic acid), hydroxycinnamic acids and flavanones. The characteristic red colour of blood oranges is an important factor for marketing purposes in order to attract more consumers to the fruit as well as the juice. In addition to blood oranges, blond oranges, also known as common orange or navel orange (because of having a navel-like structure at the bottom of the fruit), are another popular type of sweet orange. Bitter orange (*Citrus aurantium*) also belongs to the Rutaceae family and is one of the easiest accessible types of orange. Seville and biga-rade oranges are also known as sour or bitter oranges and, in India, they are recognized as important fruit crops, and are widely used in Europe for making marmalade.

The phytochemistry of oranges includes various organic acids, like citric, malic and isocitric acids, sugars such as sucrose, glucose and fructose, vitamins A, C, B1, B3 and B6, carotenoids, like carotene and xanthophyll, and flavouring compounds such as volatile hydrocarbons, phenols, esters, ketones, alcohols and hydroxycinnamic acids (Titta et al., 2010). These substances vary among dif-ferent orange types, cultivars and important environmental parameters. The functional properties of oranges that have been reported include antidiabetic, anti-obesity, anti-osteoporosis, hypocho-lesterolemic and hepatoprotective activities (Farag et al., 2020). These bioactivities are attributed to the presence of various secondary metabolites like phenols, flavonoids, carotenoids, terpenoids, etc. The ratio of acid to sugar is a major determinant factor for assessing the maturity of citrus fruits and is also a factor responsible for analysing the quality of citrus fruit flavour. The most abundant organic acids present in citrus fruits are citric and malic acids whereas others, like succinic, tartaric and oxalic acid, are less dominant in fruits. In bitter orange juice, citric acid was reported to be present at 48.79 g/L whereas sweet orange contained 11.10–15.65 g/L, a result which explains the less palatable taste of bitter oranges (Karadeniz, 2004). The major bioactive compounds reported in the leaves, roots and fruits of *C. × sinensis* are peptides, flavonoids, coumarins, steroids, fatty acids, carbamates, hydroxyamides and carotenoids, along with minerals such as potassium, cal-cium, magnesium and sodium (Favela-Hernández et al., 2016). Orange processing industries gener-ate large amounts of wastes or by-products after processing, which include peel, pulp and seeds. These wastes account for nearly 50% of the fresh fruit weight and are sources of various bioactive ingredients like essential oils, dietary fibre, flavonoids, vitamin C and phenolic compounds which exert various bioactivities, such as antimicrobial, antioxidant, anti-inflammatory, antidiabetic, cyto-toxic, and anti-obesity (Geraci et al., 2017). About 70% of the total fresh orange fruits are used by food manufacturing industries, from which nearly 50–60% of harvested fruit is converted into fruit wastes like peel, seed and cell wall residues. These generated wastes contain many nutritional and health-promoting compounds in tissues which are usually discarded (Khule et al., 2019).

6.2 CHEMISTRY OF ORANGE WASTES AND BY-PRODUCTS

The peel of sweet orange is an excellent source of orange essential oil as well as pectin and has been considered to be an important raw material from the food processing industry (Pandharipande and Makode, 2012). Peel represents about 30% of the total fresh fruit weight (Bejar et al., 2011). The citrus peel has two layers i.e., flavedo (the outer orange layer) and albedo, the inner layer. Of these two layers, the flavedo consists of small oil glands from which the oil is extracted and subsequently used by the pharmaceutical industry in the manufacturing of perfumes, cosmetics, etc. These essential oils are composed of various bioactive compounds like hydrocarbons, esters, alcohols and aldehydes. The most important components of orange essential oil are monoterpenes, such as D-limonene, which are responsible for the antimicrobial effect of orange essential effect (Geraci et al., 2017). Citrus essential oils possess many complex volatile compounds, especially terpenes. Usually, steam distillation, cold pressing and hydro-distillation are the commonly adopted oil extraction processes used on citrus peel in several food processing plants. The volatile com-pounds of sweet orange (*C. × sinensis)* peel essential oil were reported by Njoroge et al. (2005) from three varieties from Kenya, namely 'Salustiana', 'Valencia' and 'Washington'. Identification of all the volatile compounds was conducted by chromatographic techniques (gas chromatography and gas chromatography–mass spectrometry) and the authors reported a total of 56 volatile com-pounds from 'Salustiana', 73 compounds from 'Valencia' and 72 in 'Washington' peel oil. The total proportion of volatile compounds identified was 98.7, 97.8 and 97.4% for 'Salustiana', Valencia and Washington, respectively. 'Salustiana' peel oil contained 96.9% of monoterpenes followed by 'Valencia' and 'Washington' peel oil with 94.5 and 92.7%, respectively. All three peel oils were rich sources of limonene, sabinene, α-pinene and α-terpinene. Citrus oil produced by cold pressing might initiate phototoxic reactions due to the presence of furocoumarins from the coumarin family. Structurally, furocoumarins have coumarin as the base structure of the furan ring. Bergapten, bergamottin, herniarin, citropten and oxypeucedanin are well-known and established phototoxicants. Limonene represents almost 68–98% of the essential oil and is used in different industrial sectors due to its unique fragrance and antioxidant properties. It is generally recognised as a safe additive and is mostly used as a preservative by the food industry (Ozturk et al. (2019).

Pectin belongs to the group of soluble fibres and is obtained from orange processing units in the form of residues. In another study by Hosseini et al. (2019), ultrasound-assisted pectin extraction from sour orange peel was carried out and screened for different parameters. The study reported 28.07±0.67% extraction yield when 150 W ultrasound power was applied with 10 min irradiation at pH 1.5. Under these optimized condition, sour orange peel pectin (SOPP) was screened and analysed. The SOPP possessed mean±standard error moisture, ash and protein concentrations of 8.81±0.68, 1.89±0.51 and 1.45±0.23%, respectively, whereas HPLC analysis of SOPP showed the presence of galacturonic acid and neutral sugar (galactose) at concentrations of 65.3% galacturonic acid and 72% neutral sugar (galactose). The total phenolic content was reported to be 39.95±3.13 mg gallic acid equivalent per g pectin. The oil-holding and water-holding capacities were reported to be 1.32±0.21 and 3.10±0.12 g oil or water per g pectin, respectively. The surface tension was also estimated in 0.1 and 0.5% (w/v) aqueous solutions as 45.56±0.23 and 42.14±0.61 mN/m, respectively. The by-product orange pomace possesses albedo in both peel and seeds. Peel and seeds constitute 45% to 60% of total fruit weight. From a nutritional and functional point of view, orange pomace is an excellent source of dietary fibres. The presence of dietary fibres broadens the potential utilization of orange pomace in various fortified products (O'Shea et al., 2013). The by-products from oranges are also good sources of plant secondary metabolites, mainly flavonoids. These health-promoting plant chemical compounds are present in high concentrations and many publications claim that there are strong antioxidants that can be extracted from both juice and peel extracts. A study by Benelli et al. (2010) extracted antioxidants from orange pomace using supercritical fluids and compared these extracts with those obtained with other extraction techniques, such as ultrasound-assisted extraction, Soxhlet with various organic solvents and hydro-distillation. The yield from super-critical fluid extraction was found to be low but possessed good antioxidant activity using the DPPH(2,2-Diphenyl-1-Picrylhydrazyl) assay and high total phenolic concentration in comparison with Soxhlet and ultrasound-assisted extraction. The use of supercritical fluid CO_2 extraction with ethanol improved the yield and antioxidant profile. Also, antimicrobial activity was reported for a supercritical fluid extract at a pressure of 200 bar and 50°C (Benelli et al., 2010).

6.3 PROCESSING OF ORANGE WASTES AND BY-PRODUCTS

The essential oils from fruits of the citrus family are industrially relevant for their flavouring impacts. The citrus oil is mainly extracted from the oil sac which is located at the outer surface of the fruit peel. The diameter of these sacs or glands range from 0.4 to 0.6 mm. Typically, 5 kg of essential oil from orange peel can be extracted from 1000 kg of oranges. The profiling of oil showed the presence of 90% D-limonene (Ángel Siles López and Thompson, 2010), a colourless or transparent liquid at room temperature with strong, unique aroma of oranges. With the aim of improving efficiency of essential oil yield, along with energy, cost and time efficiency, many green techniques have been reported. Commonly, techniques described are ultrasound-assisted extraction, microwave-assisted extraction and supercritical fluid extraction. These techniques are green, environmentally protective extraction techniques because their main focus is to reduce waste generation and lower the consumption of both energy and solvent (Gavahian et al., 2019). Some combined techniques have also been used to improve the energy efficiency of the extraction procedure. In a study by González-Rivera et al. (2016), where the use of two different techniques i.e., microwave and ultrasound, in a combined form were used, the results showed that the simultaneous ultrasound and microwave irradiated hydro-distillation as well as the coaxial microwave-assisted hydro-distillation technique resulted in 60% greater energy efficiency compared with others. In addition to these techniques, the enzyme-assisted technique has attracted the attention of researchers, where the extraction of the essential oil from the peel was increased two-fold. Mishra et al. (2005) observed that the pre-treatment of mandarin orange peel by Xylanase enzyme resulted in higher oil yield at different enzyme concentration. Control sample yield oil as 3.72–4.41% where as 0.3% enzyme treated fruit yields oil as 4.17–4.98%. In a recent report by Bustamante et al. (2016), microwave assisted hydro-distillation (MAHD) is used for extraction of essential oil from Navel Navelate oranges. In comparison conventional treatment, MAHD treated sample showed 1.8±0.1% and hydro-distillation showed 1.7±0.1% oil yields. The cold-pressed extraction of orange peel oil resulted in 75.1% of non-hydroxylated polymethoxyflavones and 5.44% of hydroxylated polymethoxyflavones, which are responsible for inducing apoptosis in breast cancer cells (Favela-Hernández et al., 2016). Recently, deep eutectic solvents (DES) began to trend for the extraction of valuable phytochemicals due to the presence of electron donors and acceptors, as well as their properties like non-flammability, low toxicity, high biocompatibility, and negligible vapour pressure. In a study by Ozturk et al. (2018), choline chloride was used as the hydrogen bond acceptor and glycerol was used as the hydrogen bond donor for the preparation of DES. At optimum conditions of 333.15°K and a 1:10 solid-to-liquid ratio, extraction for 150 minutes resulted in the highest recovery of limonene extraction compared with ethanol extraction.

The recovery of cellulose, which is the main component of the fibre content of the peel, has been studied as it has a very important application in paper-producing industries as a filler compound. Bicu and Mustata (2011) used the sulphite pulping process in which the orange pulp was cooked in a solution of sodium sulphite after removal of the oil and pectin from the peel. Response surface methodology was used for the determination of the optimum conditions for the process. After study of the characteristics of the extracted cellulose, it was found that the cellulose extracted from this method had good purity level, bright colour, low crystallinity and good water retention. D-Limonene is the main component of the orange peel essential oil which is used by the food, pharmaceutical, cosmetic and nutraceutical industries due to its anti-oxidant and fragrance properties (Ali et al., 2015). Bio-based green solvents, like cyclopentyl methyl ether (CPME) and 2-methyl-tetrahydrofuran (2-MeTHF), have been used instead of the most commonly used solvent, hexane, to increase the yield which resulted in 80% and 40% increases in yield of D-limonene, respectively (Ozturk et al., 2019). D-Limonene causes mild irritation to eyes and the respiratory tract as it produces strong allergens, like perillyl alcohol, carvone, etc., when comes into contact with oxygen present in air (Ciriminna et al., 2014). For the extraction of pectin from orange peel, Yeoh et al. (2008) investigated different methods and solvents. In this study, microwave heating was reported to be more efficient when compared with conventional oil extraction technique i.e., Soxhlet. Microwave heating extraction for 15 min and Soxhlet extraction for 3 h recovered the same amount of extracted pectin. In the study, maximum pectin yield occurred at a strongly acidic pH of 1.5. A study by Rangarajan et al. (2010) investigated the use of orange peel solid and orange peel as substrates for the production of pectinase using *Aspergillus niger*. The submerged fermentation was carried out using peptone as a nitrogenous source with orange peel extract, which resulted in the highest yield of the pectinase enzyme. The activity of pectinase enzyme was reported to be higher when the orange peel extract was used along with peptone. The exopectinase activity was reported to be 6800 IU per g in submerged fermentation with orange peel extract. The extraction of pectinase and cellulase using the fungi *Penicillium atrovenetum*, *Aspergillus oryzae* and *Aspergillus flavus* (isolated from soil and decaying orange peels) was described by Adeleke et al. (2012). The enzyme production was carried out using orange peel in solid-state fermentation. Pectinase enzyme treatment for oil extraction results in better quality of oil compared to the oil extracted by using organic solvent due to its lower content of peroxide value, free fatty acids and colour intensity. The solid-state fermentation also reduced the refining cost by retaining the phospholipids (Rebello et al., 2017). The extraction of peroxidases from sweet orange peel has also been reported. Peroxidase activity contributes to several functional properties and hence is utilized in industries and biochemical processes such as glucose, uric acid and cholesterol quantification (Okino Delgado and Fleuri, 2016). In most non-alcoholic beverages, clouding agents are used to improve the flavour, colour and mouth feel of the beverage. Synthetic clouding agents are restricted due to the increasing demand for natural additives by consumers. Orange peel waste is one of the best-known and -studied raw materials for the production of clouding agents but the natural agent is known to flocculate and settle at the bottom over time. To improve the quality of this natural clouding agent, Espachs-Barroso et al. (2005) used polygalacturonase and cellulose. The results found that the process (addition of 90 μL/kg polygalacturonase and 69 mg/kg of cellulose at 45°C for 80 min) resulted in better cloud stability. The detailed and in-depth knowledge about orange peel has taken this process to the next level in the bio-refinery approach. Biorefinery is an integrative process which involves the biomass conversion process into fuel products, power or chemicals, using waste biomass. Ethanol production using orange peel was studied by Oberoi et al. (2010) through a two-stage hydrolysis and fermentation process at 121°C and 15 psi pressure for 15 min (Figure 6.1). Fermentation beyond 9 h did not increase the ethanol concentration, and the ultimate maximum ethanol productivity was reported to be 3.37 g/L/h in a batch fermenter. The process of bioethanol production from orange peel by using *Saccharomyces cerevisiae* for 24 hours of fermentation resulted in maximum 4.1 g/100 mL of bioethanol (Joshi et al., 2015).

Pomace generated from the orange processing industries usually contains the cell wall part from the seeds, stems and fruit which result in the production of fibre and other bioactive compounds after processing (Quiles et al., 2018). The characterisation of soluble and insoluble fibre involves a step-by-step process which includes homogenisation, protease treatment and centrifugation followed by washes. After all these processes, the supernatant and the residues are separated from each other. After dialysis, the supernatant is converted into neutral sugars and uronic acid to produce soluble dietary fibre. The residual part goes through an acid hydrolysis procedure resulting in the production of neutral sugars, uronic acid and klason lignin, and the production of insoluble dietary fibre (Gutiérrez Barrutia et al., 2019).

6.4 UTILIZATION OF ORANGE WASTES AND BY-PRODUCTS

Orange peel has different medicinal uses. The Chinese traditional medicines (chen-pi) are made with the dried mature peel of citrus fruits (*C. × sinensis and Citrus reticulata)* to treat digestive

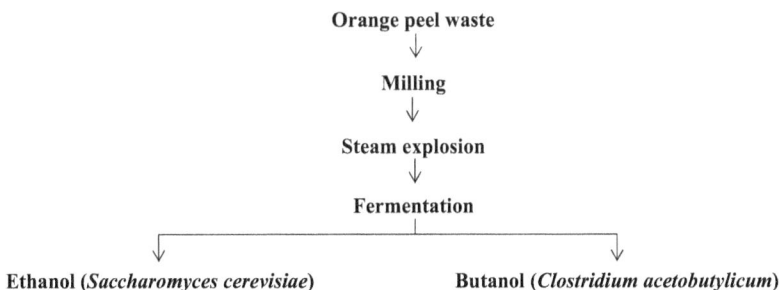

Orange peel waste
↓
Milling
↓
Steam explosion
↓
Fermentation

Ethanol (*Saccharomyces cerevisiae*)   **Butanol (*Clostridium acetobutylicum*)**

Figure 6.1 Fermentation process of orange peel waste.

and inflammation-related disorders. This medicine is also traditionally used to treat respiration-related health problems like asthma and bronchitis. In addition, several extensive studies have been conducted on its anticancer and anti-inflammation activities. Citrus flavonoids have been found to be the bioactive compounds responsible for such anti-inflammatory properties (Huang and Ho., 2010). Farag et al. (2020) compared different fruit parts of oranges for their anti-obesity activity in mice on a high-fat diet for a period of eight weeks and reported that citrus peel extract was effective at reducing total serum cholesterol, whereas citrus flesh and seed extract showed no effect. The flavonoids present in peel extract are likely to exhibit such activities. The components present in citrus peel were neo-hesperidin, poncirin, neoeriocitrin and naringin. Amines are another class of important compounds in the peel which have different physiological properties. Some amines can directly stimulate the brain by participating in neural transmission in the nervous system. Histamine is a strong amine that promotes dilation of blood vessels and which can lower blood pressure. Sipenephrine causes physiological effects characteristic of the sympathetic nervous system by promoting the stimulation of sympathetic nerves and can increase blood pressure (Okino Delgado and Fleuri, 2016). Various types of enzymes have been extracted from orange peel which are used for different functions. This mainly consisted of lipolytic enzymes, such as esterases and lipases. Lipases catalyse the reaction involving long-chain fatty acids, whereas esterases catalyse reactions involving short-chain fatty acids as well as simple esters. These enzymes are mostly used for bio-technological purposes (Lopes et al., 2011). Proteases are another group of enzymes extracted from orange peel and are mostly used in cheese industries for the coagulation of proteins (Mazorra-Manzano et al., 2013). Then there are peroxidases which have different properties and are used in industry for the quantitative measurement of uric acid, glucose and cholesterol (Vetal and Rathod, 2015). Essential oil extracted from orange is widely used in the pharmaceutical, food, cosmetic and fragrance industries. However, in recent times, essential oils have been further explored for their role in preservation, pest control and antimicrobial packaging uses. Essential oils are also incorporated into soft drinks and ice-creams as a flavouring agent. In cosmetic products, extracted orange oil is used in the development of cleansers like soaps or shampoos, perfumes, antiperspirants, room air fresheners and deodorants (Gavahian et al., 2019). Furocoumarin, a phototoxic compound found in orange essential oil, is mostly used in cosmetic products. In Europe, restrictions on the use of furocoumarin-like compounds were imposed by the Scientific Committee on Cosmetic Products in 2001 (SCCP, 2005); however, it may be used in cosmetic products within a prescribed limit of 1 ppm in the final product (Palazzolo et al., 2013).

Various studies are being undertaken to analyse the antimicrobial activity of orange essential oil on microorganisms causing food spoilage and food-borne illness. Orange essential oil has been found to be effective against the growth of common food spoilage microbes such as *Staphylococcus* and *Pseudomonas* (Schillaci et al., 2013). Orange oil has also shown strong activity against a major cause of death due to food-borne illness in most developed countries, i.e., *Listeria monocytogenes* (Vitale and Schillaci, 2016). D-Limonene in essential oil of orange peels is found to be an effective compound in inhibition of the fungus *Aspergillus niger*. Scanning electron microscope study revealed that the oil treatment was responsible for the loss of cytoplasm in fungal hyphae and for the budding of the hyphal tip, leading to the hyphal wall becoming thinner and leading to its disruption (Sharma and Tripathi, 2008). The essential oil of sweet oranges contains the bioactive compounds limonene and α-myrcene, which is reported to reduce gastric cancer risk in the host organism in response to oral intake (Toscano et al., 2017). The demand for limonene is increasing day by day due to its unique properties. In food industries, limonene is mostly used as a preservative and flavouring compound. It is also used in cosmetic industries due to its high antioxidant activities. Limonene was among the first ingredients to be used in natural pesticides to control the pest in an eco-friendly way (Hollingsworth, 2005). In food industries, pectin is utilised as an additive. The demand for pectin worldwide is reported to be nearly 30,000 tonnes, with a growing trend every year (Rezzadori et al., 2012). Pectin extracted from orange peel has great industrial opportunities in the food processing sector. The utilisation of orange peel has considerable scope for the development of valuable products, which would also lower the pollution of the environment (Maran et al., 2013) (Figure 6.2).

Dietary fibre produced from orange by-products has been shown to be rich in additional phytochemicals like flavonoids, polyphenols, carotenoids and vitamin C, and to exhibit valuable characteristics like water-binding capacity and gelling structure-building; fibre is used as a fat replacer in the ice-cream industry. The replacement of fat with dietary fibre did not cause any significant changes in flavour, colour and fragrance of the ice-cream. This results in 70% fat replacement, making the ice-cream low in calories and beneficial for gut regulation and for disease prevention (de Moraes Crizel et al., 2013). Recently, Jridi et al. (2019) reported the development of grey triggerfish skin gelatin films incorporating phenolic extracts from *Citrus* × *sinensis* peel (blood orange type). The film developed may be used as an alternative to conventional packaging films.

Comparisons were made between fresh orange peel extract and dried orange peel extract and it was concluded that fresh orange peel possessed higher antioxidant activity compared with dried orange peel extract. In addition, fresh orange peel extracts exhibited strong antibacterial properties and contained phenolics such as quinic acid (the major phenolic), rutin, *trans*-ferulic acid, naringenin and 4, 5-di-*o*-caffeoylquinate (Özcan et al., 2021). In many studies, orange peel has been reported to be a cost-effective and eco-friendly adsorbent for removing toxic and hazardous compounds from the environment (Gupta & Nayak, 2012; El-Said et al., 2013). Arami et al. (2005) showed the uses of orange peel collected from orange fields of Iran for the removal of Direct Red 23 and Direct Red 80 dyes from aqueous solution of textile dyes. The adsorption capacity was reported to be 10.72 and 21.05 mg/g for Direct Red 23 and Direct Red 80, respectively, at a pH of 2. In a similar study, the adsorption of heavy metals was performed using pre-treated banana and orange peels. The adsorption capacities of orange peel for Pb^{2+}, Ni^{2+}, Zn^{2+}, Cu^{2+} and Co^{2+} were reported to be 7.75, 6.01, 5.25, 3.65 and 1.82 mg/g, respectively. A high pH was favourable for the adsorption of Pb^{2+} and hence the orange peel can be considered to be a cost-effective adsorbent of damaging and toxic metals from synthetic solutions (Annaduari et al., 2003). Shivaraj et al. (2001) conducted an experiment to check the adsorption capacity of the dye Acid Violet 17 by utilising orange peel. The adsorption capacity was reported to be 19.88 mg/g at pH 6.3. The highest dye removal percentage was obtained at pH 2 with 87%, whereas maximum desorption was 60% when the pH was maintained at pH 10. Cellulose separated from orange peel waste is mainly used as an additive in paper-making industries and also for the retention of metal ions as an adsorbent (Bicu and Mustata., 2011). Orange peel possesses a variety of essential oils having excellent aroma and which are added to different food products. The orange peel is used in various formulations of herbal tea and marmalade. Citrus flavour is a widely used natural flavouring in cosmetic, pharmaceutical, beverage, bakery,

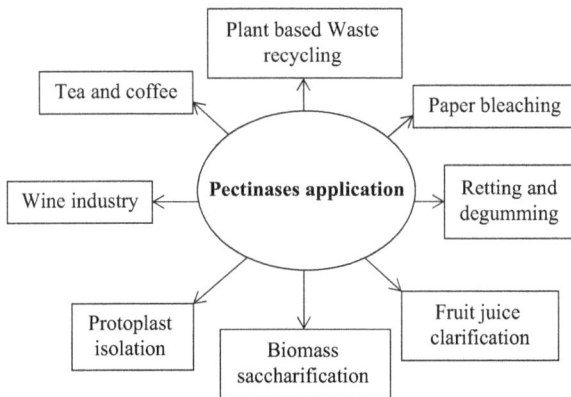

Figure 6.2 Uses of orange-based pectinase.

confectionery and perfumery industries. A study by Benjamin et al. (2007) used orange peels for the preparation of cupcakes. The grated peel transformed into sweet orange powder was produced by freezing-drying followed by oil extraction from fresh and dried peel powder by using supercritical fluid CO_2 extraction. The extracted oils were then analysed and used in cupcakes. The flavouring compounds in cupcakes were identified as limonene, linalool, myrcene, decanal and octanal. Raj and Masis (2014) developed wheat bread using orange peel at different concentration levels (5%, 7% or 10%). The orange peels were reported to have the following composition: moisture content (3.5%), protein (2.67%), ash (6.69%), fat (2.0%), fibre (2.35%) and carbohydrates (85.37%). The buns prepared with 5% orange peel powder were then analysed for moisture, fat, proteins, fibre and ash contents and found to have values of 2.3%, 9.39%, 32.65%, 1.9% and 2.3%, respectively. Whereas for control buns, the values were 2.04% for moisture, 9.51% for fat, 31.7% for protein, 1.8% for fibre and 2.1% for ash. The orange peel added to wheat flour showed increases in water absorption in the buns, whereas the dough stability decreased. Encapsulation of green tea with orange peel extracts was achieved by Rasouli Ghahroudi et al. (2017). The double emulsion was used as a microencapsulation product by the coacervation method. The essential oils from orange peel were also used as oil phase emulsions, followed by enclosure in a tea bag, thus serving as a functional tea drink. Marmalades are well-known fruit preserves prepared with high sugar content and low water activity followed by thermal cooking sterilisation of the jar to prevent any further microbial contaminants. They are very stable fruit preserves (Licciardello and Muratore, 2011). The use of natural dyes has been common since ancient times because they are anti-allergic and can easily be biodegraded. Orange peel is rich in carotenoids and is a good source of natural dyes. Dyeing of cotton fabric was studied by Edeen (2015), using the padding technique. Due to its strong antioxidant, antimicrobial and UV-protective properties, the dye extracted from orange peel has been used for wool, cotton and tencel fabric (Naveed et al., 2021).

6.5 CONCLUSIONS

Oranges are one of the most flavourful and nutritious fruits. In orange-processing plants, oranges are used to develop many products, which include jam, marmalades, jellies, fruit juice powder, fruit squash, etc. The by-products generated from orange-processing plants are pomace, seeds and peel. Among all the by-products of orange, peel contains high concentrations of essential oils and other phytochemicals. The utilisation of orange peel has been widely reported in the production of other valuable products like ethanol, pectin, enzymes and essential oils, which have been incorporated into other fortified food products. The orange peel is used in food products with the aim to enhance dietary fibre. The seeds of oranges also contain large amounts of plant secondary metabolites, like carotenoids, flavonoids, phenolics, aromatic compounds and essential oils. Pectin and pectinase enzymes are used extensively in food processing plants either of which can be extracted from orange by-products in considerable amounts. The orange essential oil is commonly extracted by the cold-press method but recently many green extraction techniques were adapted to assess the extraction efficiency and yield. Oranges are an excellent source of antioxidants and also possess many functional properties which include anti-obesity, antidiabetic, antiosteoporosis, hypocholesterolemic, hepatoprotective and antimicrobial activities and many more.

REFERENCES

Adeleke, A. J., Odunfa, S. A., Olanbiwonninu, A., & Owoseni, M. C. (2012). Production of cellulase and pectinase from orange peels by fungi. *Natural Sciences*, *10*, 107–112.
Ali, B., Al-Wabel, N. A., Shams, S., Ahamad, A., Khan, S. A., & Anwar, F. (2015). Essential oils used in aromatherapy: A systemic review. *Asian Pacific Journal of Tropical Biomedicine*, *5*(8), 601–611.

Ángel Siles López, J., Li, Q., & Thompson, I. P. (2010). Biorefinery of waste orange peel. *Critical Reviews in Biotechnology*, *30*(1), 63–69.

Annadurai, G., Juang, R. S., & Lee, D. J. (2003). Adsorption of heavy metals from water using banana and orange peels. *Water Science and Technology*, *47*(1), 185–190.

Arami, M., Limaee, N. Y., Mahmoodi, N. M., & Tabrizi, N. S. (2005). Removal of dyes from colored textile wastewater by orange peel adsorbent: Equilibrium and kinetic studies. *Journal of Colloid and Interface Science*, *288*(2), 371–376.

Bejar, A. K., Ghanem, N., Mihoubi, D., Kechaou, N., & Mihoubi, N. B. (2011). Effect of infrared drying on drying kinetics, color, total phenols and water and oil holding capacities of orange (Citrus sinensis) peel and leaves. *International Journal of Food Engineering*, *7*(5), 5.

Benelli, P., Riehl, C. A., Smânia Jr, A., Smânia, E. F., & Ferreira, S. R. (2010). Bioactive extracts of orange (*Citrus sinensis* L. Osbeck) pomace obtained by SFE and low pressure techniques: Mathematical modeling and extract composition. *The Journal of Supercritical Fluids*, *55*(1), 132–141.

Benjamin, A. C., Akingbala, J. O., & Baccus-Taylor, G. S. (2007). Effect of drying and storage on flavour quality of orange (*Citrus cinensis* (Linn) Osbeck) peel for cupcakes. *Journal of Food Agriculture and Environment*, *5*(2), 78.

Bicu, I., & Mustata, F. (2011). Cellulose extraction from orange peel using sulfite digestion reagents. *Bioresource Technology*, *102*(21), 10013–10019.

Bustamante, J., van Stempvoort, S., García-Gallarreta, M., Houghton, J. A., Briers, H. K., Budarin, V. L., Matharu, A. S., & Clark, J. H. (2016). Microwave assisted hydro-distillation of essential oils from wet citrus peel waste. *Journal of Cleaner Production*, *137*, 598–605.

Ciriminna, R., Lomeli-Rodriguez, M., Cara, P. D., Lopez-Sanchez, J. A., & Pagliaro, M. (2014). Limonene: A versatile chemical of the bioeconomy. *Chemical Communications*, *50*(97), 15288–15296.

de Moraes Crizel, T., Jablonski, A., de Oliveira Rios, A., Rech, R., & Flôres, S. H. (2013). Dietary fiber from orange byproducts as a potential fat replacer. *LWT-Food Science and Technology*, *53*(1), 9–14.

Edeen, A. B. (2015). Dyeing of Egyptian cotton fabrics with orange peel using padding technique. *International Design Journal*, *5*(3), 733–744.

El-Said, A. G., Gamal, A. M., & Mansour, H. F. (2013). Potential application of orange peel as an eco-friendly adsorbent for textile dyeing effluents. *Research Journal of Textile and Apparel*, *17*(4), 31–39.

Espachs-Barroso, A., Soliva-Fortuny, R. C., & Martín-Belloso, O. (2005). A natural clouding agent from orange peels obtained using polygalacturonase and cellulase. *Food Chemistry*, *92*(1), 55–61.

Etebu, E., & Nwauzoma, A. B. (2014). A review on sweet orange (*Citrus sinensis* L Osbeck): Health, diseases and management. *American Journal of Research Communication*, *2*(2), 33–70.

European Commission. Scientific Committee on Consumer Products (SCCP). (2005). Opinion on furocoumarins in cosmetic products. Report No. SCCP/0942/0905.

Farag, M. A., Abib, B., Ayad, L., & Khattab, A. R. (2020). Sweet and bitter oranges: An updated comparative review of their bioactives, nutrition, food quality, therapeutic merits and biowaste valorization practices. *Food Chemistry*, *331*, 127306.

Favela-Hernández, J. M. J., González-Santiago, O., Ramírez-Cabrera, M. A., Esquivel-Ferriño, P. C., & Camacho-Corona, M. D. R. (2016). Chemistry and pharmacology of *Citrus sinensis*. *Molecules*, *21*(2), 247.

Ganatsios, V., Terpou, A., Bekatorou, A., Plessas, S., & Koutinas, A. A. (2021). Refining citrus wastes: From discarded oranges to efficient brewing biocatalyst, aromatic beer, and alternative yeast extract production. *Beverages*, *7*(2), 16.

Gavahian, M., Chu, Y. H., & Mousavi Khaneghah, A. (2019). Recent advances in orange oil extraction: An opportunity for the valorisation of orange peel waste a review. *International Journal of Food Science & Technology*, *54*(4), 925–932.

Geraci, A., Di Stefano, V., Di Martino, E., Schillaci, D., & Schicchi, R. (2017). Essential oil components of orange peels and antimicrobial activity. *Natural Product Research*, *31*(6), 653–659.

González-Rivera, J., Spepi, A., Ferrari, C., Duce, C., Longo, I., Falconieri, D., … & Tine, M. R. (2016). Novel configurations for a citrus waste based biorefinery: From solventless to simultaneous ultrasound and microwave assisted extraction. *Green Chemistry*, *18*(24), 6482–6492.

Gupta, V. K., & Nayak, A. (2012). Cadmium removal and recovery from aqueous solutions by novel adsorbents prepared from orange peel and Fe_2O_3 nanoparticles. *Chemical Engineering Journal*, *180*, 81–90.

Gutiérrez Barrutia, M. B., Curutchet, A., Arcia, P., & Cozzano, S. (2019). New functional ingredient from orange juice byproduct through a green extraction method. *Journal of Food Processing and Preservation*, *43*(5), e13934.

Hollingsworth, R. G. (2005). Limonene, a citrus extract, for control of mealybugs and scale insects. *Journal of Economic Entomology*, *98*(3), 772–779.

Hosseini, S. S., Khodaiyan, F., Kazemi, M., & Najari, Z. (2019). Optimization and characterization of pectin extracted from sour orange peel by ultrasound assisted method. *International Journal of Biological Macromolecules*, *125*, 621–629.

Huang, Y. S., & Ho, S. C. (2010). Polymethoxy flavones are responsible for the anti-inflammatory activity of citrus fruit peel. *Food Chemistry*, *119*(3), 868–873.

Joshi, S. M., Waghmare, J. S., Sonawane, K. D., & Waghmare, S. R. (2015). Bio-ethanol and bio-butanol production from orange peel waste. *Biofuels*, *6*(1–2), 55–61.

Jridi, M., Boughriba, S., Abdelhedi, O., Nciri, H., Nasri, R., Kchaou, H., ... & Nasri, M. (2019). Investigation of physicochemical and antioxidant properties of gelatin edible film mixed with blood orange (*Citrus sinensis*) peel extract. *Food Packaging and Shelf Life*, *21*, 100342.

Karadeniz, F. (2004). Main organic acid distribution of authentic citrus juices in Turkey. *Turkish Journal of Agriculture and Forestry*, *28*(4), 267–271.

Khule, G. D., Khupase, S. P., & Giram, K. K. (2019). Development and quality evaluation of orange pomace fortified biscuits. *Journal of Pharmacognosy and Phytochemistry*, *8*(3), 3695–3701.

Licciardello, F., & Muratore, G. (2011). Effect of temperature and some added compounds on the stability of blood orange marmalade. *Journal of Food Science*, *76*(7), C1094–C1100.

Lopes, D. B., Fraga, L. P., Fleuri, L. F., & Macedo, G. A. (2011). Lipase and esterase: To what extent can this classification be applied accurately?. *Food Science and Technology*, *31*(3), 603–613.

Maran, J. P., Sivakumar, V., Thirugnanasambandham, K., & Sridhar, R. (2013). Optimization of microwave assisted extraction of pectin from orange peel. *Carbohydrate Polymers*, *97*(2), 703–709.

Mazorra-Manzano, M. A., Moreno-Hernández, J. M., Ramírez-Suarez, J. C., de Jesús Torres-Llanez, M., González-Córdova, A. F., & Vallejo-Córdoba, B. (2013). Sour orange *Citrus aurantium* L. flowers: A new vegetable source of milk-clotting proteases. *LWT-Food Science and Technology*, *54*(2), 325–330.

Mishra, D., Shukla, A. K., Dixit, A. K., & Singh, K. (2005). Aqueous enzymatic extraction of oil from mandarin peels. *Journal of Oleo Science*, *54*(6), 355–359.

Naveed, T., Babar, A. A., Rashdi, S. Y., Rehman, F., Naeem, M. A., Wang, W., ... & Ramzan, M. B. (2021). Dyeing and colorfastness properties of tencel fabric treated with natural dye extracted from orange peel. *Surface Review and Letters*, *28*(03), 2050055. https://doi.org/10.1142/S0218625X20500559.

Njoroge, S. M., Koaze, H., Karanja, P. N., & Sawamura, M. (2005). Essential oil constituents of three varieties of Kenyan sweet oranges (*Citrus sinensis*). *Flavour and Fragrance Journal*, *20*(1), 80–85.

Oberoi, H. S., Vadlani, P. V., Madl, R. L., Saida, L., & Abeykoon, J. P. (2010). Ethanol production from orange peels: Two-stage hydrolysis and fermentation studies using optimized parameters through experimental design. *Journal of Agricultural and Food Chemistry*, *58*(6), 3422–3429.

Okino Delgado, C. H., & Fleuri, L. F. (2016). Orange and mango by-products: Agro-industrial waste as source of bioactive compounds and botanical versus commercial description—A review. *Food Reviews International*, *32*(1), 1–14.

O'Shea, N., Doran, L., Auty, M., Arendt, E., & Gallagher, E. (2013). The rheology, microstructure and sensory characteristics of a gluten-free bread formulation enhanced with orange pomace. *Food & Function*, *4*(12), 1856–1863.

Özcan, M. M., Ghafoor, K., Al Juhaimi, F., Uslu, N., Babiker, E. E., Mohamed Ahmed, I. A., & Almusallam, I. A. (2021). Influence of drying techniques on bioactive properties, phenolic compounds and fatty acid compositions of dried lemon and orange peel powders. *Journal of Food Science and Technology*, *58*(1), 147–158.

Ozturk, B., Parkinson, C., & Gonzalez-Miquel, M. (2018). Extraction of polyphenolic antioxidants from orange peel waste using deep eutectic solvents. *Separation and Purification Technology*, *206*, 1–13.

Ozturk, B., Winterburn, J., & Gonzalez-Miquel, M. (2019). Orange peel waste valorisation through limonene extraction using bio-based solvents. *Biochemical Engineering Journal*, *151*, 107298.

Palazzolo, E., Laudicina, V. A., & Germanà, M. A. (2013). Current and potential use of citrus essential oils. *Current Organic Chemistry*, *17*(24), 3042–3049.

Pandharipande, S., & Makode, H. (2012). Separation of oil and pectin from orange peel and study of effect of pH of extracting medium on the yield of pectin. *Journal of Engineering Research and Studies, 3*(2), 06–09.

Quiles, A., Campbell, G. M., Struck, S., Rohm, H., & Hernando, I. (2018). Fiber from fruit pomace: A review of applications in cereal-based products. *Food Reviews International, 34*(2), 162–181.

Raj, A., & Masih, D. (2014). Physico chemical and rheological properties of wheat flour bun supplemented with orange peel powder. *International Journal of Science and Research, 3*(8), 391–394.

Rangarajan, V., Rajasekharan, M., Ravichandran, R., Sriganesh, K., & Vaitheeswaran, V. (2010). Pectinase production from orange peel extract and dried orange peel solid as substrates using *Aspergillus niger. International Journal of Biotechnology & Biochemistry, 6*(3), 445–453.

Rasouli Ghahroudi, F., Mizani, M., Rezaei, K., & Bameni Moghadam, M. (2017). Mixed extracts of green tea and orange peel encapsulated and impregnated on black tea bag paper to be used as a functional drink. *International Journal of Food Science & Technology, 52*(7), 1534–1542.

Rebello, S., Anju, M., Aneesh, E. M., Sindhu, R., Binod, P., & Pandey, A. (2017). Recent advancements in the production and application of microbial pectinases: An overview. *Reviews in Environmental Science and Bio/Technology, 16*(3), 381–394.

Rezzadori, K., Benedetti, S., & Amante, E. R. (2012). Proposals for the residues recovery: Orange waste as raw material for new products. *Food and Bioproducts Processing, 90*(4), 606–614.

Schillaci, D., Napoli, E. M., Cusimano, M. G., Vitale, M., & Ruberto, G. (2013). *Origanum vulgare* subsp. *hirtum* essential oil prevented biofilm formation and showed antibacterial activity against planktonic and sessile bacterial cells. *Journal of Food Protection, 76*(10), 1747–1752.

Sharma, N., & Tripathi, A. (2008). Effects of *Citrus sinensis* (L.) Osbeck epicarp essential oil on growth and morphogenesis of *Aspergillus niger* (L.) Van Tieghem. *Microbiological Research, 163*(3), 337–344.

Sivaraj, R., Namasivayam, C., & Kadirvelu, K. (2001). Orange peel as an adsorbent in the removal of acid violet 17 (acid dye) from aqueous solutions. *Waste Management, 21*(1), 105–110.

Titta, L., Trinei, M., Stendardo, M., Berniakovich, I., Petroni, K., Tonelli, C., & Rapisarda, P. (2010). Blood orange juice inhibits fat accumulation in mice. *International Journal of Obesity, 34*(3), 578–588.

Toscano-Garibay, J. D., Arriaga-Alba, M., Sánchez-Navarrete, J., Mendoza-García, M., Flores-Estrada, J. J., Moreno-Eutimio, M. A., & Ruiz-Pérez, N. J. (2017). Antimutagenic and antioxidant activity of the essential oils of *Citrus sinensis* and *Citrus latifolia. Scientific Reports, 7*(1), 1–9.

Vetal, M. D., & Rathod, V. K. (2015). Three phase partitioning a novel technique for purification of peroxidase from orange peels (*Citrus sinenses*). *Food and Bioproducts Processing, 94*, 284–289.

Vitale, M., & Schillaci, D. (2016). *Food Processing and Foodborne Illness Reference Module in Food Science.* Elsevier, 1–9.

Yeoh, S., Shi, J. T. A. G., & Langrish, T. A. G. (2008). Comparisons between different techniques for water-based extraction of pectin from orange peels. *Desalination, 218*(1–3), 229–237.

CHAPTER 7

Guava Wastes and By-Products
Chemistry, Processing, and Utilization

Krishan Kumar, Naseer Ahmed, Qurat-Ul-Eain Hyder Rizvi,
Sumaira Jan, Priyanka Thakur, Divya Chauhan, and Jaspreet Kaur

CONTENTS

7.1 INTRODUCTION

Guava (*Psidium guajava* L.) is known as the apple of the tropics, having its origins in Mexico and Central America (Somogyi, 1996). The genus *Psidium* includes about 150 species, of which the common guava, pear guava (*Psidium pyriferum* L.), Cattley guava (*Psidium cattleianum* Sabine), and apple guava (*Psidium pomiferum* L.) are the most commonly cultivated species. Being a climacteric fruit, guava matures at a fast rate and is highly perishable, with a shelf-life of only 2–3 days at ambient temperature (25–30°C) conditions. The total production of guava around the world was 55 million tonnes in 2019. India was the major producer, contributing 45% of total production, followed by other producers like China, Thailand, Indonesia, and Pakistan (Bassetto et al., 2005; FAOSTAT, 2019). Guava fruit possesses a pleasant flavor and aroma, in addition to excellent nutritional value with vitamins, dietary fibers, minerals, and antioxidants, the latter including carotenoids such as lycopene as well as β-carotenoids present in red and pink varieties of guava (Flores et al., 2015; Rodriguez-Amaya et al., 2005). Guava fruit is round with a diameter of 3–10 cm and has a yellowish or pinkish peel in some guava varieties (Lee et al., 2010). It is very nutritious but delicate, being highly susceptible to chilling injury as well as to diseases. Its storage below 10°C can result in symptoms of severe chilling injury, such as surface pitting as well as browning of the flesh and skin (Singh and Pal, 2008). Guava fruits of some varieties are sweet while others are astringent in taste (Lim et al., 2006). The fruit has a characteristic musky flavor, which decreases somewhat after the processing operations. The weight of the fruit varies between 150 and 250 g (Ayub et al., 2005). The fruits are produced twice a year but the fruits harvested in the winter are of higher value

DOI: 10.1201/9781003164463-7

(Bal and Dhaliwal, 2004). Guava is a highly productive crop as well as a highly profitable one and is well regarded by farmers due to its extensive adaptability and high yields (Hassan et al., 2012).

The guava fruit contains various sugars like sucrose, glucose, and fructose in addition to other bioactive components such as minerals, pectin, antioxidants, and dietary fibers (Bhat et al., 2015; Martínez et al., 2012; Mahattanatawee et al., 2006). The concentrations of these compounds vary with the degree of fruit ripeness (Rashida et al., 1997). The fruit is commercially processed into products like juices, jellies, and soft drinks, as well as baby food (Campoli et al., 2018; Singh, 2011). The fruits contain numerous antioxidants, such as vitamin C and carotenoids. The concentration of vitamin C in guava fruit is four times greater than that present in an orange. Also, it contains biologically active secondary metabolites such as triterpenoids, flavonoids, and other polyphenolic components (Hassimotto et al., 322005; Flores et al., 2012). Guava is widely cultivated in most parts of India and fruit is available in the rainy as well as the winter seasons, with the quality of fruits obtained in the winter season being superior to that of the fruits harvested in the rainy season. The lack of transportation and processing conditions leads to a 20–25% loss of guava fruit (Kanwal et al., 2016). Guava fruit is very easy to process and various processed products, such as juices, puree, nectar, concentrates, jam, jellies, fruit bars, canned fruit, and dehydrated powders, can be prepared from guava powder (Singh, 2011).

Processing of guava into valuable processed products generates numerous wastes and by-products of guava such as skin, seeds, pomace, etc. During the production of guava pulp, a residue is produced that constitutes about 30% by weight of the fruit, and 6% of this residue is accounted for by the seeds (Gupta et al., 2019; da Silva Lima, et al., 2019; Hernández-Acosta et al., 2011). Peel as a waste of guava processing constitutes about 20% of the fruit fresh weight (Medina and Herrero, 2016; Rojas-Garbanzo et al., 2019). The residue of guava pulp obtained during commercial juice extraction can be converted into nutritionally superior fermented products (da Silva Lima et al., 2019). In addition to the use of guava by-products for livestock feed, these can also be processed into valuable products or can be subjected to different types of extraction techniques for extracting bioactive components. This chapter deals with the generation of wastes and by-products by guava processing industries, their nutritional composition, and utilization for the extraction of bioactive components as well as their use in the development of value-added food products.

7.2 CHEMISTRY OF GUAVA WASTES AND BY-PRODUCTS

Guava fruit possesses exceptional nutritional value and is rich in valuable bioactive compounds. Guava contains 74 to 87% moisture, 0.43 to 1.39% ash content, 5.0 to 14.32% carbohydrate, 0.1 to 0.95% lipid, 2.8 to 5.5% fiber, and 0.35 to 2.55% protein. Peel has been reported to contain about 8.82 to 8.93 mg/g of phenolic compounds on a dry weight basis (Medina and Herrero, 2016; Rojas-Garbanzo et al., 2019). Guava residues obtained after juice extraction contain about 90.81% moisture, 10% crude protein, 11.71% ether extract, 75% carbohydrates, and 1.25% ash content (Muniz et al., 2020). The seeds comprise about 5–13% of oil, having significant amounts of essential fatty acids (Arvanitoyannis and Varzakas, 2008). Wastes of guava, such as peel and seeds, can be used as a constituent in preventing oxidative stress and other relevant disorders because of higher concentrations of phenolic components, insoluble dietary fibers, carotenoids, saponins, phytosterols, and fatty acids. The phenolic components present in guava residue can achieve high antioxidant and anti-radical scavenging activities (Amaya-Cruz et al., 2015; Rojas-Garbanzo et al., 2017). Being a rich source of numerous phenolic compounds, guava finds wide uses in the pharmaceutical and other industries. In addition to phenolic components, guava seeds contain dietary fiber and unsaturated fatty acids (Villacís-Chiriboga et al., 2020). The nutrients present in guava seeds on a dry weight basis include proteins (7.9–9.6%), lipids (10.5–16%), crude fibers (53.6–67.7%), and ash content

(0.9–1.2%). Guava seeds are also rich in numerous amino acids such as tryptophan and methionine (Nicanor et al., 2001; Prasad and Azeemoddin, 1994; El-Din and Yassen, 1997). In addition to its nutritional characteristics, guava also exhibits numerous medicinal properties. It has potential hepatoprotective, antioxidant, and other health-promoting benefits (Gutiérrez, Mitchell, and Solis, 2008). It is rich in several phytochemicals, vitamins, minerals, and dietary fiber, and is widely used in various folk medicines (Jiménez-Escrig et al., 2001).

Guava is a potent source of dietary fibers as reported in various studies. The leading dietary fibers in guava fruits include pectic substances and celluloses (El Tinay, Saeed, and Bedri, 1979). Simple carbohydrates, such as glucose, fructose, etc., provide nutrients and energy whereas oligo-saccharides can act as prebiotics and can help to boost the growth of probiotics in the colon of the host. Dietary fibers possess numerous health-improving characteristics, such as regulation of bowel movement in the intestine and the prevention as well as control of diabetes by slowing down the absorption of sugar in the small intestine, and prevention of cardiovascular disorders and colon cancer (Palafox-Carlos, Ayala-Zavala, and González-Aguilar, 2011). Dietary fibers can bind to fatty substances and bile acids, as well as glucose, subsequently reducing their absorption (Zhang, Zhu, and Jiang, 2014).

Lim et al. (2018) analyzed the chemical composition of three different types of pink guava by-product, i.e., refiner fraction (coarse materials and seeds with a particle size of 1.2 mm), siever fraction (skin pieces, coarse fibers, or other impurities with a particle size of 0.8 mm), and decanter fraction (unwanted pulp pieces with particle size <0.8 mm). The samples were frozen to −20°C, freeze-dried, ground to fine particles then sieved to less than 0.5 mm before being analyzed for prebiotic activity and other physico-chemical characteristics. The refiner fraction contained 7.08% protein, 7.7% lipid, high phytate content, and a high score for prebiotic activity. The siever fraction contained a large concentration of carbohydrates with lower concentrations of crude fiber and cel-lulose. Furthermore, it had the highest binding capacity with the bile acid chenodeoxycholic acid (74.8%), and the highest DPPH (2,2-diphenyl-1-picrylhydrazyl) radical scavenging (antioxidant) activity. The decanter fraction was found to be higher in cellulose content and exhibited a high score for prebiotic activity (Lim et al., 2018). Guava containing pink pulp is rich in the powerful anti-oxidant carotenoid, lycopene (Wilberg and Rodriguez-Amaya, 1995). Marina and Noriham (2014) reported a ferric reducing antioxidant power (FRAP) of 400 ppm in guava peel extract. In Malaysia, about 25% of pink guava used for puree preparation is discarded as a by-product. Antioxidants like lycopene have been found to prevent various lifestyle diseases such as oxidative stress and cancer (Basu and Imrhan, 2007). Lycopene has a radical scavenging ability greater than β-carotene and α-tocopherol (Di Mascio et al., 1989). Lycopene protects against injury to the mammalian cell membrane and DNA as well as provides protection against tumor-related oxidative damage (Matos et al., 2000). Uchôa-thomaz et al. (2014) studied the biochemical composition of guava seeds and found that these are rich in dietary fibers, protein, iron as well as zinc, and have a low calorific value. The lipid profile analysis depicted that guava seeds contain unsaturated fatty acids such as oleic, and linoleic acids. The bioactive components, like ascorbic acid, carotenoids, and insoluble dietary fiber, were found in significant amounts. The phenolic compounds were extracted at the highest concentration with 60% methanol solution, with further purification through ultrafiltration membranes (Sukeksi et al., 2016). Chang et al. (2014) quantified the total phenolic components, flavonoids, and antioxidant activity of different parts such as seeds, peel, and pulp of white-fleshed guava fruit by high-performance liquid chromatography–quantitative time-of-flight–mass spectro-metric (HPLC-QTOF-MS) analysis of an ethanol extract. They detected 69 phenolic compounds, phenolic acid derivatives, flavonoids, phenylethanoids, lignans, stilbenoids, and dihydrochalcones, while nine polar compounds like triterpenoids and iridoids were also detected. Denny et al. (2013) quantified the anti-inflammatory as well as the antinociceptive potential of phenolic compounds in guava pomace extract (GPE). The total phenolic components were estimated as 3.40 mg gallic

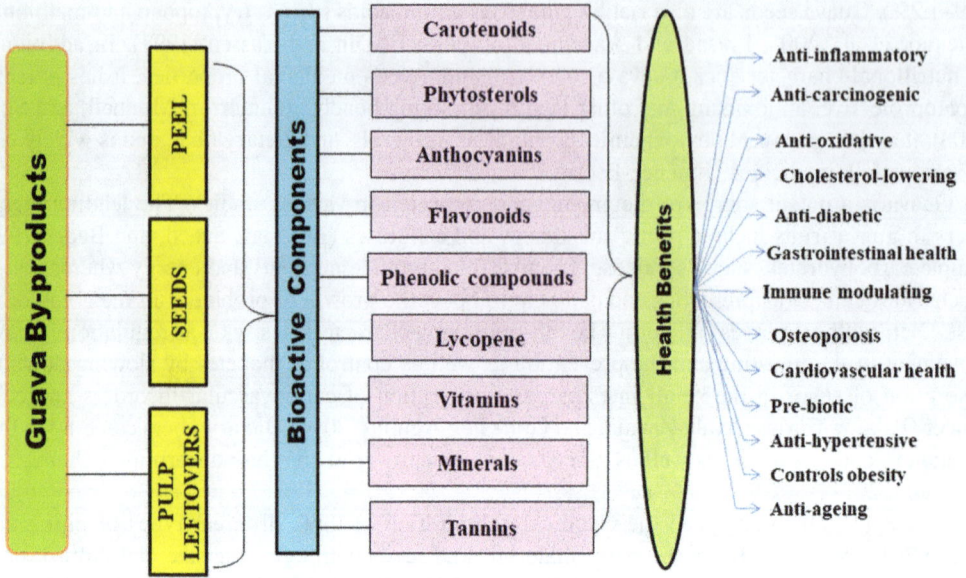

Figure 7.1 Bioactive compounds present in guava by-products and their health benefits.

acid-equivalents (GAE)/g and myricetin, quercetin, epicatechin, isovaleric, and gallic acids were the major phenolic components identified by gas chromatography-mass spectrometric (GC/MS) analysis. Therefore, the value-added utilization of guava wastes and by-products can help in recycling residues generated from the guava processing industry. The bioactive compounds present in guava by-products as well as their health benefits are depicted in Figure 7.1.

The worldwide production of guava is approximately 500,000 metric tons. The processing of guava generates considerable amounts of guava wastes and by-products and these are simply thrown away, contaminating the environment. Utilization of these wastes and by-products rather than disposing of them can help to solve the problems relevant to environmental pollution. The pomace remaining after the extraction of juice represents a major source of waste and by-products and is considered to be a major source of bioactive components (Denny et al., 2013). There has been rapid growth in the guava processing industry because of the growing demand by consumers for processed fruit products with longer shelf-life. Large amounts of waste and by-products are generated due to the mass processing of fruit, approximately 80 kg of waste per metric ton of fruit during the processing of fresh guava fruits (Souza et al., 2016). Depending upon the variety of guava, about 4–30% of fruit weight is converted into wastes and by-product residues. This solid agro-industrial residue includes a mixture of peel, pulp, and seeds, with high concentrations of phenolic compounds and antioxidant activity. This guava residue contain 15–50% seeds by weight, and about 15% peel as guava waste (Mantovani et al., 2004; Melo et al., 2011; Jiménez-Escrig et al., 2001; Medina and Herrero, 2016; Rojas-Garbanzo et al., 2019). Guava is mainly processed for guava puree in Malaysia, a leading guava-producing country, and about 25% of pink guava is discarded as waste during the preparation of puree. The use of guava by-products in the extraction of chemical compounds can be helpful in the alleviation of undesirable pollution problems (Makkar and Becker, 1999) as well as the production of economically viable compounds having a high antioxidant capacity (Kong and Ismail, 2011). Guava pomace, a major by-product generated in the guava juice industry, is a key source of bioactive components. Fruit residues extracted from processing industries are rich in bioactives and many food processing industries are adopting various techniques to incorporate the functional ingredients into their products (Assis et al., 2009).

7.3 PROCESSING OF GUAVA WASTES AND BY-PRODUCTS

7.3.1 Extraction of Bioactive Components

Processing for the extraction of bioactives from guava wastes and by-products involves the use of huge volumes of several different organic solvents, like methanol (Ademiluyi et al., 2016; Contreras-Calderón et al., 2011; Rojas-Garbanzo et al., 2017), ethyl acetate (Ongphimai et al., 2013), or formic acid (Flores et al., 2015), harming the environment as well as affecting the health of human operators. Several studies have involved the use of traditional techniques, like maceration (Musa et al., 2011; El Anany, 2015; Ademiluyi et al., 2016) and Soxhlet extraction (Ongphimai et al., 2013) for extracting the bioactives from the industrial residue. These techniques, being time-consuming, involve laborious activities which cannot be applied on a commercial level. In contrast, the innovative, green extraction techniques such as pressurized liquid extraction, ultrasound-assisted extraction, and supercritical fluid extraction, diminish the use of organic solvents, which are toxic, and reduce extraction time. These techniques involve mild extraction conditions, thereby preventing the loss of bioactive compounds, and can be utilized on a commercial scale (Selvamuthukumaran and Shi, 2017). The literature explains the use of various extraction techniques, such as solvent extraction, alcoholic extraction, hydro-alcoholic extraction, ultrasound-assisted extraction, and supercritical carbon dioxide (SC-CO$_2$) extraction, are the main techniques utilized for the extraction of bioactive components from guava wastes and by-products (Table 7.1). Ultrasound-assisted extraction is the most widely used extraction technique, with several benefits, such as improved extraction yield, reduced usage of solvents, reduced power consumption, and shorter period of extraction (Zhang et al., 2009). The transmission of ultrasound pressure waves causes cavitation, thereby resulting in higher extraction efficiency (Yolmeh et al., 2014). Additionally, the process can be further improved by disrupting the cell walls and enhancing the liberation of the cellular materials (Vilkhu et al., 2008). The collapsing of microbubbles in the cellular contents disrupts the cell wall and helps to achieve the easy release of components, to be extracted into the solvent (Patist and Bates, 2008). Da Silva et al. (2020) extracted the β-carotenoids and lycopene contents from the guava pulp and waste powders using the maceration technique and ultrasonic processing. The extracts were assessed for carotenoid concentrations, color characteristics, antioxidant activity, lycopene and β-carotenoid concentrations by HPLC with diode-array detection (HPLC-DAD). Extraction using ultrasound processing was found to be the best technique, with maximum extraction of lycopene, i.e., 135 mg/100 g from guava pulp and 76.64 mg/100 g from waste guava powder, with reduced extraction time and minimum use of solvents. Sukeksi et al. (2016) extracted phenolic compounds from guava seeds using 60% methanol solution and pure water, with further refining through ultrafiltration. In another study, da Silva et al. (2014) found that encapsulated extracts of guava peel, seeds, and unused pulp were found to exhibit higher inhibitory activities against *Listeria monocytogenes* and *Escherichia coli* than the non-encapsulated by-products (Díaz-de-Cerio et al., 2017). Kong et al. (2010c) studied the effects of various treatments, such as solvent extraction and SC-CO$_2$ extraction, on the antioxidant activity as well as the cytotoxic activities of lycopene-rich pink guava by-products. The yield of samples extracted using SC-CO$_2$ extraction was higher than *via* the solvent extraction method. Amid et al. (2016) optimized the conditions for the ultrasound-assisted extraction of pectinase enzyme from guava peel to achieve maximum recovery. The extraction was conducted under varying conditions such as sonication time (10–30 min), the temperature of the ultrasound (30–50°C), pH (2.0–8.0), and solvent-to-sample ratio (2:1 ml/g–6:1 ml/g). They achieved a high yield of 96.2%, good specific activity of 18.2 U/mg, temperature stability of 88.3%, and storage stability of 90.3% under optimum conditions. The conditions optimized included the sonication time of 20 min, the temperature of 40 °C, and a pH of 5.0.

Table 7.1 Extraction of Bioactive Components from Guava By-Products

Guava Wastes and By-Products	Bioactive/Nutritional Compounds	Extraction Method	Bioactivity/Uses	Reference
Peel, seed, pulp	Phenolics	Ultrasound-assisted hydro-alcoholic extraction	Antioxidant activity	da Silva Lima et al. (2019); Rojas-Garbanzo et al. (2019)
Peel	Pectin	Water extraction at 80°C	Soluble dietary fiber, used as a gelling agent in jams, jelly, marmalades	Chakraborty and Ray (2011)
Peel	Pectin	Using 10% deep eutectic solvent and citric acid solution	Gelling agent	Yusof, Zaini, and Azman (2020)
Peel	Pectin	Acid extraction using citric acid and HCl at 85°C for 60 min at pH 2.0	Jelly preparation	Bhat and Singh (2014)
Peel, seed, pulp	Phenolics, flavonoids	Ethanol extraction and characterization by high pressure liquid chromatography coupled with quadrupole time-of-flight mass spectrometry (HPLC-QTOF-MS)	Antioxidant activity	Chang et al. (2014)
Peel, seeds, and pulp residue	Phenolics, flavonoids	Extraction for 5 h at 60°C and pH 2.	Antioxidant activity	Kong et al. (2010b)
Pulp and wastes	Total phenolic components, flavonols, flavonoids, and condensed tannins	Extraction using probe-type ultrasound at 25°C for 2 min using ethanol: water (30:70, v/v)	Antioxidant activity	da Silva Lima et al. (2019)
Pomace	Pectin	Boiling with 0.1% citric acid solution and then precipitation with ethanol	Gelling agent in orange marmalade	Hlaing (2019)
Decanted pink guava by-product fractions	Lycopene	Supercritical fluid extraction	Antioxidant activity	Kong et al. (2010a)

7.3.2 Extraction of Pectin

Pectin is a polysaccharide present in all plant tissues. Calcium pectate exists between the plant cell walls to serve as a strengthening or building agent. It is widely used in food processing industries as a thickening agent in sauces, soups, and puddings, as a stabilizer in ice cream, margarine, mayonnaise, salad dressings, etc., and as a gel-forming agent in jams, jellies, candies, desserts, and yogurts (Figure 7.2). The conditions optimized for pectin extraction include heating the pomace at a temperature of 100°C for 30 minutes. The pectin from the water extract was then precipitated with ethanol and the paste was dried at 70°C for 150 minutes in a hot-air oven. Pectin powder was employed in the preparation of orange marmalade. Chakraborty and Ray (2011) reported that the spent guava extract was the most appropriate by-product for the extraction of pectin as they obtained a higher yield of pectin from spent guava extract at a temperature of 80°C in comparison with lime peel or apple pomace. The pectin extracted from guava residue was used in the formulation of different types of processed fruit products, such as jam, jelly, and marmalade as a thickening, stabilizing, and gelling agent. The environment friendly deep eutectic solvent (DES) and citric acid solution were utilized for the extraction of pectin components from guava peel. The pectin extracted was subjected to evaluation of chemical characteristics, with further comparison to evaluate the capacity of two different DESs as substitutes for the conventional solvent. It observed that the pectin extracted using 10% choline chloride/ethylene glycol (ChCl/EG) a higher gel-forming ability due to lower ash contents (2.02%) as compared to that extracted using 10% citric acid. It had high methoxyl content (15.6%) and a higher degree of esterification (98.47%) than pectin extracted with 10% citric acid. However, the foaming capacity of pectin extracted with 10% citric acid was higher than that of pectin extracted using 10% choline chloride/ethylene glycol (ChCl/EG) (Yusof, Zaini, and Azman, 2020). Bhat and Singh (2014) optimized the process for acid extraction of pectin from guava peel

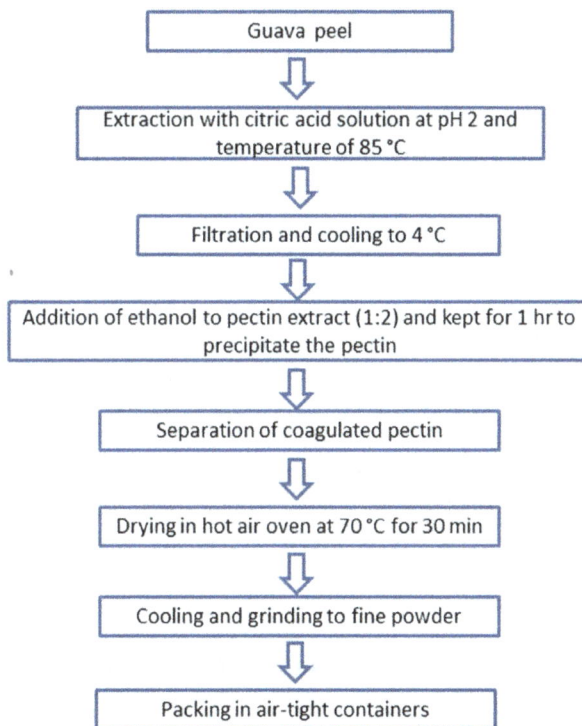

Guava peel
⇩
Extraction with citric acid solution at pH 2 and temperature of 85 °C
⇩
Filtration and cooling to 4 °C
⇩
Addition of ethanol to pectin extract (1:2) and kept for 1 hr to precipitate the pectin
⇩
Separation of coagulated pectin
⇩
Drying in hot air oven at 70 °C for 30 min
⇩
Cooling and grinding to fine powder
⇩
Packing in air-tight containers

Figure 7.2   Flow chart for preparation of pectin from guava peel.

powder by using two acids, i.e., citric acid and HCl, at three different times (30, 45, or 60 min), temperatures (65, 75, or 85°C), and pH conditions (2.0, 2.5, or 3.0). The yield of pectin using HCl ranged between 3.87 and 16.8% and that using citric acid varied from 2.65 to 11.12%. The best condition for the extraction was the use of HCl at 85°C for 60 min at a pH of 2. The pectin isolated using HCl or citric acid had equivalent weights of 685.3 or 345.4, methoxyl contents of 4.25 and 3.50%, and anhydro-uronic acid contents of 67.4 and 82.1%, respectively. The ash and moisture contents were 1.8% and 3.2% in HCl-extracted pectin, respectively, and 8.2% and 6.0% in citric acid-extracted pectin, respectively. However, organoleptic characteristics of jelly prepared by using pectin powder extracted using citric acid were better than those prepared using the HCl extracted pectin.

7.4 UTILIZATION OF GUAVA WASTES AND BY-PRODUCTS

Guava wastes and by-products, being rich in nutritional components, have great potential to be utilized for incorporation into different processed products to increase their functional and nutraceutical properties. The high pectin content of guava products can be utilized for edible coatings and they can be incorporated in the pulp of fruits lacking pectin for the preparation of fruit jam, jellies, etc. Their high sugar content can be helpful in the preparation of fermented products from guava by-products. In addition, guava wastes and by-products can be processed by various heat treatments for the inactivation of anti-nutritional components, thereby enhancing the bioavailability of nutrients. Chang et al. (2014) studied the impact of processing techniques like boiling or autoclaving and different periods of germination on the nutritional contents and phytochemical composition of guava seeds. It has been reported that, following various processing treatments, there was a decrease in the concentrations of anti-nutritional components such as tannins and saponins. The heat treatment did not exert any effect on the total dietary fiber concentration nor on ash contents. The boiling of guava seeds did not decrease the phytic acid content, but it was reduced to 91 % by autoclaving. Germination for 14 days significantly reduced the phytic acid concentration in the seed by 90 %. Thus, after thermal treatments or germination, guava seeds can be used in the human food and animal feed industries. Srivastava et al. (1997) studied the production of ethanol by three yeast strains by utilization of guava pulp containing 10% total sugars and a pH of 4.1 for ethanol production. Fermentation was conducted at 30°C and pH 4.1, by using 1.5, 3.6, or 3.9% concentrations of yeast, i.e., *Saccharomyces cerevisiae* MTCC 1972, Isolate-1 and Isolate-2, for 60 h. The highest yield of alcohol was 5.8% at pH 5.0 by Isolate-2 after 36 h of fermentation. Todisco et al. (2018) evaluated the impact of edible pectin-based coatings comprising up to 50% of guava by-products on the nutritional characteristics of dried slices of red guava. The use of pectin in edible coatings enhanced the efficiency of the edible film by acting as a barrier to the oxidation of carotenoids. The highest retention of carotenoids and total phenolic compounds was attained at a temperature of about 60°C. Bertagnolli et al. (2014) optimized the process for the preparation of cookies by incorporating different levels of guava peel flour (GPF) at levels of 30, 50, or 70% and evaluated the nutritional composition, and concentrations of total phenolic compounds, beta-carotenoids, and lycopene in the cookies. The results revealed that the cookies were found to be rich in dietary fibers, ash content, total polyphenols, and beta-carotenoids. The organoleptic evaluation of the cookies indicated that the cookies were acceptable in terms of flavor, aroma, texture, and taste attributes at the level of incorporating 30% GPF. Therefore, GPF can be substituted partially for wheat flour for cookie preparation to improve the nutritional attributes without a marked effect on the sensory characteristics of the cookies. Khalifa et al. (2016) incorporated both guava seeds and guava pomace at concentrations of 5–20% to formulate cupcake mixtures. Substitution of wheat flour with 5 or 10% guava residues did not significantly affect the sensory and thermo-mechanical characteristics. Addition of either the guava seeds and pomaces resulted in higher levels of total phenolic components, crude fiber, and antioxidant components than did pure wheat flour. The cupcake at 20% level of pomace incorporation contained the highest polyphenolics (2.34 mg GAE/g dry weight [dw]) as well as antioxidant activity (5.18 μmol Trolox Equivalent [TE]/g dw). The acceptability of cupcakes at 5 or 10% levels of incorporation did not vary significantly compared with wheat flour cupcakes in organoleptic characteristics. The guava residue obtained from guava processing industries was subjected to fermentation with *Lactobacillus plantarum* WU-P19. The product obtained contained high concentrations of conjugated linoleic acid (CLA) and vitamin B_{12}. The pulp after fermentation was used to prepare the gummy jelly which consists of 100 g of fermented guava pulp, 11 g glucose syrup, 20 g sucrose, 6 g gelatine, and 0.5 g citric acid. One gram of this jelly can meet the recommended daily requirement of vitamin B_{12}, which is deficient in many processed foods (Palachum et al., 2020).

7.5 CONCLUSIONS

Guava is a delicious fruit containing a large number of nutritional and bioactive components. It is widely used as fresh fruit or converted to various processed products such as jam, jellies, guava cheese, drinks, puree, etc. The processing of guava generates a lot of wastes and by-products by the guava-processing industries. The main wastes and by-products of guava-processing industries include pulp leftovers, peel, and seed which are also natural sources of nutritional and bioactive components. These residues contain antioxidant components such as vitamin C, carotenoids, and biologically active compounds like flavonoids, triterpenoids, and other polyphenolic components. These wastes and by-products are generally used as cattle feed or thrown away as waste material, resulting in environmental pollution. Guava wastes and by-products can be converted into value-added processed food products or these can be subjected to different extraction techniques for extraction of valuable bioactive components which can prove to be highly useful in the prevention as well as the treatment of various lifestyle diseases. Guava-based wastes and by-products can be utilized as a constituent for the development of functional food products with elevated dietary fiber and bioactive potential. Such types of products can be helpful for the control of various lifestyle diseases prevalent in society.

REFERENCES

Ademiluyi, Adedayo O, Ganiyu Oboh, Opeyemi B Ogunsuyi, and Funmilayo M Oloruntoba. 2016. A comparative study on antihypertensive and antioxidant properties of phenolic extracts from fruit and leaf of some guava (*Psidium guajava* L.) varieties. *Comparative Clinical Pathology* 25 (2):363–374.

Amaya-Cruz, Diana María, Sarahí Rodríguez-González, Iza F Pérez-Ramírez, Guadalupe Loarca-Piñaa, Silvia maya-Llanoa, Marco Gallegos-Corona, and Rosalía Reynoso-Camachoa. 2015. Juice by-products as a source of dietary fibre and antioxidants and their effect on hepatic steatosis. *Journal of Functional Foods* 17:93–102.

Amid, Mehrnoush, Fara Syazana Murshid, Mohd Yazid Manap, and Zaidul Islam Sarker. 2016. Optimization of ultrasound-assisted extraction of pectinase enzyme from guava (*Psidium guajava*) peel: Enzyme recovery, specific activity, temperature, and storage stability. *Preparative Biochemistry and Biotechnology* 46 (1):91–99.

Arvanitoyannis, Ioannis S, and Theodoros H Varzakas. 2008. Fruit/fruit juice waste management: Treatment methods and potential uses of treated waste. *Waste Management for the Food Industries* 2:569–628.

Assis, Leticia Marques de, Elessandra da Rosa Zavareze, André Luiz Radünz, Á Dias, Luiz Carlos Gutkoski, and Moacir Cardoso Elias. 2009. Propriedades nutricionais, tecnológicas e sensoriais de biscoitos com substituição de farinha de trigo por farinha de aveia ou farinha de arroz parboilizado. *Alimentos e Nutrição Araraquara* 20 (1):15–24.

Ayub, Mohammad, Alam Zeb, and Javid Ullah. 2005. Effect of various sweeteners on chemical composition of guava slices. *Sarhad Journal of Agriculture (Pakistan)* 1:23–29.

Bal, JS, and GS Dhaliwal. 2004. Distribution and quality characteristics of graded guava fruits. *Haryana Journal of Horticultural Sciences* 33 (1&2):53–54.

Bassetto, Eliane, Angelo Pedro Jacomino, Ana Luiza Pinheiro, and Ricardo Alfredo Kluge. 2005. Delay of ripening of 'Pedro Sato'guava with 1-methylcyclopropene. *Postharvest Biology and Technology* 35 (3):303–308.

Basu, Antik, and Vicky Imrhan. 2007. Tomatoes versus lycopene in oxidative stress and carcinogenesis: Conclusions from clinical trials. *European Journal of Clinical Nutrition* 61 (3):295–303.

Bertagnolli, Silvana Maria Michelin, Márcia Liliane Rippel Silveira, Aline de Oliveira Fogaça, Liziane Umann, and Neidi Garcia Penna. 2014. Bioactive compounds and acceptance of cookies made with Guava peel flour. *Food Science and Technology* 34 (2):303–308.

Bhat, Ravish, Lakshminarayana Chikkanayakanahalli Suryanarayana, Karunakara Alageri Chandrashekara, Padma Krishnan, Anil Kush, and Puja Ravikumar. 2015. *Lactobacillus plantarum* mediated fermentation of *Psidium guajava* L. fruit extract. *Journal of Bioscience and Bioengineering* 119 (4):430–432.

Bhat, SA, and ER Singh. 2014. Extraction and characterization of pectin from guava fruit peel. *International Journal of Research in Engineering & Advanced Technology* 2 (3):447–454.

Campoli, Stephanie Suarez, Meliza Lindsay Rojas, Jose Eduardo Pedroso Gomes do Amaral, Solange Guidolin Canniatti-Brazaca, and Pedro Esteves Duarte Augusto. 2018. Ultrasound processing of guava juice: Effect on structure, physical properties and lycopene in vitro accessibility. *Food Chemistry* 268:594–601.

Chakraborty, A, and S Ray. 2011. Development of a process for the extraction of pectin from citrus fruit wastes viz. lime peel, spent guava extract, apple pomace etc. *International Journal of Food Safety* 13:391–397.

Chang, Ying Ping, May Ping Tan, Wai Li Lok, Suganthi Pakianathan, and Yasoga Supramaniam. 2014. Making use of guava seed (*Psidium guajava* L): The effects of pre-treatments on its chemical composition. *Plant Foods for Human Nutrition* 69 (1):43–49.

Contreras-Calderón, José, Lilia Calderón-Jaimes, Eduardo Guerra-Hernández, and Belén García-Villanova. 2011. Antioxidant capacity, phenolic content and vitamin C in pulp, peel and seed from 24 exotic fruits from Colombia. *Food Research International* 44 (7):2047–2053.

da Silva Lima, Renan, Itaciara Larroza Nunes, and Jane Mara Block. 2020. Ultrasound-assisted extraction for the recovery of carotenoids from Guava's pulp and waste powders. *Plant Foods for Human Nutrition* 75 (1):63–69.

da Silva Lima, Renan, Sandra Regina Salvador Ferreira, Luciano Vitali, and Jane Mara Block. 2019. May the superfruit red guava and its processing waste be a potential ingredient in functional foods? *Food Research International* 115:451–459.

da Silva, Larissa Morais Ribeiro, Evania Altina Teixeira De Figueiredo, Nagila Maria Pontes Silva Ricardo, Icaro Gusmao Pinto Vieira, Raimundo Wilane de Figueiredoa Isabell, and Montenegro Brasil Carmen L Gomes. 2014. Quantification of bioactive compounds in pulps and by-products of tropical fruits from Brazil. *Food Chemistry* 143:398–404.

Denny, Carina, Priscilla S Melo, Marcelo Franchin, Adna P Massarioli, Keityane B Bergamaschi, Severino M de Alencar, and Pedro L Rosalen. 2013. Guava pomace: A new source of anti-inflammatory and analgesic bioactives. *BMC Complementary and Alternative Medicine* 13 (1):1–7.

Di Mascio, Paolo, Stephan Kaiser, and Helmut Sies. 1989. Lycopene as the most efficient biological carotenoid singlet oxygen quencher. *Archives of Biochemistry and Biophysics* 274 (2):532–538.

Díaz-de-Cerio, Elixabet, Alba Rodríguez-Nogales, Francesca Algieri, Miguel Romero, Vito Verardo, Antonio Segura-Carretero, Juan Duarte, and Julio Galvez. 2017. The hypoglycemic effects of guava leaf (*Psidium guajava* L.) extract are associated with improving endothelial dysfunction in mice with diet-induced obesity. *Food Research International* 96:64–71.

El Anany, Ayman Mohammed. 2015. Nutritional composition, antinutritional factors, bioactive compounds and antioxidant activity of guava seeds (*Psidium Myrtaceae*) as affected by roasting processes. *Journal of Food Science and Technology* 52 (4):2175–2183.

El Tinay, AH, AR Saeed, and MF Bedri. 1979. Fractionation and characterization of guava pectic substances. *International Journal of Food Science & Technology* 14 (4):343–349.

El-Din, MHA Shams, and AAE Yassen. 1997. Evaluation and utilization of guava seed meal (*Psidium guajava* L.) in cookies preparation as wheat flour substitute. *Food/Nahrung* 41 (6):344–348.

FAOSTAT. 2019. Guava production in 2018. Tridge. 2019. Retrieved 16 November 2020.

Flores, Gema, Keyvan Dastmalchi, and Sturlainny Paulino. 2012. Anthocyanins from *Eugenia brasiliensis* edible fruits as potential therapeutics for COPD treatment. *Food Chemistry* 134 (3):1256–1262.

Flores, Gema, Shi-Biao Wu, Adam Negrin, and Edward J Kennelly. 2015. Chemical composition and antioxidant activity of seven cultivars of guava (*Psidium guajava*) fruits. *Food Chemistry* 170:327–335.

Gupta, Neha, Kasturi Poddar, Debapriya Sarkar, Nitya Kumari, Bhagyashree Padhan, and Angana Sarkar. 2019. Fruit waste management by pigment production and utilization of residual as bioadsorbent. *Journal of Environmental Management* 244:138–143.

Gutiérrez, Rosa Martha Pérez, Sylvia Mitchell, and Rosario Vargas Solis. 2008. *Psidium guajava*: A review of its traditional uses, phytochemistry and pharmacology. *Journal of Ethnopharmacology* 117 (1):1–27.

Hassan, Ishtiaq, Wasif Khurshid, and Khalid Iqbal. 2012. Factors responsible for decline in guava (*Psidium guajava*) yield. *Journal of Agriciculture Research*, 50 (1):129–134.

Hassimotto, Neuza Mariko Aymoto, Maria Inés Genovese, and Franco Maria Lajolo. 2005. Antioxidant activity of dietary fruits, vegetables, and commercial frozen fruit pulps. *Journal of Agricultural and Food Chemistry* 53 (8):2928–2935.

Hernández-Acosta, Mile A, Henry I Castro-Vargas, and Fabián Parada-Alfonso. 2011. Integrated utilization of guava (*Psidium guajava* L.): Antioxidant activity of phenolic extracts obtained from guava seeds with supercritical CO_2-ethanol. *Journal of the Brazilian Chemical Society* 22 (12):2383–2390.

Hlaing, S. S. 2019. Extraction of pectin powders from guava and apple and their applications in jam making processes (Doctoral dissertation, MERAL Portal).

Jiménez-Escrig, Antonio, Mariela Rincón, Raquel Pulido, and Fulgencio Saura-Calixto. 2001. Guava fruit (*Psidium guajava* L.) as a new source of antioxidant dietary fiber. *Journal of Agricultural and food Chemistry* 49 (11):5489–5493.

Kanwal, Nida, Muhammad Atif Randhawa, and Zafar Iqbal. 2016. A review of production, losses and processing technologies of guava. *Asian Journal of Agriculture and Food Sciences* 4 (2):96–101.

Khalifa, I, H Barakat, HA El-Mansy, and SA Soliman. 2016. Influencing of guava processing residues incorporation on cupcake characterization. *Journal of Nutrition Food Science*, 6 (513):2.

Kong, Kin Weng, Amin Ismail, Chin Ping Tan, and Nor Fadilah Rajab. 2010a. Optimization of oven drying conditions for lycopene content and lipophilic antioxidant capacity in a by-product of the pink guava puree industry using response surface methodology. *LWT-Food Science and Technology* 43 (5):729–735.

Kong, Kin-Weng, Abdul Razak Ismail, Seok-Tyug Tan, Krishna Murthy Nagendra Prasad, and Amin Ismail. 2010b. Response surface optimisation for the extraction of phenolics and flavonoids from a pink guava puree industrial by-product. *International Journal of Food Science & Technology* 45 (8):1739–1745.

Kong, Kin-Weng, Nor Fadilah Rajab, K Nagendra Prasad, Amin Ismail, Masturah Markom, and Chin-Ping Tan. 2010c. Lycopene-rich fractions derived from pink guava by-product and their potential activity towards hydrogen peroxide-induced cellular and DNA damage. *Food Chemistry* 123 (4):1142–1148.

Kong, KW, and A Ismail. 2011. Lycopene content and lipophilic antioxidant capacity of by-products from *Psidium guajava* fruits produced during puree production industry. *Food and Bioproducts Processing* 89 (1):53–61.

Lee, Sarah, Hyung-Kyoon Choi, Somi Kim Cho, and Young-Suk Kim. 2010. Metabolic analysis of guava (*Psidium guajava* L.) fruits at different ripening stages using different data-processing approaches. *Journal of Chromatography B* 878 (29):2983–2988.

Lim, Si Yi, Paik Yean Tham, Hilary Yi Ler Lim, Wooi Shin Heng, and Ying Ping Chang. 2018. Potential functional byproducts from guava puree processing. *Journal of Food Science* 83 (6):1522–1532.

Lim, Yau Yan, Theng Teng Lim, and Jing Jhi Tee. 2006. Antioxidant properties of guava fruit: Comparison with some local fruits. *Sunway Academic Journal* 3:9–20.

Mahattanatawee, Kanjana, John A Manthey, Gary Luzio, Stephen T Talcott, Kevin Goodner, and Elizabeth A Baldwin. 2006. Total antioxidant activity and fiber content of select Florida-grown tropical fruits. *Journal of Agricultural and Food Chemistry* 54 (19):7355–7363.

Makkar, HPS, and K Becker. 1999. Plant toxins and detoxification methods to improve feed quality of tropical seeds-Review. *Asian-Australasian Journal of Animal Sciences* 12 (3):467–480.

Mantovani, José Ricardo, Márcio Cleber de Medeiros Corrêa, Mara Cristina Pessôa da Cruz, Manoel Evaristo Ferreira, and William Natale. 2004. Uso fertilizante de resíduo da indústria processadora de goiabas. *Revista Brasileira de Fruticultura* 26 (2):339–342.

Marina, Z, and A Noriham. 2014. Quantification of total phenolic compound and in vitro antioxidant potential of fruit peel extracts. *International Food Research Journal* 21 (5):1925–1929.

Martínez, Ruth, Paulina Torres, Miguel A Meneses, Jorge G Figueroa, José A Pérez-Álvarez, and Manuel Viuda-Martos. 2012. Chemical, technological and *in vitro* antioxidant properties of mango, guava, pineapple and passion fruit dietary fibre concentrate. *Food Chemistry* 135 (3):1520–1526.

Matos, Humberto R, Paolo Di Mascio, and Marisa HG Medeiros. 2000. Protective effect of lycopene on lipid peroxidation and oxidative DNA damage in cell culture. *Archives of Biochemistry and Biophysics* 383 (1):56–59.

Medina, Narciso Nerdo Rodríguez, and Juliette Valdés-Infante Herrero. 2016. Guava (*Psidium guajava* L.) cultivars: An important source of nutrients for human health. In *Nutritional composition of fruit cultivars*. Edited by: Monique SJ Simmonds and Victor R Preedy. Cambridge, MA: Academic Press and Elsevier, pp. 287–315.

Melo, Priscilla Siqueira, Keityane Boone Bergamaschi, Ana Paula Tiveron, Adna Prado MassarioliI, Tatiane Luiza Cadorin OldoniI, Mauro Celso ZanusII, Giuliano Elias Pereira, and Severino Matias de Alencar. 2011. Phenolic composition and antioxidant activity of agroindustrial residues. *Ciência Rural* 41 (6):1088–1093.

Muniz, Cecília Elisa S, Ângela Maria Santiago, Thaisa Abrantes Souza Gusmão, Hugo Miguel Lisboa Oliveira, Líbia de Sousa Conrado, and Rennan Pereira de Gusmão. 2020. Solid-state fermentation for single-cell protein enrichment of guava and cashew by-products and inclusion on cereal bars. *Biocatalysis and Agricultural Biotechnology* 25:101576.

Musa, Khalid Hamid, Aminah Abdullah, Khairiah Jusoh, and Vimala Subramaniam. 2011. Antioxidant activity of pink-flesh guava (*Psidium guajava* L.): Effect of extraction techniques and solvents. *Food Analytical Methods* 4 (1):100–107.

Nicanor, A Bernardino, A Ortíz Moreno, AL Martinez Ayala, and G Dávila Ortíz. 2001. Guava seed protein isolate: Functional and nutritional characterization. *Journal of Food Biochemistry* 25 (1):77–90.

Ongphimai, Nattaya, Supathra Lilitchan, Kornkanok Aryusuk, Akkarach Bumrungpert, and Kanit Krisnangkura. 2013. Phenolic acids content and antioxidant capacity of fruit extracts from Thailand. *Chiang Mai Journal of Science* 40 (4):636–642.

Palachum, Wilawan, Wanna Choorit, Supranee Manurakchinakorn, and Yusuf Chisti. 2020. Guava pulp fermentation and processing to a vitamin B12-enriched product. *Journal of Food Processing and Preservation* 44 (8):e14566.

Palafox-Carlos, Hugo, Jesús Fernando Ayala-Zavala, and Gustavo A González-Aguilar. 2011. The role of dietary fiber in the bioaccessibility and bioavailability of fruit and vegetable antioxidants. *Journal of Food Science* 76 (1):R6–R15.

Patist, Alex, and Darren Bates. 2008. Ultrasonic innovations in the food industry: From the laboratory to commercial production. *Innovative Food Science & Emerging Technologies* 9 (2):147–154.

Prasad, NBL, and G Azeemoddin. 1994. Characteristics and composition of guava (*Psidium guajava* L.) seed and oil. *Journal of the American Oil Chemists' Society* 71 (4):457–458.

Rashida, E, E Babiker El Fadil, and Abdullahi H El Tinay. 1997. Changes in chemical composition of guava fruits during development and ripening. *Food Chemistry* 59 (3):395–399.

Rodriguez-Amaya, DB, OM Porcu, and CH Azevedo-Meleiro. 2005. Variation in the carotenoid composition of fruits and vegetables along the food chain. Paper read at *I International Symposium on Human Health Effects of Fruits and Vegetables* 744.

Rojas-Garbanzo, Carolina, Benno F Zimmermann, Nadine Schulze-Kaysers, and Andreas Schieber. 2017. Characterization of phenolic and other polar compounds in peel and flesh of pink guava (*Psidium guajava* L. cv.'Criolla') by ultra-high performance liquid chromatography with diode array and mass spectrometric detection. *Food Research International* 100:445–453.

Rojas-Garbanzo, Carolina, Julia Winter, María Laura Montero, Benno F Zimmermann, and Andreas Schieber. 2019. Characterization of phytochemicals in Costa Rican guava (*Psidium friedrichsthalianum*-Nied.) fruit and stability of main compounds during juice processing-(U) HPLC-DAD-ESI-TQD-MSn. *Journal of Food Composition and Analysis* 75:26–42.

Selvamuthukumaran, M, and John Shi. 2017. Recent advances in extraction of antioxidants from plant by-products processing industries. *Food Quality and Safety* 1 (1):61–81.

Singh, SP. 2011. Guava (*Psidium guajava* L.). In *Postharvest biology and technology of tropical and subtropical fruits*. Edited by Elhadi M Yahia. Woodhead Publisher, Cambridge, United Kingdom, (pp. 213–246e).

Singh, SP, and RK Pal. 2008. Controlled atmosphere storage of guava (*Psidium guajava* L.) fruit. *Postharvest Biology and Technology* 47 (3):296–306.

Somogyi, Laszlo P. 1996. *Major processed products*. Technomic Publishing Company, Incorporated.

Souza, Henrique A, Serge-Étienne Parent, Danilo E Rozane, Daniel A Amorim, Viviane C Modesto, William Natale, and Leon E Parent. 2016. Guava waste to sustain guava (*Psidium guajava*) agroecosystem: Nutrient "balance" concepts. *Frontiers in Plant Science* 7:1252.

Srivastava, Suchi, DR Modi, and SK Garg. 1997. Production of ethanol from guava pulp by yeast strains. *Bioresource Technology* 60 (3):263–265.

Sukeksi, Lilis, Che Rosmani Che Hassan, Nik Merian Nik Sulaiman, Hamidreza Rashidi, and Sina Davazdah Emami. 2016. Polyphenols recovery from tropical fruits (pink guava) wastes via ultra-filtration membrane technology application by optimum solvent selection. *Iranian Journal of Chemistry and Chemical Engineering (IJCCE)* 35 (3):53–63.

Todisco, KM, NS Janzantti, AB Santos, FS Galli, and MA Mauro. 2018. Effects of temperature and pectin edible coatings with guava by-products on the drying kinetics and quality of dried red guava. *Journal of Food Science and Technology* 55 (12):4735–4746.

Uchôa-thomaz, Ana Maria Athayde, Eldina Castro Sousa, José Osvaldo Beserra Carioca, Selene Maia de Morais, Alessandro de Lima, Clécio Galvão Martins, Cristiane Duarte Alexandrino,Pablito Augusto Travassos Ferreira, Ana Livya Moreira Rodrigues, Suliane Praciano Rodrigues, José Celso de Albuquerque Thomaz, Jurandy do Nascimento Silva, and Larissa Lages Rodrigues. 2014. Chemical composition, fatty acid profile and bioactive compounds of guava seeds (*Psidium guajava* L.). *Food Science and Technology* 34 (3):485–492.

Vilkhu, Kamaljit, Raymond Mawson, Lloyd Simons, and Darren Bates. 2008. Applications and opportunities for ultrasound assisted extraction in the food industry—A review. *Innovative Food Science & Emerging Technologies* 9 (2):161–169.

Villacís-Chiriboga, José, Kathy Elst, John Van Camp, Edwin Vera, and Jenny Ruales. 2020. Valorization of byproducts from tropical fruits: Extraction methodologies, applications, environmental, and economic assessment: A review (Part 1: General overview of the byproducts, traditional biorefinery practices, and possible applications). *Comprehensive Reviews in Food Science and Food Safety* 19 (2):405–447.

Wilberg, Viktor C, and Delia B Rodriguez-Amaya. 1995. HPLC quantitation of major carotenoids of fresh and processed guava, mango and papaya. *LWT-Food Science and Technology* 28 (5):474–480.

Yolmeh, Mahmoud, Mohammad B Habibi Najafi, and Reza Farhoosh. 2014. Optimisation of ultrasound-assisted extraction of natural pigment from annatto seeds by response surface methodology (RSM). *Food Chemistry* 155:319–324.

Yusof, Rizana, Siti Zawani Ahmad Zaini, and Mohd Azhar Azman. 2020. Characterization of pectin extracted from guava peels using deep eutectic solvent and citric acid. In *Charting the sustainable future of ASEAN in science and technology*. Edited by Robin Haring, Ilona Kickbusch, Detlev Ganten, and Matshidiso Moeti. Springer Nature, Singapore, pp. 421–433.

Zhang, Hua-Feng, Xiao-Hua Yang, Li-Dong Zhao, and Ying Wang. 2009. Ultrasonic-assisted extraction of epimedin C from fresh leaves of Epimedium and extraction mechanism. *Innovative Food Science & Emerging Technologies* 10 (1):54–60.

Zhang, W-L, L Zhu, and J-G Jiang. 2014. Active ingredients from natural botanicals in the treatment of obesity. *Obesity Reviews* 15 (12):957–967.

CHAPTER 8

Mangosteen Wastes
Chemistry, Processing, and Utilization

Rahul Islam Barbhuiya, Sushil Kumar Singh, and Poonam Singha

CONTENTS

8.1 INTRODUCTION

Mangosteen (*Garcinia mangostana* L.) is one of the most popular, exotic, tropical fruits and is known as the "Queen of Fruits" because of its unique tangy-sweet taste. It belongs to the family Clusiaceae and is found in many countries, such as Sri Lanka, Thailand, the Philippines, Malaysia, Myanmar, and India (Ji et al., 2007; Moopayak & Tangboriboon, 2020; Pedraza-Chaverri et al., 2008). A fully ripe mangosteen fruit is rounded, 5–7 cm in diameter, and weighs 55–90 g. The fruit is juicy and slightly fibrous with an inedible, deep purple-reddish-coloured rind. The pericarp (peel) is approximately 6–10 mm thick (Li et al., 2007). The fruit comprises about 65% pericarp, 31% fresh mangosteen and 4% cap (Chaovanalikit et al., 2012). The inner part of the fruit contains soft and juicy edible white pulp enclosed typically within 6–8 arils. The arils contain "seeds", which are technically not true seeds as they bear a non-sexual embryo (Palakawong & Delaquis, 2018).

The fruit's pulp is consumed fresh, while the peel and seeds are discarded as waste. It is reported that mangosteen peel holds a variety of bioactive compounds with potential health-beneficial effects, such as anthocyanins (Palapol et al., 2009), xanthones (Zarena & Sankar, 2009), tannins (Pothitirat et al., 2009), phenolic acids (Zadernowski et al., 2009) and others. These individual bioactive compounds possess numerous biological activities such as anti-inflammatory (Chen et al., 2008), anti-microbial (Suksamrarn et al., 2002) and antioxidant activities (Jung et al., 2006). Mangosteen seed is also composed of various compounds such as moreollin, oils, lipid, carbohydrate and gambolic acid. Due to such compounds, the seed has antibacterial, antioxidant and anticancer qualities, and could be used in health care, and as detergents, lubricants, soaps and for joint treatments (Moopayak & Tangboriboon, 2020; Pedraza-Chaverri et al., 2008). This chapter discusses different types of mangosteen wastes, their biochemistry, processing and possible utilization opportunities. The waste

DOI: 10.1201/9781003164463-8

Figure 8.1 Different parts of mangosteen: (a) fruit, (b) aril, (c) seeds, and (d) peel.

materials obtained during processing of mangosteen fruits, including peel and seeds, are shown in Figure 8.1.

8.2 COMPOSITIONAL ANALYSIS OF MANGOSTEEN WASTE

The mineral composition of mangosteen processing wastes has not been studied much. However, according to Parthsarathy and Nandakishore (2014), mangosteen peel has a total mineral concentration of 163.6 mg/100 g dry peel weight with 2.58 mg/100 g sodium, 78.3 mg/100 g potassium, 5.82 mg/100 g calcium, 60.43 mg/100 g magnesium, 9.02 mg/100 g iron and 7.45 mg/kg phosphorus. Similarly, Ajayi et al. (2007) reported the mineral composition of mangosteen seeds to be 26.0 mg/kg sodium, 707 mg/kg potassium, 454 mg/kg calcium, 865 mg/kg magnesium, 90.0 mg/kg iron, 19.0 mg/kg zinc, and 18.0 mg/kg manganese (all values are in mg/kg dry matter). Both mangosteen seed and peel are shown to contain considerable concentrations of minerals which can play important roles in fulfilling the needs of the human body. For instance, calcium is required for bone formation and muscle contraction, as well as blood clotting, while potassium is required for blood pressure regulation. Magnesium and calcium work together to keep bones healthy, while calcium is also essential for maintaining healthy heart operation.

Similarly, Joseph et al. (2017) investigated the chemical composition of African mangosteen (*Garcinia livingstonei* T. Anderson). The elemental composition of dry matter of different parts of the fruits, such as the seed, epicarp, mesocarp and endocarp, was reported. The composition of the epicarp was 11880 mg/kg nitrogen, 489.65 mg/kg phosphorus, 10830 mg/kg potassium, 21.64 mg/kg sulphur, 5600 mg/kg calcium, 2280 mg/kg magnesium, 21 mg/kg zinc, 246.8 mg/kg iron, 18.1 mg/kg manganese and 10.1 mg/kg copper. The composition of the mesocarp was 6800 mg/kg nitrogen, 450.65 mg/kg phosphorus, 5738 mg/kg potassium, 4.48 mg/kg sulphur, 7600 mg/kg calcium, 1680 mg/kg magnesium, 14.2 mg/kg zinc, 156.6 mg/kg iron, 246.8 mg/kg manganese and 354.6 mg/kg copper. Similarly, the composition of the endocarp was 12940 mg/kg nitrogen, 391.72 mg/kg phosphorus, 2753 mg/kg potassium, 12.47 mg/kg sulphur, 6200 mg/kg calcium, 1440 mg/kg magnesium, 30.4 mg/kg zinc, 354.6 mg/kg iron, 7.7 mg/kg manganese and 9.7 mg/kg copper. For seed, the elemental composition was 8400 mg/kg nitrogen, 577.79 mg/kg phosphorus, 10367 mg/kg potassium, 16.53 mg/kg sulphur, 4200 mg/kg calcium, 2900 mg/kg magnesium, 40.6 mg/kg zinc, 344 mg/kg iron, 29.5 mg/kg manganese and 9.8 mg/kg copper. The micro-elements manganese and copper were found in greatest abundance in the mesocarp. As a result, mesocarp could be a good source of micronutrients. Potassium, together with sodium, is usually a crucial mineral for maintaining appropriate water balance, acid-base balance, and osmotic balance. Hence, a more detailed analysis of elements present in mangosteen waste materials is needed.

Furthermore, the organic phytochemical composition of mangosteen-processing residues has shown the presence of polyphenolic compounds such as xanthones, flavonoids, tannins and other

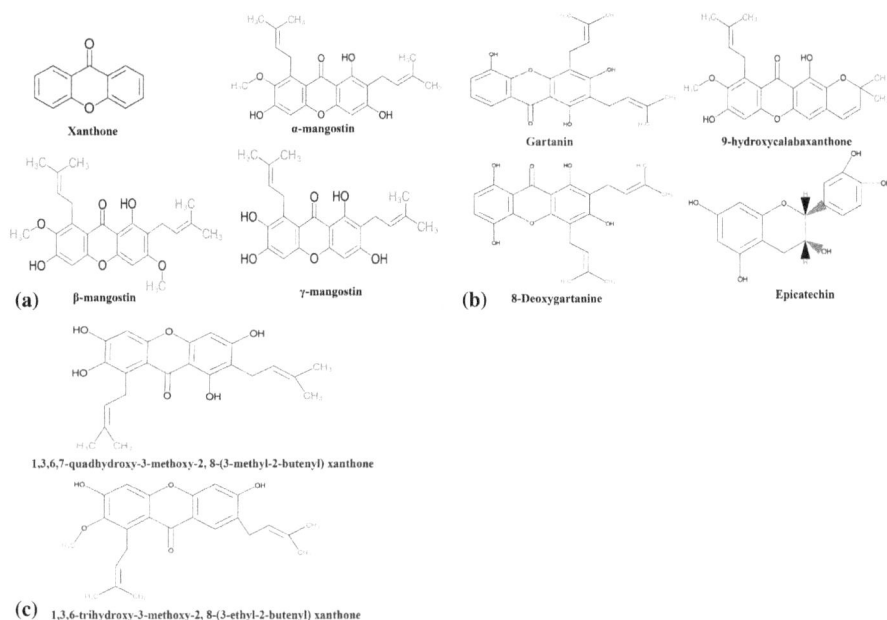

Figure 8.2 Some chemical compounds isolated from mangosteen fruit processing residues.

bioactive substances, which exhibit medicinal properties (Pothitirat et al., 2009). Figure 8.2 (a–c) shows some major bioactive compounds isolated from mangosteen waste materials. Several *in-vitro* studies have shown that the xanthones possess anti-carcinogenic, anti-inflammatory, apoptotic and antioxidant activities (Gutierrez-Orozco & Failla, 2013). To date, approximately 70 xanthones have been characterised from mangosteen extract (Aizat et al., 2019; Ovalle-Magallanes et al., 2017). The xanthone structure is mainly composed of three consecutive aromatic rings distinguished by side chains; modifying these side chains is known to influence xanthone bioactivities (Buravlev et al., 2018; Karunakaran et al., 2018). The most abundant xanthones found in mangosteen peel are α-mangostin, β-mangostin and γ-mangostin (Chen et al., 2008; Pedraza-Chaverrí et al., 2009; Zarena & Sankar, 2011). The other xanthones present in mangosteen wastes are epicatechin (Yu et al., 2007), 2,4,6,3′,5′-penta hydroxy benzophenone, maclurin-6-*O*-β-D-glucopyranoside, 2,4,3′ trihydroxy benzophenone-6-*O*-β-glucopyranoside, 1,2,4,5-tetrahydroxy benzene, isogarcinol, aromadendrin-8-*O*-β-D-glucopyranoside (Chen et al., 2017), garcimangosone D, mangostana xanthone IV, V and VI (Mohamed et al., 2017), 1,7-dihydroxy-3- methoxy 2-(3-methyl but-2-enyl) xanthone, 1,3,7-trihydroxy-2,8-di-(3-methyl but-2-enyl) xanthone, garcinone E, 1,3,6-trihydroxy-7-m ethoxy-2,8-(3-methyl-2-butenyl) xanthone, 1,3,6,7-tetrahydroxy-2,8-(3-methyl-2-butenyl), gartanin, 8-deoxygartanin and 9-hydroxycalaba xanthone (Wittenauer et al., 2012). Also, two highly oxygenated prenylated xanthones, namely mangostingone [7-methoxy-2-(3-methyl-2-butenyl)-8-(3-methyl -2-oxo-3-butenyl)-1,3,6 trihydroxyxanthone, 2] and 8-hydroxycudraxanthone G, along with garcimangosone B, 8-deoxy gartanin, and cudra xanthone G are also present in mangosteen peel (Jung et al., 2006).

Mangosteen peel is also rich in phenolic acids. For instance, the flavedo (outer part) and albedo (inner part) of mangosteen peel contains eight and six phenolic acids, respectively (Zadernowski et al., 2009). Among these, 3,4-dihydroxybenzoic acid (protocatechuic acid) is the major phenolic acid in both the outer and inner peels. The outer mangosteen peel also contains ferulic acid, caffeic acid, veratric acid, and *m*-hydroxybenzoic acid. The inner peel contains 3,4-dihydroxymandelic acid and *p*-coumaric acid (Zadernowski et al., 2009). Anthocyanins, proanthocyanidins, and (-)

epicatechin are also found in mangosteen peel. Anthocyanins are the primary-coloured compounds that indicate the maturation stages of mangosteen. Cyanidin-3-glucoside and cyanidin-3-sophoroside have been identified as the main anthocyanins in the mangosteen peel (Du & Francis, 1977). The concentration of such anthocyanins changes during the maturation of mangosteen. For example, the concentration of cyanidin-3-sophoroside and cyanidin-3-glucoside increases as the mangosteen matures. Cyanidin-3-sophoroside was found to be the most abundant anthocyanin (76.1%), followed by cyanidin-3-glucoside (13.4 %) and pelargonidin-3-glucoside (6.2 %) (Zarena & Sankar, 2012). Other anthocyanins present in mangosteen peel are cyanidin-X, cyanidin-X2, cyanidin-glucoside-X, cyanidin-glucoside-pentoside (where X denotes a residue of m/z 190, which is unified atomic mass units) (Palapol et al., 2009).

The occurrence of condensed tannins or proanthocyanidins is related to the high concentrations of anthocyanins in the peel (Chen et al., 2009). Similarly, procyanidin A-2 and procyanidin B-2 were shown to be active principal compounds in mangosteen peel (Yoshikawa et al., 1994). (−)Epicatechin is one of the flavan-3-ols (Shahidi & Naczk, 1995), which is a principal monomeric unit of mangosteen proanthocyanidins; therefore, they are often found together in the mangosteen peel (Yoshikawa et al., 1994; Yu et al., 2007). Bioactive compounds, such as anthocyanins and xanthones in the fruit peel, are responsible for the fruit's high antioxidant and anti-inflammatory properties. The content and type of bioactives present in mangosteen peel depends on several factors: extraction solvents, cultivation area conditions and maturity stage. For example, the ethanolic (95%) extracts of young mangosteen peel contain α-mangostin at a concentration of $8.07 \pm 0.11\%$ w/w of extract, while the mature mangosteen peel contains $13.63 \pm 0.06\%$ w/w of extract (Pothitirat et al., 2009). Ripe mangosteen peel also has higher tannin, flavonoid and phenolic concentrations than the young ones. Mangosteen peel collected from four different locations in Indonesia (Tasikmalaya, Subang, Purwakarta and Bogor) showed different levels of α-mangostin, γ-mangostin, and gartanin in ethanolic (70%) extracts of mangosteen peel. The peel from Bogor was shown to contain 13.87% α-mangostin, 8.28% γ-mangostin and 10.44% gartanin. The peel from Purwakarta contained 10.0 % of α-mangostin, 6.33% of γ-mangostin and 8.76% of gartanin whereas the peel from Subang contained 10.88% α-mangostin, 6.01% γ-mangostin and 8.08% gartanin, and the peel from Tasikmalaya contained 8.53% α-mangostin, 6.07% γ-mangostin and 17.28% gartanin (Muchtaridi et al., 2017). As a result of the presence of such bioactive compounds, the fruit has good medicinal properties. Numerous reviews have comprehensively addressed the medicinal properties of the fruit (Aizat et al., 2019; Ming-Hui et al., 2017; Ovalle-Magallanes et al., 2017), such as antidiabetic, antiperiodontitic (Milovanova-Palmer & Pendry, 2018), antioxidant (Ibrahim et al., 2017), antibiofilm (Agarwal & Gayathri, 2017) and anticarcinogenic (Iqbal et al., 2018) properties. Properties such as regulation of melanogenesis (Pillaiyar et al., 2017) and amelioration of metabolic disorders (Tousian Shandiz et al., 2017) have also been reported.

Additionally, during the development and germination phases of mangosteen seeds, the concentration of flavonoids and xanthones in the seeds increases, possibly as a defensive attempt to safeguard seed viability (Mazlan et al., 2018; 2019). During early development, the mangosteen seed has no embryo. The endosperm of a growing ovule is liquid, and the seed is formed by the expansion of integument cells (Yapwattanaphun et al., 2011). The outer integuments, embedded with tannins, form the seed coat (or testa), which can be up to 0.25 mm thick (Noor et al., 2016). Only about two well-developed seeds may be found in a fully ripened mangosteen fruit (Osman & Milan, 2006). When compared to other *Garcinia* species, mangosteen seeds have a significantly higher proportion of starch (Noor et al., 2016). However, studies on mangosteen seeds are still limited and more research is needed.

Mangosteens are beneficial for our health, but most nutrients are present in the mangosteen peel and very limited information is available on the composition of mangosteen seeds. Generally, these fruit-processing residues are thrown away as waste; however, for thousands of years, such products have been used in folk medicine for their health-beneficial compounds (Hiranrangsee et al., 2016). For example, the peel contains resin, which helps to stop wound infection, dysentery and diarrhoea and is widely used in Southeast Asia (Wang et al., 2012). Apart from these benefits, several other therapeutic benefits are connected with mangosteen wastes such as antiglaucoma, anti-Alzheimer, antidepressant, cytotoxic, antiviral, antifungal, antibacterial, anti-allergenic, antioxidant, anti-carcinogenic, anti-inflammatory and cardioprotective activities (Alsultan et al., 2016; Gutierrez-Orozco et al., 2013; Johnson et al., 2012). Researchers with a vision of developing efficient and economically feasible ways to use mangosteen wastes have discovered different methods by which to recover value-added substances. These alternative ways of mangosteen waste material utilization may also reduce the adverse environmental impact associated with mangosteen waste disposal.

8.3 ISOLATION OF BIOACTIVE COMPOUNDS FROM MANGOSTEEN WASTE

Studies have shown several ways to identify and isolate the main bioactive compounds of mangosteen-processing residues. As discussed earlier, mangosteen-processing residues contain a number of polyphenolic compounds such as xanthones, flavonoids and other bioactive substances. These compounds are often more soluble in organic solvents than in water and require non-polar solvents for extraction and solubilisation. However, the solvent extraction method is the most common method used to isolate bioactive compounds from the mangosteen peel, and solvents like aqua dest, acetic acid, ethyl acetate, hexane, acetone, ethanol or methanol are used commonly at different times (24, 36, or 48 hours) for extraction (Rohman et al., 2019).

According to the previous study, for extraction of xanthones, acetone was considered to be the best solvent used with an extraction time of 36 h (Kusmayadi et al., 2018). Similarly, ethyl acetate was the best solvent capable of extracting the highest concentration of α-mangostin, followed by dichloromethane, ethanol and water, from mangosteen peel (Ghasemzadeh et al., 2018). Likewise, the ultrasound-assisted extraction technique and the microwave-assisted method have been optimised to extract anthocyanins and flavonoids, respectively, from mangosteen (Hasan et al., 2016; Hiranrangsee et al., 2016). Several xanthones, such as 9-hydroxycalaba xanthone, 8-deoxy gartanin, mangostanol, garcinone E, gartanin, 3-isomangostin, α-mangostin and β-mangostin, were extracted from mangosteen peel using the liquefied dimethyl ether extraction method (Nerome et al., 2016).

However, most solvents (non-polar or semi-polar) are hazardous for topical applications and human consumption (Bundeesomchok et al., 2016). Therefore, the water-based extraction method is considered to be desirable for wider biocompatibility. One exclusive property of water is that it can boil past its boiling temperature but retain its liquid form under a high-pressure condition, a state termed "subcritical water". This process permits non-polar compounds to be dissolved in water while performing an extraction process (Machmudah et al., 2018). Another, quicker alternative green approach to isolating valuable xanthones from mangosteen peel is using a mild thermo-induced aqueous, micellar, biphasic system (Ng et al., 2018; Tan et al., 2017). This method allows for effective xanthone extraction without employing sophisticated instrumentation and with fewer chemicals (Ng et al., 2018; Tan et al., 2017). Supercritical carbon dioxide (SC-CO$_2$) combined with hydrothermal extraction can effectively extract phenolic compounds from mangosteen peel (Chhouk et al., 2016).

Furthermore, several methods such as coupling centrifugal partition chromatography with electrospray ionization mass spectrometry (CPC-ESI-MS) (Destandau et al., 2009), high-performance liquid chromatography with electrospray ionization mass spectrometry (HPLC-ESI-MS) (Zarena & Sankar, 2011, 2012) and gas chromatography with mass spectrometry (GC-MS) (Zadernowski et al., 2009) have been used to detect and quantify these bioactive compounds.

8.4 UTILIZATION OF MANGOSTEEN WASTE

Mangosteen-processing residues are utilised to treat various diseases, including arthritis, hypertension, bacterial infections, diabetes and tumours (Aizat et al., 2019; Tousian Shandiz et al., 2017). These applications highlight the value of the extracts from these processing residues in pharmaceutical and medicinal contexts. These days, the utilisation of mangosteen wastes, particularly the seed and peel, has increased exponentially. For example, mangosteen peel was used as alternative precursor for the production of a carbon-based adsorbent for ethylene removal (Mukti et al., 2018). Similarly, sulphuric acid-modified mangosteen shell (or peel) was used as an alternative for high-cost adsorbents to remove Ni (II) ions from aqueous solutions (Anitha et al., 2020). A large number of h products containing mangosteen peel extracts are available in herbal markets, such as Mangosteen powder (R) (Thailand), Mangosteen pericarp Acne Cream(R), Mangosteen Xango (Malaysia), Mastin (R) and SidoMuncul SARI KULIT MANGGIS(R) (Indonesia) (Rohman et al., 2019).

Pharmaceutical companies have developed some modern pharmaceutical formulations containing mangosteen peel extract, which include the encapsulation of mangosteen peel extract bioactives using nano-emulsion technique and intended as a topical formulation (Mulia et al., 2018). Encapsulation technique is a promising strategy by which to utilise the active components present within the fruit waste, particularly α-mangostin. For example, a topical formulation of nano-emulgel mangosteen extract in virgin coconut oil exhibits better penetration ability than its nanoemulsion (Mulia et al., 2018). Nano-emulgel is a homogeneous milky white gel formed by mixing the nanoemulsion with an aqueous solution of xanthan gum and phenoxyethanol, the phenoxyethanol being added as a preservative. Similarly, foot gangrene can be treated topically with a self-nano-emulsifying drug delivery system with enhanced solubilisation of an ethanol extract from mangosteen peel (Rohman et al., 2019).

Conventionally, mangosteen waste (seed and peel), in the form of decoctions and infusions, can be used to treat gastrointestinal infections, urinary tract infections and skin infections. They can also act as an anti-fever, anti-scorbutic and laxative agent (Ovalle-Magallanes et al., 2017). Also, as discussed earlier, mangosteen-processing wastes can be used to treat chronic ulcers, wound infection, suppuration, dysentery, diarrhoea, abdominal pain (Cui et al., 2010; Gorinstein et al., 2011), arthritis, food allergies and acne (Ming-Hui et al., 2017). These activities indeed need to be correlated with the chemical composition contained in mangosteen seeds and peel (Genovese et al., 2016). Table 8.1 presents some health benefits of different extracts/compounds from mangosteen wastes.

Table 8.1   Health Benefits of Different Extracts/Compounds from Mangosteen By-Products

By-Products	Extract/ Compound Name	Health Benefits	References
Mangosteen pericarp	Garcinone E	Garcinone E exerts anticancer activities by inducing apoptosis and suppressing migration and invasion in ovarian cancer cells, indicating its therapeutic potential for ovarian cancer.	Xu et al. (2017)
Mangosteen pericarp	Garcixanthones B and C	Garcixanthones B and C showed significant cytotoxic potential activity against MCF7 (human breast adenocarcinoma) and A549 (lung carcinoma) cell lines.	Ibrahim et al. (2018)
Mangosteen bark	Garmoxanthone	Garmoxanthone isolated from bark of *Garcinia mangostana* showed antibacterial activity against MRSA ATCC 43300 and MRSA CGMCC 1.12409.	Wang et al. (2018)
Mangosteen seedcases	β-mangostin	β-mangostin from seedcases of *Garcinia mangostana* inhibited α-melanocyte-stimulating hormone (αMSH)-mediated melanogenesis in B16F10 melanoma cells and a three-dimensional human skin model.	Lee et al. (2017)
Mangosteen leaves	Ethyl acetate extract of leaf (lower activity in hexane and methanol extract)	Highest antimicrobial activity was recorded in ethyl acetate extract among hexane, methanol and ethyl acetate which was subjected to GC-MS analysis revealing the presence of squalene (17.09%).	Lalithaet al. (2017)

In addition to these cases, many studies have explored the usage of mangosteen wastes in animal feed supplementation, food and functional food products and the determination of food shelf-life. For instance, the addition of mangosteen rind juice as a natural colourant into a sugar palm fruit jam named "Kolang-Kaling" improved its flavour, texture and red colour as selected by trained panellists (Sayuti et al., 2017). Similarly, the powder of mangosteen peel has been used as feed supplements for dairy steers and lactating cows without any adverse impact on the livestock's diets (Foiklang et al., 2016; Polyorach et al., 2016). Such supplementation improved various aspects of the steers' digestion, rumen fermentation, and microbiome composition (Foiklang et al., 2016; Polyorach et al., 2016). Furthermore, broiler chicken feed fortified with mangosteen peel was shown to increase the chickens' weight during heat stress (Hidanah et al., 2017). Such observations may be attributed to mangosteen peel's bioactive components, such as xanthones, that may improve chicken tolerance to stress. Bioactive compounds from mangosteen waste have been used in prolonging and detecting food shelf-life. For instance, a biofilm coated with anthocyanin extract from mangosteen waste could detect chicken nuggets' spoilage *via* colour indication (Ismed et al., 2016).

8.5 CONCLUSIONS

Food industries generate a vast number of wastes and by-products annually from a variety of sources. Such wastes can be utilised and valorised by extracting bioactive compounds, which can be reused for various food and pharmaceutical applications. Mangosteen wastes (peel and seeds) can be used as a food/feed component or a functional food having health benefits. The underutilised parts of mangosteens contain many phenolic compounds, such as xanthones, gartanin and mangostin, which are believed to be responsible for their antioxidant activities. Mangosteen-processing residues are also utilised to treat various diseases, including arthritis, hypertension, bacterial infections, diabetes and tumours. This comprehensive utilisation of mangosteen wastes, ranging from biomedical and technological applications to advanced materials, deserves researchers' utmost attention to promote the fruit and its cultivation further. In the future, mangosteen-derived products are envisioned to benefit various communities, including farmers, local growers, consumers and the biomaterial and biomedical industries. Further study is required so that, in the future, mangosteen-processing wastes can become sources of new lead compounds for pharmaceutical drugs that have fewer side effect.

REFERENCES

Agarwal, H., & Gayathri, M. (2017). Biological synthesis of nanoparticles from medicinal plants and its uses in inhibiting biofilm formation. *Asian Journal of Pharmaceutical and Clinical Research, 10*(5), 64–68.

Aizat, W. M., Jamil, I. N., Ahmad-Hashim, F. H., & Noor, N. M. (2019). Recent updates on metabolite composition and medicinal benefits of mangosteen plant. *PeerJ, 7*, e6324.

Ajayi, I., Oderinde, R., Ogunkoya, B., Egunyomi, A., & Taiwo, V. (2007). Chemical analysis and preliminary toxicological evaluation of *Garcinia mangostana* seeds and seed oil. *Food Chemistry, 101*(3), 999–1004.

Alsultan, Q. M. N., Sijam, K., Rashid, T. S., & Ahmad, K. B. (2016). GC-MS analysis and antibacterial activity of mangosteen leaf extracts against plant pathogenic bacteria. *American Journal of Plant Sciences, 7*(7), 1013–1020.

Anitha, D., Ramadevi, A., & Seetharaman, R. (2020). Biosorptive removal of nickel (II) from aqueous solution by Mangosteen shell activated carbon. *Materials Today: Proceedings*, 718–722.

Bundeesomchok, K., Filly, A., Rakotomanomana, N., Panichayupakaranant, P., & Chemat, F. (2016). Extraction of α-mangostin from *Garcinia mangostana* L. using alternative solvents: Computational predictive and experimental studies. *LWT-Food Science and Technology, 65*, 297–303.

Buravlev, E. V., Shevchenko, O. G., Anisimov, A. A., & Suponitsky, K. Y. (2018). Novel Mannich bases of α- and γ-mangostins: Synthesis and evaluation of antioxidant and membrane-protective activity. *European Journal of Medicinal Chemistry, 152*, 10–20.

Chaovanalikit, A., Mingmuang, A., Kitbunluewit, T., Choldumrongkool, N., Sondee, J., & Chupratum, S. (2012). Anthocyanin and total phenolics content of mangosteen and effect of processing on the quality of mangosteen products. *International Food Research Journal, 19*(3), 1047.

Chen, L.-G., Yang, L.-L., & Wang, C.-C. (2008). Anti-inflammatory activity of mangostins from *Garcinia mangostana*. *Food and Chemical Toxicology, 46*(2), 688–693.

Chen, S., Han, K., Li, H., Cen, J., Yang, Y., Wu, H., & Wei, Q. (2017). Isogarcinol extracted from *Garcinia mangostana* L. ameliorates imiquimod-induced psoriasis-like skin lesions in mice. *Journal of Agricultural and Food Chemistry, 65*(4), 846–857.

Chen, W., Fu, C., Qin, Y., & Huang, D. (2009). One-pot depolymerizative extraction of proanthocyanidins from mangosteen pericarps. *Food Chemistry, 114*(3), 874–880.

Chhouk, K., Quitain, A. T., Pag-asa, D. G., Maridable, J. B., Sasaki, M., Shimoyama, Y., & Goto, M. (2016). Supercritical carbon dioxide-mediated hydrothermal extraction of bioactive compounds from *Garcinia Mangostana* pericarp. *The Journal of Supercritical Fluids, 110*, 167–175.

Cui, J., Hu, W., Cai, Z., Liu, Y., Li, S., Tao, W., & Xiang, H. (2010). New medicinal properties of mangostins: Analgesic activity and pharmacological characterization of active ingredients from the fruit hull of *Garcinia mangostana* L. *Pharmacology Biochemistry and Behavior, 95*(2), 166–172.

Destandau, E., Toribio, A., Lafosse, M., Pecher, V., Lamy, C., & André, P. (2009). Centrifugal partition chromatography directly interfaced with mass spectrometry for the fast screening and fractionation of major xanthones in Garcina mangostana. *Journal of Chromatography A, 1216*(9), 1390–1394.

Du, C., & Francis, F. (1977). Anthocyanins of mangosteen, *Garcinia mangostana*. *Journal of Food Science, 42*(6), 1667–1668.

Foiklang, S., Wanapat, M., & Norrapoke, T. (2016). Effect of grape pomace powder, mangosteen peel powder and monensin on nutrient digestibility, rumen fermentation, nitrogen balance and microbial protein synthesis in dairy steers. *Asian-Australasian Journal of Animal Sciences, 29*(10), 1416.

Genovese, S., Fiorito, S., Taddeo, V. A., & Epifano, F. (2016). Recent developments in the pharmacology of prenylated xanthones. *Drug Discovery Today, 21*(11), 1814–1819.

Ghasemzadeh, A., Jaafar, H. Z., Baghdadi, A., & Tayebi-Meigooni, A. (2018). Alpha-mangostin-rich extracts from mangosteen pericarp: Optimization of green extraction protocol and evaluation of biological activity. *Molecules, 23*(8), 1852.

Gorinstein, S., Poovarodom, S., Leontowicz, H., Leontowicz, M., Namiesnik, J., Vearasilp, S., Haruenkit, R., Ruamsuke, P., Katrich, E., & Tashma, Z. (2011). Antioxidant properties and bioactive constituents of some rare exotic Thai fruits and comparison with conventional fruits: In vitro and in vivo studies. *Food Research International, 44*(7), 2222–2232.

Gutierrez-Orozco, F., Chitchumroonchokchai, C., Lesinski, G. B., Suksamrarn, S., & Failla, M. L. (2013). α-Mangostin: Anti-inflammatory activity and metabolism by human cells. *Journal of Agricultural and Food Chemistry, 61*(16), 3891–3900.

Gutierrez-Orozco, F., & Failla, M. L. (2013). Biological activities and bioavailability of mangosteen xanthones: A critical review of the current evidence. *Nutrients, 5*(8), 3163–3183.

Hasan, A., Nashrianto, H., Juhaeni, R., & Artika, I. (2016). Optimization of conditions for flavonoids extraction from Mangosteen (*Garcinia mangostana* L.). *Der Pharmacia Lettre, 8*(18), 114–120.

Hidanah, S., Warsito, S. H., Nurhajati, T., Lokapirnasari, W. P., & Malik, A. (2017). Effect of mangosteen peel (*Garcinia mangostana*) and ginger rhizome (*Curcuma xanthorrhiza*) on the performance and cholesterol levels of heat-stressed broiler chickens. *Pakistan Journal of Nutrition, 16*(1), 28–32.

Hiranrangsee, L., Kumaree, K. K., Sadiq, M. B., & Anal, A. K. (2016). Extraction of anthocyanins from pericarp and lipids from seeds of mangosteen (*Garcinia mangostana* L.) by Ultrasound-assisted extraction (UAE) and evaluation of pericarp extract enriched functional ice-cream. *Journal of Food Science and Technology, 53*(10), 3806–3813.

Ibrahim, S. R., Abdallah, H. M., El-Halawany, A. M., Radwan, M. F., Shehata, I. A., Al-Harshany, E. M., Zayed, M. F., & Mohamed, G. A. (2018). Garcixanthones B and C, new xanthones from the pericarps of *Garcinia mangostana* and their cytotoxic activity. *Phytochemistry Letters, 25*, 12–16.

Ibrahim, U., Kamarrudin, N., Suzihaque, M., & Abd Hashib, S. (2017). Local fruit wastes as a potential source of natural antioxidant: An overview. *IOP Conference Series: Materials Science and Engineering*, 012040.

Iqbal, J., Abbasi, B. A., Batool, R., Mahmood, T., Ali, B., Khalil, A. T., Kanwal, S., Shah, S. A., & Ahmad, R. (2018). Potential phytocompounds for developing breast cancer therapeutics: Nature's healing touch. *European Journal of Pharmacology, 827*, 125–148.

Ismed, Sylvi, D., Rahmi, I. D., & Wilianda, C. (2016). Effects of temperature and storage time on film with mangosteen (*Garcinia mangostana* L.) peel extract as smart packaging in detecting spoilage on chicken nugget. *Research Journal of Pharmaceutical Biological and Chemical Sciences*, *7*(5), 1470–1478.

Ji, X., Avula, B., & Khan, I. A. (2007). Quantitative and qualitative determination of six xanthones in *Garcinia mangostana* L. by LC–PDA and LC–ESI-MS. *Journal of Pharmaceutical and Biomedical Analysis*, *43*(4), 1270–1276.

Johnson, J. J., Petiwala, S. M., Syed, D. N., Rasmussen, J. T., Adhami, V. M., Siddiqui, I. A., Kohl, A. M., & Mukhtar, H. (2012). α-Mangostin, a xanthone from mangosteen fruit, promotes cell cycle arrest in prostate cancer and decreases xenograft tumor growth. *Carcinogenesis*, *33*(2), 413–419.

Joseph, K. S., Bolla, S., Joshi, K., Bhat, M., Naik, K., Patil, S., Bendre, S., Gangappa, B., Haibatti, V., & Payamalle, S. (2017). Determination of chemical composition and nutritive value with fatty acid compositions of African mangosteen (*Garcinia livingstonei*). *Erwerbs-Obstbau*, *59*(3), 195–202.

Jung, H.-A., Su, B.-N., Keller, W. J., Mehta, R. G., & Kinghorn, A. D. (2006). Antioxidant xanthones from the pericarp of *Garcinia mangostana* (Mangosteen). *Journal of Agricultural and Food Chemistry*, *54*(6), 2077–2082.

Karunakaran, T., Ee, G. C. L., Ismail, I. S., Mohd Nor, S. M., & Zamakshshari, N. H. (2018). Acetyl-and O-alkyl-derivatives of β-mangostin from *Garcinia mangostana* and their anti-inflammatory activities. *Natural Product Research*, *32*(12), 1390–1394.

Kusmayadi, A., Adriani, L., Abun, A., Muchtaridi, M., & Tanuwiria, U. H. (2018). The effect of solvents and extraction time on total xanthone and antioxidant yields of mangosteen peel (*Garcinia mangostana* L.) extract. *Drug Invent. Today*, *10*(12), 2572–2576.

Lalitha, J. L., Clarance, P. P., Sales, J. T., & Archana, M. A. (2017). Biological activities of *Garcinia mangostana. Asian Journal of Pharmaceutical and Clinical Research*, *10*(9), 272–278. https://doi.org/10.22159/ajpcr.2017.v10i9.18585.

Lee, K. W., Ryu, H. W., Oh, S., Park, S., Madhi, H., Yoo, J., Park, K. H., & Kim, K. D. (2017). Depigmentation of α -melanocyte-stimulating hormone-treated melanoma cells by β -mangostin is mediated by selective autophagy. *Experimental Dermatology*, *26*(7), 585–591.

Li, S., Lambros, T., Wang, Z., Goodnow, R., & Ho, C.-T. (2007). Efficient and scalable method in isolation of polymethoxyflavones from orange peel extract by supercritical fluid chromatography. *Journal of Chromatography B*, *846*(1–2), 291–297.

Machmudah, S., Lestari, S. D., Kanda, H., Winardi, S., & Goto, M. (2018). Subcritical water extraction enhancement by adding deep eutectic solvent for extracting xanthone from mangosteen pericarps. *The Journal of Supercritical Fluids*, *133*, 615–624.

Mazlan, O., Aizat, W. M., Baharum, S. N., Azizan, K. A., & Noor, N. M. (2018). Metabolomics analysis of developing *Garcinia mangostana* seed reveals modulated levels of sugars, organic acids and phenylpropanoid compounds. *Scientia Horticulturae*, *233*, 323–330.

Mazlan, O., Aizat, W. M., Zuddin, N. S. A., Baharum, S. N., & Noor, N. M. (2019). Metabolite profiling of mangosteen seed germination highlights metabolic changes related to carbon utilization and seed protection. *Scientia Horticulturae*, *243*, 226–234.

Milovanova-Palmer, J., & Pendry, B. (2018). Is there a role for herbal medicine in the treatment and management of periodontal disease? *Journal of Herbal Medicine*, *12*, 33–48.

Ming-Hui, W., Zhang, K.-J., Qin-Lan, G., Xiao-Ling, B., & Jin-Xin, W. (2017). Pharmacology of mangostins and their derivatives: A comprehensive review. *Chinese Journal of Natural Medicines*, *15*(2), 81–93.

Mohamed, G. A., Al-Abd, A. M., El-Halawany, A. M., Abdallah, H. M., & Ibrahim, S. R. (2017). New xanthones and cytotoxic constituents from *Garcinia mangostana* fruit hulls against human hepatocellular, breast, and colorectal cancer cell lines. *Journal of Ethnopharmacology*, *198*, 302–312.

Moopayak, W., & Tangboriboon, N. (2020). Mangosteen peel and seed as antimicrobial and drug delivery in rubber products. *Journal of Applied Polymer Science*, *137*(37), 49119.

Muchtaridi, M., Puteri, N. A., Milanda, T., & Musfiroh, I. (2017). Validation analysis methods of α-mangostin, ϒ-mangostin and gartanin mixture in mangosteen (*Garcinia mangostana* L.) fruit rind extract from west java with HPLC. *Journal of Applied Pharmaceutical Science*, *7*(10), 125–130.

Mukti, N. I. F., Prasetyo, I., & Mindaryani, A. (2018). Preparation of porous carbon as ethylene adsorbent by pyrolysis of extraction waste Mangosteen rinds. *MATEC Web of Conferences*, *154*, 01032.

Mulia, K., Putri, G. A., & Krisanti, E. (2018). Encapsulation of mangosteen extract in virgin coconut oil based nanoemulsions: Preparation and characterization for topical formulation. In A. Yatim (Ed.), 15th International Conference on Quality in Research, QiR 2017 (pp. 243–242). Nusa Dua, Bali, Indonesia: Materials Science Forum.

Mulia, K., Ramadhan, R. M., & Krisanti, E. A. (2018). Formulation and characterization of nanoemulgel man-gosteen extract in virgin coconut oil for topical formulation. *MATEC Web of Conferences, 156*, 01013.

Nerome, H., Hoshino, R., Ito, S., Esaki, R., Eto, Y., Wakiyama, S., Sharmin, T., Goto, M., Kanda, H., & Mishima, K. (2016). Functional ingredients extraction from *Garcinia mangostana* pericarp by liquefied dimethyl ether. *Engineering Journal, 20*(4), 155–162.

Ng, H.-S., Tan, G. Y. T., Lee, K.-H., Zimmermann, W., Yim, H. S., & Lan, J. C.-W. (2018). Direct recovery of mangostins from *Garcinia mangostana* pericarps using cellulase-assisted aqueous micellar biphasic system with recyclable surfactant. *Journal of Bioscience and Bioengineering, 126*(4), 507–513.

Noor, N. M., Aizat, W. M., Hussin, K., & Rohani, E. R. (2016). Seed characteristics and germination proper-ties of four Garcinia (Clusiaceae) fruit species. *Fruits, 71*(4), 199–207.

Osman, M. B., & Milan, A. R. (2006). *Mangosteen: Garcinia mangostana L.* University of Southampton, International Centre for Underutilised Crops.

Ovalle-Magallanes, B., Eugenio-Pérez, D., & Pedraza-Chaverri, J. (2017). Medicinal properties of mangosteen (*Garcinia mangostana* L.): A comprehensive update. *Food and Chemical Toxicology, 109*, 102–122.

Palakawong, C., & Delaquis, P. (2018). Mangosteen processing: A review. *Journal of Food Processing and Preservation, 42*(10), e13744.

Palapol, Y., Ketsa, S., Stevenson, D., Cooney, J., Allan, A., & Ferguson, I. (2009). Colour development and quality of mangosteen (*Garcinia mangostana* L.) fruit during ripening and after harvest. *Postharvest Biology and Technology, 51*(3), 349–353.

Parthsarathy, U., & Nandakishore, O. P. (2014). A study on nutrient and medicinal compositions of selected Indian Garcinia species. *Current Bioactive Compounds, 10*(1), 55–61.

Pedraza-Chaverri, J., Cárdenas-Rodríguez, N., Orozco-Ibarra, M., & Pérez-Rojas, J. M. (2008). Medicinal properties of mangosteen (*Garcinia mangostana*). *Food and Chemical Toxicology, 46*(10), 3227–3239.

Pedraza-Chaverrí, J., Reyes-Fermín, L. M., Nolasco-Amaya, E. G., Orozco-Ibarra, M., Medina-Campos, O. N., González-Cuahutencos, O., Rivero-Cruz, I., & Mata, R. (2009). ROS scavenging capacity and neuroprotective effect of α-mangostin against 3-nitropropionic acid in cerebellar granule neurons. *Experimental and Toxicologic Pathology, 61*(5), 491–501.

Pillaiyar, T., Manickam, M., & Jung, S.-H. (2017). Recent development of signaling pathways inhibitors of melanogenesis. *Cellular Signalling, 40*, 99–115.

Polyorach, S., Wanapat, M., Cherdthong, A., & Kang, S. (2016). Rumen microorganisms, methane produc-tion, and microbial protein synthesis affected by mangosteen peel powder supplement in lactating dairy cows. *Tropical Animal Health and Production, 48*(3), 593–601.

Pothitirat, W., Chomnawang, M. T., Supabphol, R., & Gritsanapan, W. (2009). Comparison of bioactive com-pounds content, free radical scavenging and anti-acne inducing bacteria activities of extracts from the mangosteen fruit rind at two stages of maturity. *Fitoterapia, 80*(7), 442–447.

Rohman, A., Rafi, M., Alam, G., Muchtaridi, M., & Windarsih, A. (2019). Chemical composition and anti-oxidant studies of underutilized part of mangosteen (*Garcinia mangostana* L.) fruit. *Journal of Applied Pharmaceutical Science, 9*(08), 047–052.

Sayuti, K., Yenrina, R., & Anggraini, T. (2017). Characteristics of "Kolang-kaling" (Sugar palm fruit jam) with added natural colorants. *Pakistan Journal of Nutrition, 16*, 69–76.

Shahidi, F., & Naczk, M. (1995). *Food phenolics*. Technomic Pub. Co.

Suksamrarn, S., Suwannapoch, N., Ratananukul, P., Aroonlerk, N., & Suksamrarn, A. (2002). Xanthones from the green fruit hulls of *Garcinia mangostana*. *Journal of Natural Products, 65*(5), 761–763.

Tan, G. Y. T., Zimmermann, W., Lee, K.-H., Lan, J. C.-W., Yim, H. S., & Ng, H. S. (2017). Recovery of mangostins from *Garcinia mangostana* peels with an aqueous micellar biphasic system. *Food and Bioproducts Processing, 102*, 233–240.

Tousian Shandiz, H., Razavi, B. M., & Hosseinzadeh, H. (2017). Review of *Garcinia mangostana* and its xanthones in metabolic syndrome and related complications. *Phytotherapy Research, 31*(8), 1173–1182.

Wang, J. J., Shi, Q. H., Zhang, W., & Sanderson, B. J. (2012). Anti-skin cancer properties of phenolic-rich extract from the pericarp of mangosteen (*Garcinia mangostana* Linn.). *Food and Chemical Toxicology, 50*(9), 3004–3013.

Wang, W., Liao, Y., Huang, X., Tang, C., & Cai, P. (2018). A novel xanthone dimer derivative with antibacterial activity isolated from the bark of *Garcinia mangostana*. *Natural Product Research, 32*(15), 1769–1774.

Wittenauer, J., Falk, S., Schweiggert-Weisz, U., & Carle, R. (2012). Characterisation and quantification of xanthones from the aril and pericarp of mangosteens (*Garcinia mangostana* L.) and a mangosteen con-taining functional beverage by HPLC–DAD–MSn. *Food Chemistry, 134*(1), 445–452.

Xu, X.-H., Liu, Q.-Y., Li, T., Liu, J.-L., Chen, X., Huang, L., Qiang, W.-A., Chen, X., Wang, Y., & Lin, L.-G. (2017). Garcinone E induces apoptosis and inhibits migration and invasion in ovarian cancer cells. *Scientific Reports*, *7*(1), 1–13.

Yapwattanaphun, C., Kobayashi, S., Yonemori, K., & Ueda, J. (2011). Hormone analysis in the locule of mangosteen fruit during apomictic seed development. In *International Symposium on Tropical and Subtropical Fruits* 1024.

Yoshikawa, M., Harada, E., Miki, A., Tsukamoto, K., Liang, S., Yamahara, J., & Murakami, N. (1994). Antioxidant constituents from the fruit hulls of mangosteen (*Garcinia mangostana* L.) originating in Vietnam. *Yakugaku Zasshi= Journal of the Pharmaceutical Society of Japan*, *114*(2), 129–133.

Yu, L., Zhao, M., Yang, B., Zhao, Q., & Jiang, Y. (2007). Phenolics from hull of *Garcinia mangostana* fruit and their antioxidant activities. *Food Chemistry*, *104*(1), 176–181.

Zadernowski, R., Czaplicki, S., & Naczk, M. (2009). Phenolic acid profiles of mangosteen fruits (*Garcinia mangostana*). *Food Chemistry*, *112*(3), 685–689.

Zarena, A., & Sankar, K. U. (2009). Screening of xanthone from mangosteen (*Garcinia mangostana* L.) peels and their effect on cytochrome c reductase and phosphomolybdenum activity. *Journal of Natural Products (India)*, *2*, 23–30.

Zarena, A., & Sankar, K. U. (2011). Xanthones enriched extracts from mangosteen pericarp obtained by supercritical carbon dioxide process. *Separation and Purification Technology*, *80*(1), 172–178.

Zarena, A., & Sankar, K. U. (2012). Isolation and identification of pelargonidin 3-glucoside in mangosteen pericarp. *Food Chemistry*, *130*(3), 665–670.

CHAPTER 9

Jackfruit Wastes and By-Products
Chemistry, Processing, and Utilization

Yogesh Gat, Renu Sharma, and Shafiya Rafiq

CONTENTS

9.1 INTRODUCTION

Jackfruit (*Artocarpus heterophyllus* Lam.) is a multi-purpose tree which provides food, fodder, fuel, timber, industrial and medicinal products. It is believed to have originated in the southwestern rainforests of India. It is native to Malaysia, Western Ghats of India, Brazil and southeastern Asia. Bangladesh and Indonesia has recognized jackfruit as their national fruit. Other countries, like

DOI: 10.1201/9781003164463-9

Malaysia, the Philippines, Vietnam, South China and Thailand, also cultivate jackfruit on a large scale. It is an evergreen tree when grown properly under a warm and humid climate. In the plains, jackfruit requires warmer and more arid conditions which are generally available in the southern part of India. The jackfruit is the largest tree-borne fruit in the world. Economically, both the mature and the immature fruits are used and are of great importance. The raw, unripe or semi-ripe fruit is eaten as a vegetable, but, when it is mature (ripe), it is utilized as a fruit. It is a multiple fruit, formed from the fusion of ovaries of many flowers, having a cylindrical shape with an average length of 30–40 cm. The outer peel of the fruit has a color range from green to yellow-brown. The flesh of the jackfruit is a good source of dietary fibers. Seeds of jackfruit are enclosed in a sweet, yellow covering which has a taste similar to that of pineapple/banana. This sheath is acidic when the fruit is young and sweet-ish when the fruit is ripened. A single jackfruit contains almost 500 seeds. Seeds of jackfruit cannot withstand freezing and drying when extracted from the fruit but can be stored for one month under cool and humid conditions. Jackfruit and its derivatives can be recognized as functional foods as they provide more than simply nutrition. Different parts of the fruit contain several valuable, bioactive compounds that can contribute to medicinal and functional properties (Figure 9.1).

The jackfruit is rich in many chemicals, such as artocarpesin, cycloartinone, isoartocarpesin, cyloartocarpin, oxydihydroartocarpesin, artocarpanone, morin, artocarpin and norartocarpetin, as well as many flavones and other pigments. The plant also contains essential amino acids such as cysteine, lysine, arginine, methionine, tryptophan, and many more. Experiments have shown that the heartwood contains 59% cellulose, 38% glucosides, 6.7% moisture, 1.7% albumin, and 0.7% lipids. Jackfruit seed is rich in a specific protein, jacalin, which is useful for the assessment of the immune status of patients infected by the human immune deficiency virus HIV-1 (Prakash et al., 2009). *A. heterophyllus* fruits contain α-carotene, β-carotene, α-zeacarotene, β-zeacarotene, dicarboxylic carotenoids and β-carotene-5, 6-epoxide carotenoids. Jackfruit is rich in minerals like potassium and calcium. The fruit provides energy and therefore is used by athletes in their diets. It can also be used to treat muscle weakness, stress and physical fatigue. In comparison with other fruits, the jackfruit has the highest protein content. The jackfruit also supplies high levels of calcium, magnesium, sodium, niacin, thiamine and vitamin B_6 to the body (Waghmare et al., 2019; Jagadeesh et al., 2007). In addition to its laxative properties, ripe fruit also provides cooling that makes it a good tonic for the brain. Dried latex exhibits androgenic action. Jackfruit extracts are effective against many strains of fungi and bacteria (Swami et al., 2012).

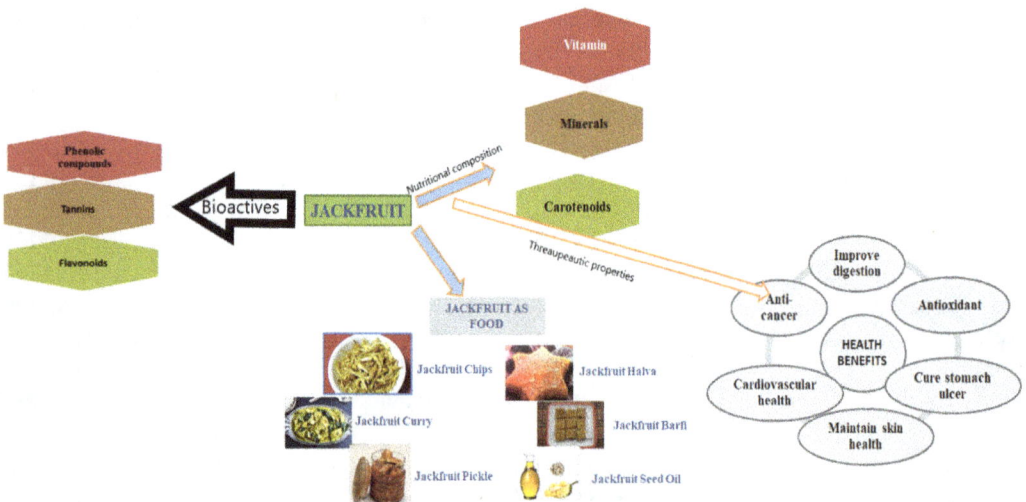

Figure 9.1   Application of jackfruit and its by-products.

There is a wide range of cultivars and species of jackfruit across the world. It can be classified into two types: soft or firm. There are many recognized varieties in the United States like 'Lemon Gold', 'Dang Rasimi', 'NS1', 'Golden Nugget', 'Black Gold' and 'Honey Gold'. There are domesticated species like *Artocarpus integer* (Thunb.) Merr., which is specific to Indonesia and Malaysia, and *Artocarpus rigidus* Blume from the Malay archipelago. There is a wide range of wild varieties of jackfruit that are freshly consumed, such as *Artocarpus lakoocha* Roxb., found in Nepal and peninsular India (Wester, 1921). The pulp of jackfruit consists of high concentrations of natural sugars so that jam prepared from it does not require the addition of sugar from external sources. Jackfruit jam is also low in calories which makes it an ideal food source for those wishing to decrease body weight. Dried sheets of pulp are used to make jackfruit leather having a sweet taste and a soft, rubbery texture. It can be used as an ingredient in ice cream, cookies and cakes. The addition of other fruits, nuts, or spices makes it a snack. Jackfruit pulp is processed to produce fruit wine, snacks and fruit juice (Nair et al., 2018).

9.2 JACKFRUIT WASTES AND BY-PRODUCTS

Of a jackfruit, only 25–35% is edible and 65–75% is retained as waste and by-products. Despite being nutritional and having economic value, there is not much focus on the utilization of these wastes or by-products as valuable compounds. Jackfruit can be segregated into three parts: pulp, seeds and peel. The peel represents more than half of the jackfruit by weight, whereas pulp and seeds contribute 30–32% and 18% of the whole fruit, respectively. Jackfruit processing always generates a notable amount of bio-waste. Greater processing leads to increased waste production, which contributes potentially to water and air pollution. The major source of these bio-wastes is the food processing industry. Fruit juice processing industries produce bio-waste in large amounts. To prevent the bioaccumulation of this bio-waste, it is very important to utilize or recycle it. These wastes are rich in essential elements, phytochemical compounds, bioactive compounds, antimicrobial activity and antioxidants (Baliga et al., 2011). The wastes and by-products of jackfruit are also processed in different ways to be utilized as food and in other product development.

The first and most important waste fraction is seeds which are widely used. The fruit's popularity is due to its sweet taste and several associated health benefits. Seeds of jackfruit are much less recognized compared to jackfruit's importance in the food sector and the properties that it exhibits. A single jackfruit contains approximately 100–500 seeds which are nutrient-rich and are edible. Seeds represent 10–15% of the fruit weight and offer nutritional benefits for the human body. Seed shape is oblong and oval with a waxy and firm nature, being individually 2.5 to 14 g in weight. There is a thin and brown-colored inner membrane of the seed coat (Gunasena et al., 1996). The seed is rich in protein, minerals, antioxidants, starch and vitamins. Carbohydrates and proteins are present in large concentrations. Various parts of the tree are also used in medicine. The major problem associated with the seeds is their processing and storage. As jackfruit is a perishable fruit, a high proportion is wasted annually along with the seeds. Furthermore, seeds germinate readily so it is difficult in less-developed countries to store the seeds in an appropriate manner (Ranasinghe, Maduwanthi & Marapana, 2019). Seeds are generally taken from the ripe fruit, dried, and stored for further use in many parts of South India. The seed shelf-life can be extended by storing seeds in a moist and cool environment. Further increases can be achieved by roasting the seeds and converting them into a powder form that is utilized in different products (Hossain, 2014). Seeds of jackfruit are used traditionally for the treatment of digestive issues. Seed powder is rich in starch and elements such as magnesium and manganese. It also contains non-reducing sugars and phenolic components. Seed powder is prebiotic and can be used to improve and maintain the microbial balance of the intestine. It is an excellent alternative to cocoa (Waghmare et al., 2019). Roots of jackfruit have been found to be helpful for patients suffering from asthma. The peel is an under-utilized waste

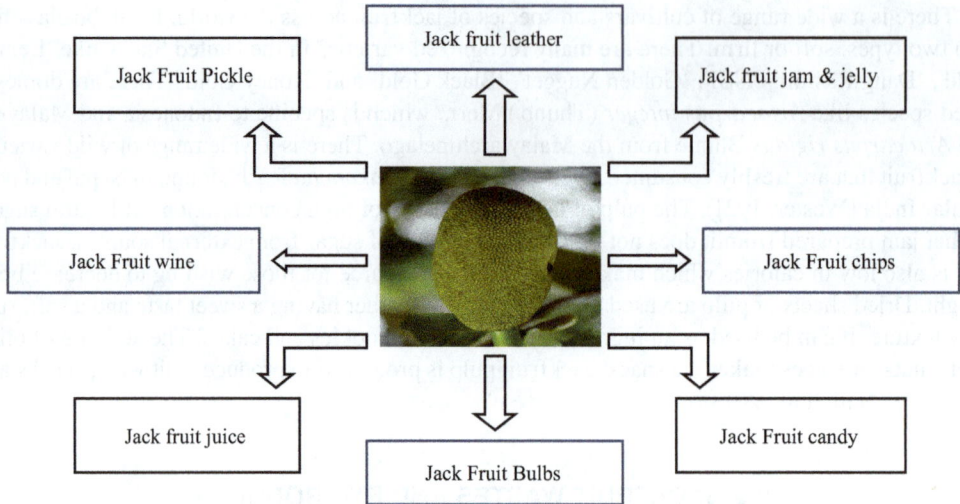

Figure 9.2   Different value-added products from jackfruit.

that is discarded by many food processing industries and in the domestic situation as kitchen waste. Jackfruit is used in the production of a number of products that generate huge amounts of peel, which is discarded as agricultural waste. Approximately 59% of the ripe fruit is composed of peel, and is rich in fibrous material, containing considerable amounts of pectin and calcium (Chadha, 1985). It is a thick outer fruit covering that tastes bitter and is therefore discarded every time. The reported annual peel waste production is in the range 2,714–11,800 kg/tree. Data submitted by the Ministry of Agriculture and Agro-Based Industry Malaysia in 2011 revealed that around 33,979 MT jackfruit peel is produced in Malaysia each year as a by-product. The peel has the potential to be utilized in many different ways as a valuable resource. It has the highest concentration of total phenols and flavonoids in the jackfruit, possibly because the peel is exposed to the external environment, particularly sunlight. External environment stresses can lead to the elevated synthesis of these compounds (Gat & Ananthanarayan, 2015). Jackfruit peel consists of unfertilized fruit perianths and is used to manufacture jellies and syrups. The reason for its use in these products is its high pectin content. At maturity, the thickness of the peel ranges from 0.4 cm to 2 cm. It is also used as a nourishing feed for livestock (Haq, 2002) (Figure 9.2).

9.3 CHEMISTRY OF JACKFRUIT-BASED WASTES AND BY-PRODUCTS

9.3.1 Seeds

Seeds are the ripened fertilized ovules of a plant, that have the potential to germinate. Seeds of jackfruit are of great importance and are used in several industries for different applications. Seeds represent 15% of the total weight of a jackfruit. Seeds have a high moisture content of around 42% and also contain a high percentage of protein but quite low ash and crude fat contents (Gupta et al., 2011). The carbohydrate and protein concentrations found in jackfruit seeds are reported to be 79.3% and 13.5%, respectively, as well as being rich in soluble and insoluble dietary fiber. Seeds are also rich in elements like calcium, copper, nitrogen, potassium, phosphorus, sulfur, and many more, of which potassium was present at the highest concentration, followed by sodium and calcium. The sulfur and S-containing metabolites found in jackfruit seeds have antimicrobial activity. Seeds also contain 12–17% starch which can be utilized in different applications, and they also contain considerable amounts of flavonoids, phenolics, saponins, steroids, triterpenes, tannins, isoflavones and lignin (Abedin et al., 2012), of which saponins were present in particularly high concentrations. As jackfruit is a seasonal fruit, the seeds cannot be stored for a long time. Therefore, seeds are processed to form products that can be stored for a longer period (Hossain et al., 2020).

Jackfruit seed contains two lectins, artocarpin and jacalin. The major protein jacalin found in jackfruit seeds is responsible for the seeds' immunological properties. The seeds also contain two B vitamins in high concentrations, namely riboflavin and thiamine. Both assist energy release to the body and are involved in other important functions of the body. Resistant starch and fiber present in seeds are important substrates for gut bacteria, which contribute to healthy digestion and immune function. Jackfruit seeds are used in many medicines for their anticancer and antispasmodic activities, as well as antioxidant and antimicrobial actions. All these properties have potential for preparing functional food and balanced diets. The medicinal role of jackfruit seeds has a particularly nutraceutical approach because the seeds are rich in phytonutrients (Maurya and Mogra, 2016).

9.3.2 Peel

The peel of jackfruit represents 46% of the fresh weight of the whole fruit, and it is discarded as waste. Peel extract contains 2.8 mg quercetin equivalent per gram of dry matter (DM) of flavonoids which are a major compound class in jackfruit peel. It also has a high concentration of other polyphenols (Meera et al.,, 2017). Jackfruit peel extract also has a strong inhibitory activity towards α-glucosidase and has strong radical scavenging (antioxidant) activity. Peel has the highest concentration of total phenolics in the fruit, containing 18 flavonoids, 12 phenolic acids, eight organic acids, eight glycosides, and three oxylipins. Peel contains several other classes of phytochemicals like tannins, saponins, proteins and carbohydrates. Ethanolic extraction provides a clear understanding of all the constituents present in peel. The preliminary screening of peel extract also showed the presence of flavonoids, alkaloids and triterpenoids. Leaves of jackfruit exhibit antimicrobial properties due to the presence of alkaloids, whereas the tannins and flavonoids are major antioxidants. Gas Chromatography–Mass Spectrometry (GC–MS) identified hexadecanoic acid (CAS) as the major compound in peel followed by squalene; CAS is the most common saturated fatty acid found in plants, animals and microorganisms, while squalene is a natural hydrocarbon that is utilized by animals and plants for the synthesis of sterol and steroid hormones, vitamin D and cholesterol in human beings. Jackfruit peel also contains calophyllolide that provides few biological activities (Ranganathan & Sundarraj, 2017). The peel of jackfruit contains 27.7% cellulose with 19.7% of total sugars. This waste provides higher amounts of cellulose than other naturally available sources. Jackfruit peel is also a good source of pectin, utilized in different industries as an important ingredient, being most commonly used in the food industries, especially in the preparation of jams and jellies (Begum et al., 2014). The peel can be utilized for the extraction of valuable components like low-methoxyl pectin and cellulose. These ingredients can be used in therapeutic or functional foods (Sundarraj & Ranganathan, 2017). The peel contains 60% of dietary fiber, of which around 56% is insoluble and 4% is soluble.

9.4 PROCESSING AND UTILIZATION OF JACKFRUIT WASTES AND BY-PRODUCTS

9.4.1 Processing and Utilization of Jackfruit Seed

9.4.1.1 Processing of Jackfruit Seed Flour

Dried seeds are also used in curries and roasted seeds can be used as snacks. Flour of jackfruit seeds is of considerable importance in the food industry. It can be used alone or can be blended to make different products like biscuits, cake and breads while maintaining its sensory and functional properties; the use of jackfruit flour in baked goods is an important one. It is used as a binding agent and thickener in different food systems. In deep-fried products, the absorption of fat is decreased to some extent when jackfruit seed flour is added (Ocloo et al., 2010). The flour is prepared by three main processes that are mechanical peeling, lye peeling and heat processing. The absorption capacities of seed flour for oil and water are important for baked goods. Fat enhances the mouthfeel and flavor, whereas good water absorption is beneficial for the making of bread. The bulk density of the seed flour is low, making packaging and transport easier. The forming properties shown by jackfruit seed flour determine important qualities like softness and swelling of bakery products. The shelf-life of the jackfruit seed flour can be improved by storing it in foil-coated aluminum and polyethylene pouches. The expected extended shelf-life is six months under cool conditions. The longer period of storage increases the moisture content and a decrease in crude fiber content is noticeable. Jackfruit seed flour is considered to be a good replacement for wheat flour. It can be used in several food products such as buttered biscuits, vada, noodles, cakes, pancakes, and chapattis (Figure 9.3).

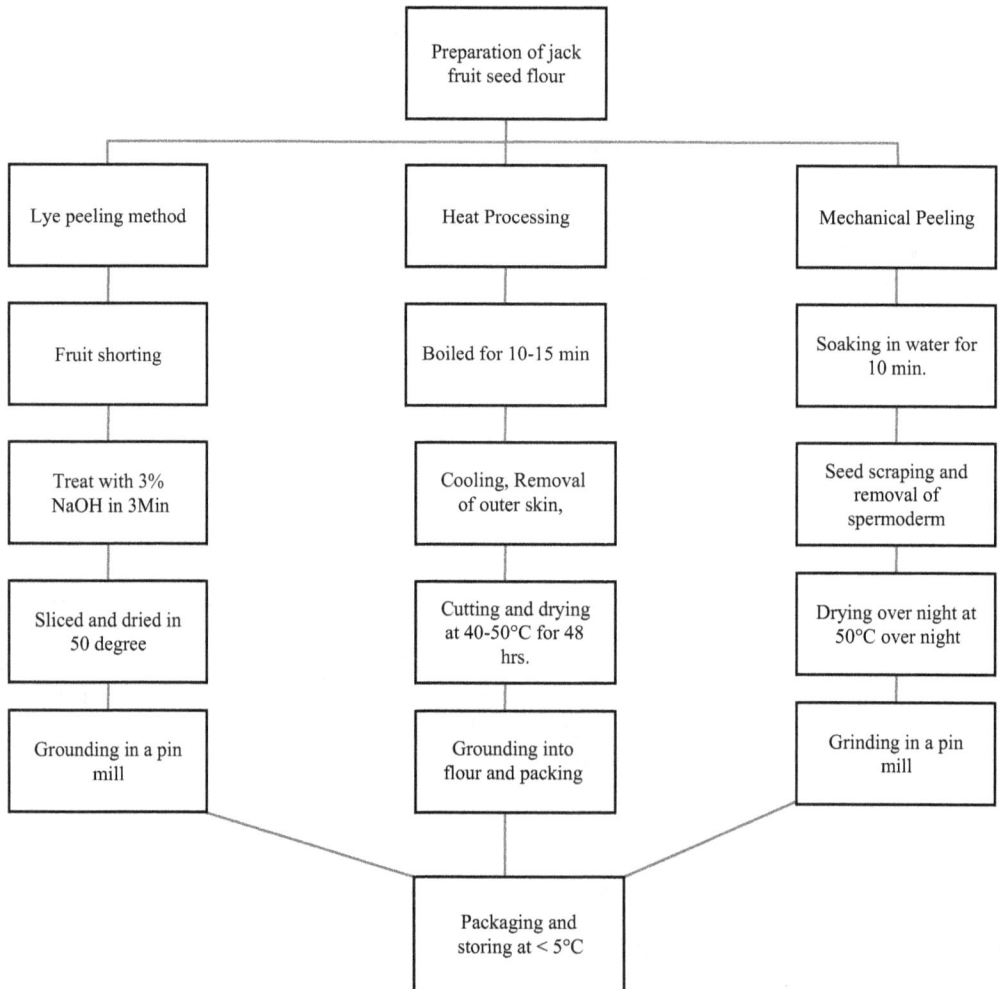

Figure 9.3 Preparation of jackfruit seed flour.

9.4.1.2 *Utilization of Jackfruit Seed Flour*

9.4.1.2.1 *Cocoa Substitute*

The demand for cocoa is increasing rapidly but the production is static, due to the greater sensitivity of the cocoa crop to changing environmental conditions. To overcome this increased demand/static production conundrum, roasted jackfruit seeds are potential alternatives as they provide the aroma and flavor of chocolates. Sensory evaluations have demonstrated that roasted seeds develop the aroma of chocolate under the influence of processing and roasting conditions. Many compounds, like 2-phenyl ethyl acetate, 3-methylbutanal and 2,3-diethyl-5-methylprazine, which are associated with chocolate aromas are present in jackfruit seeds.

9.4.1.2.2 *Extruded Products*

Extruded products have attracted the attention of customers widely because of their convenience, value, texture, appearance and taste. Products that are usually produced by this process include breakfast cereals, crispy flatbreads, cereal-based baby foods, snacks, etc. The addition of jackfruit seed flour to rice flour improves the nutraceutical as well as its nutritional properties. Noodles are mainly consumed as a staple food in many Asian countries. Instant noodles are a new trend in these fast-moving lives and gaining a lot of attention worldwide. So, it is very important to make noodles healthier (Kumari et al., 2018). Jackfruit seed flour blended with pearl millet, wheat flour, and soy flour is used to prepare noodles. Noodles prepared from this blend are rich in fat, protein, iron, calcium, fiber and energy. The noodles produced are similar to the standard available noodles on the market. Pasta prepared with the addition of jackfruit seed flour improved the nutrient content with rheological and cooking qualities being unaffected. There is a slight impact of jackfruit seed flour on the color of pasta but all other properties are acceptable and similar to the standard pasta available on the market (Kumari et al., 2015).

9.4.1.2.3 *Bakery Products*

The bakery industry attracts the attention and interest of customers because of the versatile tastes and textures of its products. Bakery products can be enhanced with improved nutritional values by the addition of specific components. The ease of incorporation of functional ingredients into bakery products is the major reason for its increased popularity. Recently, many bakery products, like cakes, biscuits, muffins, bread, and cookies, have had jackfruit seed flour incorporated to enhance the nutritional content of the product. Around 25% incorporation of jackfruit seed flour into bread increases the contents of carbohydrates, protein, crude fiber and fat. Similar increases in protein, dietary fiber and antioxidant activity have been reported in the cases of cakes and biscuits. In the case of chocolate cake, the fat content is decreased when jackfruit seed flour is added (Waghmare et al., 2019).

The use of jackfruit seed flour has primarily been studied in the preparation of bread and other confectionery products. The higher water absorption property of jackfruit seed flour decreases the dough stability time and the dough peak time. The appearance of bread from blended bread is similar to that of brown bread prepared from wheat flour. The blend also decreases the gluten percentage and enhances the crude fiber content in the prepared bread. There is increased demand for cakes with low calorific and higher fiber content and fortification of the flour with jackfruit seed flour is fulfilling this demand. Cookies prepared with addition of jackfruit seed flour exhibit good texture and organoleptic properties (Butool & Butool, 2015).

9.4.1.2.4 Snack Bars

Consumers increasingly demand healthy and nutritional products that are extracted from plant sources. There are many products available in the market that are nutritionally beneficial and are made of cereals, milk, meat, etc. Recently, nutritional bars have rapidly attracted the interest of consumers. A snack bar is a product developed by food industries to fulfill the demand for nutritional and healthy snacks for children. Jackfruit seed flour is added to ragi (finger millet) flour to prepare snack bars. This blend lowers the moisture content of the product, resulting in better shelf-life. The protein content of the snack bar is also increased by the addition of jackfruit seed flour (Chowdhury et al., 2012).

9.4.1.2.5 Traditional Food Products

Snacks are generally consumed between meals to balance the diet plan. These are generally processed and preserved for a longer period. They contain fewer calories as compared with a full-fledged meal. Jackfruit seed flour is incorporated into products like cereal bars, chapattis, and chocolate milkshakes. Fortification (15%) of cereal bars with jackfruit seed flour sustains the crispness and hardness of the product with a rise in the protein content. Cereal bars involve incorporation of 30–40% jackfruit seed meal to generate higher nutritional value, fiber content and improved sensory characteristics (Waghmare et al., 2019).

9.4.1.2.6 Protein Isolate

Protein isolated from jackfruit seeds has the potential to be used in different food industries. It can be utilized as an ingredient in the preparation of bakery products, beverages, toppings, meat products, sausages and salad dressing. Protein isolated from jackfruit seed by ultrasound treatment showed acceptable functional properties, generating a large amount of foaming and emulsifying properties, oil- and water-binding capacity, and gelation (Ulloa et al., 2017).

9.4.1.2.7 Pharmaceutical Uses of Jackfruit Seeds

Malnutrition is a growing problem in India. Jackfruit seeds, which are rich in protein, have not been explored a lot. Seeds are the optimum source to provide a sufficient amount of protein. Increased awareness among customers leads to a higher demand for jackfruit seeds. Calcium absorption is enhanced in the presence of magnesium, which is present at a high concentration in jackfruit seeds. Calcium is important for maintaining blood pressure and healthy and strong bones. Seeds contain phytochemicals that are beneficial for human health and can be added to a weight-loss diet plan. Anti-microbial activity possessed by seeds is valuable for the prevention of foodborne diseases. Resistant starch present in the jackfruit seeds is useful for controlling blood sugar. In countries like China and India, jackfruit seeds are considered to be an important component of an antidote prepared for heavy drinkers as it reduces the toxicity of alcohol. Obesity can be treated with jackfruit seeds as the absorption capacity of the fat and water by the seed flour is quite low (Rao et al., 2014). Starch is extracted from jackfruit seeds and is utilized in the preparation of fast-dissolving tablets. The starch is stable, a little granular, crisp and free flowing. Drugs prepared with jackfruit seed starch formulations are stable as they do not show any significant changes in physical parameters and drug release (Suryadevara et al., 2017).

9.4.2 Processing and Utilization of Jackfruit Peel

9.4.2.1 Extraction of Cellulose

Jackfruit peel is also an under-utilized waste. Cellulose is one of the compounds present in the peel. The jackfruit peel contains 27 g/100 g DM cellulose by mercerization. It is extracted from peel following de-pectination; this chemical process is done after the peel is treated with alkali. The extracted cellulose has good oil- and water-holding capacity. It is similar to the commercially available cellulose and can be used in the cosmetic, food and pharmaceutical industries (Sundarraj & Ranganathan, 2018b). Extracted cellulose exhibits some properties, like greater stability and emulsion-forming activity, which makes it an interesting ingredient for different formulations. For example, the particle size of cellulose extracted from jackfruit peel is larger than that of commercially available cellulose. It can be used in cake batters, salad dressings and ice cream. Peel of jackfruit is a potential source for cellulose and its derivatives like microcrystalline cellulose, α-cellulose, and carboxymethyl cellulose. These are used in pharmaceutical industries as thickening agents, tablet binders and rheology control agents (Sundarraj & Ranganathan, 2018a). Cellulose is used to produce fibers from the local biomass as it cuts down the cost of transporting the raw material. Many agriwastes are used as commercial sources of cellulose, including wheat, mulberry bark, husk fibers, corn bran, orange peel, oat hull, straw, beet pulp, tomato peel and jackfruit peel.

9.4.2.2 Extraction of Natural Antioxidants

The human body contains and generates metabolites like reactive oxygen species (ROS). Unwanted or excess ROS will lead to the development of medicinal conditions like diabetic complications and some chronic diseases. These include a weakened immune system, brain dysfunction, aging, cardiovascular disease and diabetes. Therefore, natural antioxidants which can quench ROS without side-effects can be used in food supplements, pharmaceuticals and cosmetics. Antioxidants present in the peel of jackfruit can be used in several ways and α-glucosidase can also be inhibited by peel extract (Zhang et al., 2017).

9.4.2.3 Extraction of Bio-Oil

Petroleum, natural gas and charcoal are the fossil fuels presently used as energy sources. These sources are non-renewable, harmful to the environment, and very less is left. Therefore, bio-fuel production is an alternative and promising source to provide energy, especially for the transport sector. Bio-fuel like biodiesel, bio-oil, and bio-ethanol are renewable and eco-friendly energy sources. Of these, bio-diesel and bio-ethanol use foodstuffs as raw material for their production, which can increase food prices. Jackfruit peel is used as a feedstock in the pyrolysis step to form bio-oil. The waste from jackfruit has a high concentration of volatile matter which makes it a potential resource to be utilized for the production of bio-oil. The process uses a fixed-bed reactor and the maximum yield reported is 52.6% (Soetardji et al., 2014). The optimum conditions for maximum yield are 550°C, a range of nitrogen flow of 2–4 l/min and heated at a rate of 10–50°C/min. Gas chromatography-mass spectrometry revealed that bio-oil formed from jackfruit peel contains acids, furans, nitrogen compounds, ketones, sugars, oxygenated cyclic compounds, esters, alcohols, phenols, ethers and hydrocarbon derivatives (Soetardji et al., 2014).

9.4.2.4 Extraction of Pectin

Pectin is a carbohydrate present in the form of heterogeneous polysaccharides. It has a backbone of D-galacturonic acid residues esterified with acid groups like methanol on the carboxylic group. Pectin has a wide range of applications in the pharmaceutical, food and cosmetic industries. It is used due to its versatile properties as a stabilizer, emulsifier and gelling agent. Pectin extracted from jackfruit waste is highly esterified. The solvent and the extraction conditions used affects the yield percentage and physicochemical properties of the pectin. The pectin extracted from jackfruit has poor solubility and greater ash content than commercially available pectin (Begum et al., 2014; Naik et al., 2020).

9.4.3 Processing and Utilization of Jackfruit Latex

9.4.3.1 Alternative Source of Natural Rubber

Natural rubber is used in the manufacture of a number of specialized products. It possesses superior qualities and can act as a crucial raw material for different and highly specialized products. Natural rubber is a polymer with some special attributes like high elasticity, impact resistance, resilience, malleability and abrasion resistance that cannot be shown by (or substituted by) any synthetic polymer. It is used in the manufacture of airplane tires and personal protection products used with pharmaceuticals. Latex is a natural plant-derived material that be used for the processing of natural rubber. *Cis*-1,4-polyisoprene is the main chemical element of natural rubber. The natural source explored and used for the production of natural rubber is the *Hevea brasiliensis* rubber tree (Belcher et al., 2004). The major problem associated with this natural rubber is the increasing demand but decreasing production, the latter being hampered by the limited resource of the raw material. Sources are limited as the trees bearing latex are generally grown in tropical regions. Therefore, there is a shift toward searching for better alternatives to meet the demand for natural rubber. Jackfruit produces a highly stretchable and sticky latex. Jackfruit is cultivated in many countries over a wide range of climatic conditions. The jackfruit latex has similar characteristics and chemical composition to natural rubber. The molecular weight of jackfruit latex is less than that of natural rubber and there could be other chemical constituents in jackfruit latex that can act as impurities in the process. The jackfruit latex cannot replace the natural rubber completely but it can be used to improve the wet skid resistance of tires and the dispersion of carbon black in tires (Bhadra et al., 2019).

9.4.3.2 Pharmaceutical Applications

Latex is a type of liquid emulsion produced by trees, containing many compounds like rubbers, lipids, proteins, sugars, tannins, resins, glycosides, and some proteolytic enzymes. Traditionally, it has been used in medicine due to its antimicrobial property and clot-forming ability. The latex of the jackfruit tree is valuable for treating ophthalmic disorders, dyspepsia and pharyngitis. It is also used as an antibacterial and anti-inflammatory agent. Antimicrobials can be prepared from jackfruit latex against pathogenic fungi and bacteria. Jackfruit latex contains a protein identified as a serine-centered protease having both antifungal and antibacterial properties. It possesses protease activity and other conditions do not affect its caseinolytic and gelatinolytic activities (Prakash et al., 2009). Resin present in jackfruit latex also has a potential future role in dental care. It can be used in several forms of luting agents and dental cements. It has the potential to permanently replace discolored and damaged teeth. It is also valuable in the preparation of dental varnishes. The resin exhibits a hydrophobic nature which is why it is used for the preparation of oral sustained-release tablets. Abscesses, glandular swellings and snake bites can also be healed by a mixture of latex and vinegar (Rao et al., 2014).

9.4.4 Processing and Utilization of Jackfruit Peel

9.4.4.1 Jackfruit Peel Powder

Jackfruit peel is a by-product of the jackfruit-processing industry. Jackfruit peel powder (JPP) is utilized in the processing of other products. The JPP is a rich source of dietary fiber that is used in different bakery products. JPP can be utilized depending on the protein, fat, mineral and fiber contents. The JPP is an important ingredient for introducing dietary fiber into different food products.

9.4.4.2 Incorporation into Bread

Bread is one of the most popular and daily-consumed products. The shift towards health-promoting bread has occurred as consumers are attracted to consuming healthy food. Bread can be made healthier by reducing the caloric value and enhancing the dietary fiber and protein in the product. Incorporation of JPP, a by-product of the jackfruit-processing industry with a high concentration of dietary fiber, into bread is a recent development. JPP is added as a functional ingredient into the bread to enhance its functional properties. The soluble, insoluble and total dietary fiber content in the JPP-augmented dough and bread is also higher than in bread prepared only from wheat flour. Properties like water-holding capacity, pasting properties and oil-holding capacity are significantly improved by addition of JPP. JPP is highly suitable for the preparation of functional foods, especially in case of functional bread and bread products (Felli et al., 2018).

9.4.4.3 Biodegradable Matrix Film

Jackfruit peel-based cellulose is utilized for the formation of biodegradable film. Cellulose is extracted from jackfruit peel by a bleaching process followed by alkali treatment. The cellulose presence is confirmed by Scanning Electron Microscopy (SEM). This cellulose is mixed with a mixture of glycerol and gelatine. The composite formed of gelatine/glycerol/cellulose increases the biodegradability of the matrix film. The matrix solubility is decreased, along with the moisture uptake of the film. The solubility of the film is believed to decrease because cellulose from jackfruit peel is hard to dissolve in water, requiring a long time. The matrix film has a more porous structure and the tensile strength and Young's modulus are decreased when cellulose is added (Razak et al., 2018).

9.4.4.4 Jackfruit Peel Flour (JPF)

To prepare JPF, processing is required before drying and milling of the peel. Firstly, the peel is rinsed under running tap water followed by deionized water. To soften the texture, the peel pieces are soaked in boiling water for 10 minutes. Two further chemical solutions are added, one after another, for the treatment of jackfruit peel. These chemicals are sodium bisulfite and sodium bicarbonate. After rinsing, the peel pieces are dried in a convection dryer for one day. Dried peel is then ground in a mill and sieved. The jackfruit peel flour is stored at 4°C. Jackfruit peel flour has a high concentration of crude fiber and fat but a low concentration of crude protein and moisture content, compared with commercially available wheat flour (Feili et al., 2013) and wheat flour augmented with jackfruit peel flour is used for the preparation of dietary fiber-rich bakery products.

Jackfruit peel flour is added to bakery products like bread and cookies to enhance the dietary fiber content. The addition of jackfruit peel flour to wheat flour for the preparation of bread manipulates the texture quality. There is a significant increase in the crust hardness that is directly affected by the ratio of amylose and amylopectin matrix present in the mixture (Schiraldi & Fessas, 2000). The interaction between fibrous material and gluten determines the hardness of the bread. The presence of gluten and starch in the dough of the bread makes it more elastic. This elastic behavior makes the bread a continuous sponge structure. Cohesiveness is decreased with addition of jackfruit peel flour whereas adhesiveness is unaffected. The color of the bread also changes with addition of JPF. Substitution (5%) of JPF in wheat flour for the preparation of bread proved to be the most acceptable blend among the consumers (Feili et al., 2013) (Figure 9.4).

9.4.4.5 Cookie Formulation

To produce high-fiber cookies, wheat flour is partially substituted with JPF. Cookies were made by using a standard recipe with different ratios of JPF substitution. Similar to bread, cookies also

```
┌─────────────────────────────────────┐
│  Rinse jackfruit rind with tap water │
│  followed by deionized water         │
└─────────────────────────────────────┘
                    │
                    ▼
┌─────────────────────────────────────┐
│  Soak in boiling water (10 min) and rinse │
└─────────────────────────────────────┘
                    │
                    ▼
┌─────────────────────────────────────┐
│  Soak in boiling solution of sodium  │
│  bisulfite (0.1% w/w, 1o min)        │
└─────────────────────────────────────┘
                    │
                    ▼
┌─────────────────────────────────────┐
│  Again soak in boiling solution of sodium │
│  bicarbonate (15 min)                │
└─────────────────────────────────────┘
                    │
                    ▼
┌─────────────────────────────────────┐
│  Dry for 24 hours (50°C)             │
└─────────────────────────────────────┘
                    │
                    ▼
┌─────────────────────────────────────┐
│  Milling and sieving (355-μm mesh)   │
└─────────────────────────────────────┘
                    │
                    ▼
┌─────────────────────────────────────┐
│  Store at 4°C                        │
└─────────────────────────────────────┘
```

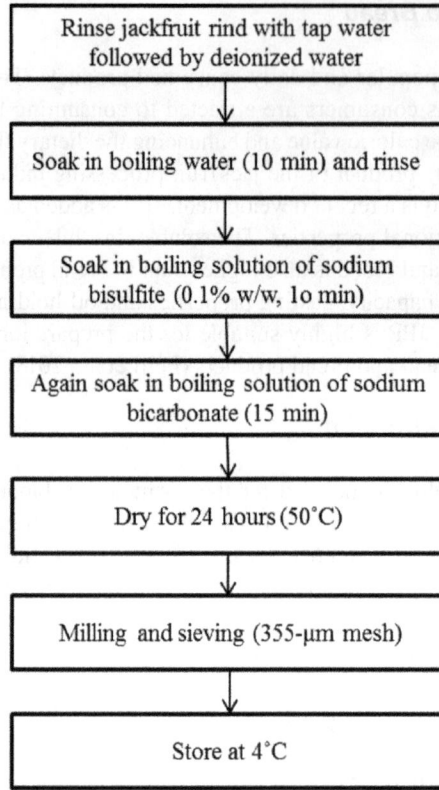

Figure 9.4 Steps involved in the preparation of jackfruit peel flour.

showed denser texture and darker color with the addition of JRF. The 5% JPF substitution was acceptable for consumption in all aspects as compared with other substitutions. The moisture content decreased and the ash content increased with the addition of JPF in cookies. Replacement also lowered the total carbohydrate and fat content of the product. Substitution of JPF in different ratios significantly changed the sensory and physicochemical attributes of the cookies (Ramya et al., 2020) (Table 9.1).

9.5 CONCLUSIONS

Jackfruit is a perishable fruit and cannot be stored for a long period hence its use is not widespread from its site of cultivation. The majority of the fruit is waste, which contains several valuable compounds. These wastes are generally produced from jackfruit-processing industries and households. Waste generated from households is quite negligible as compared with the food industries. Recent studies have evaluated the potential of jackfruit waste or by-products to be utilized in different industries. It also provides another income source to the industries and improves economic conditions as the raw material is waste and processing of it requires very less expenditure. The use of these wastes in different applications will reduce the waste and protect the environment. It can be used as an alternative to commercially available components like pectin, cellulose, and many more, and as a potential functional ingredient in different food and non-food products. It is also used in the treatment of wastewater for the removal of dyes. Jackfruit by-products are used by pharmaceutical companies and possess several health benefits. It is a promising natural agri-based resource for

Table 9.1 Processing and Utilization of By-Products from Jackfruit

	Properties	Application	References
By-products from jackfruit peel			
Cellulose extraction	Greater stability, more emulsion activity, improved water- and oil-holding capacity	Ice cream, salad dressing and cake batters, tablet binders and rheology control agents	Sundarraj & Ranganathan (2018a)
Pectin extraction	Lower solubility, higher ash content, gelling agent, stabilizer and emulsifier	Jam, jellies, confectionery, medicines, cosmetics	Begum et al. (2014)
Bio-oil extraction	Renewable and eco-friendly fuel	As transport, energy source	Soetarji et al. (2014)
Natural adsorbent	Removal of Cd(II) dyes from water	Wastewater treatment	Jayarajan et al. (2011)
By-products from jackfruit seed			
Extruded products	Increased nutrition and energy, more protein, fat and fiber Decreased color quality	Incorporated to make healthier noodles and pasta (staple food)	Nandkule et al. (2015)
Bakery products	Increased crude fiber Decreased gluten content and calorific value	Specifically used in cake, bread, biscuit preparation	Akter & Haque (2018)
Protein isolate	Good emulsifying and foaming properties, improved gelling property, increased oil- and water-binding capacity	Protein isolate is used in preparation of toppings, beverages, meat products and confectionery	Ulloa et al. (2017)
Pharmaceutical products	Free flowing, stable, granular and crisp products	Used for the preparation of stable and fast-dissolving tablets	Suryadevara et al. (2017)

the production of renewable energy. A lot more studies can be done to explore different methods to study valuable compounds present in by-products and to use them in different sectors.

REFERENCES

Abedin, M.S., Nuruddin, M.M., Ahmed, K.U., and Hossain, A. (2012). Nutritive compositions of locally available jackfruit seeds (*Artocarpus heterophyllus*) in Bangladesh. *International Journal of Biosciences*, 2(8), 1–7.

Akter, B., and Haque, A. (2018). Utilization of jackfruit (*Artocarpus heterophyllus*) seed's flour in food processing: A review. *The Agriculturists*, 16(2), 131–142.

Baliga, M.S., Shivashankara, A.R., Haniadka, R., Dsouza, J., and Bhat, H.P. (2011). Phytochemistry, nutritional and pharmacological properties of *Artocarpus heterophyllus* Lam (jackfruit): A review. *Food Research International*, 44, 1800–1811.

Begum, R., Aziz, M.G., Uddin, M.B., and Yusof, Y.A. (2014). Characterization of jackfruit (*Artocarpus heterophyllus*) waste pectin as influenced by various extraction conditions. *Agriculture and Agricultural Science Procedia*, 2, 244–251.

Belcher, B., Rujehan, N.I., and Achdiawan, R. (2004). Rattan, rubber, or oil palm: Cultural and financial considerations for farmers in Kalimantan. *Economic Botany*, 58, 77–87.

Bhadra, S., Mohan, N., Parikh, G., and Nair, S. (2019). Possibility of *Artocarpus heterophyllus* latex as an alternative source for natural rubber. *Polymer Testing*, 79, 106066.

Butool, S., and Butool, M. (2015). Nutritional quality on value addition to jackfruit seed flour. *International Journal of Scientific Research*, 4, 2406–2411.

Chadha, Y.R. (Ed.). (1985). *The Wealth of India-Raw Materials*. Publications and Information Directorate. New Delhi: CSIR, 450.

Chowdhury, A.R., Bhattacharyya, A.K., and Chattopadhyay, P. (2012). Study on functional properties of raw and blended jackfruit seed flour (a non-conventional source) for food application. *Indian Journal of Natural Products and Resources*, 3(3), 347–353.

Feili, R., Zzaman, W., Abdullah, W.N.W., and Yang, T.A. (2013). Physical and sensory analysis of high fiber bread incorporated with jackfruit rind flour. *Food Science and Technology*, 1(2), 30–36.

Felli, R., Yang, T.A., Abdullah, W.N.W., and Zzaman, W. (2018). Effects of incorporation of jackfruit rind powder on chemical and functional properties of bread. *Tropical Life Sciences Research*, 29(1), 113–126.

Gat, Y., and Ananthanarayan, L. (2015). Physicochemical, phytochemical and nutrimental impact of fortified cereal based extrudate snacks: Effect of jackfruit seed flour addition and extrusion cooking. *Advance Journal of Food Science and Technology*, 8(1), 59–67.

Gunasena, H.P.M., Ariyadas, K.P., Wikramasinghe, A., Herath, H.M.W., Wikramsinghe, P., Rajakaruna, S.B. (1996). *Manual of Jackfruit Cultivation in Sri Lanka*. Sri Lanka: Forest information service. Department of Forest Publication, 48.

Gupta, R.K., Gupta, D., Mann, S., and Sood, A. (2011). Phytochemical, nutritional and antioxidant activity evaluation of seeds of jackfruit (*Artocarpus heterophyllus* Lam.). *International Journal of Pharma and Bio Sciences*, 2(4), 336–345.

Haq, N. (2002). *Improvement of Underutilized Fruits in Asia*. Annual Report. UK: Community Fund.

Hossain, M.A., Evan, M.S.S., Moazzem, M.S., Roy, M., and Zzaman, W. (2020). Response surface optimization from antioxidant extraction from jackfruit (*Artocarpus heterophyllus* Lam.) seed and pulp. *Journal of Scientific Research*, 12(3), 397–409.

Hossain, M.T. (2014). Development and quality evaluation of bread supplemented with jackfruit seed flour. *International Journal of Nutrition and Food Sciences*, 3(5), 484.

Jagadeesh, S.L., Reddy, B.S., Swamy, G.S.K., Gorbal, K., Hegde, L., and Raghavan, G.S.V. (2007). Chemical composition of jackfruit (*Artocarpus heterophyllus* Lam.) selections of Western Ghats of India. *Food Chemistry*, 102, 361–365.

Jayaranjan, M., Arunachalam, R., and Annadurai, G. (2011). Agricultural wastes of jackfruit peel nano-porous adsorbent for removal of rhodamine dye. *Asian Journal of Applied Sciences*, 4(3), 263–270.

Kumari, S., Prasad, R., and Gupta, A. (2018). Processing and utilization of jackfruit seeds, pearl millet and soybean flour for value addition. *Journal of Pharmacognosy and Phytochemistry*, 7(6), 569–572.

Kumari, V., Divakar, S., Ukkru, M., and Nandini, P.V. (2015). Development of raw jackfruit based noodles. *Food Science Research Journal*, 6(2), 326–332.

Maurya, P., and Mogra, R. (2016). Assessment of consumption practices of jackfruit (*Artocarpus heterophyllus* lam.) seeds villages of Jalalpur block district Ambedarnagar (U.P.) India. *Remarking*, 2, 73–75.

Meera, M., Ruckmani, A., Saravanan, R., and Prabhu, R.L. (2017). Anti-inflammatory effect of ethanolic extract of spine, skin and rind of jackfruit peel – A comparative study. *Natural Product Research*, 32(22), 1–5.

Naik, M., Rawson, A., and Rangarajan, J.M. (2020). Radio frequency-assisted extraction of pectin from jackfruit (*Artocarpus heterophyllus*) peel and its characterization. *Journal of Food Process Engineering*, 43(6), 1–11.

Nair, P.N., Palanivel, H., and Kumar, R. (2018). Jackfruit (*Artocarpus heterophyllus*), a versatile but underutilized food source. *Fiji Agricultural Journal*, 57(1), 5–18.

Nandkule, V.D., Masih, D., Sonkar, C., and Patil, D.D. (2015). Development and quality evolution of jackfruit seed and soy flour noodles. *International Journal of Science, Engineering and Technology*, 3(3), 802–806.

Ocloo, F.C.K., Bansa, D., Boatin, R., Adom, T., and Agbemavor, W.S. (2010). Physico-chemical, functional and pasting characteristics of flour produced from jackfruit (*Artocarpus heterophyllus*) seeds. *Agriculture and Biology Journal of North America*, 1(5), 903–908.

Prakash, O., Kumar, R., Mishra, A., and Gupta, R. (2009). *Artocarpus heterophyllus* (jackfruit): An overview. *Pharmacognosy Reviews*, 3(6), 353–358.

Ramya, H.N., Anitha, S., and Ashwini, A. (2020). Nutritional and sensory evaluation of jackfruit rind powder incorporated with cookies. *International Journal of Current Microbiology and Applied Sciences*, 9(11), 3305–3312.

Ranasinghe, R.A.S.N., Maduwanthi, S.D.T., and Marapana, R.A.U.J. (2019). Nutritional and health benefits of jackfruit (*Artocarpus heterophyllus* Lam.): A review. *International Journal of Food Science*, 1–12.

Ranganathan, T.V., and Sundarraj, A.A. (2017). Phytochemical screening and spectroscopy analysis of jackfruit. *International Research Journal of Pharmacy*, 8(9), 151–159.

Rao, J., Singh, K., Singh, S., Mishra, S.M., and Bajpai, M. (2014). *Artocarpus heterophyllus* (jackfruit) potential unexplored in dentistry- an overview. *Universal Journal of Pharmacy*, 3(1), 50–55.

Razak, S.F.A., Rahman, W.A., and Majid, N.A. (2018). Effect of jackfruit rind-based cellulose (JR-CEL.) on physical and mechanical properties of the biodegradable glycerol/gelatin matrix film. *AIP Conference Proceedings*, 20075, 1–8.

Schiraldi, A., and Fessas, D. (2000). Mechanism of staling. In C. Pavinee, Vodovotz, (eds). *Bred Staling*. New York: CRC Press, Inc., 2–10.

Soetardji, J.P., Widjaja, C., Djojorahardjo, Y., Soetaredjo, F.E., and Ismadji, S. (2014). Bio-oil from jackfruit peel waste. *Procedia Chemistry*, 9, 158–164.

Sundarraj, A.A., and Ranganathan, T.V. (2017). Physicochemical characterization of jackfruit (*Artocarpus integer* (Thumb.).) peel. *Research Journal of Pharmaceutical, Biological and Chemical Sciences*, 8(3), 2285–2295.

Sundarraj, A.A., and Ranganathan, T.V. (2018a). Characterization of cellulose from jackfruit (*Artocarpus heterophyllus*) peel. *Journal of Pharmacy Research*, 12(3), 311–315.

Sundarraj, A.A., and Ranganathan, T.V. (2018b). Extraction and characterization of cellulose from Jackfruit (*Artocarpus heterophyllus*) peel. *Journal of Experimental Biology and Agricultural Sciences*, 6(2), 414–424.

Suryadevara, V., Lankapalli, S.R., Danda, L.H., Pendyala, V., and Katta, V. (2017). Studies on jackfruit seed strach as a novel natural superdisintegrant for the design and evaluation of irbesartan fast dissolving tablets. *Integrative Medicine Research*, 6, 280–291.

Swami, S.B., Thakor, N.J., Haldankar, P.M., and Kalse, S.B. (2012). Jackfruit and its many functional components as related to human health: A review. *Comprehensive Reviews in Food Science and Food Safety*, 11, 565–576.

Ulloa, J.A., Barbosa, M.C.V., Vazquez, J.A.R., Ulloa, P.R., Ramirez-Ramirez, J.C., Carrillo, Y.S., and Torres, L.G. (2017). Production, physico-chemical and functional characterization of a protein isolate from jackfruit (*Artocarpus heterophyllus*) seeds. *CYTA-Journal of Food*, 15(4), 497–507.

Waghmare, R., Memon, N., Gat, Y., Gandhi, S., Kumar, V., and Panghal, A. (2019). Jackfruit seed: An accompaniment to functional foods. *Brazilian Journal of Food Technology*, 22, 1–9.

Wester, P.J. (1921). The food plants of Philippines. *Philippine Journal of Agricultural Scientist*, 14, 211–384.

Zhang, L., Tu, Z-c., Xie, X., Wang, H., Wang, H., Wang, Z-x., Sha, X-m., and Lu, Y. (2017). Jackfruit (*Artocarpus heterophyllus* Lam.) peel: A better source of antioxidant and α-glucosidase inhibitors than pulp, flake and seed, and phytochemical profile by HPLC-QTOF-MS/MS. *Food Chemistry*, 234, 1–39.

CHAPTER 10

Banana Wastes
Chemistry, Processing, and Utilization

Yogesh Kumar, Samandeep Kaur, Saptashish Deb, and D.C. Saxena

CONTENTS

DOI: 10.1201/9781003164463-10

10.1 INTRODUCTION

Bananas are an important fruit from the Musaceae family, grown extensively in tropical and sub-tropical regions and considered to be the first cultivated food crop (Ploetz, 2015; Emaga et al., 2007). It is important to note that the most edible bananas (including plantains) come from two species and their hybrids: *Musa acuminata* and *Musa balbisiana*. The cultivars more closely related to *Musa balbisiana* tend to be the starchy cooking bananas commonly known as plantains; cultivars closer to *Musa acuminata* tend to be sweet dessert varieties of banana. The possible uses of banana are determined by its chemical, physicochemical and functional properties (Rodriguez-Ambriz et al., 2008). The chemical composition of banana changes during the maturation period. Banana is an abundant source of flavonoid, phenolic and antioxidant compounds such as dopamine (Alothman et al., 2009). These bioactive compounds are known to prevent certain diseases such as heart disease and cancer. Due to the aforementioned nutritive and health-promoting properties, the rate of banana consumption has been steadily increasing. In addition to starch, sugar and antioxidants, banana is also a rich source of potassium, calcium, phosphorus, and fiber which are equally beneficial to the human body (Mota et al., 2000). Banana fibers are considered to be one of the unexplored potential bio-resources (Pappu et al., 2015). Owing to these beneficial properties, banana is recommended to be part of our daily diet. Various scientific research and genetic studies have been performed to improve the production of banana as well as to improve quality, output, and resistance to disease.

Banana ranks fourth among the most important food crops, after wheat, rice, and maize. In 2017 alone, 22.7 million tonnes of bananas, excluding plantains, were traded, representing almost 20% of global production that year. The Asia-Pacific region leads the world banana market and shares 61% of total global consumption, with India ranked top in banana production, contributing about 25.7% to global production (Figure 10.1). Globally, Ecuador is the highest exporter of bananas, covering 24.7% of the total world exports (Mordor Intelligence, 2020). In India, banana is the second-largest cultivated fruit next to mango. Tamil Nadu is the largest banana-producing state in India, followed by Maharashtra.

More than one thousand varieties of bananas have been cultivated and consumed locally all over the world. Some popularly grown commercialized banana varieties are listed in Table 10.1. 'Cavendish' is the dominant commercial banana variety with global production around 50 billion

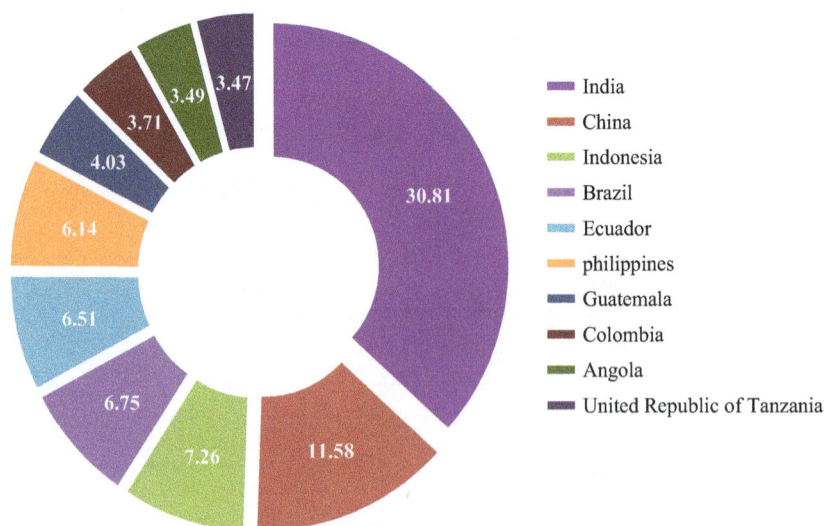

Figure 10.1 Contribution of various countries to global production (million tons) of bananas in 2018.

Table 10.1   List of Popular Banana Varieties Grown around the World

Variety	Binomial	Use
Cavendish banana	*Musa acuminata* 'Cavendish' subgroup	Baking, fruit salad, desserts, ice cream, smoothies
Manzano banana (also called Apple, Latundan, and Silk banana)	*Musa acuminata* × *M. balbisiana* (AAB Group) 'Silk'	Making desserts
Dwarf Cavendish Banana	*Musa acuminata* 'Dwarf Cavendish'	Eating raw or cooked
Dwarf plantain	*Musa* × *paradisiaca*	Cooking
Ice Cream banana	*Musa acuminata* × *balbisiana* 'Blue Java'	Eaten as fresh or cooked, making desserts, salads
Gros Michel	*Musa acuminata* (AAA Group) 'Gros Michel'	Eating raw, food flavoring
Red banana	*Musa acuminata* 'Red Dacca'	Eating raw, making desserts and fruit salads, baking, frying
Lady Finger banana	*Musa acuminata* 'Lady Finger'	Eating raw, baked goods or smoothies
Pisang Raja	*Musa acuminata* × *Musa balbisiana* (AAB Group) 'Pisang Raja'	Eating raw, making desserts

tonnes, which accounts for around 47% of global production. The global production of bananas increased from 67 MT in 2000 to 114 MT in 2017 at a compound annual growth rate of 3.25%. However, overall production of bananas has been projected to slow to 1.5% per annum globally, reaching 135 MT in 2028 (FAO, 2020a,b).

Banana peels are the main residue of the banana fruit, comprising about 30–40% of the total fruit fresh weight. With normalized composition, the banana peels can be utilized to obtain various products of industrial and domestic use (Emaga et al., 2007). Because of the very high consumption rate of bananas, huge weights of banana peel are generated by various industries and by domestic households. In most cases, this vast resource of banana peel is disposed of in fields, on wastelands or in water bodies, which may cause environmental problems. A small proportion of banana peels are used as cattle feed (Bakry et al., 1997). Like banana pulp, the chemical composition of banana peel also changes with maturation. Depending on the chemical composition, physicochemical and functional properties, some countries have started using banana peel under the waste utilization sector (Rodriguez-Ambriz et al., 2008). With the development of modern processing techniques, numerous food processing industries have started producing various value-added products using banana peels to overcome the environment related issues, generate employment opportunities, introduce new products and generate new income streams (Zhang et al., 2005). Banana peels have been used for the manufacture of, activated carbon and adsorbents for purification of water (Annadurai et al., 2002), for flavoring oil and wine (Faturoti et al., 2006), ethanol by fermentation (Ibrahim, 2015), nano-fertilizer from peel ash (Hussein et al., 2019), ash-derived alkali solution for soap production (Udosen and Enang, 2000), biomethane (Bardiya and Somayaji, 1996; Nathoa et al., 2014), and cellulose nanofibers (Khawas and Deka, 2016).

Banana peels consist of carbohydrates, vitamins, lignocellulosic compounds and pigments as well as phytochemicals (mainly phenolic compounds) with antioxidant properties. Around forty different phenolic compounds have already been identified in the peel (Vu et al., 2019a). A comparative study of antioxidant activity in banana peels and banana pulp has shown that banana peel antioxidant content is greater than that in banana pulp (Someya et al., 2002). Therefore, there is great potential of banana peel in food processing industries to produce novel products without compromising the final product's quality. Moreover, banana peel is a rich source of other chemical constituents such as 6–9% of crude protein, 3.8–11% of crude fat and 43.2–49.7% of dietary fiber (Mohapatra et al., 2010). These constituents of peel can be utilized as fortification/supplementation ingredients to develop new products. Extracted minerals, fiber and monosaccharides from banana peel waste can be used as ingredients in various food systems (Mukhopadhyay et al., 2008; Oliveira et al., 2007). Banana peels show different nutritional composition at different stages of maturity. A mature banana peel contains 6–9% of protein, which indicates huge potential of banana peel to be an economically viable and abundant source of protein. Apart from the high amount of protein, fat and dietary fiber, mature banana peels are rich in free sugars (Emaga et al., 2011).

However, banana peel also contains some anti-nutritional elements too, with tannins being the main anti-nutritional compounds present in banana peel. Apart from anti-nutritional disadvantage, tannins impart an astringent taste to unripe banana peels making them unpalatable (Li et al., 2005). The lower tannin content in the peel of ripe bananas peel is due to the migration of tannins from peel to pulp during the maturation period and its degradation by peroxidases and polyphenol oxidases (Emaga et al., 2008). The stage of maturity of a banana directly influences the composition of the peel and the quality of products manufactured from peel. Thus, ripe banana peel is safer from an anti-nutritional point of view, compared with unripe banana peel, for further utilization in food processing sectors. The effect of ripeness on the chemical composition of banana peel varies according to variety (Emaga et al., 2007). When choosing a perfectly ripened banana peel, to get its best proximate composition, it is always better to choose yellow-colored banana peel without any dark spots. Due to the numerous chemical changes during ripening, synthesis of volatile components and changes in the banana pulp and peel texture take place (Drury et al., 1999), resulting in uneven changes in banana peel color and brown spots appearing all over a yellow background. Dark spots over the banana peels convert the color of the banana peel from yellow to dark brown or black (Choehom et al., 2004). Dark-spotted banana peel imparts a 'muddy' brown-color when dried and ground to flour, while the chocolaty aroma disappears.

During the time of making banana peel flour for further utilization, drying is the foremost and crucial step. During drying, stress is generated in the cellular structure of the peel which disintegrates the natural peel structure, followed by physical and chemical changes of the product (Pan et al., 2008). For this reason, the drying temperature is considered to be a critical parameter. High temperatures may increase the drying rate of banana peel but may degrade the color, texture and overall quality of the final banana peel flour (Baini and Langrish, 2009). Keeping in view the above points, this chapter elaborates on the chemistry of banana peel, its processing, and the various possible applications.

10.2 CHEMISTRY OF BANANA PEEL

10.2.1 Nutritional Composition

Peels are the major waste of banana/plantain (cooking banana) processing. According to reports, banana peel contains 3.8–11% crude fat, polyunsaturated fatty acids (44.5–52.4% of FA), especially linoleic acid (21.8–26.9%) and α-linolenic acid (19.7–29.3%), 6–9% crude protein, 3% starch, 43.2–49.7% total dietary fiber, 6–12.1% lignin, 10–21% pectin, 7.5–9.6% cellulose, 6.4–8.4% hemicelluloses, at different ripeness stages. Various pectin fractions of dried banana peels (genotype AAA, maturity 5) contained galacturonic acid (42–66.4%), rhamnose (0.2–0.5%), arabinose (1.4–2.7%), xylose (0.3–0.8%), mannose (2–2.7%), glucose (11–16.5%), and galactose (2.5–12.9%) (Emaga et al., 2007, 2008; Davey et al., 2009). Whereas the plantain peel contains 8.1–8.6% crude protein, 2.2–3.7% crude fat, 24–39.3% starch, 6.4–7.5%ash, 35.7–37.3% total dietary fiber, 6.1–7.1% cellulose, 4.5–6.3% hemicellulose, and 7.9–15% lignin at various maturity stages (Emaga et al., 2007).

Green banana peel has high starch and carbohydrate content, but, as the fruit ripens, a decrease in starch content is observed (Zhang et al., 2005; Eshak, 2016). The mineral analysis of banana peel showed that peel contains plenty of minerals such as potassium, sodium, phosphorus, calcium, zinc, manganese and magnesium (Table 10.2) (Abubakar et al., 2016). Bananas peels also contain

Table 10.2 Nutritional Composition of Banana Peel of Different Banana Varieties

Variety	Proximate (% Dry Weight)							Reference
	Moisture	Ash	Fiber	Lipid	Protein	Carbohydrate		
Banana peel	13.49	9.83	14.83	23.93	5.53	32.39		Abubakar et al. (2016)
Plantain peel	11.43	8.63	9.43	37.53	5.79	27.18		
Musa sapientum peel	6.70	8.50	31.70	1.70	0.90	59.00		Anhwange et al. (2009)
Organic banana (USA) peel, color index=4	22.06	12.35	11.51	6.65	9.42	38.06		Puraikalan (2018)
Poovan banana (India) peel	9.52	12.95	14.41	3.64	11.63	47.85		
Unripe banana peel	10.0	7.6	-	4.7	8.4	69.4		Waghmare & Arya (2016)
Banana peel ('Ethiopia')	20.0	-	-	6.0	6.0	62.0		Gebregergs et al. (2016)
Banana peel ('Mohali')	3.25	10	-	5.1	4.79	56.1		Budhalakoti (2019)

	Mineral Content (mg/100 g)									
	K	Fe	Na	Mg	P	Zn	Mn	Ca	Cr	
Banana peel	527.19	41.03	812.87	27.28	1460	150	-	4113.72	71.33	Abubakar et al. (2016)
Plantain peel	854.49	55.08	672.96	64.12	2977	41.82	-	1702.38	73.91	
Musa sapientum peel	7810	61	2430	-	7810	-	7620	1920	-	Anhwange et al. (2009)

carotenoids like lutein, β-carotene, violaxanthin, auroxanthin, neoxanthin, isolutein, and β- and α-cryptoxanthin, which contribute to the specific color of the peel. Banana peel (*Musa × paradisiaca*) also contains antibacterial compounds like β-sitosterol, 12-hydroxy stearic acid and malic acid, which are effective against microorganisms like *Bacillus cereus, Escherichia coli, Staphylococcus aureus*, and *Salmonella enteritidis* (Mokbel and Hashinaga, 2005).

As the fruit ripens, there are changes in the nutritional content of the peel and pulp. In a study of banana peel, it was observed that a decrease in chlorophyll (90%) and an increase in carotenoid (50%), and flavonoid (27%) content was observed during the ripening process. However, the antioxidant capacity and phytochemical content of the peel decrease in response to overripening of fruit (Vu et al., 2019a). Compositional analysis of banana peel confirms that it is a potential source of nutrients and can be exploited to obtain nutritional/ nutraceutical products of plant origin.

10.2.2 Antioxidant Potential

Banana peels contain plenty of bioactive compounds like phenolics, which have antioxidant potential (Vijayakumar et al., 2008). A chemical compound is said to have antioxidant properties if it can delay or prevent free radical chain reactions by scavenging/neutralizing free radicals. The antioxidant potential of banana peels is attributed to the presence of polyphenols, flavonoids or flavanols like gallocatechin (Vijayakumar et al., 2008; Someya et al., 2002); but such activity is not limited to phenolics; other compounds, like vitamin C (ascorbic acid), tocopherols (vitamin E), carotenes (β-carotene), dopamine, rutin and catecholamines, also enhance its antioxidant potential (Sulaiman et al., 2011; Kanazawa & Sakakibara, 2000). An ethanolic extract of banana peel possessed an IC_{50} value of 19.10 µg/ml, against DPPH, which means that banana peel exhibit a strong antioxidant potential (Dahham et al., 2015). Studies have proven that banana peel extracts inhibit lipid peroxidation and are also capable of scavenging DPPH and ABTS free radicals (González-Montelongo et al., 2010). Although it has been observed that, due to a decrease in content of biogenic amines (like dopamine) on ripening, the antioxidant activity of banana peel also decreases upon ripening but remains significant (10 mg/100 g) at stages of maturation (Kanazawa & Sakakibara, 2000). Total Trolox equivalent (TE) antioxidant capacity (TEAC) of 5.67 mg TE/g dry weight (DW) and 83% DPPH radical scavenging activity were observed in banana peel by Babbar et al. (2011). Gallocatechin, a phenolic compound, was isolated from peel and pulp extracts of bananas (*Musa acuminata 'cavendish'*), using HPLC: 158 mg/100 g DW of gallocatechin was observed in banana peel whereas 29.6 mg/100 g DW of gallocatechin was observed in banana pulp (Someya et al., 2002). Rutin, a flavanol glycoside, has been reported in plantain peel in the range 0.242–0.619 mg/g DW (Tsamo et al., 2015). Banana peels also contain several catecholamines and phenethylamine. In a study, 80 to 560 mg/100 g of dopamine was reported in ripened '*cavendish*' peel (Kanazawa & Sakakibara, 2000). These studies indicate that banana peel can be utilized as a source of antioxidants.

10.2.3 Extraction and Evaluation of Bioactive Components

Bioactive compounds are chemical compounds that provide health benefits to the consumer. Luckily, the banana peel contains a lot of these important bioactive components as shown in Figure 10.2. These compounds can be extracted and isolated from different food matrices and incorporated into food materials. Extraction is an essential step in analyzing the bioactive compounds of fruit and vegetable wastes and by-products, which need to be separated from the fruit cells for further analysis. For the extraction of phytochemicals from these residues, the raw material is cleaned, dried, and ground under conditions that do not distort or destroy the target functional compounds (Fabricant & Farnsworth, 2001; Kaur et al., 2021). The solvents used for extraction are selected according to the target compound's chemical properties, using the concept 'like dissolves

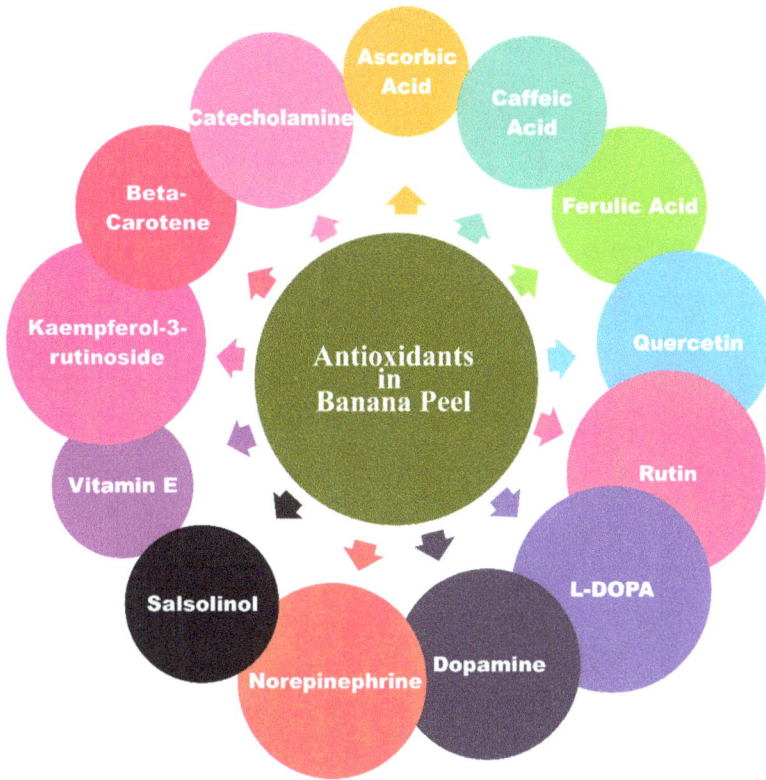

Figure 10.2 Some important bioactive compounds observed in banana peel.

like'. When the target compound is hydrophilic, solvents with higher polarity (i.e., methanol, ethanol, etc.) are used, whereas, when the target compounds are lipophilic, solvents like dichloromethane are selected (Cos et al., 2006). Pure aqueous solutions of methanol, ethanol, acetone, and ethyl acetate or water are common extraction solvents used on plant matrices. Since the target compounds can be of varied nature (polar/ non-polar, heat-labile/heat-resistant), the suitability of extraction techniques should also be considered.

During extraction (at 25°C) of dopamine from banana peel, using methanol, with an increase in extraction time from 1 to 120 min, an increase in dopamine concentration was observed (González-Montelongo et al., 2010). Phytosterols like β-sitosterol (0.3–0.6 g/kg DM), campesterol (0.1–0.7 g/kg DM), cycloeucalenol (5–86 mg/g DM), Cycloeucalenone (0.8–9.4 g/kg DM), cycloartenol (0–37 mg/kg DM), and stigmasterol (0.2–0.4 g/ kg DM) were also reported in unripe banana peels of several *M. balbisiana* cultivars (Villaverde et al., 2013). An increase in L-DOPA and tyramine levels occurs as the fruit ripens, whereas, upon overripening of the fruit, a decline in the levels of these chemicals is observed (Romphophak et al., 2005), which is related to the formation of melanin upon oxidation and polymerization of dopamine. The optimum conditions for extraction of polyphenols/bioactives from banana peel using different extraction methods have been compiled in Table 10.3.

10.3 PROCESSING OF BANANA PEEL

The industrial processing of bananas produces around 40% of the total fruit weight as peel, which is biodegradable in nature. However, simply dumping this peel is neither environmentally

Table 10.3 Optimum Conditions for Extraction of Polyphenols/Bioactives from Banana Peel Using Different Extraction Methods

Variety/Tissue	Extraction Method	Target Compound	Results	Reference
Banana (*Musa acuminata* 'Cavendish') peel	Ultrasound-assisted extraction (UAE)	Phenolic and antioxidant compounds	**Optimum conditions:** T=30°C, time =5 min, ultrasonic power =150 W, sample =8 g/100 mL, acetone conc. =60% **Optimum responses:** phenolic compounds =23.49 mg/g peel, flavonoids =39.46 mg/g peel, proanthocyanidins =13.11 mg/g	Vu et al. (2017)
Banana peel	Microwave-assisted extraction and water	Phenolic compounds	**Optimum conditions:** pH =1, Sample =2 g/100 mL, irradiation time =6 min, microwave power =960 W **Optimum response:** phenolics =50.55 mg/g dried peel	Vu et al. (2019b)
Musa acuminata ('Cavendish') unripe peel	UAE	Total tannins, antioxidant activity	**Optimum conditions:** T=60°C, time =30 min, sample =5 g/100 mL **Optimum responses:** extract yield =14.9%, total tannin content =119.2 mg TAE/g sample, flavonoids =29.0 mg RE/ g sample, DPPH activity =80.8%, ABTS scavenging activity =84.7%	Ishak et al., (2020)
Dwarf banana peel (Brazil)	Microwave-assisted extraction	Polyphenols	**Optimum Conditions:** ethanol conc. =50%, microwave power =380W, time =100 s, sample =1 g/35 mL **Optimum Response:** polyphenols =2.16%	Gu et al. (2014)
Banana peel ('Grande Naine' and 'Gruesa' cultivar)	Maceration using different solvents	Bioactive compounds	Maximum extraction was observed in aqueous acetone; extraction yield decreased at high temperature and in long-term extractions. At 25 and 55°C and 1 min extraction, phenolics =3.3±0.8%, anthocyanins =0.434±97 mg cyanidin 3-glucoside equivalents/100 g freeze-dried banana peel	González-Montelongo et al. (2010)
Peel and pulp of different plantain cultivars	Maceration at 40°C	Phenolics	Myricetin-deoxyhexose-hexoside, rutin and kaempferol-deoxyhexose-hexoside were observed in peels of *Musa* spp.; ferulic acid-dihexoside, myricetin-deoxyhexose-hexoside, and quercetin-deoxyhexose-hexoside were dominant in pulp	Tsamo et al. (2015)
Cavendish banana (*Musa acuminata* 'Cavendish') peel at different ripening stages	Comparison of different extraction methods	Carotenoids	Maximum carotenoids were observed at stage 5 ripening; carotenoids =1.86 μg; (xanthophyll =0.57 μg, β-carotene =0.84 μg); saponification method provided maximum β-carotene yield, and acetone provided higher yield of β-carotene than xanthophylls	Yan et al. (2016)
Dwarf Cavendish (*Musa acuminata* 'Dwarf Cavendish') peel	Solvent extraction at 27 and 35°C	Carotenoids	Higher antioxidant activity was observed in ethyl acetate extracts at 27°C, using FRAP method.	Sheikhzadeh et al. (2015)

Abbreviations: RE: Rutin Equivalent; DPPH: 2, 2-Diphenyl-1-picrylhydrazyl; ABTS: 2,2'-azino-bis(3-ethylbenzothiazoline-6-sulfonic acid; TAE: Tannic Acid Equivalent; FRAP: Ferric Reducing Antioxidant Power

friendly nor economical. The peel can be processed into a variety of products like starch, pectin, dietary fiber, and protein, and to produce non-food products like biochar and biofuels (green diesel, ethanol, and biogas (methane)) by anaerobic digestion. However, banana peel is highly perishable and bulky, with moisture representing 80–90% of the total peel weight. Therefore, the peels from the banana processing plant must be preserved to prevent any deterioration of nutrients and organic components if they are to be utilized in food-based applications. The preservation of peels could be achieved by adopting a series of unit operations like blanching, drying, grinding and packaging, with storage before further utilization. The processing of peels through these unit operations has significantly improved over the years due to innovations and the arrival of alternate technologies and methods. Hence, it is essential to study the recent trends in processing technology of generated peel prior to further utilization without compromising nutritional and organic components.

The first step in the processing of banana peel should be grading and sorting according to the banana fruits' maturity (Figure 10.3). For food-based applications, the target should be peel which is rich in nutrients and bioactive compounds. The peel from overripe fruit should be avoided as they have already lost or converted the important components like starch, fibers, and antioxidants into other forms. However, they contain high concentrations of sugars and could be used in non-food-based applications like the production of biofuels (ethanol, biodiesel, or methane) or solid biomass like biochar.

After grading and sorting, for food-based applications, the next step is blanching to minimize quality loss over time. In general, blanching inactivates the enzymes, reduces the numbers of surface microorganisms and retards the loss of bioactive components like vitamins and phytochemicals. By doing so, blanching greatly preserves the color, flavor and nutritional values of peel. Moreover, several studies have also claimed that blanching could reduce the concentration of antinutrients in peel as well as pesticide residue concentrations on the fruit surface. Traditionally, blanching is performed as high-temperature short-term blanching using hot water or steam at temperatures of 90–95°C for a few minutes. However, in recent decades, lower-temperature (50–60°C) blanching for 30–40 min has been intensively studied. The effectiveness of lower-temperature hot-water blanching can be improved using chemicals like potassium metabisulfite, citric acid, sodium bicarbonate, sodium chloride and calcium chloride. Unfortunately, leaching of water-soluble nutrients during blanching is a major concern that could be avoided by using steam. Alternatively, rapid innovations in blanching have been noticed in recent years by introducing novel technologies like high-pressure processing, ultrasound, microwave, Ohmic, infrared, and radiofrequency blanching. However, these novel technologies are still not viable commercially due to lower capacity with higher economic inputs.

After blanching, peel should be dried to remove moisture, reduce water activity and bring a substantial reduction in weight and volume. Drying minimizes the overall packaging and storage as well as transportation cost of peel during movement from the site of generation to the site of processing. Moreover, drying also restricts spoilage and decay due to microorganisms, and inactivates enzymes to decrease chemical changes. In recent decades, drying techniques have advanced due to the arrival of new drying methods and developments in hybrid drying. However, the new methods like microwave and microwave-assisted, ultrasound-assisted, and ohmic drying are not economical due to higher establishment and operational costs. Nevertheless, these techniques have significantly reduced the drying time without any detrimental effect on quality parameters. However, one method which seems promising for the economical drying of peel is utilizing solar energy with existing drying methods. Normal sun drying is one of the oldest methods and requires prolonged drying times to remove the moisture to a safe level. Hence, external inputs like solar panels, electrical fans, electrical coils and drying chambers improve the overall heat utilization, losses and efficiency with fast drying. Though our primary target is to reduce the moisture to a safe level, the drying temperatures should be selected so that most of the bioactive components will not be affected. After drying, size reduction is an important unit operation that decides the particle size distribution of the dried

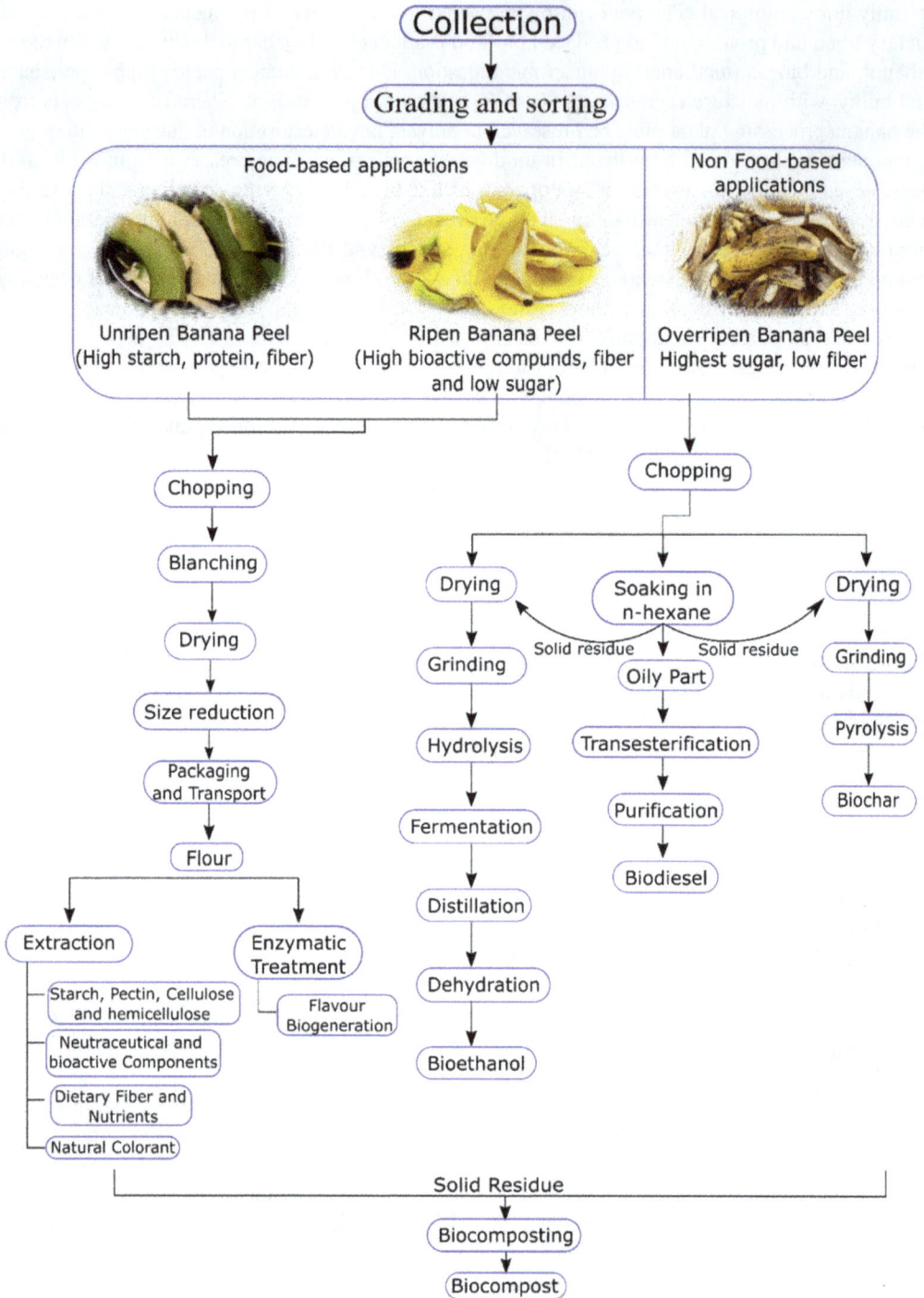

Figure 10.3 Flow chart for processing banana peel for food- and non-food-based applications.

banana peel. Depending on the purpose of utilization, the particle size range varies and hence the size reduction method should be selected in such a way that flour generated must fulfill the required particle size distribution.

10.3.1 Extraction of Starch, Pectin, Protein, Cellulose and Fiber from Banana Peel

Starch extraction from banana peel is a simple process in which the sliced banana peel is boiled in water at 98°C followed by filtration using a muslin cheesecloth. Furthermore, addition of a starter culture for the fermentation process is used to produce wine. The various properties of the peel starch were determined, and it was found that the gelatinization temperature was more than the banana flesh starch and that the content of resistant starch was greater in peel than in the flesh of the fruit (Byarugaba-Bazirake et al., 2014).

Banana peel also contains high concentrations of pectin which is extracted from the peel powder by homogenizing it with hydrochloric acid at pH 1.5 and citric acid at pH 1.7. Continuous heating and stirring on a hot plate for 4 h is followed by filtration through cheesecloth, which is later washed using ethanol. This extraction process of pectin from banana peel was found to be similar to that of commercial citrus pectin in terms of sensory basis. This concludes that the banana peel pectin can be used for various food applications (Castillo-Israel et al., 2015).

Different methods can be used to extract protein from banana peel, such as phenol extraction, trichloroacetic acid-acetone precipitation and trichloroacetic acid preparation (Zhang et al., 2012). Budhalakoti (2021) extracted protein from banana peels, where pieces of banana peel were soaked in acetic acid, followed by homogenization, foam separation and a centrifugation process to separate the final protein sample. However, very few experiments regarding the extraction of protein from banana peels have been reported to date.

The optimum extraction of cellulose from banana peel can be achieved by removing fat using 90% ethanol for 16 h followed by treating with NaOH at pH 11.6 for 24 h to remove protein and bleaching in 15% hydrogen peroxide for 3 h. The cellulose obtained using this procedure was light yellow. It had good emulsifying and oil retention capacity, i.e., 40.70 g oil/g dry matter and 0.08 g oil/g dry matter, respectively. Cellulose extracted from banana peels can be utilized as a natural ingredient in various food systems (Singanusong et al., 2014).

Banana peel is a good source of dietary fiber which shows around 50 g of dietary fiber/100 g dry matter. Various extraction methods have been used to extract the dietary fiber from banana peel, among which wet milling concentrates have been found to achieve the highest yields. In wet milling, washing with hot tap water resulted in removal of the protein fraction and extracting dietary fiber concentrates with good water- and oil-holding capacities (Wachrasiri et al., 2009).

10.3.2 Production of Ethanol, Biodiesel, and Biochar from Banana Peel

Ethanol demand has increased in recent times due to high demands for ethanol by industries. Ethanol is produced by chemical synthesis of petrochemical substrates or by microbial conversion of carbohydrates. Banana peel is pretreated before undergoing acid hydrolysis of banana peel starch and sugar. The starch present is hydrolyzed and converted to fermentable sugar and then transformed microbially to ethanol followed by cooling and filtration of the ethanol produced. Brewer's yeast (*Saccharomyces cerevisiae*) is utilized during the process as an inoculant to achieve a higher yield of ethanol (Bhatia & Paliwal, 2010).

In the past 30 years, the global production of bananas has escalated by 175% g. Therefore, production of biodiesel from banana peel has also increased. Fruit- and vegetable-based waste, such as banana peel, are among the prominent sources for biodiesel production and banana is considered to be one of the most important tropical fruits in many Asian countries. Mostly, banana peel is used as a catalyst to produce biodiesel which gives a very high yield of biodiesel of around 98.95% (Gohain et al., 2017; Pathak et al., 2018).

Biochar is used to increase crop productivity in agriculture, and to reduce leaching and nutrient losses. Biochar is now being produced by the banana peel pyrolysis process under reduced O_2 conditions. The biochar produced from banana peel has high potassium content and, when applied to a crop field, the plant growth was healthier, stronger and darker green in appearance. This suggests that banana peel biochar can serve as an alternative to chemical fertilizer as well as being a way to use agricultural waste (Islam et al., 2019).

10.3.3 Production of Lactic Acid from Banana Peel

To produce lactic acid, large amounts of raw materials are required for the saccharification fermentation process. Banana peel can be an abundant source of raw material, as banana is a tropical fruit and can produced throughout the year. Banana peel can also be readily utilized for the production of lactic acid as they are rich in reducing sugars and can be converted to fermentable sugars through enzymatic hydrolysis known as saccharification. Subsequent fermentation of the resistant starch produces lactic acid (Martinez-Trujillo et al., 2020). Co-culture of lactic acid bacteria, such as *Bacillus licheniformis*, and the fungus *Aspergillus awamori* has been utilized for the production of lactic acid (Mufidah et al., 2017).

10.3.4 Production of Jelly from Banana Peel

Banana peel can be utilized for making a jelly which contains dietary fiber and is a rich source of antioxidants. The process flow for the production of the jelly from banana peel powder is shown in Figure 10.4. It has been found that banana peel jelly has a high content of phenolics and has good antioxidant activity (Lee et al., 2010; Rasidek et al., 2016). The superior antioxidant properties of jellies make them suitable for use as functional foods and nutraceuticals.

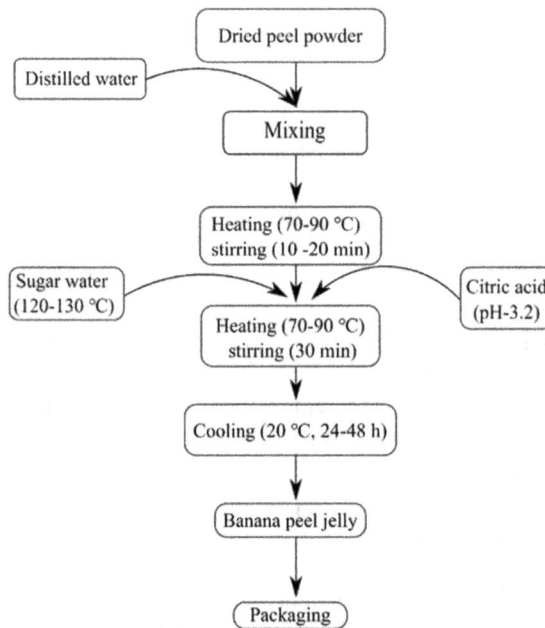

Figure 10.4 Flow chart for preparation of banana peel jelly.

10.4 UTILIZATION OF BANANA PEEL

Due to being rich sources of micronutrients and bioactive compounds, banana peel has been explored for various food and non-food-based uses (Figure 10.5).

10.4.1 Food-Based Utilization

Because of the rich nutritional profile of banana peel, it can be directly utilized in various food formulations. Moreover, the important biocomponents can be extracted with modern extraction methods and used in food supplements and nutraceuticals to protect against multiple nutrient chronic disorders in a sustainable way. Food-based utilization either involves extraction of biocomponents like phytochemicals, antioxidants, starch, proteins, dietary fibers or essential oils or the direct utilization of banana peel in food formulations.

10.4.1.1 Bakery Industry

Banana peel is an excellent source of dietary fiber and contains around 50% dietary fiber (Wachirasiri et al., 2009). The ratio of insoluble to soluble dietary fiber in banana peel varies from 4.48: 1 to 5.46: 1 (Sharma et al., 2016). Studies have reported higher dietary fiber content in banana peel than in wheat, rice, oats, and barley (Sudha et al., 2007). Therefore, the use of dietary fiber extracted from banana peel to produce low-caloric and high-dietary-fiber content products such as biscuits can be a promising option for the bakery industry. It has been reported that incorporation of banana peel at a level of 10% during biscuit production does not affect the color, aroma or taste of the biscuits (Joshi, 2007).

10.4.1.2 Meat Industry

The uses of banana peel-derived components are not restricted to use as a food ingredient, but also have the potential to be used as a food preservative and these preservatives are attracting enormous attention from researchers and industrial scientists due to the increased awareness of the drawbacks of chemical preservatives (Padam et al., 2014). Various antimicrobial compounds, such as β-sitosterol, 12-hydroxystrearic acid and malic acid, extracted from banana peel of *M.* × *paradisiaca* showed an inhibitory effect against various food-borne pathogens like *Staphylococcus aureus*, *Escherichia coli*, *Bacillus cereus* and *Salmonella* sp. Owing to its antioxidant properties, banana peel extract also exhibited preservative capability by reducing the lipid oxidation of raw meat (Devatkal et al., 2014; Aminzare et al., 2019). Therefore, in future, these components could be very good preservatives in industrial applications to replace chemical preservatives (Mokbel and Hashinaga, 2005). Devatkal et al. (2014) conducted a study on banana peel of *M.* × *paradisiaca* and

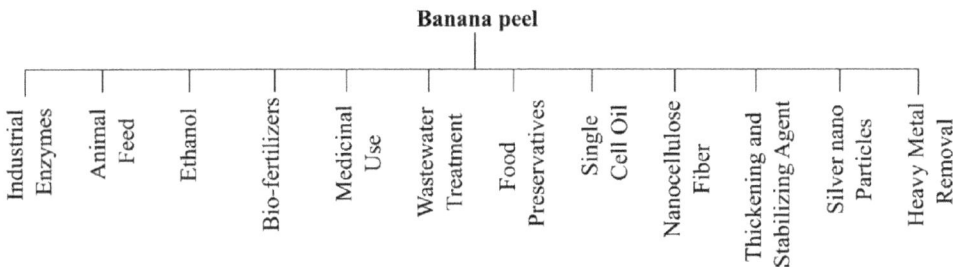

Figure 10.5 Utilization of banana peel.

found that water extracted from the peel of *M. × paradisiaca* minimized the lipid oxidation process in raw meat and could be a promising alternative to a synthetic antioxidant (butylated hydroxy toluene).

10.4.1.3 Jam and Jelly Production

Compared with green bananas, peels of ripen bananas are rich sources of pectin (Mohapatra et al., 2011). Pectin extracted from banana peel is an economic source of pectin for banana-producing countries and it also minimizes the dependency for pectin import (Padam et al., 2014). Pectin extracted from banana peel can be used for the production of jams and jellies (Mohapatra et al., 2011).

10.4.2 As an Animal Feed

Banana peel is rich in micronutrients which can improve the health of animals. Incorporation of approximately 15–30% of banana peel into animal feed has been seen to increase milk production from cows. A growth study reveals that addition of banana peel to a broiler chicken diet acted as a growth enhancer. Banana peel can serve as a complete replacement for maize in the diet for cattle, pigs, goats and chickens without showing any adverse effects (Pathak et al., 2016).

The protein and sugar content of banana peel can be enriched by 34% and 142%, respectively, through solid-state fermentation, by processing banana peel as a mycological medium on which to grow microfungi (Akinyele and Agbro, 2007; Yabava and Ado, 2008). The high concentrations of protein and sugar present in processed banana peel can increase the nutritional quality of animal feeds, which could be a comparable alternative to imported animal feeds such as soyabean meal (Padam et al., 2014).

10.4.3 Biofertilizer

Due to their high alkaline nature, banana peel can be used to reduce the acidity of soil. Vermicompost produced by decomposition of banana peel using earthworms can serve as an effective fertilizer that will encourage nutrient uptake and improve plant growth and yield (Achsah and Prabha, 2013).

10.4.4 Medicinal Use

Banana peel has been used to fight cancer and heart diseases due to the presence of gallocatechin in high concentrations (Khoozani et al., 2019). Arabinoxylans in banana peels may be used as a potential nutrient supplement. 7,8-dihydroxy-3-methyl-isochromanone-4 extracted from banana peel can be used as a cell adhesive and for treatment of atherosclerosis and endothelium injury (Ranjha et al., 2022). The lipid, phenolics and tannins in banana peel have good antimicrobial properties and anti-allergic effects against disease caused by both Gram-positive and Gram-negative bacteria. Banana peel extract can also suppress the growth of ventral prostrates and seminal vesicles, increasing testosterone activity; hence it can be used for prostate hyperplasia.

10.4.5 Production of Single Cell Protein and Industrial Enzymes

Single cell protein has been produced from *Aspergillus oryzae*, *Rhizopus* sp. and *S. cerevisiae* by growing them over banana peels as the carbon source. Mycelial protein has also been produced using the fungus *Aspergillus niger* with banana peel as a substrate (Yabaya & Ado, 2008). Alpha-amylase has been produced using banana peels as a substrate by solid-state fermentation with bacteria such as *Bacillus subtilis* and fungi such as *Aspergillus flavus*. (Adeniran et al., 2010).

10.4.6 Wastewater Treatment

Banana peels have been used as a bio-adsorbent to remove contamination by dyes and organic pollutants from wastewaters. The presence of carboxyl, hydroxyl and amino groups allows a high adsorption capacity of organic compounds in banana peel (Khan et al., 2017; Darge & Mane, 2015).

10.4.7 Banana Peel-Based Nanocomposites

Cellulose nanofiber extracted from banana peels can be used in biomedical industry and paper industry as a strengthening component in composite materials (Tibolla et al., 2014). Moreover, nanocomposites have been used to reinforce biodegradable films to improve the mechanical properties. Furthermore, cellulose nanofibers have been used as a Pickering emulsifier to stabilize oil-in-water emulsions. Cellulose nano fiber has many advantages like they are strong, stiff and highly crystalline. Cellulose nano fiber can be extracted from different methods like chemical or enzymatic. Readers can find details about production of nanofibers from banana peels in studies conducted by Tibolla et al. (2014), and Khawas et al. (2016).

10.4.8 Biofuel from Banana Peel

The growing world population is increasing the demand for ethanol in various industries and has resulted in searches for alternative sources of energy (Santa-Maria et al., 2013). The negative impact of fossil fuels on the environment has provided the momentum to search for alternative sources, such as biofuels. Banana peel contains cellulose and hemicelluloses in significant amounts, and hence can be used in the production of bioethanol or biochar.

10.4.9 Production of Bioethanol from Banana Peel

A study by Waghmare and Arya (2016) optimized the fermentation conditions for production of ethanol from banana peel. In this study, 35.5 g/L ethanol was obtained by hydrolyzing green banana peel powder at optimum conditions of 1.5% H_2SO_4 at 120°C for 20 min and subsequent fermentation by *S. cerevisiae* strain NCIM 3095. Moreover, oily parts of the peel extract can be readily separated using organic solvents like n-hexane for production of biodiesel by transesterification.

10.4.10 Briquettes Made from Banana Peel

A mixture of banana peel and molasses treated using a high-pressure press was reported to generate briquettes using a low-cost raw material (Wilaipon, 2009). These briquettes can be a promising alternative to solid fuel, such as coal, and these briquettes were found to exhibit a low burning rate compared with other agricultural waste-based briquettes, such as rice husks, coconut fiber, sawdust, palm fiber, etc. (Padam et al., 2014).

10.5 CONCLUSIONS

Banana peel represents a major waste material obtained from banana fruits that contain important organic components like phytochemicals, cellulose, hemicelluloses, lignin, starch, sugars and protein. However, the concentration of these components greatly depends on the ripening stage of banana fruits. Therefore, the processing approach may vary depending on the utilization. Banana peel has been found to be a very convenient and economic alternative biomaterial for extraction of bioactive compounds for functional or nutraceutical food, and for production of enzymes and lactic acid. Furthermore, banana peel can be utilized in non-food-based applications. Despite its many potential food- and non-food-based uses, the processing and utilization of banana peel is still restricted.

ACKNOWLEDGMENT

One of the authors, Mr. Yogesh Kumar, is thankful to AICTE, Department of Higher Education, India for supporting him with a National Doctoral Fellowship, 2019.

REFERENCES

Abubakar, U. S., Yusuf, K. M., Safiyanu, I., Abdullahi, S., Saidu, S. R., Abdu, G. T., & Indee, A. M. (2016). Proximate and mineral composition of corn cob, banana and plantain peels. *International Journal of Food Science and Nutrition, 1*(6), 25–27.

Achsah, R. S., & Prabha, M. L. (2013). Potential of vermicompost produced from banana waste (*Musa paradisiaca*) on the growth parameters of solanum lycopersicum. *International Journal of ChemTech Research, 5*, 2141–2153.

Adeniran, H. A., Abiose, S. H., & Ogunsua, A. O. (2010). Production of fungal β-amylase and amyloglucosidase on some Nigerian agricultural residues. *Food Bioprocess Technology, 3*, 693–698.

Akinyele, B. J., & Agbro, O. (2007). Increasing the nutritional value of plantain wastes by the activities of fungi using the solid-state fermentation technique. *Research Journal of Microbiology, 2*(2),117–124.

Alothman, M., Bhat, R., & Karim, A. A. (2009). Antioxidant capacity and phenolic content of selected tropical fruits from Malaysia, extracted with different solvents. *Food Chemistry, 115*, 785–788.

Aminzare, M., Hashemi, M., Ansarian, E., Bimkar, M., Azar, H. H., Mehrasbi, M. R., ... & Afshari, A. (2019). Using natural antioxidants in meat and meat products as preservatives: A review. *Advances in Animal and Veterinary Sciences, 7*(5), 417–426.

Anhwange, B. A., Ugye, T. J., & Nyiaatagher, T. D. (2009). Chemical composition of Musa sapientum (banana) peels. *Electronic Journal of Environmental, Agricultural and Food Chemistry, 8*(6), 437–442.

Annadurai, G., Juang, R. S., & Lee, D. J. (2002). Use of cellulose-based wastes for adsorption of dyes from aqueous solutions. *Journal of Hazardous Materials, 92*, 263–274.

Babbar, N., Oberoi, H. S., Uppal, D. S., & Patil, R. T. (2011). Total phenolic content and antioxidant capacity of extracts obtained from six important fruit residues. *Food Research International, 44*(1), 391–396.

Baini, R., & Langrish, T. A. G. (2009). Assessment of colour development in dried bananas–measurements and implications for modelling. *Journal of Food Engineering, 93*(2), 177–182.

Bakry, F., Carreel, F., Caruana, M. L., Côte, F. X., Jenny, C., & Têzenas du Montcel, H. (1997). Les bananiers. In Charrier, A., Hamon, S., Jacquot, M., & Nicolas, D., eds. *L'amélioration des plantes tropicales* (pp. 109–139). Montpellier, France: CIRAD/ORSTOM.

Bardiya, N., & Somayaji, K. (1996). Biomethanation of banana peel and pineapple waste. *Bioresource Technology, 58*, 73–76.

Bhatia, L., & Paliwal, S. (2010). Banana peel waste as substrate for ethanol production. *International Journal of Biotechnology and Bioengineering Research, 1*(2), 213–218.

Budhalakoti, N. (2019). Formulation and standardisation of banana peel extracted insoluble dietary fibre based buns. *Current Journal of Applied Science and Technology, 32*, 1–9.

Budhalakoti, N. (2021). Extraction of protein from banana by-product and its characterization. *Journal of Food Measurement and Characterization, 15*(3), 2202–2210.

Byarugaba-Bazirake, G. W., Byarugaba, W., Tumusiime, M., & Kimono, D. A. (2014). The technology of producing banana wine vinegar from starch of banana peels. *African Journal of Food Science and Technology, 5*(1), 1–5.

Castillo-Israel, K. A. T., Baguio, S. F., Diasanta, M. D. B., Lizardo, R. C. M., Dizon, E. I., & Mejico, M. I. F. (2015). Extraction and characterization of pectin from Saba banana [Musa'saba'(*Musa acuminata* x *Musa balbisiana*)] peel wastes: A preliminary study. *International Food Research Journal, 22*(1), 202–207.

Choehom, R., Ketsa, S., & van Doorn, W. G. (2004). Senescent spotting of banana peel is inhibited by modified atmosphere packaging. *Postharvest Biology and Technology*, *31*(2), 167–175.

Cos, P., Vlietinck, A. J., Berghe, D. V., & Maes, L. (2006). Anti-infective potential of natural products: How to develop a stronger in vitro 'proof-of-concept'. *Journal of Ethnopharmacology*, *106*(3), 290–302.

Dahham, S. S., Mohamad, T. A., Tabana, Y. M., & Majid, A. M. S. A. (2015). Antioxidant activities and anticancer screening of extracts from banana fruit (Musa sapientum). *Academic Journal of Cancer Research*, *8*(2), 28–34.

Darge, A., & Mane, S. J. (2015). Treatment of industrial wastewater by using banana peels and fish scales. *International Journal of Science and Research (IJSR)*, *4*(7), 600–604.

Davey, M. W., Van-den Bergh, I., Markham, R., Swennen, R., & Keulemans, J. (2009). Genetic variability in musa fruit provitamin A carotenoids, lutein and mineral micronutrient contents. *Food Chemistry*, *115*, 806–813.

Devatkal, S. K., Kumboj, R., & Paul, D. (2014). Comparative antioxidant effect of BHT and water extracts of banana and sapodilla peels in raw poultry meat. *Journal of Food Science and Technology*, *51*(2), 387–391.

Drury, R., Hortesteiner, S., Donnsion, I., Bird, C. R., & Seymour, G. B. (1999). Chlorophyll catabolism and gene expression in the peel of ripening banana fruits. *Physiologia Plantarum*, *107*, 32–38.

Emaga, T. H., Andrianaivo, R. H., Wathelet, B., Tchango, J. T., & Paquot, M. (2007). Effects of the stage of maturation and varieties on the chemical composition of banana and plantain peels. *Food Chemistry*, *103*(2), 590–600.

Emaga, T. H., Bindelle, J., Agneesens, R., Buldgen, A., Wathelet, B., & Paquot, M. (2011). Ripening influences banana and plantain peels composition and energy content. *Tropical Animal Health and Production*, *43*(1), 171–177.

Emaga, T. H., Robert, C., Ronkart, S. N., Wathelet, B., & Paquot, M. (2008). Dietary fibre components and pectin chemical features of peels during ripening in banana and plantain varieties. *Bioresource Technology*, *99*(10), 4346–4354.

Eshak, N. S. (2016). Sensory evaluation and nutritional value of balady flat bread supplemented with banana peels as a natural source of dietary fiber. *Annals of Agricultural Sciences*, *61*(2), 229–235.

Fabricant, D. S., & Farnsworth, N. R. (2001). The value of plants used in traditional medicine for drug discovery. *Environmental Health Perspectives*, *109*(suppl 1), 69–75.

FAO. (2020a). Banana market review. *February 2020 Snapshot*, 1–4.

FAO. (2020b). MEDIUM-TERM OUTLOOK. Prospects for global production and trade in bananas and tropical fruits. http://www.fao.org/3/ca7568en/ca7568en.pdf.

Faturoti, B. O., Emah, G. N., Isife, B. I., Tenkouano, A., & Lemchi, J. (2006). Prospects and determinants of adoption of IITA plantain and banana-based technologies in three Niger Delta States of Nigeria. *African Journal of Biotechnology*, *5*, 1319–1323.

Gebregergs, A., Gebresemati, M., & Sahu, O. (2016). Industrial ethanol from banana peels for developing countries: Response surface methodology. *Pacific Science A Review: Natural Science and Engineering*, *18*(1), 22–29.

Gohain, M., Devi, A., & Deka, D. (2017). Musa balbisiana colla peel as highly effective renewable heterogeneous base catalyst for biodiesel production. *Industrial Crops and Products*, *109*, 8–18.

González-Montelongo, R., Lobo, M. G., & González, M. (2010). Antioxidant activity in banana peel extracts: Testing extraction conditions and related bioactive compounds. *Food Chemistry*, *119*(3), 1030–1039.

Gu, S., Zhu, K., Luo, C., Jiang, Y., & Xu, Y. (2014). Microwaves-assisted extraction of polyphenols from banana peel. *Medicinal Plant*, *5*(1), 21.

Hussein, H. S., Shaarawy, H. H., Hussien, N. H., & Hawash, S. I. (2019). Preparation of nano-fertilizer blend from banana peels. *Bulletin of the National Research Centre*, *43*(1), 1–9.

Ibrahim, H. M. M. (2015). Green synthesis and characterization of silver nanoparticles using banana peel extract and their antimicrobial activity against representative microorganisms. *Journal of Radiation Research and Applied Sciences*, *8*(3), 265–275.

Ishak, N. A., Razak, N. A. A., Dek, M. S. P., & Baharuddin, A. S. (2020). Production of high tannin content and antioxidant activity extract from an unripe peel of *Musa acuminata* (Cavendish) using ultrasound-assisted extraction (UAE). *BioResources*, *15*(1), 1877–1893.

Islam, M., Halder, M., Siddique, M. A., Razir, S. A. A., Sikder, S., & Joardar, J. C. (2019). Banana peel bio-char as alternative source of potassium for plant productivity and sustainable agriculture. *International Journal of Recycling of Organic Waste in Agriculture*, *8*(1), 407–413.

Joshi, R. V. (2007). Low calorie biscuits from banana peel pulp. *Journal of Solid Waste Technology & Management, 33*(3), 142–147.

Kanazawa, K., & Sakakibara, H. (2000). High content of dopamine, a strong antioxidant, in cavendish banana. *Journal of Agricultural and Food Chemistry, 48,* 844–848.

Kaur, S., Panesar, P. S., & Chopra, H. K. (2021). Citrus processing by-products: An overlooked repository of bioactive compounds. *Critical Reviews in Food Science and Nutrition,* 1–20.

Khan, A., Naqvi, H. J., Afzal, S., Jabeen, S., Iqbal, M., & Riaz, I. (2017). Efficiency enhancement of banana peel for waste water treatment through chemical adsorption. *Proceedings of the Pakistan Academy of Sciences: A. Physical and Computational Sciences, 54*(3), 329–335.

Khawas, P., & Deka, S. C. (2016). Isolation and characterization of cellulose nanofibers from culinary banana peel using high-intensity ultrasonication combined with chemical treatment. *Carbohydrate Polymers, 137,* 608–616.

Khoozani, A., Birch, J., & Bekhit, A. E. D. A. (2019). Production, application and health effects of banana pulp and peel flour in the food industry. *Journal of Food Science and Technology, 56*(2), 548–559.

Lee, E. H., Yeom, H. J., Ha, M. S., & Bae, D. H. (2010). Development of banana peel jelly and its antioxidant and textural properties. *Food Science and Biotechnology, 19*(2), 449–455.

Li, S., Zhang, H. Q., Tony, J. Z., & Hsieh, F. (2005). Textural modification of soya bean/corn extrudates as affected by moisture content, screw speed and soya bean concentration. *International Journal of Food Science and Technology, 40,* 731–741.

Martínez-Trujillo, M. A., Bautista-Rangel, K., García-Rivero, M., Martínez-Estrada, A., & Cruz-Díaz, M. R. (2020). Enzymatic saccharification of banana peel and sequential fermentation of the reducing sugars to produce lactic acid. *Bioprocess and Biosystems Engineering, 43*(3), 413–427.

Mohapatra, D., Mishra, S., Singh, C. B., & Jayas, D. S. (2011). Post-harvest processing of banana: Opportunities and challenges. *Food and Bioprocess Technology, 4*(3), 327–339.

Mohapatra, D., Mishra, S., & Sutar, N. (2010). Banana and its by-product utilisation: An overview. *Journal of Scientific and Industrial Research, 69,* 323–329.

Mokbel, M. S., & Hashinaga, F. (2005). Antibacterial and antioxidant activities of banana (Musa, AAA cv. Cavendish) fruits peel. *American Journal of Biochemistry and Biotechnology, 1*(3), 125–131.

Mordor Intelligence. (2020). Banana market size, share, analysis – growth, trends, and forecast (2020–2025). https://www.mordorintelligence.com/industry-reports/banana-market.

Mota, R. V., Lajolo, F. M., Ciacco, C., Cordenunsi, B. R., & Paulo, S. (2000). Composition and functional properties of banana flour from different varieties. *Starch Staerke, 52,* 63–68.

Mufidah, E., Prihanto, A. A., & Wakayama, M. (2017). Optimization of l-lactic acid production from banana peel by multiple parallel fermentation with *Bacillus licheniformis* and *Aspergillus awamori*. *Food Science and Technology Research, 23*(1), 137–143.

Mukhopadhyay, S., Fangueiro, R., Arpac, Y., & Senturk, U. (2008). Banana fibers variability and fracture bahaviour. *Engineered Fibers and Fabrics, 3*(2), 39–45.

Nathoa, C., Sirisukpoca, U., & Pisutpaisal, N. (2014). Production of hydrogen and methane from banana peel by two phase anaerobic fermentation. *Energy Procedia, 50,* 702–710.

Oliveira, L., Cordeiro, N., Evtuguin, D. V., Torres, I. C., & Silvestre, A. J. D. (2007). Chemical composition of different morphological parts from 'Dwarf Cavendish' banana plant and their potential as a non-wood renewable source of natural products. *Industrial Crops Products, 26,* 163–172.

Padam, B. S., Tin, H. S., Chye, F. Y., & Abdullah, M. I. (2014). Banana by-products: An under-utilized renewable food biomass with great potential. *Journal of Food Science and Technology, 51*(12), 3527–3545.

Pan, Z., Shih, C., McHugh, T. H., Wood, D., & Hirschberg, E. (2008). Sequential infrared radiation and freeze-drying method for producing crispy strawberries. *Transactions of the ASABE, 51*(1), 205–216.

Pappu, A., Patil, V., Jain, S., Mahindrakar, A., Haque, R., & Thakur, V. K. (2015). Advances in industrial prospective of cellulosic macromolecules enriched banana biofibre resources: A review. *International Journal of Biological Macromolecules, 79,* 449–458.

Pathak, G., Das, D., Rajkumari, K., & Rokhum, L. (2018). Exploiting waste: Towards a sustainable production of biodiesel using *Musa acuminata* peel ash as a heterogeneous catalyst. *Green Chemistry, 20*(10), 2365–2373.

Pathak, P. D., Mandavgane, S. A., & Kulkarni, B. D. (2016). Valorization of banana peel: A biorefinery approach. *Reviews in Chemical Engineering, 32*(6), 651–666.

Ploetz, R. C. (2015). Fusarium Wilt of Banana. *Phytopathology, 105*(12), 1512–1521.

Puraikalan, Y. (2018). Characterization of proximate, phytochemical and antioxidant analysis of banana (*Musa sapientum*) peels/skins and objective evaluation of ready to eat/cook product made with banana peels. *Current Research in Nutrition and Food Science Journal*, *6*(2), 382–391.

Ranjha, M. M. A. N., Irfan, S., Nadeem, M., & Mahmood, S. (2022). A comprehensive review on nutritional value, medicinal uses, and processing of banana. *Food Reviews International*, *38*(2), 199–225.

Rasidek, N. A. M., Nordin, M. F. M., & Shameli, K. (2016). Formulation and evaluation of semisolid jelly produced by *Musa acuminata* Colla (AAA Group) peels. *Asian Pacific Journal of Tropical Biomedicine*, *6*(1), 55–59.

Rodriguez-Ambriz, S. L., Islas-Hernandez, J. J., Agama-Acevedo, E., Tovar, J., & Bello-Perez, L. A. (2008). Characterization of a fiber-rich powder prepared by liquefaction of unripe banana flour. *Food Chemistry*, *107*, 1515–1521.

Romphophak, T., Siriphanich, J., Ueda, Y., Abe, K., & Chachin, K. (2005). Changes in concentrations of phenolic compounds and polyphenol oxidase activity in banana peel during storage. *Food Preservation Science*, *31*, 111–115.

Santa-Maria, M., Ruiz-Colorado, A. A., Cruz, G., & Jeoh, T. (2013). Assessing the feasibility of biofuel production from lignocellulosic banana waste in rural agricultural communities in Peru and Colombia. *Bioenergy Research*, *6*(3), 1000–1011.

Sharma, S. K., Bansal, S., Mangal, M., Dixit, A. K., Gupta, R. K., & Mangal, A. K. (2016). Utilization of food processing by-products as dietary, functional, and novel fiber: A review. *Critical Reviews in Food Science and Nutrition*, *56*(10), 1647–1661.

Sheikhzadeh, M., Hakimzadeh, V., & Abedi, M. R. (2015). Carotenoids extraction optimization of lutein-based banana peel. *Journal of Applied Environmental and Biological Sciences*, *4*(11S), 213–217.

Singanusong, R., Tochampa, W., Kongbangkerd, T., & Sodchit, C. (2014). Extraction and properties of cellulose from banana peels. *Suranaree Journal of Science and Technology*, *21*(3), 201–213.

Someya, S., Yoshiki, Y., & Okubo, K. (2002). Antioxidant compounds from bananas (*Musa Cavendish*). *Food Chemistry*, *79*, 351–354.

Sudha, M. L., Vetrimani, R., & Leelavathi, K. (2007). Influence of fibre from different cereals on the rheological characteristics of wheat flour dough and on biscuit quality. *Food Chemistry*, *100*(4), 1365–1370.

Sulaiman, S. F., Yusoff, N. A. M., Eldeen, I. M., Seow, E. M., Sajak, A. A. B., & Ooi, K. L. (2011). Correlation between total phenolic and mineral contents with antioxidant activity of eight Malaysian bananas (*Musa* sp.). *Journal of Food Composition and Analysis*, *24*(1), 1–10.

Tibolla, H., Pelissari, F. M., & Menegalli, F. C. (2014). Cellulose nanofibers produced from banana peel by chemical and enzymatic treatment. *LWT-Food Science and Technology*, *59*(2), 1311–1318.

Tsamo, C. V. P., Herent, M., Tomekpe, K., Emaga, T. H., Quetin-Leclercq, J., Rogez, H., Larondelle, Y., & Andre, C. (2015). Phenolic profiling in the pulp and peel of nine plantain cultivars (*Musa* sp.). *Food Chemistry*, *167*, 197–204.

Udosen, E. O., & Enang, M. I. (2000). Chemical composition and soaping characteristics of peels from plantain (*Musa paradisiaca*) and banana (*Musa sapientum*). *Global Journal of Pure and Applied Sciences*, *6*(1), 79–82.

Vijayakumar, S., Presannakumar, G., & Vijayalakshmi, N. R. (2008). Antioxidant activity of banana flavonoids. *Fitoterapia*, *79*(4), 279–282.

Villaverde, J. J., Oliveira, L., Vilela, C., Domingues, R. M., Freitas, N., Cordeiro, N., Freire, C. S. R., & Silvestre, A. J. (2013). High valuable compounds from the unripe peel of several Musa species cultivated in Madeira Island (Portugal). *Industrial Crops and Products*, *42*, 507–512.

Vu, H. T., Scarlett, C. J., & Vuong, Q. V. (2017). Optimization of ultrasound-assisted extraction conditions for recovery of phenolic compounds and antioxidant capacity from banana (*Musa cavendish*) peel. *Journal of Food Processing and Preservation*, 41(5), e13148.

Vu, H. T., Scarlett, C. J., & Vuong, Q. V. (2019a). Changes of phytochemicals and antioxidant capacity of banana peel during the ripening process; with and without ethylene treatment. *Scientia Horticulturae*, *253*, 255–262.

Vu, H. T., Scarlett, C. J., & Vuong, Q. V. (2019b). Maximising recovery of phenolic compounds and antioxidant properties from banana peel using microwave assisted extraction and water. *Journal of Food Science and Technology*, *56*(3), 1360–1370.

Wachirasiri, P., Julakarangka, S., & Wanlapa, S. (2009). The effects of banana peel preparations on the properties of banana peel dietary fibre concentrate. *Songklanakarin Journal of Science and Technology*, *31*(6), 605–611.

Waghmare, A. G., & Arya, S. S. (2016). Utilization of unripe banana peel waste as feedstock for ethanol production. *Bioethanol*, 2, 146–156.

Wilaipon, P. (2009). The effects of briquetting pressure on banana-peel briquette and the banana waste in Northern Thailand. *American Journal of Applied Sciences*, 6(1), 167–171.

Yabaya, A., & Ado, S. A. (2008). Mycelial protein production by *Aspergillus niger* using banana peels. *The Scientific World Journal*, 3, 9–12.

Yan, L., Fernando, W. M., Brennan, M., Brennan, C. S., Jayasena, V., & Coorey, R. (2016). Effect of extraction method and ripening stage on banana peel pigments. *International Journal of Food Science & Technology*, 51(6), 1449–1456.

Zhang, L. L., Feng, R. J., & Zhang, Y. D. (2012). Evaluation of different methods of protein extraction and identification of differentially expressed proteins upon ethylene-induced early-ripening in banana peels. *Journal of the Science of Food and Agriculture*, 92(10), 2106–2115.

Zhang, P., Whistler, R. L., BeMiller, J. N., & Hamaker, B. R. (2005). Banana starch: Production, physicochemical properties, and digestibility—A review. *Carbohydrate Polymers*, 59(4), 443–458.

CHAPTER 11

Peach Wastes and By-Products
Chemistry, Processing, and Utilization

Saadiya Naqash and Haroon Naik

CONTENTS

11.1 INTRODUCTION

Peach (Prunus persica L.) falls under the classification of fruits that thrive under mild climatic conditions, with Persia as its suggested country of origin, from whence it has derived its species name persica, although its actual country of origin is China (Alipasandi et al., 2013). Taxonomically, peach has been categorized as a member of the family Rosaceae and the subfamily Prunoideae. The genus for this fruit is Prunus, subgenus is Amygdalus and it falls under section Euamygdalus. The peach fruit is widely cultivated over a broad range of climatic conditions. Peach trees have a strong tolerance to humid conditions as compared with other drupes and prefer sunny and warm weather during the fruiting season and requiring about 24°C to achieve ripening. Peach cultivation requires proper soil drainage because of its high sensitivity to waterlogging; however, it thrives best in foothills, high hills and mid hills situations. Peach cultivation and fruit production is an emerging industry and is highly active amongst other fruits of the same category, with new cultivars/varieties being produced around the year (Elsadr, 2016) with an annual increase in production rate of 3.34%. As reported by FAO (2013), the total global peach production in 2013 was about 21 MT.

The peach fruit crop is ready to harvest for the export markets when it reaches an appearance of good colour with a hard, outer skin. However, if the fruits are intended to be consumed domestically, they are generally harvested at the ripe stage. The yield of the crop depends strongly on the pre- and post-harvest management practices, seed quality/variety and soil type. There is a huge number of peach varieties distributed across the globe, reflecting a huge phenotypic diversity and providing varieties with a range of aromas, sweetness/acidity ratios and textures for the consumers (Okie, Bacon, & Bassi, 2008). An average peach weighs between 145 g and 200 g, with an edible weight in the range 85–100 g depending upon the variety. The apparent skin colour of peach fruit is not a prime parameter for the varieties utilized in the peach-processing industry, although, for the varieties cultivated for whole-fruit consumption, colour is a fundamental factor for consumer appeal (Byrne et al., 2007). A uniform cream-yellow colour of skin and flesh is ideal especially in the delayed-ripening varieties cultivated in Spain (Espada et al., 2009).

11.2 PHYSICAL AND CHEMICAL COMPOSITION OF PEACH

Mechanical properties of different peach varieties fall across a broad range. Average geometric properties, namely length, breadth, thickness, volume, polar diameter, surface area, density, sphericity, arithmetic diameter and equivalent diameter are in the range between 50–65 (mm)±0.65, 40–70 (mm)±0.57, 25–68 (mm)±0.87, 40–161 (cm^3)±1.72, 4.5–9.9 (mm), 4,365–13,984 (mm^2), 0.81–1.19 (g/mm^3), 79–104.2 (%), 42.3–66.8 (mm) and 42.02–66.5 (mm) respectively. Cumulative concentration of acids in peaches range from 5.59 to 18.50 g/L, identified as malic acid (42.92%–84.30% of total acids), citric acid (3.72%–31.61%), quinic acid (14.56%–57.54%) and shikimic acid (0.14%–1.89%), but the level of L-ascorbic acid (vitamin C) in peaches is quite low as compared with kiwis and oranges. (Remorini, 2008). Like almost all other fruits and vegetables, the nutritional value and the post-harvest quality of a peach is estimated by its biochemical composition (Wang et al., 2009, Colaric et al., 2005). Generally, peach is considered to be a nutritionally rich fruit, which is full of major and minor nutrients (Iordǎnescu et al., 2015, Manzoor et al., 2012). It contains a negligible amount of fat but a large amount of water (88.87% of total fresh fruit weight). The average total sugar concentration in peaches ranges from 89 g/L to 185 g/L, wherein approximately 55%–73%, 6.5%–15.5%, 6.8%–17% and 1.1%–10.8% are represented by sucrose, glucose, fructose and sorbitol, respectively (Reig et al., 2013). In the case of ripe and mature peaches, a trace amount of other sugars like maltose, isomaltose, raffinose, xylose, trehalose, etc. are also found (Cirilli et al., 2016). Whole peaches represent an excellent source of fibre, containing approximately 1.5% fibre including both soluble and insoluble dietary fibres. Peaches contain approximately 0.25 g/100 g fresh weight of fat, although the fatty acids present in peaches (linoleic, linolenic and palmitic acid) are considered to have a foremost protective role with respect to a number of coronary diseases (Moreno et al., 2006, Duan et al., 2013). Major vitamins in peach are the B- complex vitamins, vitamin C and vitamin E. The concentration of vitamin C in peaches has been reported to be in the range 10.4–33.3 mg/100 g (Saidani et al., 2017), while the mineral content identified in peach consists mainly of potassium, calcium, nitrogen, phosphorus, magnesium, manganese, iron, copper, zinc, chromium, nickel, lead, cobalt, selenium and fluoride (USDA, 2016, Başar et al., 2006, Melo et al., 2016); of all the minerals present, the most abundant mineral in peach is potassium (190 mg/100 g), followed by phosphorus (24 mg/g), calcium (9 mg/g) and magnesium (6 mg/g) (Wills, 1983, USDA, 2016).

Peach is also a storehouse of health-promoting phytochemicals that have been reported to exhibit antioxidant, antimicrobial (Umar et al., 2005, Raturi et al., 2011) antiparasitic (Kumar et al., 2015), antidiabetic (Bahadoran, 2013, Scalbert, 2005) and anti-inflammatory activities (Gasparotto et al., 2014), as well as protectants against cardiovascular diseases (Kono et al., 2013), neurodegenerative diseases (Kozłowska et al., 2014) and cancers (Noratto et al., 2014). The phytochemicals have been classified into different groups, i.e., phenolic compounds, volatile compounds and carotenoids (Holst, 2008, Liu, et al., 2004). As reported in the literature, more than one hundred bioactive compounds are present in peaches (Wang et al., 2009). The major phenolic compounds present in peaches are the phenylpropanoids and their derivatives, chlorogenic acid and neochlorogenic acid (an isomer of chlorogenic acid), although gallic acid is also present in lesser concentrations (Aubert et al., 2014, Davidović, 2013, Campbell et al., 2013, Liu et al., 2015). The major carotenes identified in peach fruit are β-cryptoxanthin and β-carotene, in the concentration range 6×10^{-4} g/kg–3.6×10^{-3} g/kg in the peel and 6×10^{-4} g /kg–1.60×10^{-3} g/kg in the flesh, (Gil et al., 2002), In contrast, the subtypes of flavonoids present in peach are flavonols, flavan-3-ols and anthocyanins (Zhao et al., 2015). The flavonols found in peaches include quercetin and its derivatives that include its glucosides, quercetin galactosides and quercetin rutinosides (Aubert et al., 2014, Davidović et al., 2013, Campbell et al., 2013, Liu et al., 2015, Zhao et al., 2015; Scordino et al., 2012, Tomas-Barberan et al., 2001, Bento et al., 2018). The amount of flavonols in white-fleshed peaches is in the range 1.5 ± 0.01 to 6.3 ± 0.2 mg/100 g fresh weight and in the case of yellow-flesh peaches is in the range 1.5 ± 0.1 to 8.3 ± 0.3 mg/100 g fresh weight (Ceccarelli et al., 2016). The major anthocyanins reported to be present in peach are cyanidin-3-O-glucoside and cyanidin-3-O-rutinoside (Aubert, 2014, Campbell et al., 2013, Zhao et al., 2015, Tomas-Barberan et al., 2001, Bento et al., 2018) among which cyanidin-3-O-glucoside was the most abundant, with concentrations that range from 52.4 ± 12.9 to 325.1 ± 82.4 mg/kg fresh weight in peel and 2.7 ± 1.2 to 17.6 ± 4.9 mg/kg fresh weight in pulp. In the case of peel, little or no cyanidin-3-O-rutinoside was detected, with concentrations ranging from 1.8 ± 0.8 to 12.2 ± 0.4 mg/kg fresh weight (Tomas-Barberan et al., 2001) As detailed by Ceccarelli and collaborators (2016), three major flavan-3-ols are in peach, namely catechin, epicatechin and proanthocyanidin B. That study identified the white-fleshed peaches to be rich in epicatchin with values ranging from 11.7 ± 0.5 to 20.6 ± 0.9 mg/100 g fresh weight in the exocarp and 1.0 ± 0.1 to 8.8 ± 0.2 mg/100 g fresh weight in the mesocarp, followed by procyanidin B1 and catechin present in the range of 0.7 ± 0.5 to 13.3 ± 1.3 mg/ 100 g fresh weight in the exocarp and 1.1 ± 0.1 to 9.2 ± 0.6 mg/100 g fresh weight in the mesocarp, and 0.6 ± 0.1 to 3.8 ± 0.2 mg/100 g fresh weight in the exocarp and 1.0 ± 0.1 to 5.5 ± 0.2 mg/100 g fresh weight in the mesocarp (Ceccarelli et al., 2016). The major phytochemicals present in the peach and their structure is presented in Figure 11.1.

11.3 CHEMISTRY OF PEACH WASTES AND BY-PRODUCTS

In the current age targeting zero waste, scientists and researchers are focusing on the valorisation of industrial wastes that can help to improve the overall sustainability of the food chain. Waste generation in food industries across the globe has attracted attention owing to its social, economic and environmental effects, with 39% of food waste and by-products in developed countries being generated from the food processing industries. This waste from food industries is expected to be a potential source of high-value bioactives. Peach-processing industries generate peel, seeds and some fruit chunks. Ripeness of peaches is a major indicator of waste production, depending upon which approximately 10% of peach pulp is lost as waste. On average, 100 g of waste peach pulp contains 0.1 g sugar, 0.15 g protein, 0.03 g raw fat, 0.15 g potassium, 0.01 g calcium, 0.01 g phosphorus, 0.25 g cellulose and 0.03 g pectin (Argun and Dao, 2017). Owing to their nutrient-dense composition and the high concentrations of phytochemicals, processing of these wastes can help to recover the components like carotenoids and phenolic compounds that can provide an economic boost to the industry and achieve environmental protection, the bioactives also having broad area of application in the food, pharmaceutical and cosmetic industries.

11.3.1 Peach Stone

The peach stone (or pit) comprises a seed that contains a large amount of protein which is yet to be explored in terms of characterisation and exploitation (Lima et al., 2014, Pelentir et al., 2011). Peach stones form a major part of the peach processing waste, and therefore can be taken as a

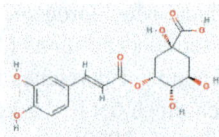

Neochlorogenic acid
CID: 5280633

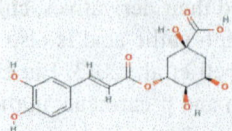

Chlorogenic Acid
CID: 1794427

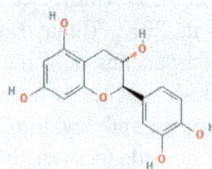

Catechin
CID: 9064

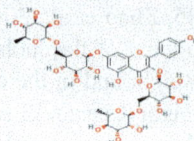

Kaempferol 3-O-rutinoside 7-O-rutinoside
CID: 102180242

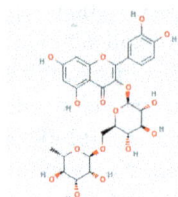

Quercetin-3-*O*-rutinoside
CID: 5280805

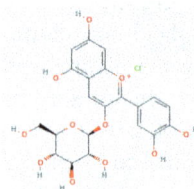

Cyanidin-3-*O*-glucoside
CID: 197081

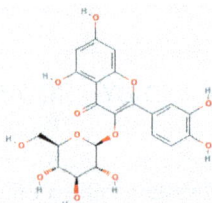

Quercetin-3-*O*-glucoside
CID: 5280804

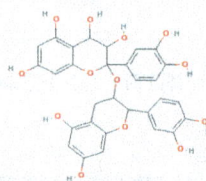

Procyanidin
CID: 107876

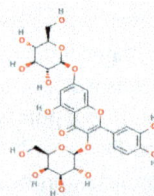

Quercetin-3-*O*-galactcoside
CID: 14130922

Figure 11.1   Structural representation of the main phytochemicals found in peaches.

Table 11.1 Proximate Analysis of Peach Stone

S. No.	Component	Concentration (%) Dry Basis
1	Moisture	4.77
2	Volatile matter	75.27
3	Fixed carbon	18.43
4	Ash	1.53
5	Carbon	51.35
6	Hydrogen	6.01
7	Nitrogen	0.58
8	Oxygen	40.32
9	Sulphur	0.14

Table 11.2 Quality Description of Peach Pomace

S. No.	Component	Fresh Pomace Composition
1	Moisture	7.0 ± 0.5 g/100 g
2	Fat	<3.0 g/100 g
3	Protein	7.5 ± 0.3 g/100 g
4	Ash	3.0 ± 0.2 g/100 g
5	Total fibre	54.2 ± 2.3 g/100 g
6	SDF	19.1 ± 1.0 g/100 g
7	IDF	35.4 ± 1.5 g/100 g
8	Calorific value	150–170 cal/100 g
9	Water Holding capacity	3.5–4.3 water/g
10	Pectinesterase activity	ND
11	Polygalacturonase activity	ND

Source: Adapted from Pagan and Ibraz (2001)
ND = Not detected

potential, economic source of proteins and peptides (Vásquez-Villanueva et al., 2015). Proximate analysis of peach stone is shown in Table 11.1. Bioactive peptides obtained from foods are considered to be explicit peptide chains that have a remarkable effect on the functional quality of the human body, ultimately influencing its overall health (Kitts & Weiler, 2003). All these properties make peach stones a rich and cheap source of peptides and proteins.

11.3.2 Peach Pomace

Pomace is a by-product of peach which is produced after juice extraction from peaches and tends to be a good source of polyphenols which can be added to foods as antioxidants and pigments. Peach pomace accounts for approximately 10% of the original fruit fresh weight (Argun and Dao, 2017). Biocompounds present in peach are largely present in the pulp and skin tissues, which represent the components of peach pomace, thereby making it an important by-product resource for the extraction of several important compounds. The major composition compounds of peach pomace are shown in Table 11.2.

11.3.3 Peach Kernel

Kernel is a stone that is enclosed in a seed and forms another major waste stream of peach processing that approximates up to 20% (w/w) of the fruit fresh weight. Approximately 8×10^6 kg/year

Table 11.3   Composition of Peach Kernel

S. No.	Component	mg/100 g Dry eight
1	Crude oil	48
2	Crude protein	26.7
3	Crude fibre	5.4
4	Ash	4
5	Carbohydrate	16
6	Reducing sugars	7.1
7	Potassium	691
8	Phosphorus	565
9	Magnesium	102
10	Calcium	17.3
11	Sodium	16
12	Aluminium	1

of peach waste is produced from the peach processing industries of Brazil. Major compositional constituents of peach kernels are shown in Table 11.3. The peach kernel is full of oil that has significant therapeutic properties and has useful nutritional qualities due to the presence of large amounts of oleic and linoleic acids. Peach kernels contain about 50% by weight of oils (Yolanda et al., 2009). Wu et al. (2011) identified compounds like dithiothreitol, hydroxycinnamic acid, quercetin, B-type proanthocyandin, phenylpropanoic acid, dihydroxybenzoic acid(s), catechins, gentisic acid, kuromanin chloride, vanillic acid, (–)-epicatechin gallate, among others. The kernels are composed of an outer covering and an almond that contains a large amount of oil, and they are generally used as animal fodder or incinerated as burnable materials (Calgaroto et al., 2005). In addition to kernel oil, dehulled kernels are used as a raw material to produce persipan that employs hydrolysis of the cyanogenic glycosides in order to avert the bitterness. It has been reported that pulverizing, steeping and cooking decreases the concentration of cyanogenic glycosides, but a complete hydrolysis of cyanogenic glycosides is likely to be attained by the use of β-glucosidase (Schieber et al., 2001). These kernels are also regarded as excellent sources of protein approximating up to 28% in concentration, although the core predicament for their utilization in foods is the existence of the cyanogenic glucoside amygdalin (Abd El-Aal et al., 1986, Rahma and Abd El-Aal, 1988).

11.4   PROCESSING OF PEACH WASTE AND BY-PRODUCTS

Like other fruits and vegetable wastes, peach processing wastes are usually disposed of either in landfills or decomposed anaerobically, although these methods do not provide anything of significant importance and also tend to be a burden on the environment, thus incurring a huge economic loss to the processing industries. Various studies have identified extraction methods for important compounds from industrial peach waste. Extraction of these biological compounds could represent an effective valorisation strategy for peach pomace waste. As reported by Kurz et al. (2008), pectin is a major polysaccharide in peaches. The isolation of pectin from peach pomace has been reported in several experimental studies. Pagan et al. (1999) reported the production and characterisation of pectin obtained from pomace at different pH values. The extractions were carried out in aqueous samples that were continuously mixed at 150 rpm in an insulated reactor attached to a stable heating source in a discontinuous process at 40°, 60° or 80°C. Varying quantities of 70% (w/w) nitric acid were added into the reactor to achieve pH values of 2.53, 2.05, 1.79, 1.54, 1.40 or 1.20. Sample extraction from each condition combination was performed between 10 and 80 minutes at intervals of 10 minutes. The pectin obtained was washed thoroughly with ethanol to remove the monosaccharides and disaccharides and was then vacuum dried in an oven at 50°C until a constant weight was attained, followed by fine grinding to 30-mesh size. In a method developed by Faravash and Ashtiani (2007), a 10 g aliquot of pomace powder sample was mixed with 250 mL of distilled water, followed by acidification with different volumes of 0.1 N HCl until it attained a pH of 1.5, 2, 2.3, 2.5 or 3.5. Extraction was done over four different times (30, 60, 120 or 180 min). The slurry was filtered twice and the final filtrate was reduced to one-fifth of its original volume at a temperature of 52 ± 1°C (Pagan and Ibarz, 1999, Pagan et al., 2001). Following this, the whole set-up was kept at room temperature for 4 hours in order to allow the pectin to separate (Kalapathy and Proctor, 2001) and attain a colloid–liquid state equilibrium. As reported by Pagan and Ibarz (1999) and Pagan et al. (2001), precipitated material was rinsed with 45% ethanol to eliminate any contaminants. Re-centrifugation at 1000 rpm for 60 min was done to separate the pectic substances from the liquid. Vigorous rinsing was repeated until a complete clarification of the liquid phase was achieved. The pectin obtained from the peach pomace at pH 2.05 at 60°C has been reported to be of high quality with a degree of esterification of 95.4%.

Conventional techniques used in the extraction of antioxidants from fruits involve the use of organic solvents at elevated temperatures (Li et al., 2006). These extraction techniques incur elevated costs and climatic effects, and result in thermal degradation of important biocompounds. Extraction of carotenoids from peach pomace has been reported by Lalas et al. (2019). The peach pomace was dried, using an optimized drying process, in order to retain the maximum carotenoid yield. An aliquot (30 g) of pomace was spread over an aluminium tray in a 5-mm thick layer. The tray was placed in an oven at 100°C for 48 hours. Dried peach pomace was then ground in a blender and carotenoid extraction was carried out in an extraction solvent mixture (50% hexane, 25% acetone and 25% ethanol [v/v]). An aliquot (0.5 g) dried and ground pomace material was placed in a 25 mL glass vial with a screw cap closure and was covered with aluminium foil. To this, 10 mL of extraction solvent was added. The mix was blended at a room temperature for 30 minutes over a magnetic stirrer at 300 rpm, followed by the addition of 1.5 mL distilled water. The mix was again stirred for five minutes and then allowed to stand for five minutes in order to achieve phase separation, after which a desired volume of the organic phase was diluted 1:20 with acetone and transferred into a 1-cm quartz cuvette followed by the measurement of the absorbance at wavelength 450 nm. Extraction of compounds from fruit tissues is a mass transfer process, that primarily depends on relocation resistance of these compounds into the extraction solvent, and the compartmentalisation of these compounds into the tissues of fruit matrices amplify this resistance (Donsì et al., 2010). Therefore, production techniques with assisting technologies are used to disorder the integrity of cellular tissue, increasing the production effectiveness, while conserving the biological compounds in the extract. Such approaches utilize microwaves, high pressures and other emerging techniques like ultrasound and pulsed electric fields (PEF). In case of PEF, the fruit, or part of it, is exposed to external electric fields (1–10 kV/cm) for a fraction of time that results in electroporation of the cell membranes leading to an increased cell permeability, favouring the extraction of biocompounds (Donsì et al., 2010). Utilisation of pulse electric field has several benefits over other assisting techniques such as lower energy costs and increased recovery of the extraction solvent, due to negligible disintegration of the cell wall (Redondo et al., 2018). Moreover, due to no increase in temperature, the loss of biocompounds is less than that achieved by microwaves or ultrasound (Cacace and Mazza, 2003). Due to these benefits, pulsed electric field has been used in the extraction of marketable compounds from the matrices of different vegetables (Kumari et al., 2018). The high moisture content of peach pomace renders it highly prone to microbial spoilage which is one of the major management issues (Ajila et al., 2012). Therefore, it is necessary to freeze or air-dry the pomace prior to any extraction process, either of which might result in compositional changes as well as reducing the extraction efficiency of bioactive compounds. Some studies have reported that the pectin obtained from fresh peach pomace is highly methoxylated and has excellent gelling capacity as compared with the pectin obtained from stored pomace (Pagan et al., 1999).

Peach kernels are a rich source of oil. De Campos et al. (2008) extracted oil from peach kernels by employing the low-pressure extraction technique. The peach almonds were ground and then filled in a sealed unit placed inside the 250 mL extractor. An aliquot (5 g) of the sample was placed in a 250 mL flask filled with 150 mL of solvent for 6 hours at the boiling point of the solvent, in a Soxhlet apparatus. Extraction was replicated using varying solvents that include n-hexane, dichloromethane, ethyl acetate and ethanol (Byers, 2007). Solvent mixtures comprising distilled water and ethanol as one mix, and hexane and dichloromethane in a 1:1 ratio (v/v). The extraction involves grinding of the peach almond under cold conditions so as to avoid any thermal degradation of the target bioactives. The extraction involved use of a 50 g macerated sample in a flask filled with 200 mL of ethanol for seven days at ambient temperature. The extract obtained was allowed to evaporate to 10% of the original volume, resulting in the crude extract, followed by addition of 1 mL of each solvent (Kitzberger et al., 2007). The organic solvents used were of 99% purity and were used in sequence in order of increasing polarity of 0, 4.4 and 9.0 (Byers, 2007). The hydro-distillation consisted of placing 50 g of ground peach almond in a 2 L Clevenger-type flask with 700 mL of distilled water as described by Martínez et al. (2004). Afterwards, 2 mL hexane was added in the decantation part of the Clevenger-type flask to dissolve the volatile oil during the extraction process that continued for 3 hours. The supercritical fluid extraction of peach almond oil has been reported by Zetzel et al. (2003), where extraction was carried out in a dynamic extraction unit. A co-solvent pump (Constametric, 3200, EUA), was attached to the extraction line so as to supply the modifier at a previously estimated flow rate, to mix with the CO_2 flow before going to the extraction container. According to Michielin et al. (2005), the extraction procedure consisted of using a 3 g sample of dried and ground peach almond inside the extractor vessel in order to make a fixed bed of the particles. This step is taken forwards by the controlling temperature and pressure throughout the process. Once the extraction is over, the solute is collected in dark bottles after 150 minutes and weighed. The supercritical fluid extraction assays, carried out to obtain the ratio between extracted oil quantity and the amount of raw material on a dry weight basis, were performed in duplicate and segregated into two groups, i.e., pure carbon dioxide assays, where CO_2 is used as a solvent at a temperature of 30°C, 40°C or 50°C and maintaining a pressure of 100, 200 or 300 bars with a constant solvent flow rate of 8.3 ± 0.8 g/min. The second group was the co-solvent assay, where ethanol was mixed with supercritical carbon dioxide in concentrations of 2% or 5% (w/w); in this group, temperature and pressure were maintained at 50°C and 300 bars and the solvent was separated from the extract by conventional techniques. The 99.9% pure CO_2 used in the process was delivered at a pressure of up to 60 bars, and the density of solvent was obtained for each operating condition (Angus et al., 1976).

11.5 UTILIZATION OF PEACH WASTES AND BY-PRODUCTS IN FOOD APPLICATIONS

11.5.1 Peach Pomace

The use in jellies and jams with a high sugar content is the major industrial application of pectin, although pectin is also used in the pharmaceutical industry, dentistry and cosmetics sector. Pectic substances comprise of 0.5–4.0% of the fresh weight of plant material (Sakai et al., 1993, Kashyap et al., 2001) and peaches contain about 0.1–0.9% fresh weight of pectic substances (Jayani, et al., 2005). Although production of peach pomace is less than that of apple pomace and approximates up to only 10% of the apple pomace production, it still represents a potential commercially significant source of pectin and is widely used in development of bakery products enriched with dietary fibre. Carotenoids are red, orange and yellow pigments present in high concentrations and large numbers in plant tissues. In addition to acting as pigments to facilitate photosynthesis and attract pollinators or seed dispersal agents, carotenoids also function as antioxidants, protecting living tissues from damage caused by over-accumulation of reactive oxygen species, known carcinogenic factors (Boon et al., 2010). Several carotenoids that are generally found in foods play important roles that are linked with the maintenance of the fundamental functions of the body and prevention of various diseases. The most common carotenoids include carotenes, lycopene, lutein and zeaxanthin that have health benefits, like decreasing the risk of oncogenesis. Amongst the peach processing wastes, peach pomace acts as a major source of carotenoids.

11.5.2 Peach Kernel Oil

Peach kernels also form a significant industrial residue in peach processing. Peach kernels are a potential source of an oil having significant therapeutic and medicinal values (Mezommo et al., 2010) as a result of the composition of unsaturated fatty acids and high oleic acid concentration (55–77%) (El Saadany et al., 1993; Calgaroto et al., 2005). The lucrative market for peach almond oil paves the way for the peach industry to use the kernel residue to produce high-value products. Peach kernel oil is extensively used in the cosmetics industry as an ingredient in various products, because of its light and penetrating nature that absorbs easily without leaving a sticky residue after use. Peach kernel oil also possesses a rich nutritional profile that provides an opportunity for producing high-value products from the biowaste from the peach industry owing to the presence of saturated fatty acids (Saadany et al., 1993) and antioxidant constituents (Saadany et al., 1993). Therefore, peach kernels are well regarded as a potential source of essential oil for the food and pharmaceutical industries alike.

11.5.3 Peach Kernel Flour

The attractive and rich composition and the high bioactive values of peach kernels make them an important, potential ingredient for use in various food products. One such product is the preparation of peach kernel flour that is rich in bioactive compounds. Peach kernel flour has the potential to be used for the fortification and enrichment of human food and animal feed (Pelentir et al., 2011). The flour has an oil absorption capacity (OAC) higher than its water absorption capacity (WAC); the OAC is reported to be twice the WAC. This property makes the peach kernel flour a potential alternative to traditional flour in products like cakes or biscuits that require flour with a high OAC. Apricot kernel flour has been reported to show a contradictory trend, regarding OAC and WAC, when compared with peach kernel flour (Abd El-Aal et al., 1986). Therefore, amalgamation of both flours may result in a product that has the potential of having an improved ability to absorb and retain more water and oil, which could be advantageous in many consumables. The emulsification property of peach kernel flour has also been reported to be high, thereby increasing its viability in the products like sausages and other meat analogues. The foaming properties of peach kernel flour are reported to be fairly good, thereby suiting its use in different beverages and soft drinks and thus improving their nutritional quality. (Abd El-Aal et al., 1986).

11.6 CONCLUSIONS

A significant proportion of peaches produced worldwide is processed, resulting in the generation of huge quantities of wastes and by-products that include peel, pomace and stone. Peach stone is a high-fibre by-product that is used as fuel. This is also used to produce activated carbon and fungicidal oil, in addition to the production of inhibitory peptides and fibres. Peach peel is also a rich bioresource of carotenoids, phenolics and flavonols. Peach pomace tends to be a good source of industrial pectin, and it is also a sustainable source of corrosion inhibitors. Peach kernel is a potential source of peach kernel flour that can have a wide usage in human food and animal feed fortification. Peach kernel oil is also an important by-product of peach that has a potential usage in the food, medicinal and cosmetic industries. Extraction of bioactive compounds from peach wastes is an important strategy to bring down the environmental pollution caused by industrial waste production and disposal, and is also the cheapest and most sustainable strategy for the production of various bioactive and functional products. Traditional as well as novel technically assisted processes have been developed or are emerging in order to have continuous extraction and greater exploitation of valuable products from material which is otherwise regarded as waste.

BIBLIOGRAPHY

Abd El-Aal, M. H., Hamza, M. A., & Rahma, E. H. (1986). In vitro digestibility, physico-chemical and functional properties of apricot kernel proteins. Food Chemistry, 19(3), 197–211.

AIJN European Fruit Juice Association. (2016). Liquid fruit market report. Brussels, Belgium: AIJN European Fruit Juice Association.

Ajila, C. M., Brar, S. K., Verma, M., & Prasada Rao, U. J. S. (2012). Sustainable solutions for agro processing waste management: An overview. Environmental Protection Strategies for Sustainable Development, 4, 65–109.

Alipasandi, A., Ghaffari, H., & Alibeyglu, S. Z. (2013). Classification of three varieties of peach fruit using artificial neural network assisted with image processing techniques. International Journal of Agronomy and Plant Production, 4(9), 2179–2186.

Angus, S., Armstrong, B., & De Reuck, K. M. (1976). International thermodynamic tables of the fluid state: Carbon dioxide. Oxford: Pergamon Press.

Argun, H., & Dao, S. (2017). Bio-hydrogen production from waste peach pulp by dark fermentation: Effect of inoculum addition. International Journal of Hydrogen Energy, 42(4), 2569–2574.

Arvanitoyannis, I. S., & Varzakas, T. H. (2008). Vegetable waste treatment: Comparison and critical presentation of methodologies. Critical Reviews in Food Science and Nutrition, 48(3), 205–247.

Aubert, C., Bony, P., Chalot, G., Landry, P., & Lurol, S. (2014). Effects of storage temperature, storage duration, and subsequent ripening on the physicochemical characteristics, volatile compounds, and phytochemicals of western red nectarine (Prunus persica L. Batsch). Journal of Agricultural and Food Chemistry, 62(20), 4707–4724.

Bahadoran, Z., Mirmiran, P., & Azizi, F. (2013). Dietary polyphenols as potential nutraceuticals in management of diabetes: A review. Journal of Diabetes & Metabolic Disorders, 12(1), 1–9.

Başar, H. (2006). Elemental composition of various peach cultivars. Scientia Horticulturae, 107(3), 259–263.

Bento, C., Goncalves, A. C., Silva, B., & Silva, L. R. (2018). Assessing the phenolic profile, antioxidant, antidiabetic and protective effects against oxidative damage in human erythrocytes of peaches from Fundão. Journal of Functional Foods, 43, 224–233.

Boon, C. S., McClements, D. J., Weiss, J., & Decker, E. A. (2010). Factors influencing the chemical stability of carotenoids in foods. Critical Reviews in Food Science and Nutrition, 50(6), 515–532.

Byers, J. A. (2007). Catálogo phenomenex. http://www.phenomenex.com/phen/Doc.

Byrne, D. H., Noratto, G., Cisneros-Zevallos, L., Porter, W., & Vizzotto, M. (2007, October). Health benefits of peach, nectarine and plums. In II International Symposium on Human Health Effects of Fruits and Vegetables: FAVHEALTH 2007 841 (pp. 267–274).

Cacace, J. E., & Mazza, G. (2003). Mass transfer process during extraction of phenolic compounds from milled berries. Journal of Food Engineering, 59(4), 379–389.

Calgaroto, C., Pilecco, J., Oliveira, M. P., Furlan, L., & Zambiazi, R. (2005). Extração e caracterização do óleo de amêndoa de pêssego. In Proceeding of XIV Congresso de Iniciação Científica da Universidade Federal de Pelotas, Pelotas.

Campbell, O. E., & Padilla-Zakour, O. I. (2013). Phenolic and carotenoid composition of canned peaches (Prunus persica) and apricots (Prunus armeniaca) as affected by variety and peeling. Food research international, 54(1), 448–455.

Ceccarelli, D., Simeone, A. M., Nota, P., Piazza, M. G., Fideghelli, C., & Caboni, E. (2016). Phenolic compounds (hydroxycinnamic acids, flavan-3-ols, flavonols) profile in fruit of Italian peach varieties. Plant Biosystems-An International Journal Dealing with all Aspects of Plant Biology, 150(6), 1370–1375.

Cirilli, M., Bassi, D., & Ciacciulli, A. (2016). Sugars in peach fruit: A breeding perspective. Horticulture Research, 3(1), 1–12.

Colaric, M., Veberic, R., Stampar, F., & Hudina, M. (2005). Evaluation of peach and nectarine fruit quality and correlations between sensory and chemical attributes. Journal of the Science of Food and Agriculture, 85(15), 2611–2616.

Davidović, S. M., Veljović, M. S., Pantelić, M. M., Baošić, R. M., Natić, M. M., Dabić, D. C., ... & Vukosavljević, P. V. (2013). Physicochemical, antioxidant and sensory properties of peach wine made from redhaven cultivar. Journal of Agricultural and Food Chemistry, 61(6), 1357–1363.

De Campos, L. M., Leimann, F. V., Pedrosa, R. C., & Ferreira, S. R. (2008). Free radical scavenging of grape pomace extracts from Cabernet sauvingnon (Vitis vinifera). Bioresource Technology, 99(17), 8413–8420.

Donsì, F., Ferrari, G., & Pataro, G. (2010). Applications of pulsed electric field treatments for the enhancement of mass transfer from vegetable tissue. Food Engineering Reviews, 2(2), 109–130.

Duan, Y., Dong, X., Liu, B., & Li, P. (2013). Relationship of changes in the fatty acid compositions and fruit softening in peach (Prunus persica L. Batsch). Acta Physiologiae Plantarum, 35(3), 707–713.

El Saadany, R. M. A., Kalaf, H. H., & Soliman, M. (1993, August). Characterization of lipids extracted from peach kernels. In International Symposium on Postharvest Treatment of Horticultural Crops 368 (pp. 123–131).

Elsadr, H. (2016). A Genome Wide Association Study of Flowering and Fruit Quality Traits in Peach [(Prunus persica (L.) Batsch] (Doctoral dissertation, University of Guelph).

Espada, J. L., Romero, J., & Alonso, J. M. (2009). Preview of the second clonal selection from the autochthonous peach population "Amarillos Tardíos de Calanda" (late yellow peaches of Calanda). Acta Horticulturae, 1, 251–254.

FAO stat. (2010). http://faostat.fao.org/site/567/DesktopDefault.aspx? PageID¼567#ancor.

FAO stat. (2013). http://faostat.fao.org/site/567/DesktopDefault.aspx? PageID¼567#ancor.

Faravash, R. S., & Ashtiani, F. Z. (2007). The effect of pH, ethanol volume and acid washing time on the yield of pectin extraction from peach pomace. International Journal of Food Science & Technology, 42(10), 1177–1187.

Felipe, A. J. (2009). 'Felinem', 'Garnem', and 'Monegro'almond× peach hybrid rootstocks. HortScience, 44(1), 196–197.

Gasparotto, J., Somensi, N., Bortolin, R. C., Moresco, K. S., Girardi, C. S., Klafke, K., ... & Gelain, D. P. (2014). Effects of different products of peach (Prunus persica L. Batsch) from a variety developed in southern Brazil on oxidative stress and inflammatory parameters in vitro and ex vivo. Journal of Clinical Biochemistry and Nutrition, 55(2), 110–119.

Gil, M. I., Tomás-Barberán, F. A., Hess-Pierce, B., & Kader, A. A. (2002). Antioxidant capacities, phenolic compounds, carotenoids, and vitamin C contents of nectarine, peach, and plum cultivars from California. Journal of Agricultural and Food Chemistry, 50(17), 4976–4982.

Gradziel, T. M. (2002, August). Interspecific hybridizations and subsequent gene introgression within Prunus subgenus Amygdalus. In XXVI International Horticultural Congress: Genetics and Breeding of Tree Fruits and Nuts 622 (pp. 249–255).

Grimi, N., Mamouni, F., Lebovka, N., Vorobiev, E., & Vaxelaire, J. (2011). Impact of apple processing modes on extracted juice quality: Pressing assisted by pulsed electric fields. Journal of Food Engineering, 103(1), 52–61.

Holst, B., & Williamson, G. (2008). Nutrients and phytochemicals: From bioavailability to bioefficacy beyond antioxidants. Current Opinion in Biotechnology, 19(2), 73–82.

Iordănescu, O. A., Alexa, E., Radulov, I., Costea, A., Dobrei, A., & Dobrei, A. (2015). Minerals and amino acids in peach (Prunus persica L.) cultivars and hybrids belonging to world germoplasm collection in the conditions of West Romania. Agriculture and Agricultural Science Procedia, 6, 145–150.

Jayani, R. S., Saxena, S., & Gupta, R. (2005). Microbial pectinolytic enzymes: A review. Process Biochemistry, 40(9), 2931–2944.

Kalapathy, U., & Proctor, A. (2001). Effect of acid extraction and alcohol precipitation conditions on the yield and purity of soy hull pectin. Food Chemistry, 73(4), 393–396.

Kashyap, D. R., Vohra, P. K., Chopra, S., & Tewari, R. (2001). Applications of pectinases in the commercial sector: A review. Bioresource Technology, 77(3), 215–227.

Kaynak, B., Topal, H., & Atimtay, A. T. (2005). Peach and apricot stone combustion in a bubbling fluidized bed. Fuel Processing Technology, 86(11), 1175–1193.

Kitts, D. D., & Weiler, K. (2003). Bioactive proteins and peptides from food sources. Applications of biopro-
cesses used in isolation and recovery. Current Pharmaceutical Design, 9(16), 1309–1323.

Kitzberger, C. S. G., Smânia Jr, A., Pedrosa, R. C., & Ferreira, S. R. S. (2007). Antioxidant and antimicrobial
activities of shiitake (Lentinula edodes) extracts obtained by organic solvents and supercritical fluids.
Journal of Food Engineering, 80(2), 631–638.

Kono, R., Okuno, Y., Nakamura, M., Inada, K. I., Tokuda, A., Yamashita, M., ... & Utsunomiya, H. (2013).
Peach (Prunus persica) extract inhibits angiotensin II-induced signal transduction in vascular smooth
muscle cells. Food Chemistry, 139(1–4), 371–376.

Kozlowska, A., & Szostak-Wegierek, D. (2014). Flavonoids-food sources and health benefits. Roczniki
Państwowego Zakładu Higieny, 65(2), 79–85.

Kumar, D. R., Kumar, M. A., & Naidu, P. B. (2015). Evaluation of anthelmintic activity of prunus persica (L.).
Asian Journal of Pharmaceutical and Clinical Research, 4(5), 163–165.

Kumari, B., Tiwari, B. K., Hossain, M. B., Brunton, N. P., & Rai, D. K. (2018). Recent advances on application
of ultrasound and pulsed electric field technologies in the extraction of bioactives from agro-industrial
by-products. Food and Bioprocess Technology, 11(2), 223–241.

Kurz, C., Carle, R., & Schieber, A. (2008). Characterisation of cell wall polysaccharide profiles of apricots
(Prunus armeniaca L.), peaches (Prunus persica L.), and pumpkins (Cucurbita sp.) for the evaluation of
fruit product authenticity. Food Chemistry, 106, 421–430.

Lalas, S., Alibade, A., Bozinou, E., & Makris, D. P. (2019). Drying optimisation to obtain carotenoid-enriched
extracts from industrial peach processing waste (pomace). Beverages, 5(3), 43.

Lazos, E. S. (1991). Composition and oil characteristics of apricot, peach and cherry kernel. Grasas y aceites,
42(2), 127–131.

Li, B. B., Smith, B., & Hossain, M. M. (2006). Extraction of phenolics from citrus peels: I. Solvent extraction
method. Separation and Purification Technology, 48(2), 182–188.

Lima, B. N. B., Lima, F. F., Tavares, M. I. B., Costa, A. M. M., & Pierucci, A. P. T. R. (2014). Determination of
the centesimal composition and characterization of flours from fruit seeds. Food Chemistry, 151, 293–299.

Liu, H., Cao, J., & Jiang, W. (2015). Evaluation of physiochemical and antioxidant activity changes during fruit
on-tree ripening for the potential values of unripe peaches. Scientia Horticulturae, 193, 32–39.

Liu, R. H. (2004). Potential synergy of phytochemicals in cancer prevention: Mechanism of action. The
Journal of Nutrition, 134(12) Supplement, 3479S–3485S.

Manzoor, M., Anwar, F., Mahmood, Z., Rashid, U., & Ashraf, M. (2012). Variation in minerals, phenolics
and antioxidant activity of peel and pulp of different varieties of peach (Prunus persica L.) fruit from
Pakistan. Molecules, 17(6), 6491–6506.

Martinez, J., Rosa, P. T., Menut, C., Leydet, A., Brat, P., Pallet, D., & Meireles, M. A. A. (2004). Valorization
of Brazilian vetiver (Vetiveria zizanioides (L.) Nash ex Small) oil. Journal of Agricultural and Food
Chemistry, 52(21), 6578–6584.

Martínez-Gómez, P., & Gradziel, T. M. (2001, May). New approaches to almond breeding at the university
of California-Davis program. In III International Symposium on Pistachios and Almonds 591 (pp.
253–256).

Melo, G. W. B. D., Sete, P. B., Ambrosini, V. G., Freitas, R. F., Basso, A., & Brunetto, G. (2016). Nutritional
status, yield and composition of peach fruit subjected to the application of organic compost. Acta
Scientiarum. Agronomy, 38, 103–109.

Mezzomo, N., Martínez, J., & Ferreira, S. R. (2009). Supercritical fluid extraction of peach (Prunus persica)
almond oil: Kinetics, mathematical modeling and scale-up. The Journal of Supercritical Fluids, 51(1),
10–16.

Michielin, E. M., Bresciani, L. F., Danielski, L., Yunes, R. A., & Ferreira, S. R. (2005). Composition profile
of horsetail (Equisetum giganteum L.) oleoresin: Comparing SFE and organic solvents extraction. The
Journal of Supercritical Fluids, 33(2), 131–138.

Moreno, M. A. (2003). Breeding and selection of Prunus rootstocs at the Estacion Experimental de Aula Dei
[Spain]. ITEA.

Moreno, C., Pascual-Teresa, S. D., Ancos, B. D., & Cano, M. P. (2006). Nutritional values of fruits. In Y. H.
Hui (Ed.), Handbook of fruits and fruit processing (pp. 29–44). Nova Jersey: Blackwell Publishing.

Noratto, G., Porter, W., Byrne, D., & Cisneros-Zevallos, L. (2014). Polyphenolics from peach (Prunus persica
var. Rich Lady) inhibit tumor growth and metastasis of MDA-MB-435 breast cancer cells in vivo. The
Journal of Nutritional Biochemistry, 25(7), 796–800.

Okie, W. R., Bacon, T., & Bassi, D. (2008). Fresh market cultivar development. The Peach: Botany, Production and Uses, 139.

Pagán, J., Ibarz, A., Llorca, M., & Coll, L. (1999). Quality of industrial pectin extracted from peach pomace at different pH and temperatures. Journal of the Science of Food and Agriculture, 79(7), 1038–1042.

Pagan, J., Ibarz, A., Llorca, M., Pagan, A., & Barbosa-Cánovas, G. V. (2001). Extraction and characterization of pectin from stored peach pomace. Food Research International, 34(7), 605–612.

Pelentir, N., Block, J. M., Monteiro Fritz, A. R., Reginatto, V., & Amante, E. R. (2011). Production and chemical characterization of peach (Prunus persica) kernel flour. Journal of Food Process Engineering, 34(4), 1253–1265.

Pinochet, J. (2009). 'Greenpac', a new peach hybrid rootstock adapted to Mediterranean conditions. HortScience, 44(5), 1456–1457.

Piotrowska, P., Zevenhoven, M., Hupa, M., Giuntoli, J., & de Jong, W. (2013). Residues from the production of biofuels for transportation: Characterization and ash sintering tendency. Fuel Processing Technology, 105, 37–45.

Plazzotta, S., Ibarz, R., Manzocco, L., & Martín-Belloso, O. (2021). Modelling the recovery of biocompounds from peach waste assisted by pulsed electric fields or thermal treatment. Journal of Food Engineering, 290, 110196.

Rahma, E. H., & Abd El-Aal, M. H. (1988). Chemical characterization of peach kernel oil and protein: Functional properties, in vitro digestibility and amino acids profile of the flour. Food Chemistry, 28(1), 31–43.

Raturi, R., Singh, H., Bahuguna, P., Sati, S. C., & Badoni, P. P. (2011). Antibacterial and antioxidant activity of methanolic extract of bark of Prunus persica. Journal of Applied and Natural Science, 3(2), 312–314.

Redondo, D., Venturini, M. E., Luengo, E., Raso, J., & Arias, E. (2018). Pulsed electric fields as a green technology for the extraction of bioactive compounds from thinned peach by-products. Innovative Food Science & Emerging Technologies, 45, 335–343.

Reig, G., Iglesias, I., Gatius, F., & Alegre, S. (2013). Antioxidant capacity, quality, and anthocyanin and nutrient contents of several peach cultivars [Prunus persica (L.) Batsch] grown in Spain. Journal of Agricultural and Food Chemistry, 61(26), 6344–6357.

Remorini, D., Tavarini, S., Degl'Innocenti, E., Loreti, F., Massai, R., & Guidi, L. (2008). Effect of rootstocks and harvesting time on the nutritional quality of peel and flesh of peach fruits. Food Chemistry, 110(2), 361–367.

Saidani, F., Giménez, R., Aubert, C., Chalot, G., Betrán, J. A., & Gogorcena, Y. (2017). Phenolic, sugar and acid profiles and the antioxidant composition in the peel and pulp of peach fruits. Journal of Food Composition and Analysis, 62, 126–133.

Sakai, T., Sakamoto, T., Hallaert, J., & Vandamme, E. J. (1993). Pectin, pectinase, and protopectinase: Production, properties, and applications. Advances in Applied Microbiology, 39, 213–294.

Scalbert, A., Manach, C., Morand, C., Rémésy, C., & Jiménez, L. (2005). Dietary polyphenols and the prevention of diseases. Critical Reviews in Food Science and Nutrition, 45(4), 287–306.

Schieber, A., Stintzing, F. C., & Carle, R. (2001). By-products of plant food processing as a source of functional compounds—recent developments. Trends in Food Science & Technology, 12(11), 401–413.

Scordino, M., Sabatino, L., Muratore, A., Belligno, A., & Gagliano, G. (2012). Phenolic characterization of Sicilian yellow flesh peach (Prunus persica L.) cultivars at different ripening stages. Journal of Food Quality, 35(4), 255–262.

Scorza, R., & W. Okie. (1990). Peaches. In J. N. Moore & J. R. Ballington Jr (Eds.), Genetic resources of temperate fruit and nut crops (pp. 175–232). Wageningen: ISHS.

Scorza, R., Sherman, W. B., & Lightner, G. W. (1988). Inbreeding and co-ancestry of low chill short fruit development period freestone peaches and nectarines produced by the University of Florida breeding program. Fruit Varieties Journal, 42, 79–85.

Tomás-Barberán, F. A., Gil, M. I., Cremin, P., Waterhouse, A. L., Hess-Pierce, B., & Kader, A. A. (2001). HPLC– DAD– ESIMS analysis of phenolic compounds in nectarines, peaches, and plums. Journal of Agricultural and Food Chemistry, 49(10), 4748–4760.

Umar Lule, S., & Xia, W. (2005). Food phenolics, pros and cons: A review. Food Reviews International, 21(4), 367–388.

USDA. (2016). National nutrient database for standard reference. https://ndb.nal.usda.gov/ndb/search (accessed June 1, 2016).

Vargas, E. F. D., Jablonski, A., Flôres, S. H., & Rios, A. D. O. (2017). Waste from peach (Prunus persica) processing used for optimisation of carotenoids ethanolic extraction. International Journal of Food Science & Technology, 52(3), 757–762.

Vásquez-Villanueva, R., Marina, M. L., & García, M. C. (2015). Revalorization of a peach (Prunus persica (L.) Batsch) byproduct: Extraction and characterization of ACE-inhibitory peptides from peach stones. Journal of Functional Foods, 18, 137–146.

Versari, A., Castellari, M., Parpinello, G. P., Riponi, C., & Galassi, S. (2002). Characterisation of peach juices obtained from cultivars redhaven, suncrest and maria marta grown in Italy. Food Chemistry, 76(2), 181–185.

Wang, Y., Yang, C., Li, S., Yang, L., Wang, Y., Zhao, J., & Jiang, Q. (2009). Volatile characteristics of 50 peaches and nectarines evaluated by HP-SPME with GC-MS. Food Chemistry 116(1), 356–364.

Watkins, R. (1995). Cherry, plum, peach, apricot and almond. In Smartt, J., & Simmonds, N. W. (Eds.), Evolution of crop plants (pp. 423–429). London: Longman Scientific & Technical.

Wills, R. B., Scriven, F. M., & Greenfield, H. (1983). Nutrient composition of stone fruit (Prunus spp.) cultivars: Apricot, cherry, nectarine, peach and plum. Journal of the Science of Food and Agriculture, 34(12), 1383–1389.

Wu, H., Shi, J., Xue, S., Kakuda, Y., Wang, D. F., & Jiang, Y. M. (2011). Essential oil extracted from peach (Prunus persica) kernel and its physicochemical and antioxidant properties. LWTFood Science and Technology, 44, 2032–2039.

Yolanda, S. V., Albertina, C., Juan, A. R., & Pando, C. (2009). Supercritical fluid extraction of peach (Prunus persica) kernel oil using carbon dioxide and ethanol. The Journal of Supercritical Fluids, 49, 167–173.

Zarrouk, O., Gogorcena, Y., Gómez-Aparisi, J., Betrán, J. A., & Moreno, M. A. (2005). Influence of peach x almond hybrids rootstocks on flower and leaf mineral concentration, yield and vigour of two peach cultivars. Scientia Horticulturae, 106(4), 502–514.

Zerva, E., Abatis, D., Skaltsounis, A. L., & Fokialakis, N. (2012). Development and application of a methodology for the recovery of high added value products from peach industry waste. Planta Medica, 78, PJ98.

Zetzel, C., Brunner, G., & Meireles, M. A. A. (2003). Standardized low-cost batch SFE units for university education and comparative research. In Proceedings of the Sixth International Symposium on Supercritical Fluids, vol. 1. Versailles, pp. 577–581.

Zhao, X., Zhang, W., Yin, X., Su, M., Sun, C., Li, X., & Chen, K. (2015). Phenolic composition and antioxidant properties of different peach [Prunus Persic (L.)batsch] cultivars in China. International Journal of Molecular Sciences, 16(3), 5762–5778.

CHAPTER 12

Papaya Wastes and By-Products
Chemistry, Processing, and Utilization

Madiha Abdel-Hay, Asmat Farooq, and Fozia Kamran

CONTENTS

12.1 INTRODUCTION

Papaya (*Carica papaya* L.) is a tropical fruit in the Caricaceae family throughout and considered to be one of the most economically valuable fruits cultivated as a food crop around the world, including tropical and subtropical regions. In 2019, papaya was cultivated on an area of 462,552 ha globally. The five highest-producing countries in that time were India with about 6,050,000 t, the Dominican Republic with 1,171, 336 t, Brazil with almost 1,161,808 t, Mexico with 1,083,133 t, and Indonesia with about 986,991 t (FAOSTAT, 2020). Papaya is a tree-like plant, classified as being semi herbaceous in nature, reaching up to about 10 m high with a hollow stem (Sharma et al., 2020). The fruits are oval, large in size, weighing from 0.5 up to 20 lbs and have a central cavity for seeds. Fruits are carried in an axillary manner on the main stem, either singly or in small clusters. The fruits are green during the immature stage and turn yellow or reddish yellow upon ripening. The edible part surrounds the seed cavity. The fruits ripen over 5–9 months, depending on cultivar and temperature.

DOI: 10.1201/9781003164463-12

The largest producer region of papaya is Asia, followed by the Americas and Africa. The global export of papaya has expanded in different regions year after year (Figure 12.1). World production increases each year, reaching up to 13,735,086 tonnes in 2019 (Figure 12.2). The global production of papaya increases by 2.1 percent each year; therefore, it is anticipated that, in 2029, it will reach up to 16.6 million tonnes according to reports (FAO, 2020). Papaya remains well known under different names, e.g., pawpaw, papye, lapaya, tapayas and kapaya papayo, and different varieties, e.g., 'Solo', 'Sekaki', 'Maradol', 'Rainbow', 'Bettina', 'Cariflora', 'Eksotika', 'Red Lady', 'Tainung', 'Sunrise' and 'Formosa' in different parts of the world (Bhattachrjee, 2001). The unripe form of the fruit contains latex, which is rich in enzymes, widely used in nutritional, industrial and therapeutic applications (Azarkan et al., 2003). These facts disclose the nutritional importance of papaya as it is an excellent source of dietary components that may be beneficial in satisfying human nutritional requirements and hence considered valuable for a complete, healthy life. Papaya fruit contains an abundant amount of bioactive ingredients, mostly phenolics, carotenoids, vitamins A, C, B, B_5, E and folate, minerals such as potassium and magnesium, and fiber, which have many powerful health impacts on the body. Other bioactive compounds in papaya include polysaccharides, glycosides, enzymes, lectins, saponins, steroids and others (Aravind et al., 2013). Papaya contains an enzyme

Figure 12.1   Papaya fruit (adapted from FAOSTAT, 2020).

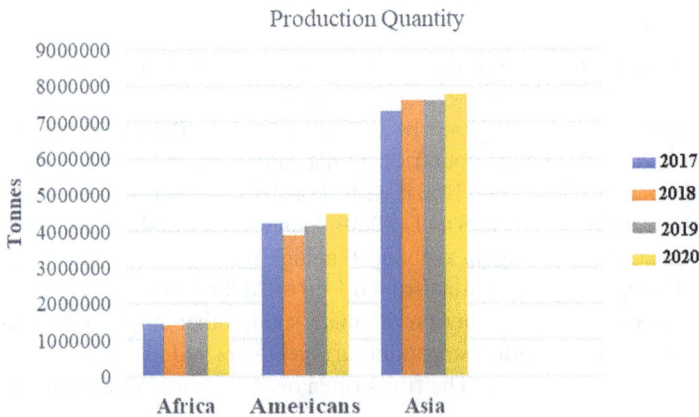

Figure 12.2   Papaya fruit production in different geographical regions (adapted from FAOSTAT, 2020).

called papain, which is required in industrial processes to tenderize meat and in other applications, including pharmaceuticals, brewing industries and skincare. Every part of the papaya fruit (seeds, peel and latex) finds an industrial application, such as dietary additives, functional foods, home remedies, animal feed and pharmaceutical products. Papaya peel contains a wide range of phyto-chemicals like phenolic compounds, carotenoids, dietary fibers and vitamins that exhibit significant antioxidant properties (Suleria et al., 2020; Samsuri et al., 2020). Seeds and peel are the important by-products of papaya processing and consist of approximately 8.5 and 12%, respectively, of the total fruit fresh weight. The waste streams of papaya are discarded, resulting in environmental pollution and health issues if not reused and utilized (Parniakov et al., 2015; Parniakov et al., 2014; Abu Qdais et al., 2019). These wastes of papaya are considered to be a valuable source of health-beneficial compounds that can be used to develop high-value highly nutritious products in an economically inexpensive manner. Therefore, it is necessary to come up with novel methods and applications that enable industry to utilize and manufacture useful and profitable products as "wealth from waste" by extracting or isolating some of the active ingredients which can be used in nutritional/functional foods, food supplements and pharmaceutical products (Ningrum & Schreiner, 2017) (Figure 12.3; Tables 12.1 and 12.2).

12.2 CHEMISTRY OF PAPAYA WASTES AND BY-PRODUCTS

12.2.1 Peel

Papaya peel is often discarded as waste and does not receive much attention in terms of re-utilization or recycling. This is due to a lack of applied research for commercial purposes (Suleria et al., 2020). Papaya peel makes up 25% of the dry matter of the fruit (Pavithra et al., 2017). The

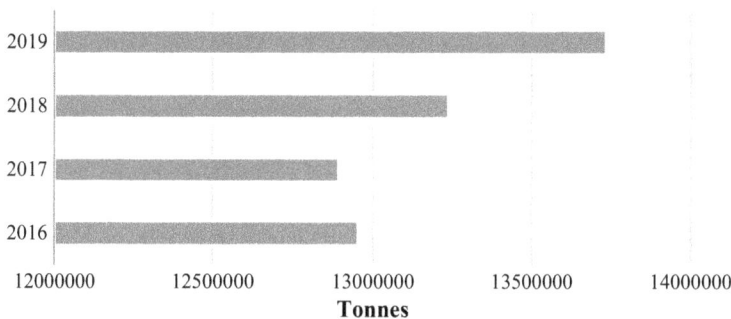

Figure 12.3 Papaya fruit production around the world.

Table 12.1 Proximate Compositions of Different Parts of Papaya (Dry Weight)

Parts	Moisture (%)	Fiber (%)	Ash (%)	Protein (%)	Fat (%)	Carbohydrates (%)	References
Ripe papaya	89.1	11.9	4.59	8.26	0.93	86.2	Pal and Mazumder (2013)
Unripe papaya	92.2	9.8	6.6	7.9	0.00	64.2	Maisarah et al. (2014)
Peel	58.22	13.67	4.84	9.04	0.31	27.87	Chukwuka et al.(2013)
Seed	11.02	8.02	5.21	27.41	28.61	19.70	Makanjuola and Makanjuola (2018)
Seed oil	6.2	21.8	2.4	28.2	27	14.4	Yanty et al. (2014)
Latex	17.76	0.67	7.0	57.2	5.21	12.16	Macalood et al. (2013)

Table 12.2   Mineral Composition of Various Parts of Papaya

Parts	Ca (mg/100 g)	K (mg/100 g)	Fe (mg/100 g)	Zn (mg/100 g)	References
Ripe papaya	146.8	1238	12.84	0.92	Pal and Mazumder (2013)
Unripe papaya	432.4	2743	8.11	0.07	Pal and Mazumder (2013)
Peel	16.22	504.33	2.73	1.94	Chukwuka et al. (2013)
Seed	6.43	720.83	4.20	6.41	Makanjuola and Makanjuola (2018)

Table 12.3   Phytochemical Components of Different Parts of Papaya (Dry Weight)

Parts	Total Phenolic Concentration (mg GAE/100 g)	Flavonoids (mg Gallic acid equivalent /100 g)	Tannins (%)	Saponins (mg/100 g)	References
Ripe papaya	272.66	92.95	0.638	2.6	Thakuria et al. (2018) Okon et al. (2017)
Unripe papaya	339.91	53.44	0.55	2.91	Nna et al. (2019)
Peel	1735.1	0.41	0.54	143	Chukwuka et al. (2013) Jamel et al., (2017)
Seed	30.32–37.34	59.54	3.8	380	Maisarah et al. (2013) Doan et al. (2020) Nna et al. (2019)

peel contains carbohydrates (~27%), ash (~5%) and fat (~1%), and representative biomolecules, such as proteinases, glutathione and cysteine (Macalood et al., 2013). Papaya peel is considered to be fiber rich since the quantities of total, soluble and insoluble dietary fiber, estimated by Calvache et al. (2016), were reported to be $59.8 \pm 0.5\%$, $19.93 \pm 0.01\%$ and $39.9 \pm 0.5\%$, respectively. A study conducted by Jamal et al. (2017) on total phenolic compound concentration in papaya peel found that were 1,735.1 mg/L gallic acid-equivalents (GAE) at 120°C and with a 5-h incubation time (Table 12.3) which indicated that papaya peel may be used as a natural source of antioxidants. Sancho et al. (2011) identified three phenolic compounds present at high concentrations in papaya peel, namely ferulic acid followed (in terms of concentration) by *p*-coumaric acid and caffeic acid (Table 12.4 and Figure 12.4). The isolated components generally extracted from peels represent a significant source of antioxidant compounds, which play a key role in nutrition and human health (Abbas et al., 2017). Papaya peel contains health-beneficial compounds with bioactive properties, which can be utilized as nutraceuticals, food additives, pharmaceuticals, novel food products, or in industrial processes (Pathak et al., 2019). These alternative sources of nutrients can be added and utilized in various food formulations (Samsuri et al., 2020).

12.2.2 Seeds

Papaya seeds are black in color and are located in the center of the ripe fruit pulp (Kadiri et al., 2016), and have a spicy and sharp taste. Papaya seeds represent nearly 20% of the weight of the fresh fruit and are known to be a protein-rich by-product source, in addition to being rich in crude fibers, fatty acids, calcium and phosphorus (Chielle et al., 2016). The total phenolic content was reported to be 37.34 mg GAE/g (Doan et al., 2020). Kadiri et al. (2017) identified bioactive phenolic compounds in papaya seeds, such as flavonoids, caffeic acid, *p*-hydroxybenzoic acid, ferulic acid, *p*-coumaric acid, quercetin-3-galactoside, myricetin and kaempferol-3-glucoside (Figure 12.4 and Table 12.4) (Adachukwu et al., 2013; Ikeyi et al., 2013). Rodrigues et al. (2019) purified 30 phenolic compounds from papaya seeds, of which the main compounds were ferulic acid, mandelic acid and vanillic acid. Kumoro et al. (2020) reported that papaya seed contained 1,945.48 mg GAE/100 g polyphenols,

Table 12.4 Bioactive Compounds in Different Parts of Papaya

Parts	Bioactive Components	References
Fruit	Kaempferol, quercetin, myricetin and carotenoids – lycopene, cryptoxanthin, zeaxanthin, β-carotene, violaxanthin. β-sitosterol, linalool, 4-terpineol and monoterpenoids	Sangsoy et al. (2017)
Unripe fruit	Hexadecanoic acid, octadecanoic acid, hexadecanoic acid Z-11 and its methyl ester	Ezekwe and Chikezie (2017)
Peel	Gallic acid, *p*-hydroxybenzoic acid, caftaric acid, syringic acid, ferulic acid, epicatechin, quercetin-3-glucoside, epicatechin gallate, kaempferol, quercetin.	Suleria et al. (2020)
Latex	Papain, caricain, chymopapain, protease omega, chitinase, cysteine endopeptidases, and glutaminyl cyclase	Macalood et al. (2013)
Seed	Cholest-4-en-3-one, 26-(acetyloxy), stigmasterol, stigmast-4-en-3-one, *N*-hexadecanoic acid, 1,19- eicosadiene, squalene, *p*-hydroxybenzoic acid, hyperoxide, genteel alcohol, triallyl glucose, salicylic acid, kaempferol hexoside, caffeic acid, ferulic acid, *p*-coumaric acid, quercetin-3-galactoside, kaempferol-3-glucoside, hexadecanoic acid, and octadecenoic acid	Seshamamba et al. (2018)Singh et al. (2020a, 2020b, 2020c), Kadiri et al. (2017), Doan et al. (2020)
Oil	Flavonoids (kaempferol, myricetin), palmitic acid, linoleic acid, oleic acid, 1,19-eicosadiene, squalene, *N*-hexadecanoic acid, stigmasterol, cholest-4-en-3-one, 26-(acetyloxy), gamma-sitosterol and stigmast-4-en-3-one	Adachukwu et al. (2013), Doan et al. (2020), Seshamamba et al. (2018)

total carotenoids 5,690 μg/100 g and flavonoids 117.48 mg catechin equivalent/100 g. Julaeha et al. (2015) identified 1,2,3,4-tetrahydropyridin-3-yl-octanoate in an extract of papaya seeds. Gas chromatography-mass spectrometric (GC-MS) studies of papaya seed extracts reported the presence of amides, fatty acids, fatty aldehydes, organic acids, nitriles and sterols (Sani et al., 2020). The GC-MS profile of papaya seed extracts detected the presence of some major bioactive compounds including hexadecanoic acid, methyl ester, 11-octadecenoic acid, *N*, *N*-dimethyl, n-hexadecanoic acid and oleic acid. Therefore, it is suggested that the seeds could be used as an effective antioxidant agent (Agadaa et al., 2021). Papaya seeds have also been vaunted for their oil content (25.6%) that may be beneficial for food and pharmaceutical purposes (Afolabi et al., 2011). PPapaya seed oil has highest concentrations of triacylglycerols (triolein >37%) and monounsaturated fatty acid (oleic acid >70 %) of total oil content respectively (Samaram et al., 2013; Panzarini et al., 2014). Furthermore, papaya seed oil exhibits great stability to oxidation reactions and exhibits antioxidant properties. The physicochemical features of papaya seed oil demonstrated its superiority and benefits for consumption (Agunbiade & Adewole, 2014). Ghosh et al. (2017) separated oleic acid from the seed by the methods of ^{1}H nuclear magnetic resonance (proton NMR), Fourier-transform infrared spectroscopy (FTIR), and carbon-13 nuclear magnetic resonance (C-13 NMR) spectroscopy, including mass spectrometry (MS). The chemical constituents of papaya seed oil were found to be oleic acid (47.7%), palmitic acid (6%) and linoleic acid (37.3%) (Anwar et al., 2018). Seshamamba et al. (2018) found that the phenolics concentration in papaya seed oil varied from 112±0.25 to 114±0.55 mg/100 g. Generally, the high concentrations of phytochemicals, carotenoids, proteins, lipids and carbohydrates in papaya seeds make this oil a good ingredient for food fortification. So, processing by extracting or isolation from papaya seeds of various compounds with good nutritional and functional properties is a novel idea and a first step toward the incorporation of papaya seeds as a functional ingredient (Kumoro et al., 2020).

12.2.3 Latex

Crude latex of papaya is well known to have great uses and potentials, specifically in the agricultural, food manufacture and pharmaceutical industries. The latex contains various kinds of proteins

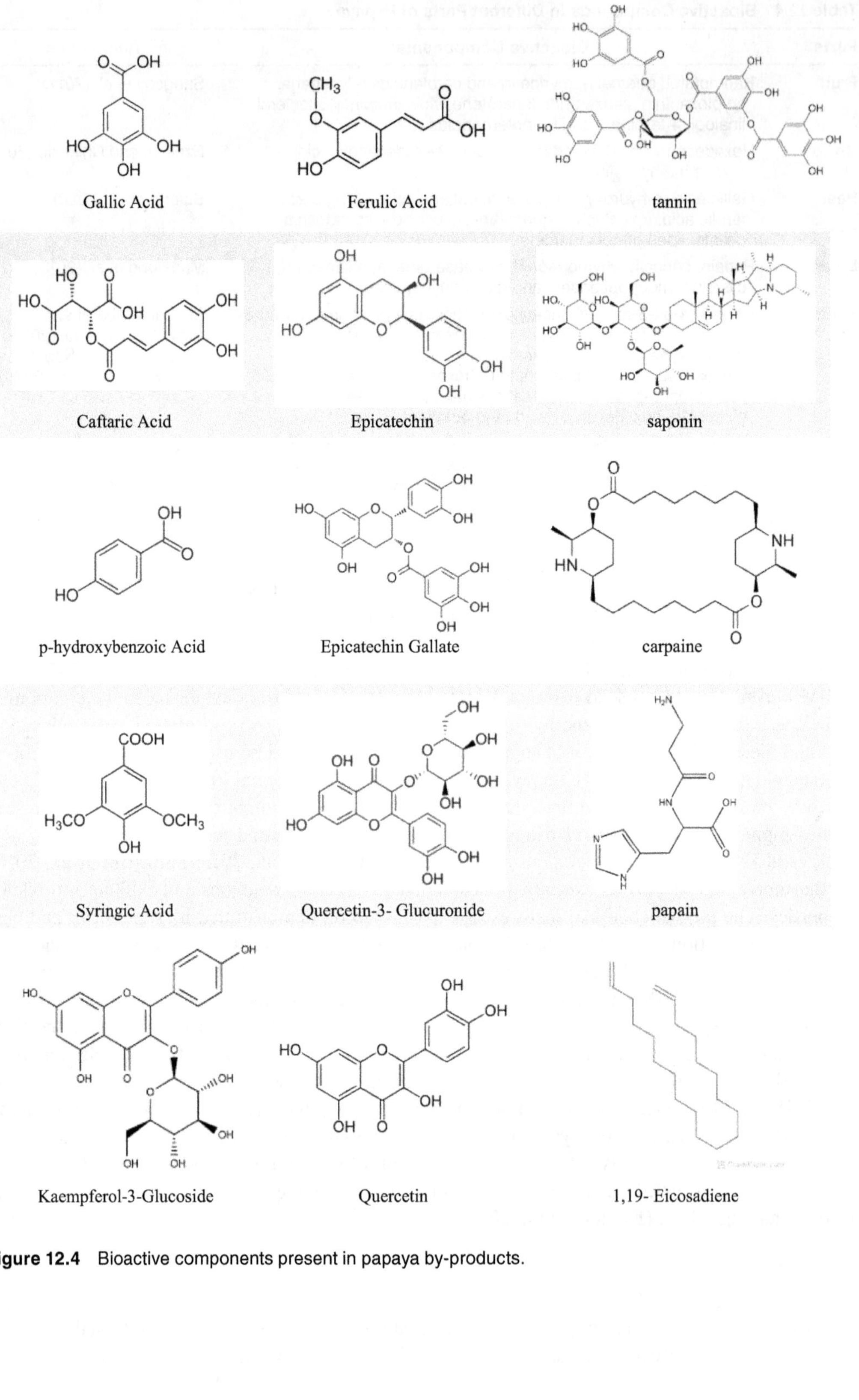

Figure 12.4 Bioactive components present in papaya by-products.

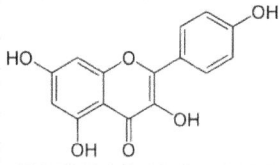
Kaempferol

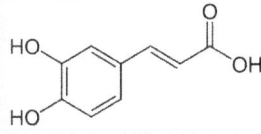
Caffeic acid

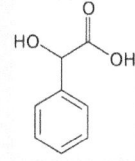
Mandelic acid

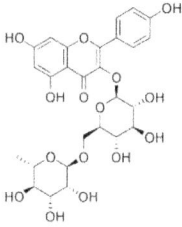
kaempferol 3-rutinoside

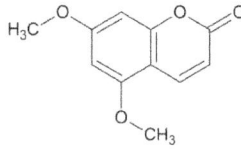
5,7-dimethoxy coumarin

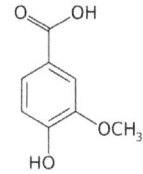
vanillic acid

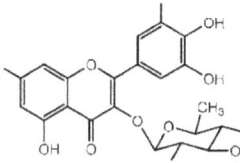
myricetin 3-rhamnoside,

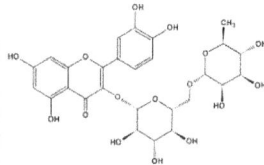
quercetin3-(2G-rham
nosylrutinoside),

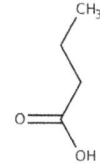
N-Butyric acid

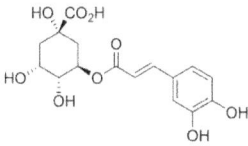
chlorogenic acid

N-Hexadecanoic Acid

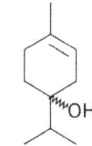
4-terpineol

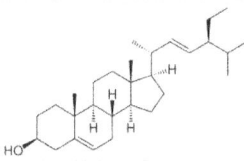
Stigmasterol,

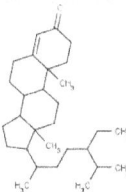
Stigmast-4-EN-3-One

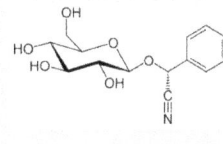
Prunasin

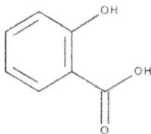
salicylic acid,

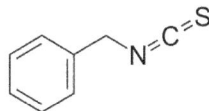
benzyl isothiocyanate

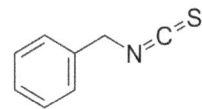
Benzyl isothiocyanate

Figure 12.4 (Continued)

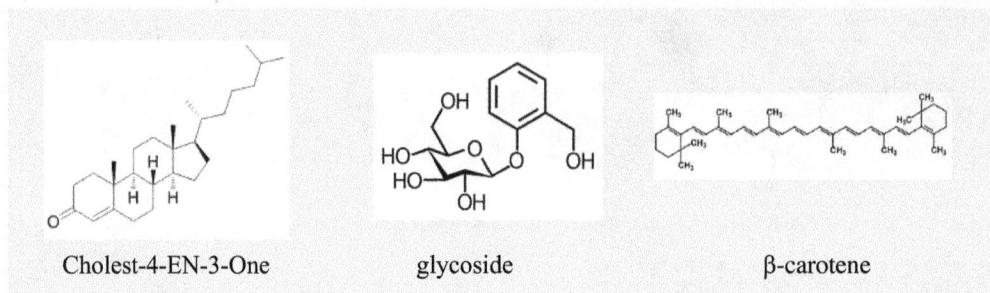

Cholest-4-EN-3-One   glycoside   β-carotene

Figure 12.4   (Continued)

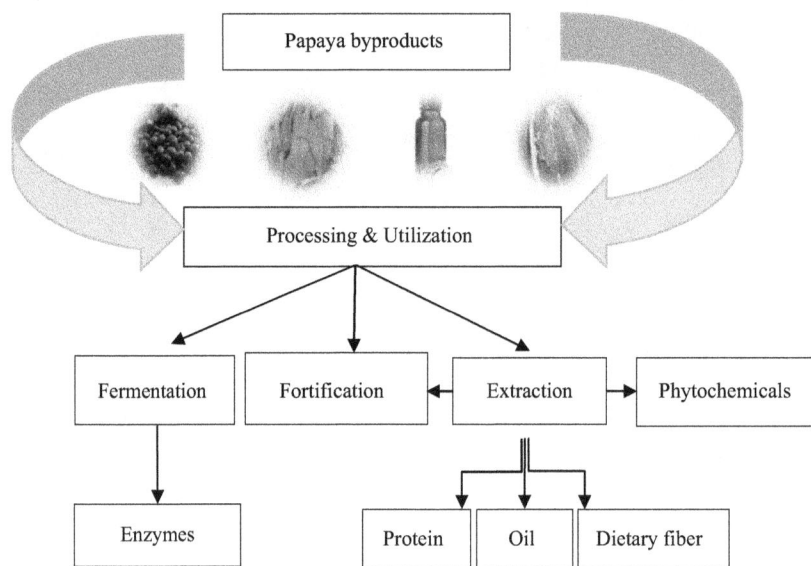

Figure 12.5   Various food applications from the processing and utilization of papaya by-products.

and enzymes (Macalood et al., 2013), such as the papain enzyme which has a characteristic milk clotting activity as a result of proteolytic activity (Rana et al., 2017). Enzymes present in papaya latex (Table 12.4), as well as in different parts of the papaya plant, include papain, caricain, chymopapain and protease omega (Dhivya et al., 2018). Pal and Mazumder (2013) and Kumar and Mishra (2019) reported that the latex of unripe papaya contained proteolytic enzymes such as chymopapain A, B and C, peptidase A and B, papain glutamine cyclotransferase, and lysozymes, while Nhat and Ha (2019) isolated a lipase from the papaya latex.

12.3 PROCESSING OF PAPAYA WASTES AND BY-PRODUCTS

During the processing of papaya, a great number of wastes and by-products (peels, seeds, oils and latex) are generated and removed. However, it is established that these by-products are novel sources of value-added biomolecules that are obtained during enzyme isolation, dietary fiber extraction, oil production and protein extraction, which can be utilized for various uses in the food industry (Figure 12.5).

12.3.1 Enzyme Production

The separation methods for isolating enzymes from plant by-products have been improved by the development of new techniques that ensure high and efficient extraction rates (Nadar et al., 2017). The by-products of papaya have been extensively studied to isolate proteolytic enzymes and many other functional products. To isolate papain from papaya latex, precipitation with ammonium sulfate and polyethylene glycol was studied (Nitsawang et al., 2006), along with polyvinyl sulfonate (Braia et al., 2013), while a papain extraction method was developed using the macro-ligand alginate (Rocha et al., 2016). Isolating proteolytic enzymes from dried papaya by-products can also be adopted by a convection method (Rojas et al., 2018). Total soluble protein was extracted with phosphate buffer (pH 7.0) as the solvent, and treating the waste products at 40°C to about 20% moisture, which allows an improved extraction and does not affect the proteolytic activity of the proteins isolated.

Dried green papaya peel contains more papain activity (914.34 ± 25.47 U/mg) than ripe peel (881.70 ± 23.69 U/mg). When 20% ammonium sulfate is utilized, it precipitates 74% of the papain (Rojas et al., 2020). Papain and pectinesterase enzyme isolation from fruits of papaya were studied by Foda et al. (2016). The optimum concentration of NaCl for pectinesterase activity was estimated to be 0.3 M. The optimum pH for crude papain activity was found to be 6.4 and that for papaya pectinesterase activity was estimated to be 7.0. Rivera et al. (2017) demonstrated that papaya lipase 1 is stable in organic solvents and this unlocks the possibilities of analyzing the catalytic reactivity of pure papaya lipase 1 in different reactions. Meanwhile, Nhat and Ha (2019) suggested an improved methodology to solubilize the latex and precipitate the lipase protein by the use of ammonium sulfate and sodium lauroyl sarcosinate. The data showed that using the freeze-drying method improved the activity of lipase in latex powder when compared with conventional methods, while freeze-drying of papaya latex could be beneficial in preserving its lipase activity and sodium lauroyl sarcosinate could achieve the separation of the lipase from the papaya latex. In a recent study by Hafid et al. (2020), a three-phase partitioning method was used as a quick, simple, inexpensive and very efficient method for papain recovery from the papaya latex. The molecular weight of the papain obtained was 23.2 kDa and showed highest activity at 50°C and pH 6.0. In the presence of many metal ions, protease activity with the four isoforms was constant over 40–80°C and a pH range of 6.0–7.5, none of the ions inactivating the protease recovered in this process. The milk-clotting activity tests showed greater stability of the papaya papain when stored at −20°C compared with 25°C or 4°C for up to 5 weeks.

12.3.2 Dietary Fiber Extraction

When structural properties of pectin from papaya peel were studied, it was found that the properties could be exploited to understand the functional characteristics of pectin as an ingredient in the food industry. In this regard, Koubala et al. (2014) extracted pectin from the peel of two papaya varieties, Solo and Local. Papaya peel is considered to be rich in pectin when isolated from the cell wall material. The peel of papaya is mainly composed of linear Ca^{2+}-cross-linked homogalacturonan, with low methyl-esterification, and high molar mass, whereas a smaller proportion of branched, water-soluble, highly methyl-esterified pectins could be recovered from the peel of 'Solo' and 'Local' papayas. Pectin consists of a heteropolysaccharide structure found in the primary cell walls of terrestrial plants. Pectin is generally added as a gelling substance to food products, especially in jams and jellies. It can also be beneficial in fillings, medicines, sweets, as a fruit juice stabilizer and milk drinks, and as a good source of soluble dietary fibers. In a study by Altaf et al. (2015), who isolated the pectin from papaya peel, the extraction process had an effect on the extraction yields of pectin. Pectin yield was high when using hydrochloric acid for extraction when compared with using citric acid. As the extraction time and temperature increase, the pectin yield increases.

Meanwhile, as the pH increases, the pectin yield decreases. The dietary fiber can be produced from papaya by-products by extraction methods using ethanol and microwave-assisted dehydration, and it was reported that a 15-minute extraction with 2.9 mL of ethanol/g of papaya pulp, followed by drying at 40°C in the next step produces dietary fiber concentrates with best properties for function. The dietary fiber concentrates obtained from papaya peel under similar conditions revealed higher glass transition temperatures (38°C) and higher levels of cell wall polysaccharides than one obtained from pulp (8°C), with the polyphenol amount being twice that yielded from pulp. Both dietary fiber concentrates revealed their potential to be used for nutritional functions and for technological purposes (Nieto et al., 2017). Furthermore, Calvache et al. (2016) used the peel papaya by-product to yield dietary fiber concentrates when treated with ethanol and analyzed after microwave drying. The fibers obtained exhibited higher antioxidant activity; therefore, they can be used in functional food preparations. The antioxidant compounds prevent lipid oxidation in foods and hence protect from food rancidity, retaining the shelf life, as well as being used in the formulation of health food products in the food industry.

In another method of extraction, Patidar et al. (2016) used a solid-state fermentation process with dried peel to isolate acidic pectin methylesterase by a fungal strain *Aspergillus tubingensis*. This method was more efficient and cheaper than the ones usually applied. Zhang et al. (2017) reported that the soluble dietary fibers were isolated from the peel of papaya by either alkaline extraction or ultrasound-assisted alkaline extraction. Under suitable conditions, a maximum yield of 36.99% was recorded by the ultrasound-assisted alkaline method. The ultrasound-assisted alkaline extraction preparations exhibited superior functional properties such as higher water-holding capacity, temperature stability, oil-holding capacity and swelling abilities than fiber extracted by alkaline extraction alone. Thus, papaya peel is potentially a rich source of natural, health-beneficial dietary fibers with reasonable functional features. Ripening of papaya is associated with gradual depolymerization of pectic polysaccharides, such as homogalacturonans, arabinogalactans, rhamnogalacturonans and their derivatives.

12.3.3 Oil Extraction

The seed oil from papaya is rich in oleic and palmitic acids, with concentrations estimated to be in the range of 66.74–76.80% and 12.80–19.70%, respectively. High concentrations of oleic acids are useful in reducing the risk of chronic problems such as cardiovascular disease (Kris-Etherton, et al., 1999). Oleic acid is also useful in the formulation of skin creams, cosmetics and plant oil-based lubricants and chemicals (Yanty et al., 2014). Plant oils with higher levels of polyunsaturated fatty acids are sensitive to oxidation and tend to deteriorate when exposed to room temperature and air, but oleic acid obtained from papaya seed oil is very stable against oxidation reactions (Tan and Ghazali, 2019; Puangsri et al., 2005). Thus, papaya seed oil containing a low level of polyunsaturated fatty acid (<5%) and high concentrations of oleic acid (>66%) could be fit for household cooking applications.

There are various methods of papaya seed oil extraction. The method most generally used for extraction of oil from papaya seed is Soxhlet extraction. The oil content obtained from the Soxhlet method is reported to be $25.3 \pm 0.7\%$ which was similar to that from oil-rich seeds such as caraway (18.9–20.1%), white mustard (25.3–28.9%) and coriander (20–22%) (Kozłowska et al., 2016). Papaya seed oil was extracted by using different polar solvents such as ethanol, acetone, ethyl acetate and hexane by Lee et al. (2011). In that study, papaya seed powder was soaked with the selected solvent for 1 week, and the oil was recovered by evaporation. The results presented that the most polar solvent (ethanol) extracted the lowest yield of papaya seed oil, compared with the less polar solvents. This is due to the fact that most of the lipids are non-polar in nature and therefore less soluble in the more polar solvents. When Samaram et al. (2013) suspended papaya seed powder in a non-polar solvent, hexane, in a ratio of 1:10 (w/v) and agitated the suspension under various experimental conditions under temperatures ranges from 25 to 50°C and for times from 3 to 12 h, the oil was extracted by evaporation. It was reported that the highest amount of papaya seed oil was obtained at 25°C and a 12-h agitation time. Furthermore, in another study, it was observed that the yield of oil was higher when the powder-to-solvent ratio gradually increased from 1:5 to 1:12 (w/v). The higher ratio of seed powder dispersed into the solvent consequently increased the mass transfer and hence oil extraction (Senrayan and Venkatachalam, 2018). Puangsri et al. (2005) reported extraction of papaya seed oil using an aqueous enzymatic method using various enzymes (pectinase, protease, α-amylase and cellulase). In that process, the seed powder was suspended in distilled water at a ratio of 1:10 (w/v) and treated with 2% of enzyme at 45°C. The protease enzyme achieved the highest yield of oil, followed by pectinase, α-amylase and cellulose. Another technique, ultrasound-assisted solvent extraction was used for the extraction of papaya seed oil by Samaram et al. (2013). In this method, papaya seed oil was extracted with the non-polar solvent hexane at a ratio of 1:8 (w/v) at 50°C under 30 min sonication in a water bath, with the oil being recovered by evaporation. This study found that the yield of papaya seed oil was 23.03%. In addition, Zhang et al. (2019) also estimated the effectiveness of ultrasound-assisted solvent extraction of seed oil from papaya by using an ultrasonic waterbath. To obtain an optimum extraction yield, Box–Behnken experimental design was used. It has been observed that ultrasonic power has a great impact on the efficiency of oil extraction. The greatest extraction was obtained at an ultrasonic power of 250 W and a sonication time of 20 min. It was concluded that, for seed oil extraction from papaya, the ultrasound-assisted solvent extraction obtained 7% higher yield than that achieved by the traditional Soxhlet extraction method. On the other hand, Briones-Labarca et al. (2015) used high hydrostatic pressure-assisted solvent extraction method for oil production from papaya seed. The papaya seed powder was extracted with hexane, sealed airtight in a polyethylene bag, and the pressure applied was 500 MPa for 15 min. The concentration of seed oil extracted from the high hydrostatic pressure solvent method (40.5%) was higher than that achieved by Soxhlet extraction (32.1%), indicating the commercialization potential of the newly developed technique for papaya seed oil extraction.

12.3.4 Protein Extraction

The crude protein in papaya peel is about 18.1% (Romelle et al., 2016). The presence of some additional peel ingredients, such as phenolic compounds, polysaccharides and lipids, can result in co-extraction with the proteins (González-García et al., 2014). Using more environmentally friendly solvents can enhance the performance and sustainability of protein extraction. Through the process of protein extraction, the most suitable method for extraction should be selected to maximize the reaction between the extracting solvent and the target compounds. Parniakov et al. (2015) compared the efficiency of aqueous extraction of papaya seeds at different pH values in the range 2.5–11 and temperatures from 20–60°C for extracting proteins from papaya seeds. The extraction was processed by pulsed electrical energy or high-voltage electric discharges. It was reported that, when the sample was pretreated initially with electric field strengths of $\approx$ 40 kV/cm or $\approx$ 13.3 kV/cm for pulsed electric fields or high-voltage electric discharges, respectively, the protein yield was significantly improved compared with conventional solid–liquid extraction.

12.4 UTILIZATION OF PAPAYA WASTES AND BY-PRODUCTS

Papaya peel rich in dietary fiber was reported by Manzoor et al. (2019) to be prepared in the form of a powder. The peel powder was incorporated into yoghurt at various concentrations (3.0 % and 1.5 % w/w) and the quality parameters viscosity, pH and color were recorded. Sensory score was also measured after fortification of yogurt with papaya peel powder. Experiments showed that the thickness of the curd increased in response to an increase in the quantity of peel powder added. It was also found that addition of the peel powder resulted in greater quality of the yogurt for a period of 21 days of storage. When the concentration of powder was increased, the pH values of the samples prepared did not change ($P > 0.05$). Other factors also remained unaffected for the storage period, except that the pH decreased. Sensor findings showed that the curd fortified with 1.5% concentration of peel powder and dried at 55°C presented the highest sensory scores. Omar et al. (2020) prepared ice cream enriched with papaya seeds by preparing 1.0%, 2.0% or 3.0% of the papaya seed formulations and compared with normal or untreated samples. It was established that the papaya seeds contain high numbers and concentrations of antioxidants, and the addition of the seeds retained the properties of the ice cream, so that the quality and sensory profiles of ice cream was acceptable to the food panelists. In a study by Maskey and Shrestha (2020), using papaya latex protease enzyme as a milk clotting agent, it was shown that most milk clotting activity was observed at pH 6.5, temperature of 70°C and an enzyme concentration of 1 g/1000 mL milk. The addition of the papaya protease enzyme to cheese enhanced the sensory properties and showed the papaya latex to be a good source of crude papaya protease which could possibly be used in the processing of soft-unripened cheese. The proteases identified and characterized in papaya latex are papain, chymopapain, glycyl endopeptidase and caricain (Azarkan et al., 2003) and are extensively utilized for the tenderization of meat. Azevedo and Campagnol (2014) estimated the consequences of the incorporation of papaya seed flour on the technical and sensory quality parameters of hamburgers. Four treatments were used for preparing hamburgers, namely 0, 1, 2 or 3% seed flour, to achieve the beneficial effects of incorporating the high protein and fiber contents of papaya seed flour. It was found that the incorporation of papaya seed flour is a practical way of enhancing the technical qualities and eliminating the negative effects of spoilage on the sensory qualities of the hamburgers. Bokaria and Ray (2016) developed flour from the peel of raw papaya for the preparation of highly nutritious cookies. Peel flour-enriched cookies were formulated by adding 5%, 7.5% or 10% papaya peel flour to regular wheat flour in that study. The result showed 5% peel flour produced cookies with the best outcomes in terms of physical, chemical and sensory quality of the cookies. The results showed that papaya peel flour-enriched cookies contained substantial higher levels of protein and antioxidant activity which improved the nutritional profile. Bhosale and Udachan (2018) used papaya seed powder and papaya peel for the formulation of cookies and conducted sensory evaluations which showed that fortified functional cookies containing 3.82% papaya peel powder or 1.24% papaya seed powder blend were acceptable and indicated that the addition of papaya peel or papaya seed powder improved nutritional properties, physicochemical characteristics and organoleptic attributes. Kadiri et al. (2017) processed papaya seed into a flour that was high in protein concentration and had exceptional reducing power as well as the potential to be used as a food additive with the capacity to inhibit Fe^{3+}–Fe^{2+} transformation. Kugo et al. (2018) formulated a maize flour enriched with dried papaya seed flour to make fortified porridge as a school meal recipe for primary school children (n=326) in Kenya. The children consumed 300 mL porridge fortified with papaya on a daily basis for two months. The results revealed that fortified porridge with papaya seed had substantial beneficial effects on the recovery of children from *Ascaris lumbricoides* (roundworm) infection. The fortified porridge resulted in improved health results, with fewer fungal infections than children treated with the synthetic drug albendazole for parasitic worm infestations. Using papaya incorporation into school meals had a potential beneficial impact in reducing malnutrition and improving the health of primary school children. In addition, Avila et al. (2020) prepared cornmeal porridges enriched with papaya seed and reported that prior treatment of the papaya seed with acetic acid and sodium bicarbonate enhanced the phenolic compound concentration and antioxidant activities, which caused improvements in the phytochemical and functional characteristics of the cornmeal porridge.

12.4.1 Formulated Edible Coatings

The application of edible coatings is an important strategy to minimize the use of conventional packaging materials and increase the shelf-life of processed fruit and other products. Ma et al. (2021) formulated papaya polysaccharide and corn starch layers by applying the solution casting method.

Consequently, the fused papaya polysaccharide and corn starch layers exhibited effective antioxidant and moisture-retaining properties and suitable antibacterial performance. The results showed that, after incorporating the papaya polysaccharide, the films showed a substantial rise in swelling and tensile strength. The incorporation of papaya polysaccharide with corn starch affected the shelf-life of fresh apples, while including the edible layer with papaya polysaccharide certainly improve the sensory acceptance of the blended materials.

12.4.2 Antimicrobial Agent

Papaya peel is an appreciable source of organic complexes that promotes the growth of microorganisms. Thus, papaya peel can be utilized as an efficient agent for the culture and isolation of microbes (Saheed et al., 2016). Han et al. (2018) developed an economical method of using culled papaya to yield different products, such as seed oil, crude myrosinase-detoxified/defatted seed meal, purée, and glucosinolates with biofumigation and antimicrobial purposes. This method depends on combination of two papaya by-products (seed oil and purée; prepared by chopping and crushing of papaya without peel and seeds as carbon substrates for the growth of *Yarrowia lipolytica* (a yeast) to generate only cell proteins and value-added recombinant proteins as functional products. *Y. lipolytica* is amenable to gene manipulation and is recognized as a useful cellular factory with several industrialized purposes. Oleic acid was isolated from papaya seed by Ghosh et al. (2017) who investigated its antibacterial activity on live fish which had been exposed to pathogenic bacteria such as *Klebsiella* strain KBSG14. The pure isolated compound showed a high chemo-preventive effect against bacterial infection in *vivo*. Another report showed antibacterial activity when a methanolic papaya seed extract was tested against *Escherichia coli*, *Pseudomonas aeruginosa*, *Bacillus subtilis* and *Staphylococcus aureus* (Sahni, 2020). In addition, Singh et al. (2020a, 2020b, 2020c) reported that the papaya seed methanolic extract was assessed for its antibacterial activity against *Pseudomonas vulgaris*, *E. coli* and *Klebsiella pneumoniae*. The extract showed greater activity against *E. coli* followed by *Ps. vulgaris* and *K. pneumoniae*.

12.5 CONCLUSIONS

Papaya is a popular fruit cultivated in various tropical and subtropical areas of the world. All parts of the papaya fruit are rich in beneficial bioactive components including phenolics, flavonoids, minerals, vitamins and enzymes which are responsible for its functionality in different applications. Papaya processing industries dispose of massive quantities of papaya waste and by-products, especially peels, seeds and other fruit residues. Different papaya parts can be utilized as good sources of functional food formulations and nutraceuticals not only to resolve waste problems but also to produce additional profits from the waste streams for the fruit processing industries. Additionally, several papaya by-products have found their applications in food fortification, enzyme production, dietary fiber extraction, oil production and protein extraction. Previous studies illustrated that papaya could play a vital role in the food system and be beneficial in improving the finances of many papaya-growing countries. Additional research is required to investigate the hidden beneficial applications in papaya fruit and its byproducts which can be utilized as nutritional, nutraceutical and functional product formulations, to benefit the food industry.

REFERENCES

Abbas, M., Saeed, F., Anjum, F. M., Afzaal, M., Tufail, T., Bashir, M. S., & Suleria, H. A. R. (2017). Natural polyphenols: An overview. *International Journal of Food Properties*, *20*(8), 1689–1699.

Abu Qdais, H., Wuensch, C., Dornack, C., & Nassour, A. (2019). The role of solid waste composting in mitigating climate change in Jordan. *Waste Management & Research*, *37*(8), 833–842.

Afolabi, I. S., Marcus, G. D., Olanrewaju, T. O., & Chizea, V. (2011). Biochemical effect of some food processing methods on the health promoting properties of under-utilized *Carica papaya* seed. *Journal of Natural Products*, *4*, 17–24.

Agada, R., Thagriki, D., Lydia, D. E., Khusro, A., Alkahtani, J., Al Shaqha, M. M., ... & Elshikh, M. S. (2021). Antioxidant and anti-diabetic activities of bioactive fractions of *Carica papaya* seeds extract. *Journal of King Saud University-Science*, *33*(2), 101342.

Agunbiade, F. O., & Adewole, T. A. (2014). Methanolysis of Carica papaya seed oil for production of biodiesel. *Journal of Fuels*, Article ID 904076, 1–6.

Altaf, U., Immanuel, G., & Iftikhar, F. (2015). Extraction and characterization of pectin derived from papaya (*Carica papaya* Linn.) peel. *International Journal of Science, Engineering and Technology*, *3*(4), 970–974.

Anwar, M., Rasul, M. G., & Ashwath, N. (2018). A systematic multivariate analysis of *carica papaya* biodiesel blends and their interactive effect on performance. *Energies*, *11*(11), 2931.

Anwar, M., Rasul, M. G., Ashwath, N., & Nabi, M. N. (2019). The potential of utilising papaya seed oil and stone fruit kernel oil as non-edible feedstock for biodiesel production in Australia—A review. *Energy Reports*, *5*, 280–297.

Aravind, G., Bhowmik, D., Duraivel, S., & Harish, G. (2013). Traditional and medicinal uses of *Carica papaya*. *Journal of Medicinal Plants Studies*, *1*(1), 7–15.

Ávila, S., Kugo, M., Hornung, P. S., Apea-Bah, F. B., Songok, E. M., & Beta, T. (2020). *Carica papaya* seed enhances phytochemicals and functional properties in cornmeal porridges. *Food Chemistry*, *323*, 126808.

Azarkan, M., El Moussaoui, A., Van Wuytswinkel, D., Dehon, G., & Looze, Y. (2003). Fractionation and purification of the enzymes stored in the latex of *Carica papaya*. *Journal of Chromatography B*, *790*(1–2), 229–238.

Azevedo, L. A., & Campagnol, P. C. B. (2014). Papaya seed flour (*Carica papaya*) affects the technological and sensory quality of hamburgers. *International Food Research Journal*, *21*(6), 2141.

Bhattachrjee, S. K. (2001). *Carica papaya*. In Shashi Jain (Ed.), *Hand book of medicinal plant* (3rd rev. ed.) (pp. 1–71). Jaipur: Pointer Publisher.

Bhosale, P., & Udachan, I. S. (2018). Studies on utilization of papaya peel and seed powder for development of fiber enriched functional cookies. *IJRAR-International Journal of Research and Analytical Reviews (IJRAR)*, *5*(4), 459–466.

Bokaria, K., & Ray, S. (2016). Development of papaya peel flour based cookies and evaluation of its quality. *Development*, *3*(12), 6393–6396.

Braia, M., Ferrero, M., Rocha, M. V., Loureiro, D., Tubio, G., & Romanini, D. (2013). Bioseparation of papain from *Carica papaya* latex by precipitation of papain–poly (vinyl sulfonate) complexes. *Protein Expression and Purification*, *91*(1), 91–95.

Briones-Labarca, V., Plaza-Morales, M., Giovagnoli-Vicuña, C., & Jamett, F. (2015). High hydrostatic pressure and ultrasound extractions of antioxidant compounds, sulforaphane and fatty acids from Chilean papaya (*Vasconcellea pubescens*) seeds: Effects of extraction conditions and methods. *LWT-Food Science and Technology*, *60*(1), 525–534.

Calvache, J. N., Cueto, M., Farroni, A., de Escalada Pla, M., & Gerschenson, L. N. (2016). Antioxidant characterization of new dietary fiber concentrates from papaya pulp and peel (*Carica papaya* L.). *Journal of Functional Foods*, *27*, 319–328.

Chandrasekaran, R., Gnanasekar, S., Seetharaman, P., Keppanan, R., Arockiaswamy, W., & Sivaperumal, S. (2016). Formulation of *Carica papaya* latex-functionalized silver nanoparticles for its improved antibacterial and anticancer applications. *Journal of Molecular Liquids*, *219*, 232–238.

Chielle, D. P., Bertuol, D. A., Meili, L., Tanabe, E. H., & Dotto, G. L. (2016). Spouted bed drying of papaya seeds for oil production. *LWT-Food Science and Technology*, *65*, 852–860.

Chukwuka, K. S., Iwuagwu, M., & Uka, U. N. (2013). Evaluation of nutritional components of *Carica papaya* L. at different stages of ripening. *IOSR Journal of Pharmacy and Biological Sciences*, *6*(4), 13–16.

Dahunsi, S. O., Oranusi, S., & Efeovbokhan, V. E. (2017). Cleaner energy for cleaner production: Modeling and optimization of biogas generation from *Carica papayas* (Pawpaw) fruit peels. *Journal of Cleaner Production*, *156*, 19–29.

Dhivya, R., Rashma, R. S., Vinothini, B., & Pavithra, R. (2018). Extraction and purification of papain enzyme from *Carica papaya* for wound debridement. *International Journal of Pure and Applied Mathematics*, *119*(15), 1265–1274.

Doan, M. T. N., Huynh, M. C., Pham, A. N. V., Chau, N. D. Q., & Le, P. T. K. (2020). Extracting seed oil and phenolic compounds from papaya seeds by ultrasound-assisted extraction method and their properties. *Chemical Engineering Transactions, 78*, 493–498.

Etim, A. O., Eloka-Eboka, A. C., & Musonge, P. (2020). Potential of *Carica papaya* peels as effective biocatalyst in the optimized parametric transesterification of used vegetable oil. *Environmental Engineering Research, 26*(4), 200299.

Ezekwe, S. A., & Chikezie, P. C. (2017). GC–MS analysis of aqueous extract of unripe fruit of *Carica papaya*. *Journal of Nutrition & Food Sciences, 7*(3), 1–5.

FAO. (2020). Major tropical fruits – preliminary market results 2019. Rome, 3–4.

FAOSTAT. (2020). http://www.fao.org/faostat/en/#data/QC.

Foda, F. F., Saad, S. M., Attia, N. Y., & Eid, M. S. (2016). Production and evaluation of papain and pectinesterase enzymes from papaya fruits. In *3rd International Conference on Biotechnology Applications in Agriculture (ICBAA)*, Benha University, Moshtohor and Sharm El-Sheikh (pp. 5–9).

Ghosh, S., Saha, M., Bandyopadhyay, P. K., & Jana, M. (2017). Extraction, isolation and characterization of bioactive compounds from chloroform extract of *Carica papaya* seed and it's in vivo antibacterial potentiality in Channa punctatus against Klebsiella PKBSG14. *Microbial Pathogenesis, 111*, 508–518.

González-García, E., Marina, M. L., & García, M. C. (2014). Plum (*Prunus domestica* L.) by-product as a new and cheap source of bioactive peptides: Extraction method and peptides characterization. *Journal of Functional Foods, 11*, 428–437.

Hafid, K., John, J., Sayah, T. M., Domínguez, R., Becila, S., Lamri, M., … & Gagaoua, M. (2020). One-step recovery of latex papain from *Carica papaya* using three phase partitioning and its use as milk-clotting and meat-tenderizing agent. *International Journal of Biological Macromolecules, 146*, 798–810.

Han, Z., Park, A., & Su, W. W. (2018). Valorization of papaya fruit waste through low-cost fractionation and microbial conversion of both juice and seed lipids. *RSC Advances, 8*(49), 27963–27972.

Ikeyi, A. P., Ogbonna, A. O., & Eze, F. U. (2013). Phytochemical analysis of paw-paw (*Carica papaya*) leaves. *International Journal of Life Sciences Biotechnology and Pharma Research, 2*(3), 347–351.

Jamal, P., Akbar, I., Jaswir, I., & Zuhanis, Y. (2017). Quantification of total phenolic compounds in papaya fruit peel. *Pertanika Journal of Tropical Agricultural Science, 40*(1), 121–135.

Julaeha, E., Permatasari, Y., Mayanti, T., & Diantini, A. (2015). Antifertility compound from the seeds of *Carica papaya*. *Procedia Chemistry, 17*, 66–69.

Kadiri, O., Akanbi, C. T., Olawoye, B. T., & Gbadamosi, S. O. (2017). Characterization and antioxidant evaluation of phenolic compounds extracted from the protein concentrate and protein isolate produced from pawpaw (*Carica papaya* Linn.) seeds. *International Journal of Food Properties, 20*(11), 2423–2436.

Kadiri, O., Olawoye, B., Fawale, O. S., & Adalumo, O. A. (2016). Nutraceutical and antioxidant properties of the seeds, leaves and fruits of *Carica papaya*: Potential relevance to humans diet, the food industry and the pharmaceutical industry-a review. *Turkish Journal of Agriculture-Food Science and Technology, 4*(12), 1039–1052.

Koubala, B. B., Christiaens, S., Kansci, G., Van Loey, A. M., & Hendrickx, M. E. (2014). Isolation and structural characterisation of papaya peel pectin. *Food Research International, 55*, 215–221.

Kozłowska, M., Gruczyńska, E., Ścibisz, I., & Rudzińska, M. (2016). Fatty acids and sterols composition, and antioxidant activity of oils extracted from plant seeds. *Food Chemistry, 213*, 450–456.

Kris-Etherton, P. M., Pearson, T. A., Wan, Y., Hargrove, R. L., Moriarty, K., Fishell, V., & Etherton, T. D. (1999). High–monounsaturated fatty acid diets lower both plasma cholesterol and triacylglycerol concentrations. *The American Journal of Clinical Nutrition, 70*(6), 1009–1015.

Kugo, M., Keter, L., Maiyo, A., Kinyua, J., Ndemwa, P., Maina, G., … & Songok, E. M. (2018). Fortification of *Carica papaya* fruit seeds to school meal snacks may aid Africa mass deworming programs: A preliminary survey. *BMC Complementary and Alternative Medicine, 18*(1), 1–7.

Kumar, A., & Mishra, S. (2019). Formulation and processing of papaya by products *International Journal of Food Science and Nutrition, 4*(5): 143–148.

Kumoro, A. C., Alhanif, M., & Wardhani, D. H. (2020). A critical review on tropical fruits seeds as prospective sources of nutritional and bioactive compounds for functional foods development: A case of Indonesian exotic fruits. *International Journal of Food Science, 2020*, 1–15, 4051475.

Lee, W. J., Lee, M. H., & Su, N. W. (2011). Characteristics of papaya seed oils obtained by extrusion–expelling processes. *Journal of the Science of Food and Agriculture, 91*(13), 2348–2354.

Ma, Y., Zhao, Y., Xie, J., Sameen, D. E., Ahmed, S., Dai, J., … & Liu, Y. (2021). Optimization, characterization and evaluation of papaya polysaccharide-corn starch film for fresh cut apples. *International Journal of Biological Macromolecules*, *166*, 1057–1071.

Macalood, J. S., Vicente, H. J., Boniao, R. D., Gorospe, J. G., & Roa, E. C. (2013). Chemical analysis of *Carica papaya* L. crude latex. *American Journal of Plant Sciences*, *4*(10), 1941.

Maisarah, A. M., Asmah, R., & Fauziah, O. (2014). Proximate analysis, antioxidant and anti proliferative activities of different parts of *Carica papaya*. *Journal of Tissue Science & Engineering*, *5*(1), 1.

Maisarah, A. M., Nurul Amira, B., Asmah, R., & Fauziah, O. (2013). Antioxidant analysis of different parts of *Carica papaya*. *International Food Research Journal*, *20*, 1043–1048.

Makanjuola, O. M., & Makanjuola, J. O. (2018). Proximate and selected mineral composition of ripe pawpaw (*Carica papaya*) seeds and skin. *Journal of Scientific and Innovative Research*, *7*(3), 75–77.

Manzoor, S., Yusof, Y. A., Chin, N. L., Tawakkal, A., Mohamed, I. S., Fikry, M., & Chang, L. S. (2019). Quality characteristics and sensory profile of stirred yogurt enriched with papaya peel powder. *Pertanika Journal of Tropical Agricultural Science*, *42*(2), 519–533.

Maskey, B., & Shrestha, N. K. (2020). Optimization of crude papaya (*Carica papaya*) protease in soft-unripened cheese preparation. *Journal of Food Science and Technology Nepal*, *12*(12), 1–8.

Mohammad, I. (2019). Gold nanoparticle: An efficient carrier for MCP I of *Carica papaya* seeds extract as an innovative male contraceptive in albino rats. *Journal of Drug Delivery Science and Technology*, *52*, 942–956.

Muazu, U., & Aliyu-Paiko, M. (2020). Evaluating the potentials of *Carica papaya* seed as phytobiotic to improve feed efficiency, growth performance and serum biochemical parameters in broiler chickens. *IOSR Journal of Biotechnology and Biochemistry*, *6*(1), 8–18.

Nadar, S. S., Pawar, R. G., & Rathod, V. K. (2017). Recent advances in enzyme extraction strategies: A comprehensive review. *International Journal of Biological Macromolecules*, *101*, 931–957.

Nhat, D. M., & Ha, P. T. V. (2019). The isolation and characterization of lipase from *Carica papaya* latex using zwitterion sodium lauroyl sarcosinate as agent. *Potravinarstvo*, *13*(1), 773–778.

Nieto Calvache, J. E., Soria, M., De Escalada Pla, M. F., & Gerschenson, L. N. (2017). Optimization of the production of dietary fiber concentrates from by-products of papaya (*Carica papaya* L. Var. Formosa) with microwave assistance. Evaluation of its physicochemical and functional characteristics. *Journal of Food Processing and Preservation*, *41*(4), e13071.

Ningrum, A., & Schreiner, M. (2017). Extensive potentiality of selected tropical fruits from Indonesia. *Indonesian Food and Nutrition Progress*, *14*(2), 85–90.

Nitsawang, S., Hatti-Kaul, R., & Kanasawud, P. (2006). Purification of papain from *Carica papaya* latex: Aqueous two-phase extraction versus two-step salt precipitation. *Enzyme and Microbial Technology*, *39*(5), 1103–1107.

Nna, P. J., Egbuje, O. J., & Don-Lawson, D. C. (2019). Determination of phytoconstituents and antimicrobial analysis of the ethylacetate extract of *carica papaya* seed. *International Journal of Research and Innovation in Applied Science (IJRIAS)*, *5*, 1–7.

Okon, W. I., Ogri, A. I., Igile, G. O., & Atangwho, I. J. (2017). Nutritional quality of raw and processed unripe *Carica papaya* fruit pulp and its contribution to dietary diversity and food security in some peasant communities in Nigeria. *International Journal of Biological and Chemical Sciences*, *11*(3), 1000–1011.

Omar, S. R., Aminuddin, F., Karim, L., Suhaimi, N., & Omar, S. N. (2020). Acceptability of novel antioxidant ice cream fortified with nutritious *Carica papaya* seed. *Journal of Academia*, *8*(1), 7–17.

Pal, A., & Mazumder, A. (2013). *Carica Papaya*, a magic herbal remedy. *International Journal of Advantages Research (IJAR)*, *5*(1), 2626–2635.

Panzarini, E., Dwikat, M., Mariano, S., Vergallo, C., & Dini, L. (2014). Administration dependent antioxidant effect of Carica papaya seeds water extract. *Evidence-Based Complementary and Alternative Medicine*, Article ID 281503, 1–13.

Parniakov, O., Barba, F. J., Grimi, N., Lebovka, N., & Vorobiev, E. (2014). Impact of pulsed electric fields and high voltage electrical discharges on extraction of high-added value compounds from papaya peels. *Food Research International*, *65*, 337–343.

Parniakov, O., Roselló-Soto, E., Barba, F. J., Grimi, N., Lebovka, N., & Vorobiev, E. (2015). New approaches for the effective valorization of papaya seeds: Extraction of proteins, phenolic compounds, carbohydrates, and isothiocyanates assisted by pulsed electric energy. *Food Research International*, *77*, 711–717.

Pathak, P. D., Mandavgane, S. A., & Kulkarni, B. D. (2019). Waste to wealth: A case study of papaya peel. *Waste and Biomass Valorization*, *10*(6), 1755–1766.

Pathak, R., Thakur, V., & Gupta, R. K. (2018). Formulation and analysis of papaya fortified biscuits. *Journal of Pharmacognosy and Phytochemistry, 7*(4), 1542–1545.

Patidar, M. K., Nighojkar, S., Kumar, A., & Nighojkar, A. (2016). Papaya peel valorization for production of acidic pectin methylesterase by *Aspergillus tubingensis* and its application for fruit juice clarification. *Biocatalysis and Agricultural Biotechnology, 6*, 58–67.

Pavithra, C. S., Devi, S. S., Suneetha, W. J., & Rani, C. V. D. (2017). Nutritional properties of papaya peel. *The Pharma Innovation Journal, 6*(7), 170–173.

Puangsri, T., Abdulkarim, S. M., & Ghazali, H. M. (2005). Properties of *Carica papaya* L. (papaya) seed oil following extractions using solvent and aqueous enzymatic methods. *Journal of Food Lipids, 12*(1), 62–76.

Puja, I. G. K., Wardana, I. N. G., Irawan, Y. S., & Choiron, M. A. (2018). The role of *Carica papaya* latex and aluminum oxide on the formation of carbon nanofibre made of coconut shell. *Advances in Natural Sciences: Nanoscience and Nanotechnology, 9*(3), 035021.

Rachmatika, R., & Prijono, S. N. (2015). Biological potency of *Carica papaya* L. seed for improving raja duck performance. *Buletin Peternakan, 39*(2), 123–128.

Rana, M. S., Hoque, M. R., Rahman, M. O., Habib, R., & Siddiki, M. S. R. (2017). Papaya (*Carica papaya*) latex-an alternative to rennet for cottage cheese preparation. *Journal of Advanced Veterinary and Animal Research, 4*(3), 249–254.

Rivera, I., Robles, M., Mateos-Díaz, J. C., Gutierrez-Ortega, A., & Sandoval, G. (2017). Functional expression, extracellular production, purification, structure modeling and biochemical characterization of *Carica papaya* lipase 1. *Process Biochemistry, 56*, 109–116.

Rocha, M. V., Di Giacomo, M., Beltramino, S., Loh, W., Romanini, D., & Nerli, B. B. (2016). A sustainable affinity partitioning process to recover papain from *Carica papaya* latex using alginate as macro-ligand. *Separation and Purification Technology, 168*, 168–176.

Rodrigues, L. G. G., Mazzutti, S., Vitali, L., Micke, G. A., & Ferreira, S. R. S. (2019). Recovery of bioactive phenolic compounds from papaya seeds agroindustrial residue using subcritical water extraction. *Biocatalysis and Agricultural Biotechnology, 22*, 101367.

Rojas, L. F., Cortés, C. F., Zapata, P., & Jiménez, C. (2018) Extraction and identification of endopeptidases in convection dried papaya and pineapple residues: A methodological approach for application to higher scale. *Waste Management, 78*, 58–68.

Rojas, R., Alvarez-Pérez, O. B., Contreras-Esquivel, J. C., Vicente, A., Flores, A., Sandoval, J., & Aguilar, C. N. (2020). Valorisation of mango peels: Extraction of pectin and antioxidant and antifungal polyphenols. *Waste and Biomass Valorization, 11*(1), 89–98.

Romelle, F. D., Rani, A., & Manohar, R. S. (2016). Chemical composition of some selected fruit peels. *European Journal of Food Science and Technology, 4*(4), 12–21.

Saheed, O. K., Jamal, P., Karim, M. I. A., Alam, M. Z., & Muyibi, S. A. (2016). Utilization of fruit peels as carbon source for white rot fungi biomass production under submerged state bioconversion. *Journal of King Saud University-Science, 28*(2), 143–151.

Sahni, S. K. (2020). Antimicrobial activity of methanolic extract of *Carica papaya* seeds. *Sustainable Humanosphere, 16*(1), 316–319.

Samaram, S., Mirhosseini, H., Tan, C. P., & Ghazali, H. M. (2013). Ultrasound-assisted extraction (UAE) and solvent extraction of papaya seed oil: Yield, fatty acid composition and triacylglycerol profile. *Molecules, 18*(10), 12474–12487.

Samsuri, S., Li, T. H., Ruslan, M. S. H., & Amran, N. A. (2020). Antioxidant recovery from pomegranate peel waste by integrating maceration and freeze concentration technology. *International Journal of Food Engineering, 16*(10), 1–7.

Sancho, L. E. G. G., Yahia, E. M., & González-Aguilar, G. A. (2011). Identification and quantification of phenols, carotenoids, and vitamin C from papaya (*Carica papaya* L., cv. Maradol) fruit determined by HPLC-DAD-MS/MS-ESI. *Food Research International, 44*(5), 1284–1291.

Sangsoy, K., Mongkolporn, O., Imsabai, W., & Luengwilai, K. (2017). Papaya carotenoids increased in oxisols soils. *Agriculture and Natural Resources, 51*(4), 253–261.

Sani, M. S. A., Bakar, J., Rahman, R. A., & Abas, F. (2020). Effects of coated capillary column, derivatization, and temperature programming on the identification of *Carica papaya* seed extract composition using GC/MS analysis. *Journal of Analysis and Testing, 4*(1), 23–34.

Senrayan, J., & Venkatachalam, S. (2018). Solvent-assisted extraction of oil from papaya (*Carica papaya* L.) seeds: Evaluation of its physiochemical properties and fatty-acid composition. *Separation Science and Technology, 53*(17), 2852–2859.

Seshamamba, B. S. V., Malati, P., Ruth, A. N. G., Mallika, A. S., & Sharma, V. (2018). Studies on physicochemical properties and proximate analysis of *Carica papaya* seed. *Journal of Pharmacognosy and Phytochemistry*, *7*(6), 1514–1519.

Sharma, A., Bachheti, A., Sharma, P., Bachheti, R. K., & Husen, A. (2020). Phytochemistry, pharmacological activities, nanoparticle fabrication, commercial products and waste utilization of *Carica papaya* L.: A comprehensive review. *Current Research in Biotechnology*, *16*, 316–319.

Singh, P. K., Dwivedi, M. K., & Sonter, S. (2020a). Antioxidant, antibacterial activity, and phytochemical characterization of *Carica papaya* flowers. *Beni-Suef University Journal of Basic and Applied Sciences*, *9*(23), 1–11.

Singh, P. G., Madhu, S. B., Shailasree, S., Gopenath, T. S., Basalingappa, K. M., & Sushma, B. V. (2020b). In vitro antioxidant, anti-inflammatory and anti-microbial activity of *Carica papaya* seeds. *Global Journal of Medical Research*, *20*, 19–38.

Singh, S. P., Kumar, S., Mathan, S. V., Tomar, M. S., Singh, R. K., Verma, P. K., ... & Acharya, A. (2020c). Therapeutic application of *Carica papaya* leaf extract in the management of human diseases. *DARU Journal of Pharmaceutical Sciences*, *28*(2), 735–744.

Singh, S. P., Mishra, A., Shyanti, R. K., Singh, R. P., & Acharya, A. (2021). Silver nanoparticles synthesized using *Carica papaya* leaf extract (AgNPs-PLE) causes cell cycle arrest and apoptosis in human prostate (DU145) cancer cells. *Biological Trace Element Research*, *199*(4), 1316–1331.

Sugiharto, S. (2020). Papaya (*Carica papaya* L.) seed as a potent functional feedstuff for poultry–A review. *Veterinary World*, *13*(8), 1613.

Suleria, H. A., Barrow, C. J., & Dunshea, F. R. (2020). Screening and characterization of phenolic compounds and their antioxidant capacity in different fruit peels. *Foods*, *9*(9), 1206.

Tan, C. X., & Ghazali, H. M. (2019). Avocado (*Persea americana* mill.) oil. In M. Ramadan (Ed.), *Fruit oils: Chemistry and functionality* (pp. 353–375). Cham: Springer.

Thakuria, P., Nath, R., Sarma, S., Kalita, D. J., Dutta, D. J., Borah, P., & Hussain, J. (2018). Quantitative analysis of phytochemicals in methanolic extract of Artocarpus heterophyllus, *Carica papaya* and *Terminalia bellerica* plant leaves. *International Journal of Chemical Studies*, *6*(2), 1229–1231.

Utama, G. L., Sidabutar, F. E., Felina, H., Wira, D. W., & Balia, R. L. (2019). The utilization of fruit and vegetable wastes for bioethanol production with the inoculation of indigenous yeasts consortium. *Bulgarian Journal of Agricultural Science*, *25*(2), 264–270.

Yanty, N. A. M., Marikkar, J. M. N., Nusantoro, B. P., Long, K., & Ghazali, H. M. (2014). Physico-chemical characteristics of papaya (*Carica papaya* L.) seed oil of the Hong Kong/Sekaki variety. *Journal of Oleo Science*, *63*(9), 885–892.

Zhang, W., Pan, Y. G., Huang, W., Chen, H., & Yang, H. (2019). Optimized ultrasonic-assisted extraction of papaya seed oil from Hainan/Eksotika variety. *Food Science & Nutrition*, *7*(8), 2692–2701.

Zhang, W., Zeng, G., Pan, Y., Chen, W., Huang, W., Chen, H., & Li, Y. (2017). Properties of soluble dietary fiber-polysaccharide from papaya peel obtained through alkaline or ultrasound-assisted alkaline extraction. *Carbohydrate Polymers*, *172*, 102–112.

CHAPTER 13

Apricot Fruit Wastes
Chemistry, Processing, and Utilization

Gülşah Çalışkan Koç

CONTENTS

13.1 INTRODUCTION

Several *Prunus* species of the Prunoidae sub-family of the Rosaceae family are cultivated as apricot (such as *Prunus armeniaca*). The Rosaceae is one of the largest plant families, having around 3400 species, such as plums, peaches, and apples (Hacıseferoğulları et al. 2007; Ali et al. 2015). Apricots are widely grown in countries with Mediterranean-like climate, which have mild, warm summers and cool winters (Hussain et al. 2010; Davarynejad et al. 2010). In 2018, 3,838,523 tonnes of apricot were produced worldwide. In 2015, apricots were mostly produced in Turkey (750,000 tonnes), Uzbekistan (493,842 tonnes), Iran (342,479 tonnes), Algeria (242,243 tonnes), Italy (229,020 tonnes), Spain (176,289 tonnes), Pakistan (128,382 tonnes), France (114,785 tonnes), Japan (112,400 tonnes), and Morocco (101,612 tonnes) (FAO 2020).

Apricots are rich in sugars (sucrose, glucose, and fructose), fibers, vitamins (pro-vitamin A, B-group vitamins, vitamins K, and E), minerals (K, P, Ca, Mg, Fe, Se, Na, etc.), bioactive phytochemicals (polyphenols (flavonols, anthocyanins, etc.) and carotenoids (β-carotene, γ-carotene, lycopene, etc.)), organic acids (malic, citric, tartaric, etc.,) (Table 13.1; Ruiz et al. 2005; Hacıseferoğulları et al. 2007; Akin et al. 2008; Leccese et al. 2010; Davarynejad et al. 2010). As a result of the presence of several phytochemicals showing antioxidant activity, apricots have many health-beneficial effects. Apricots are recommended to be consumed in cases of trace element deficiencies, stress, depression, anemia (due to their high Fe content), physical and mental fatigue, neurosis, etc.

DOI: 10.1201/9781003164463-13

Table 13.1 The Proximate Composition, Concentrations of Total Phenolics, Carotenoids, Sugars, and Flavonoids, and Antioxidant Activity of the Apricot

Antioxidant Activity	Ash Content (%)	Protein Content (%)	Crude Lipid Content (%)	Crude or Dietary Fiber Content (%)	Total Phenolics	Total Carotenoids	Total Sugars	Total Flavonoids	References
57.79±9.04–248.40±58.39 μmol eq./100 g (DPPH) 152.2±15.4–334.4±72.9 μmol eq./100 g (FRAP)	-	-	-	-	33.46±3.16–113.44±9.54 mg eq./100 g	2.23±0.71–11.58±3.00 mg eq./100 g	11.33±0.30–14.59±2.04 g/100g	16.87±6.43–41.42±8.16 mg eq./100 g	Kafkaletou et al. (2019, FW)*
0.045±0.006–0.357±0.002 μM Fe2+/ g FW	1.10±0.10–6.32±0.21	0.66±0.01–1.33±0.01	0.100±0.01–0.570±0.043	-	0.44±0.11–1.10±0.09 mg GA/g FW	-	-	-	Iordanescu et al. (2018)**
55.70±1.01–82.33±2.0% inhibition of DPPH	9.25±0.024–12.10±0.015	6.18±0.073–8.70±0.245	2.10±0.012–3.00±0.100	11.38±0.27–13.60±0.30	4591±210–7310±390 mg GAE/100 g	10.12±0.21–18.13±0.34 β-carotene/100 g	56.78±0.47–64.90±0.73	-	Ali et al. (2011, dw)***
150–520 μg AA/100 g (FRAP) 18–60 μg TE/mf FW (Lipophilic antioxidant capacity)	-	-	-	-	-	-	-	-	Davarynejad et al. (2010)****
-	0.50±0.013–0.89±0.008	-	-	-	4233.70±174.03–8180.49±380.98 mg GAE/100 g (dw)	14.83±1.47–91.89±3.66 mg β-carotene/100 g (dw)	-	-	Akin et al. (2008)*****
-	2.72–5.34	2.84–4.29	0.55–3.12	0.77–2.41	-	-	-	-	Hacıseferoğulları et al. (2007)******

* Apricot varieties: Greece (Bebecou, Diamantopoulou, Neraida, and Tyrvi) and France (Kioto, Farhial, Farely, and Farbaly).
** Apricot Varieties: Hungarian Best, Selena, Sirena, Olimp, Sulmona, Sulina, and Silvana (Depending on three ripening stages: unripe, half-ripe, and fully ripe).
*** Apricot Varieties: Alman, Habi, Khakhas, Mirmalik, Neeli, and Shai.
**** Apricot Varieties: Tom cot, Sweet cot, Goldstrike, Goldbar, Jumbo cot, Bergeron, Bergarouge, Zebra, and Yellow cot.
***** Apricot Varieties: Hacıhaliloğlu, Hasanbey, Soğancı, Kabaaşı, Çöloğlu, Çataloğlu, Hacıkız, Tokaloğlu, Alyanak, Iğdır, and Bursa.
****** Apricot Varieties: Hacıhaliloğlu, Çataloğlu, Kabaaş, Soğancı, Hasanbey, and Zerdali.
GAE: gallic acid equivalents, dw: dry weight, FW: fresh weight, eq.: equivalents, AA: ascorbic acid, TE: trolox.

Apricots are generally consumed in fresh, dried, or frozen forms, as well as processed fruit such as jam, jellies, marmalades, pulp, juices, nectars, alcoholic beverages, etc. (Hacıseferoğulları et al. 2007; Kafkaletou et al. 2019). The dry matter content of apricot is important for its commercial value and the apricots which contain a high proportion of dry matter are generally used to produce dried apricots. Low dry matter content of apricot causes some problems during transportation, processing, and drying, etc. (Akin et al. 2008). Due to the perishable nature of the apricot (3–5 days) and limited marketing opportunities (3–4 weeks), a high percentage of the apricots harvested is wasted (Kafkaletou et al. 2019; Rai et al. 2016; El-Adawy et al. 1994, FAO/DOA 2007). Drying, such as sun-drying and osmotically drying, canning, and producing jam are the methods commonly used for apricot preservation (Sharma 2018). During these processes, large volumes of apricot seeds are discarded by the processing plants, causing losses of potentially valuable resources as well as representing disposal problems (Özcan et al. 2010; Al-Juhaimi et al. 2018).

Apricots include a hard shell/stone (endocarp) with a seed inside (kernel), and the stone is surrounded by an outer fleshy part (exocarp and mesocarp) (Ruiz et al. 2005). The apricot shells, which consist of lignin and cellulose, are generally used as fuel (El-Adawy et al. 1994; Hacıseferoğulları et al. 2007). Apricot kernels are an important agricultural waste obtained in large amounts. Large amounts of oil, protein, and fiber can be obtained from the apricot kernel (Femenia et al. 1995). Valorizing the apricot waste is a promising strategy, with the advantages of increasing its economic potential, achieving sustainable utilization of these waste components, and reducing environmental pollution.

13.2 APRICOT WASTE

13.2.1 Apricot Shells

The apricot shell (~10% of the fruit mass) is a non-edible food waste from the processed apricot fruit industry. The apricot shell contains 7% moisture (fresh weight (fw)), 2% ash (fw), 35% lignin (dry weight, dw), 23.7% hemicellulose (dw), 26.5% cellulose (dw) (Popa et al. 2019). The oil, sugar, protein, ash, K, Ca, P, Mg, Na, Zn, I, Cu, Mn, phenolic, and pectin contents (dry weight basis) of the shell are in the range 16.6–22.0 g/kg, 19.3–20.0 g/kg, 12.1–17.7 g/kg, 16.0–16.7 g/kg, 2.42–2.46 g/kg, 1.04–1.34 g/kg, 240–310 mg/kg, 210–310 mg/kg, 100–140 mg/kg, 15.8–18.9 mg/kg, 9.6–32.6 mg/kg, 4.2–4.6 mg/kg, 2.5–2.6 mg/kg, 7.4–9.0 g/kg, and 6.2–8.0 g/kg, respectively (Cañellas et al. 1992).

Apricot shells are generally used as fuel due to their high gross energy value (17262 kJ/kg). They can also be used as a lignocellulosic feedstock for biorefineries (Cañellas et al. 1992; Femenia et al. 1995; Popa et al. 2019). The apricot shell can be used for adsorption of cadmium from water (Salah Azab and Peterson 1989), to carry out fast pyrolysis (Lucchesi et al. 1988), or for the production of plastic panels (Cañellas et al. 1992), activated carbon (Soleimani and Kaghazchi 2008; Petrova et al. 2010; Janković et al. 2019), and bio-oil (Demiral and Kul 2014).

The low ash content of apricot shells highlights the benefits of using them for combustion, pyrolysis, or carbonization (Corbett et al. 2015). The apricot shell can be used as a potential source of dietary fiber due to its neutral detergent fiber content (905–935 g/kg dw). The cellulose, hemicelluloses, and lignin can be extracted from the apricot shell (Cañellas et al. 1992; Popa et al. 2019). Moreover, apricot shells can be used as a raw material for the production of furfural and xylose because of their high hemicellulose content. Apricot shells cannot be used as animal feed because of their low protein content and low digestibility (Cañellas et al. 1992). They can also be used as a raw material for the production of antioxidant supplements, stabilizers, and preservatives, the latter due to their antioxidizing properties (Corbett et al. 2015).

13.2.2 Apricot Kernels

The apricot kernel is generally regarded as an unwanted part of the apricot (Tanwar et al. 2018b). Apricot kernels can be divided into two groups, namely bitter and sweet, based on their amygdalin (a cyanogenic glycoside) content (Yildirim et al. 2010). The sweet apricot kernel is consumed as a fresh, dried, or roasted snack, whereas the high-amygdalin bitter kernel is generally used for oil extraction (Dwivedi and Ram 2008). Despite its high protein, fiber, and oil content, the presence of amygdalin in bitter apricot kernels restricts its direct human consumption and limits its use in foods/feed (Femenia et al. 1995). Consumption of bitter almond kernel leads to hydrocyanic acid (HCN) production and has a toxic effect, acting predominantly on the nervous system and the thyroid gland. The toxicity symptoms are weakness, vomiting, mental confusion, headache, cardiac arrest, nausea, dizziness, abdominal cramps, circulatory and respiratory failure, coma, etc. (Gupta and Sharma 2009; Matthäus and Özcan 2009; Tanwar et al. 2018b). For this reason, removing amygdalin is an important process for the processing of bitter apricot kernels (Zhang et al. 2016).

Removal of the amygdalin is generally accomplished by soaking the bitter kernels in distilled water. Alternatively, a distilled water and 0.1 M ammonium hydroxide solution, or 25% sodium chloride solution, etc., can be used (El-Adawy et al. 1994; Tanwar et al. 2018a). The immersing of the bitter apricot kernel in the ammonium hydroxide solution resulted in a greater decrease (97%) in amygdalin content. The temperature of the immersing water is very important in order to determine the debitterizing processing time which is important for reducing the amygdalin content. The debitterizing process of the apricot kernel is shown in Figure 13.1. However, the immersing of apricot kernels has several disadvantages, such as higher moisture content and fatty acid rancidity, and transferring some water-soluble substances to the water, etc. (El-Adawy et al. 1994; Garrido et al. 2008; Zhang et al. 2016). Drying of the amygdalin-eliminated apricot kernels can be used to decrease the moisture content, to increase the self-life of the apricot kernels (Zhang et al. 2016). In addition, the recycling of water-soluble compounds like protein, polysaccharides, etc. from the wastewater of kernel processing is important to both manufacturers and for environmental protection (Zhang et al. 2018).

The chemical composition of the apricot kernel is given in Table 13.2. The apricot kernel contains large amounts of protein and fat compared with apricot fruit. For this reason, the apricot kernel has the potential for being a non-traditional and inexpensive source of protein and oil (Tanwar et al. 2018a). The crude energy of the apricot kernel ranges between 630.0 and 672.4 cal/100 g (Özcan 2000). The apricot kernel has a high concentration of phenolics (92.2–162.1 mg gallic acid-equivalent (GAE)/100 g; Korekar et al. 2011). Oleic, linoleic, and palmitic acids are the major fatty acids in the apricot kernel, and gallic acid, 3,4-dihydroxybenzoic acid, (+)-catechin, 1,2-dihydroxybenzene, etc. are the major phenolic compounds (Al-Juhaimi et al. 2018). The phenolic and flavonoid contents, and antioxidant capacity of apricot kernels can be increased by a roasting process (Durmaz and Alpaslan 2007; Al-Juhaimi et al. 2018).

The apricot kernel is a good source of proteins. For this reason, in addition to its consumption in the food industry, the apricot kernel can also be used as animal feedstuff and fertilizers (Manzoor et al. 2012). The apricot kernel has several health-beneficial effects, such as decreasing LDL cholesterol levels, reducing the colon cancer risk, protecting against cancer and cardiovascular diseases, regulating the intestinal flora, increasing the HDL cholesterol levels, antispasmodic, pectoral sedative, toning the respiratory system, etc. (Kan and Bostan 2010; Southon and Faulks 2002; Özcan 2000; Davis and Iwahashi 2001; Hyson et al. 2002; Knekt et al. 2002; Liu 2005; Durmaz and Alpaslan 2007; Korekar et al. 2011). The water and methanolic extracts of apricot seed kernels showed inhibitory activities against *Staphylococcus aureus*, *Escherichia coli*, *Proteus mirabilis*, *Salmonella typhimurium*, and *Candida albicans* (Abtahi et al. 2008; Yiğit et al. 2009). The antioxidant activity of sweet apricot kernels is high. In addition, the sweet kernel extract has higher antioxidant activities such as lipid peroxidation inhibition and DPPH radical scavenging

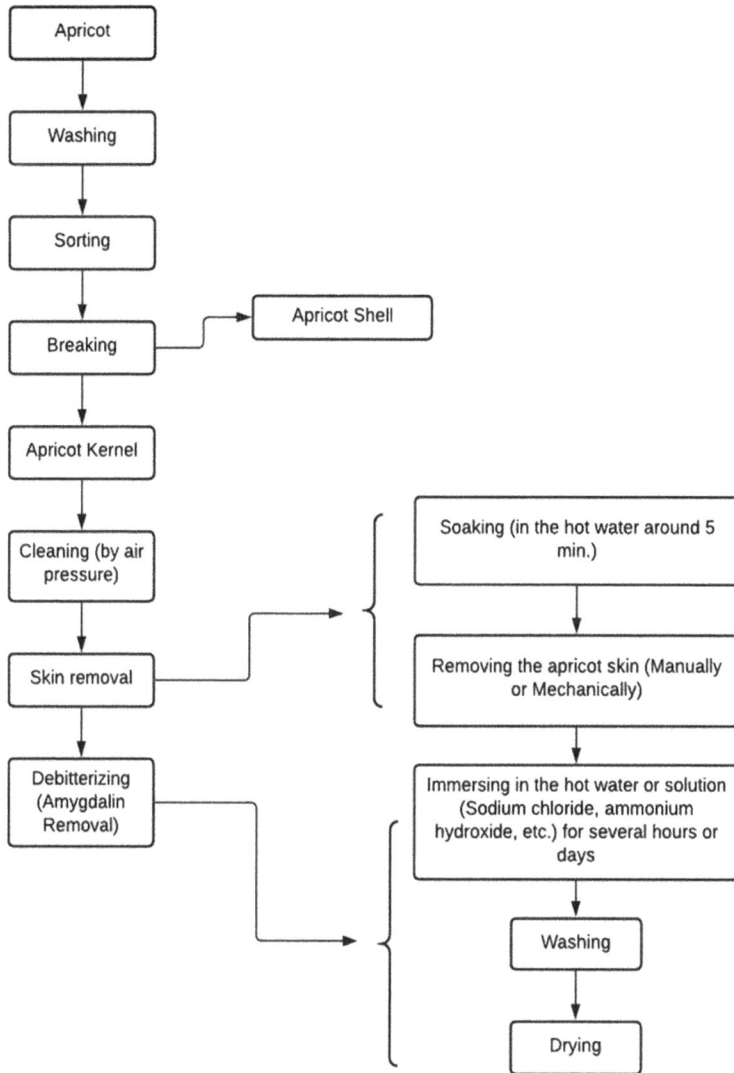

Figure 13.1 Debitterizing process of the apricot kernel (Mandalari et al., 2010; Zhang et al., 2016).

activities than bitter kernel extracts (Yiğit et al. 2009; Kostadinović Veličkovska et al. 2018). Sweet kernels can be used for food, medicinal, and cosmetic purposes because of their high nutritional value and health-beneficial effects. Generally, apricot kernel is used as an appetizer (raw or roasted), and in foods such as bakery and confectionery products; it can also be used for the production of apricot kernel oil, cosmetic products, benzaldehydes, and active carbon (Ahmadi et al. 2008; Esfahlan et al. 2010).

13.3 PROCESSING OF APRICOT KERNELS

The apricot kernel contains 32–56% (by weight) oil (Table 13.2). The sweet apricot kernel includes a higher amount of oil (average of 53%) than the bitter apricot kernel (average of 43%)

Table 13.2 Physicochemical Properties of Apricot Kernel

			Properties			
Carbohydrate (%)	Moisture (%)	Ash (%)	Protein (%)	Crude Lipid (%)	Dietary Fiber (%)	References
13.82±0.43	3.13±0.12	2.60±0.05	25.41±1.09	55.07±1.35	0.85±0.02	Tanwar et al. (2018b, wild apricot kernel)
21.47±0.49	3.34±0.19	2.41±0.07	22.10±1.69	50.77±2.06	0.61±0.06	Tanwar et al. (2018b, detoxified wild apricot kernel)
-	4.11±0.09–6.43±0.14	2.11±0.04–3.89±0.07	13.21±0.27–20.90±0.21	32.23±0.69–42.51±0.79	5.13±0.09–9.81±0.17	Manzoor et al. (2012)*
8.2	4.0–4.1	2.2–2.4	24.4	45.6–46.3	5.4	Gupta et al. (2012)**
		2.138–3.454	14.948–24.201	42.195–57.185	4.060–7.638	Özcan et al. (2010)***
21.16–35.26	-	-	17.75–22.56	45.51–54.24	0.84–4.96	Dwivedi and Ram (2008)****
-	4.91–5.12	2.10–2.67	23.58–27.70	46.3–51.4	13.49–17.98	Özcan (2000)*****
-	6.7±2.1	2.2±0.0–2.9±0.0	24.4±0.1–28.8±0.2	39.7±1.5–47.2±0.5	-	Femenia et al. (1995, Bitter kernel, dw)
-	5.4±1.7	2.0±0.1–2.9±0.4	22.4±0.6–29.3±0.3	49.8±2.1–56.1±0.7	-	Femenia et al. (1995 Sweet kernel, dw)
-	2.80	-	-	50.90 (Hexane extract) 51.30 (Chloroform-methanol extract)	-	El-Aal et al. (1986)

* Apricot variety: Nari, Halmas, Travet, and Charmagzi.
** Depending on the apricots grown in different locations (Mandi, Shimla, and Kinnaur) of Himachal Pradesh.
*** Apricot Variety: Hüdayi, Hacıhaliloğlu, Soğancı, Şahinbey, Hacıkız, Canino, Caona, Sakıt–2, Çekirge–52, Erkenağerik, 693-K, Karacabey, Ethembey, Alyanak, Kabaaşı, Hasanbey and Aprikoz.
**** Depending on the different villages in Ladakh, India [Leh, Nimo, and Saspol from the Indus valley, Kargil, and Poyon from the Suru valley, and Turtuk, Hundar, Terchey, and Sumur from the Nobra valley].
***** Apricot variety: Çataloğlu, Hacıhaliloğlu, Hasanbey, and Çöloğlu.

(Femenia et al. 1995). Although both sweet and bitter apricot kernels can be used for oil extraction, the sweet apricot kernel is generally used as a dried or roasted snack and the bitter kernel is generally used for extraction of oil (Dwivedi and Ram 2008). The kernel oil has an apricot odor and it has a light to deep yellow color (Gupta and Sharma 2009; Rai et al. 2016; Priyadarshi et al. 2018). The extraction of apricot kernel oil can be accomplished by several methods, such as physical means or solvent, supercritical fluid, and enzyme-assisted extractions, etc. (Sharma 2018; Bhanger et al. 2020). The physical process includes several unit operations such as breaking of the pits, removing of kernels, expression of kernel oil, filtration, and packing (Kostadinović Veličkovska et al. 2018; Sharma 2018; Bhanger et al. 2020). For other extraction processes, the fresh kernels are first dried, crushed, and milled. In the solvent extraction process, several solvents like petroleum ether, hexane, or chloroform-methanol mixture, etc., can be used. The solvent extraction process can be performed by using Soxhlet apparatus until the extraction is completed. Then the solvent is removed using an evaporator (Abd El-Aal et al. 1986; Erdogan Orhan et al. 2008; Özcan et al. 2010; Wang and Yu 2012; Manzoor et al. 2012; Priyadarshi et al. 2018; Stryjecka et al. 2019). The oil yield of the solvent extraction process (45.03%) is higher than that obtained by physical means (34.5% by pressing (Sharma 2018). The oil yield by solvent extraction of the apricot kernels was found to be $50.18\pm3.92\%$ (Wang and Yu 2012), 50.90%, and 51.30% for petroleum ether, hexane, and chloroform-methanol solvents, respectively (El-Aal et al. 1986). In addition to the selected solvents, the oil yield strongly depends on the apricot variety. The oil yields of hexane-extracted 'Kalecik', 'Malatya', and 'Bodrum' apricots were found to be 21.8%, 43.6%, and 41.5%, respectively (Erdogan Orhan et al. 2008). The apricot kernel oil extracted by hexane had a light yellow color and a desirable odor. It was also found free from HCN, which has a toxic effect (El-Aal et al. 1986).

The bitter apricot kernel oil press cake obtained during oil extraction contains a high concentration of essential oil (bitter almond oil; Sharma et al. 2010) and crude protein (around 34–45%, Gupta et al. 2012, Table 13.3). This cake, which does not include toxic hydrocyanic acid, can be used for the isolation of proteins or as animal feed. The protein can be extracted from apricot kernel oil press cake and it can be used as a supplement in the food industry (Sharma et al. 2010).

Apricot kernel oil is a good source of unsaturated fatty acids like oleic, linoleic, linolenic acids, etc. (El-Aal et al. 1986; Wang and Yu 2012; Matthaus et al. 2016; Priyadarshi et al. 2018; Tanwar et al. 2018a). The physicochemical composition and fatty acid profile of the apricot kernel oil are given in Table 13.3. The oil can be classified as a semi-dry oil according to the iodine number (El-Aal et al. 1986; Femenia et al. 1995). The apricot variety and harvest time affect the efficiency of apricot kernel oil extraction and the tocopherol and fatty acid composition of the oil. The oil contains α-, γ-, and δ-tocopherol isomers (Manzoor et al. 2012; Matthaus et al. 2016). Apricot kernel oil includes a high amount of γ- tocopherol and the total tocopherol concentrations of 'Hasanbey', 'Hacihaliloglu', 'Kabaasi', and 'Sogancı' kernels are in the range 38.22–64.98 mg/kg, 28.55–51.81 mg/kg, 29.87–36.71 mg/kg, and 24.18–60.09 mg/kg, respectively, depending on the maturation level (Matthaus et al. 2016). The α-, γ-, and δ- tocopherol concentrations of apricot kernel oil are in the range 14.8 ± 0.3–40.4 ± 0.8 mg/kg, 330.8 ± 6.7–520.8 ± 11.2 mg/kg, and 28.5 ± 0.5–60.2 ± 1.3 mg/kg, respectively, depending upon the apricot variety ('Nari', 'Halmas', 'Travet', and 'Charmagzi'; Manzoor et al. 2012). The peroxide value of apricot kernel oil is below the maximum limit for vegetable oils (10 meq O_2/kg). For this reason, it can be stated that apricot seed oil has good oxidative stability (Pardo et al. 2009; Matthaus et al. 2016). The content of the antioxidant phenolics and tocopherols of apricot oil contribute to its long shelf-life (Alpaslan and Hayta 2006; Ramadan et al. 2011).

Apricot kernel oil can be used in food, medicinal, and pharmaceutical applications because of its antioxidant, antimicrobial, antifungal, antibacterial, antiseptic, antiaging, etc., properties (Bhanger et al. 2020). It can be utilized in edible oils, lubricants, cosmetics, surfactants, etc. (Gupta et al. 2012; Bhanger et al. 2020). Oils such as apricot kernel oil, with high concentrations of oleic and linoleic fatty acids, are very popular because of their high stability and nutritional value. Moreover,

Table 13.3　Fatty Acid Composition and Physicochemical Properties of the Apricot Kernel Oil

Fatty Acids (%)	Wang and Yu (2012) (wb, %)	Gupta et al. (2012)**	Manzoor et al. (2012)***	Matthaus and Özcan (2009)****	Femenia et al. (1995) Bitter Kernel	Femenia et al. (1995) Sweet Kernel	Abd El-Aal et al. (1986)
Myristic (C14:0)	0.03±0.09	-	-	-	-	-	Trace
Palmitic (C16:0)	3.79±0.78	5.0–7.8	3.35±0.07–5.93±0.13	4.9–5.7	5.4±0.2	4.9±0.2	4.37
Palmitoleic (C16:1)	0.67±0.25	0.5–0.7	0.32±0.01–0.71±0.01	-	1.0±0.0	1.0±0.0	0.12
Margaric (17:0)	-	-	-	-	-	-	-
Stearic (C18:0)	1.01±0.31	0.9–2.0	1.10±0.03–1.68±0.03	0.8–2.1	1.0±0.0	0.8±0.0	0.46
Oleic (C18:1)	65.23±4.97	62.1–70.6	62.34±1.34–80.97±1.40	62.3±0.3–71.6±0.6	63.4±1.9	67.6±1.6	66.29
Linoleic (C18:2)	28.92±4.62	20.5–27.8	13.33±0.41–30.33±0.71	18.7±0.3–28.0±0.3	28.8±1.1	25.3±0.9	28.64
Linolenic or α-Linolenic (C18:3)	0.14±0.05	0.4–1.4	ND–1.03±0.04	0.1–0.2	-	-	0.12
Arachidic acid (C20:0)	0.09±0.04	-	-	ND–0.1	-	-	-
Cis-11-eicosenoic acid (C20:1)	0.11±0.02	-	-		-	-	-
Behenic (C22:0)	-	-	-		-	-	-
Lignoceric (C24:0)	-	-	-		-	-	-
Total saturated fatty acids	-	6.5–8.7	4.45–7.24		6.4	5.7	4.83
Total unsaturated fatty acids	-	91.6–92.6	92.93–94.62		93.2	93.9	95.17
Total essential fatty acids	-	-	13.33–31.06		-	-	-
Moisture content (%)	0.17	0.25–0.26	-		-	-	-
Specific gravity	-	0.914–0.915 g/cc	-		0.926±0.009 (20°C)	0.919±0.008 (20°C)	0.9136 (25°C)
Density (24°C)	-	-	0.87±0.04–0.93±0.03		-	-	-
Refractive index	-	1.4720–1.4729 (40°C)	1.4655±0.02–1.4790±0.02 (40°C)		1.4680±0.003 (20°C)	1.4665±0.004 (20°C)	1.4638 (25°C)

(Continued)

Table 13.3 (Continued)   Fatty Acid Composition and Physicochemical Properties of the Apricot Kernel Oil

Fatty Acids (%)	Wang and Yu (2012) (wb, %)	Gupta et al. (2012)**	Manzoor et al. (2012)***	Matthaus and Özcan (2009)****	Femenia et al. (1995) Bitter Kernel	Femenia et al. (1995) Sweet Kernel	Abd El-Aal et al. (1986)
Peroxide value (meq O$_2$ /kg oil)	-	5.12–5.27	1.00±0.03– 2.32±0.04	-	-	-	0.04
Saponification value/ number	-	189.8–191.3 mg KOH/g oil	189.1±3.23– 199.4±2.9 mg KOH/g oil	-	190±2	192±2	189.7
Unsaponifiable matter (%)	-	-	0.59±0.01– 0.88±0.04	-	0.70±0.04	0.76±0.05	0.86
Acid value/ number	0.46 mg KOH/g	2.27–2.78 mg KOH/g oil	-	-	-	-	0.12
Iodine value/number	-	100.2–100.4 g I$_2$/100 g oil	96.4±1.9– 106.3±2.12 g I/100 g oil	-	107±2	104±2	103.8

ND: Not detected.

* Depending on the apricot variety (Hasanbey, Hacıhaliloğlu, Kabaasi, and Soganci) and maturation level.

** Depending on the apricots grown in different locations (Mandi, Shimla, and Kinnaur) of Himachal Pradesh.

*** Apricot variety: Nari, Halmas, Travet, and Charmagzi.

**** Apricot variety: Şekerpare, Alyanak, Çataloğlu, Soğancı, Hacıhaliloğlu, Wild and Sultan.

linoleic acid has a positive effect on cardiovascular diseases and plasma cholesterol (Nicolosi et al. 2004; Manzoor et al. 2012). The dietary lipid, which includes a large amount of unsaturated fatty acids, prevents diabetes, cardiovascular diseases, inflammatory, and autoimmune disorders, etc. Alpha-tocopherol has a very high vitamin E potency and antioxidant activity (Matthäus and Özcan 2009; Manzoor et al. 2012). For this reason, it can be stated that apricot kernel oil has many health-beneficial effects because of its functional fatty acids and tocopherol contents. Apricot kernel oil shows high resistance to oxidation because of its α- and γ-tocopherol contents (Matthaus et al. 2016). In Germany and the USA, apricot kernel oil is used in oil and macaroon paste formulations (Femenia et al. 1995; Matthaus et al. 2016). According to the aforementioned health-promoting effects of apricot kernel oil, with its high unsaturated and low saturated fatty acid contents, it can be utilized as an edible oil and for other industrial purposes.

13.4 UTILIZATION OF APRICOT WASTE

13.4.1 Utilization of the Apricot Kernels

Apricot kernels are often used in food formulations as whole kernels, cracked kernels, or in flour forms. Apricot kernel flour from sweet apricots does not require much processing and it requires simply to be milled to obtain flour. In order to obtain apricot kernel flour from the bitter kernel, the first detoxification process has to be applied. For this purpose, the apricot kernel is immersed in distilled water containing 0.1 M ammonium hydroxide (1: 2 kernel weight: solution volume) at 47°C for 30 h. The dried and crushed kernels are used for oil extraction using several types of solvents, such as hexane. The defatted, crushed kernels are dried and ground to obtain fine kernel flour (El-Adawy et al. 1994). Another detoxification method for bitter apricot kernels involves immersing the apricot kernel flour in 25% NaCl solution (kernel flour: solution, 1: 5 weight: volume) for 12 h, washing, immersing again in 25% NaCl solution for another 12 h, followed by washing and drying (Tanwar et al. 2018a, b). In another method, the dried and cracked apricot kernels (1: 12 kernel: soaking solution weight: volume) are soaked in the water (40°C for 10–15 min) and soaking solutions (1% ascorbic acid, 1% citric acid, and 1% NaCl solutions), respectively for 6, 12, 18, and 24 h at room temperature; after drying, the lipids are extracted, and the defatted, dried kernels are ground into flour (Elkot et al., 2018).

The detoxification process decreases the content of protein and antinutritional factors of the apricot kernel because of the removal of bitter and anti-nutritive substances (El-Adawy et al., 1994). In addition, the leaching process causes the loss of water-soluble minerals and sugars (El-Adawy et al. 1994; Tanwar et al. 2018b). Moreover, the detoxification process improves the amino acid profile, especially the essential amino acids, because of the leaching out of the protein fractions having high non-essential amino acid content, and increases the digestibility, which may be due to removal of antinutritional substances, such as tannins (El-Adawy et al. 1994).

Apricot kernel flour is rich in minerals (K, Mg, Ca, Na, Fe, Zn, etc.), vitamins (B group, tocopherol), unsaturated fatty acids (oleic and linoleic acids), and proteins (40–48%, mainly albumins) (El-Aal et al. 1986; El-Adawy et al. 1994; Dhen et al. 2017). The high nutritional content of apricot kernel flour makes it a good food ingredient for producing innovative functional foods. Table 13.4 shows the physicochemical properties of kernel flour. Apricot kernel flour has lower moisture content and higher ash, protein, fat, fiber, and carbohydrate contents than wheat flour (moisture, carbohydrate, fiber, ash, fat, and protein concentrations of apricot kernel flour are 13.45%, 71.52%, 0.99%, 0.64%, 2.70%, and 10.69%, respectively; Dhen et al. 2017, 2018). For these reasons, apricot kernel flour is used in bakery products, pasta, pasta-like products, etc. (as whole kernels or apricot kernel flour) to enhance the nutritional content of these products. Apricot kernel flour is generally used to produce protein-enriched foods because of the high protein content of the apricot kernel.

Table 13.4 Chemical Composition of Apricot Kernel Flour and Apricot Kernel Oil Press Cake Left after Apricot Kernel Oil Extraction

	Moisture (%)	Ash (%)	Protein (%)	Crude Lipid (%)	Crude Fiber (%)	Carbohydrate (%)	References
Apricot Kernel Flour	5.52	2.76	26.93	45.37	-	19.96	Dhen et al. (2017)
	11.01±0.02	3.96±0.00	55.12±0.12	4.9±0.01	5.18±0.04	29.78±0.04	Elkot et al. (2018, defatted apricot kernel flour)
	-	2.32±0.11	26.76±0.11	51.26±0.10	3.43±0.0	16.23±0.05	El-Safy et al. (2012, dw)
	7.80	2.31	25.60	44.39	2.51	25.19	El-Demery and Elsanat (2010, dw)
	-	2.71	21.8	40.2	-	-	Özboy-Özbas et al. (2010)
	6.9	2.39	26.9	52.0	-	-	Eyidemir and Hayta (2009, dw)
	10.9	3.92	54.7	4.17	5.11	30.2	El-Adawy et al. (1994, dw)
	9.96	3.32	50.3	5.35	10.5	29.2	El-Adawy et al. (1994, dw, detoxified apricot kernel oil by distilled water)
	10.2	3.10	49.8	5.20	11.0	28.8	El-Adawy et al. (1994, dw detoxified apricot kernel flour by ammonium hydroxide)
Apricot Kernel Oil Press Cake	-	-	13.21–20.90	-	-	-	Manzoor et al. (2012)*
	4.9–7.5	4.9–5.1	34.3–44.5	5.4–9.7	7.0–10.8	27.5–32.7	Gupta et al. (2012)**
	7.2	-	34.3	9.7	10.8	27.5	Sharma et al. (2010)

*Depending on the apricot variety (Nari, Halmas, Travet, and Charmagzi).
**Depending on the apricots grown in different locations (Mandi, Shimla, and Kinnaur) of Himachal Pradesh.

It is desirable that the enriched product should be economically affordable, nutritious, and have the desired sensory properties. For this reason, the apricot kernel, a waste product obtained from apricot processing, is a cheap, natural protein source and a promising food ingredient (Eyidemir and Hayta 2009; El-Demery et al. 2010; Dhen et al. 2018; Sheikh et al. 2020). The addition of apricot kernel flour (0–20%) resulted in an increase in the protein, lipid, and ash contents of noodles (Eyidemir and Hayta 2009). The noodles with 15% apricot kernel flour were acceptable to the panelists in terms of physicochemical and sensorial properties (Eyidemir and Hayta 2009). The partial replacement of apricot kernel flour (4–24% for bread and 5–25% for spaghetti) decreased the moisture and carbohydrate contents and increased the concentrations of ash, protein, and lipid, and the caloric value of the bread and spaghetti. The partial replacement of up to 12% of wheat flour with apricot kernel flour retains the desired sensorial properties, color, loaf volume, and texture of the bread, while 10% replacement retained the desired sensorial properties for spaghetti (El-Demery et al. 2010; Dhen et al. 2018). The partial replacement of wheat flour with apricot kernel flour (wheat flour: apricot kernel flour 100: 0–75: 25) resulted in an increase in the protein, ash, fat, and crude fiber contents, radical scavenging activity, and phosphomolybdenum reduction assay, depending on the amount of the apricot kernel flour used. The desired supplementation level was found to be 10%, depending on the sensorial evaluation (Sheikh et al. 2020). The bread, spaghetti, and biscuits made with apricot kernel flour are important for nutrition, health, and weight control due to their low carbohydrate and high protein contents. The supplementation of 1% of apricot kernel flour into a yogurt formulation resulted in higher ash, total solids, protein content and acidity, and lower pH value compared to non-supplemented samples. The yogurt samples supplemented with apricot kernel flour exhibited a storage advantage because of their high protein and solids contents (Elkot et al. 2018). The partial replacement of wheat flour with apricot kernel flour improved the water-binding capacity of the wheat flour due to the higher protein content of apricot kernel flour (Dhen et al. 2017). In addition, apricot kernel flour can be used as a fat substitute in cookies (Özboy-Özbaş et al. 2010; Seker et al. 2010). Apricot kernels were also used to produce a health care tea effective against anorexia, dry mouth, disturbed sleep, etc. (Jinyi 2006).

13.4.2 Utilization of Apricot Kernel Oil in Foods

Similar to the consumption of apricot kernel flour, the kernel oil is used in bakery products (biscuits and cakes) without any deteriorative effects on their sensorial properties. The apricot kernel oil was found to be comparable with corn oil in the biscuits (sweet and salted) and cake formulations (Abd El-Aal et al. 1986). The kernel oil can be used as a substitute for other traditional oils in food product formulations. Chitosan film impregnated with apricot kernel oil (1: 0–1 weight: volume chitosan: apricot kernel oil) showed higher antioxidant and antibacterial activities against both *Escherichia coli* and *Bacillus subtilis* compared with the control. These activities of the films increased with increasing concentration of the kernel oil. Moreover, the oil inhibited mold growth on bread and enhanced the shelf-life of the bread (Priyadarshi et al. 2018).

In India, apricot oil is used for cosmetics (for hair, soap, etc.), cooking (in lamps, etc.), and religious purposes (Dwivedi and Ram 2008). The kernel oil can be used as a raw material for benzaldehyde production and it can also be used as an air disinfectant, preservative, and antimicrobial agent due to its antimicrobial activity toward bacteria and yeasts (Lee et al. 2014). Moreover, apricot kernel oil has the potential for biodiesel production (Ullah et al. 2009; Wang and Yu 2012). The biodiesel yield of Siberian apricot kernel oil was 88.7% (Wang and Yu 2012).

13.5 CONCLUSIONS

Apricot is a delicious fruit which offers functional waste materials, including apricot shell (non-edible part) and kernel (edible part). In addition to using it as fuel, the apricot shell, which is rich in fiber, is used as a raw material for biorefineries, cadmium adsorption from water, and for the production of activated carbon and bio-oil, etc. The apricot kernel has an enormous industrial potential (medicinal, food, cosmetic, etc. industries) and health-promoting effects. The apricot kernel flour is used to increase the protein content of foods such as bread, pasta, yogurt, etc. The desired supplementation level of apricot kernel flours is around 10–15%, depending on the sensorial evaluation. Apricot kernel oil, obtained by cold pressing or solvent extraction methods, from sweet apricot kernels is a potential alternative source of edible oil. it can be used in active food packaging materials, cosmetic products, or for medicinal purposes. In addition, the apricot kernel oil press cake and the water used in the debitterizing process also include valuable components. The recycling process can also be applied to the soaking water and press cake to recover water-soluble substances such as protein, sugar, etc.

REFERENCES

Abd El-Aal, M.H., Khalil, M.K.M., and E.H. Rahma. 1986. Apricot kernel oil: Characterization, chemical composition and utilization in some baked products. *Food Chemistry* 19:287–298.

Abtahi, H., Ghazavi, A., Karimi, M., Mollaghasemi, S., and G. Mosayebi. 2008. Antimicrobial activities of water and methanol extracts of bitter apricot seeds. *Journal of Medicinal Science* 8 (4):433–436.

Ahmadi, H., Fathollahzadeh, H., and H. Mobli. 2008. Some physical and mechanical properties of apricot fruits, pits and kernels (C.V *Tabarzeh*). *American-Eurasian Journal of Agricultural & Environmental Sciences* 3 (5):703–707.

Akin, E.B., Karabulut, I., and A. Topcu. 2008. Some compositional properties of main Malatya apricot (*Prunus armeniaca* L.) varieties. *Food Chemistry* 107:939–948.

Al Juhaimi, Fa, Özcan, M., Ghafoor, K., and E.E. Babiker. 2018. The effect of microwave roasting on bioactive compounds, antioxidant activity and fatty acid composition of apricot kernel and oils. *Food Chemistry* 243:414–419.

Ali, S., Masud, T., and K.S. Abbasi. 2011. Physico-chemical characteristics of apricot (*Prunus armeniaca* L.) grown in Northern Areas of Pakistan. *Scientia Horticulture* 130 (2):386–392.

Ali, S., Masud, T., Abbasi, K.S., Mahmood, T., and A. Hussain. 2015. Apricot: Nutritional potentials and health benefits-A review. *Annals. Food Science and Technology* 16 (1):175–189.

Alpaslan, M., and M. Hayta. 2006. Apricot kernel: Physical and chemical properties. *Journal of the American Oil Chemists' Society* 83:469–471.

Bhanger, M.I., Anwar, F., Memon, N., and R. Qadir. 2020. Cold pressed apricot (*Prunus armeniaca* L.) kernel oil. In: *Cold Pressed Oils: Green Technology, Bioactive Compounds, Functionality, and Applications*, M.F. Ramadan (ed.). 725–730. UK: Academic Press.

Cañellas, J., Femenia, A., Rosselló, C., and L. Soler. 1992. Chemical composition of the shell of apricot seeds. *Journal of the Science of Food and Agriculture* 59:269–271.

Corbett, D., Kohan, N., Machado, G., Jing, C., Nagardeolekar, A., and B.M. Bujanovic. 2015. Chemical composition of apricot pit shells and effect of hot-water extraction. *Energies* 8:9640–9654.

Davarynejad, G., Khorshidi, S., Nyéki, J., Szabó, Z., and J. Gal-Remennyik. 2010. Antioxidant capacity, chemical composition and physical properties of some apricot (*Prunus armeniaca* L.) cultivars. *Horticulture, Environment, and Biotechnology* 51(6):477–482.

Davis, P.A., and C.K. Iwahashi. 2001. Whole almonds and almond fractions reduce aberrant crypt foci in a rat model of colon carcinogenesis. *Cancer Letters* 165:27–33.

Demiral, İ., and Ş.Ç. Kul. 2014. Pyrolysis of apricot kernel shell in a fixed-bed reactor: Characterization of bio-oil and char. *Journal of Analytical and Applied Pyrolysis* 107:17–24.

Dhen, N., Rejeb, I.B., Boukhris, H., Damergi, C., and M. Gargouri. 2018. Physicochemical and sensory properties of wheat- Apricot kernels composite bread. *LWT- Food Research and Technology* 95:262–267.

Dhen, N., Rejeb, I.B., Martínez, M.M., Román, L. Gómez, M., and M. Gargouri. 2017. Effect of apricot kernels flour on pasting properties, pastes rheology and gels texture of enriched wheat flour. *European Food Research and Technology* 243:419–428.

Durmaz, G., and M. Alpaslan. 2007. Antioxidant properties of roasted apricot (*Prunus armeniaca* L.) kernel. *Food Chemistry* 100:1177–1181.

Dwivedi, D., and B.B. Ram. 2008. Chemical composition of bitter apricot kernels from Ladakh, India. *Acta Horticulturae* 765:335–338.

El-Aal, M.H.A., Hamza, M.A., and E.H. Rahma. 1986. In vitro digestibility, physico-chemical and functional properties of apricot kernel proteins. *Food Chemistry* 19:197–211.

El-Adawy, T.A., Rahma, E.H., El-Badawey, A.A., Gomaa, M.A., Lásztity, R., and L. Sarkadi. 1994. Biochemical studies of some non-conventional sources of proteins Part 7. Effect of detoxification treatments on the nutritional quality of apricot kernels. *Food/Nahrung* 38:12–20.

El-Demery, M., Elsanat, S., and S. Elsanat. 2010. Influence of apricot kernel flour addition on sensory characteristics of spaghetti. *Egyptian Journal of Food Science* 38:49–65.

El-safy, S., Salem, R., and M. Abd El-Ghany. 2012. Chemical and nutritional evaluation of different seed flours as novel sources of protein. *World Journal of Dairy & Food Sciences* 7:59–65.

Elkot, W., El-Nawasany, L.I., and H. Sakr. 2018. Composition and quality of stirred yoghurt supplemented with apricot kernels powder. *Journal of Agroalimentary Processes and Technologies* 23 (3):125–130.

Erdogan Orhan, I., Koca, U., Aslan Erdem, S., Kartal, M., and Ş. Küsmenoplu. 2008. Fatty acid analysis of some Turkish apricot seed oils by GC and GC-MS techniques. *Turkish Journal of Pharmaceutical Science* 5 (1):29–34.

Esfahlan, A.J., Jamei, R., and , Esfahlan, R.J. 2010. The importance of almond (*Prunus amygdalus* L.) and its by-products. *Food Chemistry* 120:349–360.

Eyidemir, E., and M. Hayta. 2009. The effect of apricot kernel flour incorporation on the physicochemical and sensory properties of noodle. *African Journal of Biotechnology* 8:85–90.

FAO. 2020. http://www.fao.org/faostat/en/#data/QC/visualize (accessed December, 2020).

FAO. DOA. 2007. Food and Agriculture Organization and Department of Agriculture, Northern Areas. *Fruit Production in Northern Areas, Survey Report.* UN-PAK/FAO/2001/003.

Femenia, A., Rossello, C., Mulet, A., and J. Canellas. 1995. Chemical composition of bitter and sweet apricot kernels. *Journal of Agricultural and Food Chemistry* 43:356–361.

Garrido, I., Monagas, M., Gómez-Cordovés, C., and B. Bartolomé. 2008. Polyphenols and antioxidant properties of almond skins: Influence of industrial processing. *Journal of Food Science* 73:C106–C115.

Gupta, A., and P.C. Sharma. 2009. Standardization of methods for apricot kernel oil Extraction, packaging and storage. *Journal of Food Science and Technology* 46:121–126.

Gupta, A., Sharma, P.C., Thilakaratne, B.M.K., and A.K. Verma. 2012. Studies on physico-chemical characteristics and fatty acid composition of wild apricot (*Prunus armeniaca* Linn.) kernel oil. *Indian Journal of Natural Products and Resources* 3:366–370.

Hacıseferoğulları, H., Gezer, İ., Özcan, M.M., and B. Murat Asma. 2007. Post-harvest chemical and physical–mechanical properties of some apricot varieties cultivated in Turkey. *Journal of Food Engineering* 79:364–373.

Hussain, A., Yasmin, A., and J. Ali. 2010. Comparative study of chemical composition of some dried apricot varieties grown in northern areas of Pakistan. *Pakistan Journal of Botany* 42 (4):2497–2502.

Hyson, D., Schneeman, B., and P. Davis. 2002. Almonds and almond oil have similar effects on plasma lipids and LDL oxidation in healthy men and women. *Journal of Nutrition* 132:703–707.

Iordanescu, O.A., Lalescu, D., Berbecea, A., Camen, D., Poiana, M.A., Moigradean, D., and M. Bala. 2018. Chemical composition and antioxidant activity of some apricot varieties at different ripening stages. *Chilean Journal of Agricultural Research* 78:266–275.

Janković, B., Manić, N., Dodevski, V., Radovic, I., Pijovic, M., Katnic, D., and G. Tasic. 2019. Physico-chemical characterization of carbonized apricot kernel shell as precursor for activated carbon preparation in clean technology utilization. *Journal of Cleaner Production* 236:117614.

Jinyi, L. 2006. *Health-care tea contg. apricot seed.* CN Patent No. 171804.

Kafkaletou, M., Kalantzis, I., Karantzi, A., Christopoulos, M.V., and E. Tsantili. 2019. Phytochemical characterization in traditional and modern apricot (*Prunus armeniaca* L.) cultivars – Nutritional value and its relation to origin. *Scientia Horticulturae* 253:195–202.

Kan, T., and S. Bostan. 2010. Changes of contents of polyphenols and vitamin A of organic and conventional fresh and dried apricot cultivars (*Prunus armeniaca* L.). *World Journal of Agriculture and Soil Science* 6 (2):120–126.

Knekt, P., Kumpulainen, J., Järvinen, R., Rissanen, H., Heliövaara, M., Reunanen, A., Hakulinen, T., and A. Aromaa. 2002. Flavonoid intake and risk of chronic diseases. *American Journal of Clinical Nutrition* 76:560–568.

Korekar, G., Stobdan, T., Arora, R., Yadav, A., and S.B. Singh. 2011. Antioxidant capacity and phenolics content of apricot (*Prunus armeniaca* L.) kernel as a function of genotype. *Plant Foods for Human Nutrition* 66:376–383.

Kostadinović Veličkovska, S., Mot, A.C., Mitrev S., Gulaboski, R., Brühl, L., Mirhosseini, H., Silaghi-Dumitrescu, R., and B. Matthäus. 2018. Bioactive compounds and "in vitro" antioxidant activity of some traditional and non-traditional cold-pressed edible oils from Macedonia. *Journal of Food Science and Technology* 55:1614–1623.

Leccese, A., Bureau, S., Reich, M., Renard, M.G.C.C., Audergon, J.M., Mennone, C., Bartolini, S., and R. Viti. 2010. Pomological and nutraceutical properties in apricot fruit: Cultivation systems and cold storage fruit management. *Plant Foods for Human Nutrition* 65:112–120.

Lee, H., Ahn, J., Kwon, A.R., Lee, E.S., Kwak, J.H., and Y.H. Min. 2014. Chemical composition and antimicrobial activity of the essential oil of apricot seed. *Phytotherapy Research* 28:1867–1872.

Liu, R. 2005. Potential synergy of phytochemicals in cancer prevention: Mechanism of action. *Journal of Nutrition* 134 Supplement:3479S–3485S.

Lucchesi, A., Maschio, G., Rizzo, C., and G. Stoppato. 1988. A pilot plant for the study of the production of hydrogen-rich syngas by gasification of biomass BT. In: *Research in Thermochemical Biomass Conversion,* Bridgwater, A.V., and Kuester, J.L. (eds). 642–654. Springer Netherlands, Dordrecht.

Mandalari, G., Tomaino, A., Arcoraci, T., Martorana, M., Turco, V.L., Cacciola, F.,Rich, G.T., Bisignano, C., Saija, A., Dugo, P., Cross, K.L., Parker, P.L., Waldron, K.W., and Wickham, M.S.J. 2010. Characterization of polyphenols, lipids and dietary fibre from almond skins (*Amygdalus communis* L.). *Journal of Food Composition and Analysis*, 23 (2):166–174.

Manzoor, M., Anwar, F., Ashraf, M., and K. Alkharfy. 2012. Physico-chemical characteristics of seed oils extracted from different apricot (*Prunus armeniaca* L.) varieties from Pakistan. *Grasas y Aceites* 63:193–201.

Matthäus, B., and M. Özcan. 2009. Fatty acids and tocopherol contents of some *Prunus* spp. kernel oils. *Journal of Food Lipids* 16:187–199.

Matthaus, B., Özcan, M.M., and F. Al Juhaimi. 2016. Fatty acid composition and tocopherol content of the kernel oil from apricot varieties (Hasanbey, Hacihaliloglu, Kabaasi and Soganci) collected at different harvest times. *European Food Research and Technology* 242:221–226.

Nicolosi, R.J., Woolfrey, B., Wilson, T.A., Scollin, P., Handelman, G., and R. Fisher. 2004. Decreased aortic early atherosclerosis and associated risk factors in hypercholesterolemic hamsters fed a high- or mid-oleic acid oil compared to a high-linoleic acid oil. *Journal of Nutritional Biochemistry* 15:540–547.

Özboy-Özbaş, Ö., Seker, I.T., and I. Gökbulut. 2010. Effects of resistant starch, apricot kernel flour, and fiber-rich fruit powders on low-fat cookie quality. *Food Science and Biotechnology* 19:979–986.

Özcan, M. 2000. Composition of some apricot (*Prunus armeniaca* L.) kernels grown in Turkey. *Acta Alimentaria* 29:289–294.

Özcan, M., Özalp, C., Ünver, A., Arslan, D., and N. Dursun. 2010. Properties of apricot kernel and oils as fruit juice processing waste. *Food Science & Nutrition* 01:31–37.

Pardo, J., Fernández, E., Rubio, M., Alvarruiz, A., and G.L. Alonso. 2009. Characterization of grape seed oil from different grape varieties (*Vitis Vinifera*). *European Journal of Lipid Science and Technology* 111:188–193.

Petrova, B., Budinova, T., Tsyntsarski, B., Kochkodan, V., Shkavro, Z., and N. Petrov. 2010. Removal of aromatic hydrocarbons from water by activated carbon from apricot stones. *Chemical Engineering Journal* 165:258–264.

Popa, V., Puitel, A., and G. Lisa. 2019. Characterization of components isolated from Algerian apricot shells (*Prunus Armeniaca* L.). *Cellulose Chemistry and Technology* 53:851–859.

Priyadarshi, R., Sauraj, Kumar, B., Deeba, F., Kulshreshtha, A., and Y.S. Negi. 2018. Chitosan films incorporated with apricot (*Prunus armeniaca*) kernel essential oil as active food packaging material. *Food Hydrocolloid* 85:158–166.

Rai, I., Bachheti, R.K., Saini, C.K., Joshi, A., and R.S. Satyan. 2016. A review on phytochemical, biological screening and importance of Wild Apricot (*Prunus armeniaca* L.). *Oriental Pharmacy and Experimental Medicine* 16:1–15.

Ramadan, M.F., Zayed, R., Abozid, M., and M. Asker. 2011. Apricot and pumpkin oils reduce plasma cholesterol and triacylglycerol concentrations in rats fed a high-fat diet. *Grasas y Aceites* 62 (4) 443–452.

Ruiz, D., Egea, J., Tomás-Barberán, F., and M. Gil. 2005. Carotenoids from new apricot (*Prunus armeniaca* L.) varieties and their relationship with flesh and skin color. *Journal of Agricultural and Food Chemistry* 53:6368–6374.

Salah Azab, M., and Peterson, P.J. 1989. The removal of cadmium from water by the use of biological sorbents. *Water Science and Technology* 21:1705–1706.

Seker, I.T., Ozboy-Ozbas, O., Gokbulut, I., Öztürk, S., and H. Köksel. 2010. Utilization of apricot kernel flour as fat replacer in cookies. *Journal of Food Processing and Preservation* 34:15–26.

Sharma, P.C., Tilakratne, B.M.K.S., and A. Gupta. 2010. Utilization of wild apricot kernel press cake for extraction of protein isolate. *Journal of Food Science and Technology* 47:682–685.

Sharma, R. 2018. Value addition of wild apricot fruits grown in North–West Himalayan regions - a review. *Journal of Food Science and Technology* 51:2917–2924.

Sheikh, M., Mudasir, A., and N. Anjum. 2020. To study the effect of apricot kernel flour (by-product) on physico-chemical, sensorial and antioxidant properties of biscuits. *Journal of Current Research in Food Science* 1 (1):16–22.

Soleimani, M., and T. Kaghazchi. 2008. Adsorption of gold ions from industrial wastewater using activated carbon derived from hard shell of apricot stones – An agricultural waste. *Bioresource Technology* 99:5374–5383.

Southon, S., and R. Faulks. 2002. Health benefits of increased fruit and vegetable consumption. In: Jongen, W. (ed). *Fruit and Vegetable Processing: Improving Quality*. 2–19. CRC Press, Washington, DC.

Stryjecka, M., Kiełtyka-Dadasiewicz, A., Michalak, M., Rachon, L., and A. Glowacka. 2019. Chemical composition and antioxidant properties of oils from the seeds of five apricot (*Prunus armeniaca* L.) cultivars. *Journal of Oleo Science* 68:729–738.

Tanwar, B., Modgil, R., and A. Goyal. 2018a. Effect of detoxification on biological quality of wild apricot (*Prunus armeniaca* L.) kernel. *Journal of the Science of Food and Agriculture* 99:517–528.

Tanwar, B., Modgil, R., and A. Goyal. 2018b. Antinutritional factors and hypocholesterolemic effect of wild apricot kernel (*Prunus armeniaca* L.) as affected by detoxification. *Food & Function* 9 (4):2121–2135.

Ullah, F., Nosheen, A., Hussain, I., and A. Bano. 2009. Base catalyzed transesterification of wild apricot kernel oil for biodiesel production. *African Journal of Biotechnology* 8 (14):3289–3293.

Wang, L., and H. Yu. 2012. Biodiesel from Siberian apricot (*Prunus sibirica* L.) seed kernel oil. *Bioresource Technology* 112:355–358.

Yiğit, D., Yigit, N., and A. Mavi. 2009. Antioxidant and antimicrobial activities of bitter and sweet apricot (*Prunus armeniaca* L.) kernels. *Brazilian Journal of Medical and Biological Research* 42:346–352.

Yildirim, F.A., and M.A. Aşkın. 2010. Variability of amygdalin content in seeds of sweet and bitter apricot cultivars in Turkey. *African Journal of Biotechnology* 9:6522–6524.

Zhang, Q.A., Song, Y., Wang, X., Zhao, W.Q., and X.H. Fan. 2016. Mathematical modeling of debittered apricot (*Prunus armeniaca* L.) kernels during thin-layer drying. *CyTA - Journal of Food* 14:509–517.

Zhang, Q.A., Wei, C.X., Fan, X.H., and F.F. Shi. 2018. Chemical compositions and antioxidant capacity of by-products generated during the apricot kernels processing. *CyTA - Journal of Food* 16:422–428.

CHAPTER 14

Plum Wastes and By-Products
Chemistry, Processing, and Utilization

Mohd Aaqib Sheikh, Nadira Anjum, Arshied Manzoor, Mohammad Ubaid,
Fozia Hameed, Charanjiv Singh Saini, and Harish Kumar Sharma

CONTENTS

14.1 INTRODUCTION

Fruit plays an important role in our daily diet, and several diseases could be prevented by adequate fruit consumption. Recently, there has been a huge increase in agricultural production and agro-industrial processing which generates wastes and by-products in large quantities. Plum (*Prunus domestica* L.) has been commercially grown around the globe, with China, Romania, Serbia, India, and the United States leading commercial cultivation. Amongst the angiosperms, the Rosaceae is one of the largest plant families, with about 3400 species associated with this family including important fruit crops such as plum, peach, almond, apricot, and cherries (Sheikh et al., 2021a). Plums are a large and botanically diverse group of stone fruits with an early harvest season and abundant fruit production. Plum was one of the first fruits domesticated by humans and they are a well-known fruit crop around the world for attributes such as delicious taste, color, and appealing aroma, with 12.61 million metric tons of global production in 2018 alone (FAOSTAT, 2020). Plum is a very old, traditional tree fruit, and is one of the most popular fruits among consumers, playing a vital role in both the economy and diet (Milosevic and Milosevic, 2012).

Plums are temperate fruit trees grown in climates with well-differentiated seasons. The plum is a drupe ("stone fruit") having a large stone ("pit") in the center of the fruit. It is a climacteric fruit which means that it will ripen off the tree, following harvest. Plum fruits vary considerably from variety to variety, but they are generally oblong in shape, mealy, juicy, and extremely delicate when ripe. The fruit may be green, red, purple, or yellow and has a delicious flavor varying from very

DOI: 10.1201/9781003164463-14

sweet to acid. Plum has been a part of the human diet due in part to several health-promoting effects (Sheikh et al., 2021c). In general, plum fruits are good for human nutrition, owing to their good nutritional profile, with high energy value and a relatively high content of carbohydrates (glucose, sucrose, fructose, sorbitol, etc.) but a low content of fat and protein as shown in Table 14.1.

Fresh plums are considered to be a healthy food due to their higher-than average concentration of sugars, dietary fiber, pectin, organic acids, vitamins, and minerals, although the nutritional composition depends largely on the variety and conditions under which the fruit was grown (Pashova, 2006). A daily diet containing a high amount of dietary fiber helps in preventing heart diseases, cancer, obesity, and other diseases like diabetes (Stacewicz-Sapuntzakis et al., 2001). The components constituting dietary fiber include pectin, cellulose, hemicelluloses, and lignin. In addition to this, plums have a high antioxidant activity because plums contain a high concentrations of phenolics and flavonoids. The high antioxidant potential, concentrations of phenolic acids, flavonoids, vitamins, minerals, and the high content of sorbitol are characteristic of the *Prunus* genus (Stacewicz-Sapuntzakis et al., 2001; Milosevic and Milosevic, 2012; Dreher, 2018; Sheikh et al., 2021c). A study conducted to measure the antioxidant activity of plums showed that plums possess three times the antioxidant potential of apples. Moreover, plum fruit can scavenge the oxygen-derived free radicals to reduce oxidative stress damage which can be involved in the development of several chronic diseases (Kim et al., 2003). The antioxidants present in plums are responsible for maintaining the oxidative as well as the microbial stability of the various products processed from the plum.

Plum is a delicious, seasonal, stone fruit that is mainly consumed during the summer season, which means that it cannot be consumed fresh throughout the year (Sheikh et al., 2021b). So, a greater proportion of plums is processed to prepare foodstuffs like jam, juice, jelly, nectar, dried fruit, etc., to provide an alternative to fresh plum consumption. The highly perishable nature of the fruit contributes to its high post-harvest losses occurring at ambient temperature. If the fruit is stored under refrigerated conditions, the storage life can be extended by only 2–4 weeks. Processing of plums into various value-added products is the best method to control the huge post-harvest losses of the fruit which are incurred. Most of the plums produced are consumed fresh while the remainder go for processing. The most important plum products are plum juice, dried plums, and canned whole plum. Dried fruits are mostly preferred in the market because of their stability. Drying also results in a reduced mass of the product that in turn decreases the packaging as well as the transportation

Table 14.1   Compositional Characteristics of Plum (*Prunus domestica*) Fruit

Moisture content (%)	78–88	Stacewicz-Sapuntzakis et al. (2001)
Carbohydrates (%)	8.76–21.06	
Protein content (%)	0.7–0.9	Walkowiak-Tomczak (2008)
Crude fat (%)	0.2–0.4	
Crude fiber (%)	1.4–2.79	Milosevic and Milosevic (2012);
Total ash (%)	4.54–4.82	
Minerals (mg/100 g)	120–208	Dreher (2018)
- K	6.0–45.0	
- Ca	4.0–8.0	
- Mg		
Vitamins	4–11	Dreher (2018)
- Vitamin C (mg/100 g)	94–780	
- Vitamin A (µg/100 g)	20–56	
- Thiamin (µg/100 g)	200–900	
- Niacin (µg/100 g)	38–50	
- Riboflavin (µg/100 g)		
Total phenolics (mg/100 g, f.w.)	221.7–769	Stacewicz-Sapuntzakis et al. (2001)
Total anthocyanins (mg/100 g, f.w.)	18–926	
Total acids (% as malic acid)	0.55	

costs. The other plum products available in the market are plum jam, juice, paste, and leather, etc. (Chang et al., 1994). Drying is considered to be a mild processing method which helps in the maintenance of the original levels of components such as dietary fiber in agricultural products like fruits and vegetables (Ki and Jong, 2000). The phytochemical components and the related health benefits of the plum have attracted the attention of researchers toward the development of health-promoting products from the plum (Sheikh et al., 2021d). Plum fruit is also considered to be an excellent source of minerals like calcium, potassium, magnesium, and iron. The plum products, including waste and by-products from plum processing, have the scope for commercial exploitation. The efficient and profitable transformation of plum fruit into value-added products not only reduces the post-harvest losses but also provides better returns to the fruit growers.

14.2 CHEMISTRY OF PLUM WASTES AND BY-PRODUCTS

The mature plum seeds (or kernels) constitute an important part of the plum fruit that is present within the fruit pit or stone. Fruits, including plum, can be converted into their dried forms used in different dishes through various dehydration processes. Plum kernels are more often referred to as non-traditional food sources and deliver many health benefits such as preventing heart diseases, microbial growth, and carcinogenesis, coupled with prevention of osteoporosis and digestive system ailments, which are credited to the presence of phenolics such as flavonols and caffeic acid derivatives, as well as dietary fiber. However, the ripening stage of the plum determines the number of bioactive compounds present in the fruit (Milala et al., 2013). A significant amount of oil is obtained from the plum kernel (>45 g/100 g) which is used in cosmetics, pharmaceuticals, and, of course, in the food industries due to the presence of many bioactive compounds (such as vanillic, chlorogenic, syringic acids, and rutin), which bestow on the kernel a significant antioxidant potential Table 14.2).

The fatty acid profile of plum shows that, among the fatty acids, linoleic (C18:2) and oleic (C18:1) acids are present in the highest concentrations in the plum kernel oil coupled with other beneficial compounds, such as phytosterols and tocopherols, the concentration of which also depends on the variety of plum used (Górnaś et al., 2017; Savic et al., 2020). To date, studies have revealed the presence in plum kernels of a cyanogenic glucoside (amygdalin), a bitter compound that can treat various degenerative diseases (He et al., 2020).

Plum skin has also been reported to contain a much higher antioxidant activity than the corresponding plum flesh, as a result of the bioactive compounds present. Díaz-Mula et al. (2009) reported that the antioxidant activity of skin of some plum varieties was 2–40 times greater than that of the plum flesh, the activity increasing along with the total phenolics, total anthocyanins, and carotenoids during storage (Table 14.2). Plum kernels of plum species, such as *P. domestica* and the cherry plum *Prunus cerasifera* Ehrh., also contain various isomers of tocopherol (α, β, γ, and δ) and the tocotrienol α-T3 (Górnaś et al., 2015).

In a recent study, it was reported that the skin of black plum contains a high amount of phenolic compounds, particularly anthocyanins, that can be used in the development of food packaging films where they provide improved antioxidant activity along with enhanced ethylene- and radical-scavenging activity (Zhang et al., 2019). However, when plums were stored under modified atmospheric conditions, the increase in concentration of total phenolics and of anthocyanins, and in total antioxidant activity was more rapid in the skin of purple plum varieties than in the yellow varieties (Díaz-Mula et al., 2011). Additionally, these compounds help in the improvement of film barrier properties against water vapor and UV-light. The extracts obtained from the skin of black plums contained higher concentrations of total phenolics, flavonoids, and anthocyanins, and total phenols than in extracts of black plum pulp. The antioxidant activity of plum pulp extract also helped in the prevention of the deteriorative oxidation process. Correia et al., (2020) demonstrated that the extract of plum pulp mixed with a black tea extract (50:50) resulted in the inhibition of biodiesel

Table 14.2   Bioactive Compounds in Plum By-Products (Fruit, Skin, and Kernel Oil)

Plum Species (Variety)	Bioactive Compound	Concentration	References
Plum fruit			
Prunus salicina (Black Diamond)	Anthocyanins	34–177 mg/kg	Díaz-Mula et al. (2009)
Prunus salicina. (Golden Globe)	β-carotene	8 ± 0.5 mg/kg	Díaz-Mula et al. (2009)
Prunus salicina (Black Amber)	Total phenolics	5210 ± 180 mg/kg	Díaz-Mula et al. (2009)
Prunus domestica (Casselman)	Anthocyanins, flavonoids, total phenolics, and ascorbic acid	N.I.	Medina-Meza et al. (2015)
Prunus salicina (Black Diamond)	Anthocyanins	0.06–0.36 mmol/kg	Díaz-Mula et al. (2008)
Prunus salicina (Black Amber)	Total phenolics Chlorogenic acid Epicatechin	0.32–1.93 mg/g 210.20 mg/g 0.32 mg/g	Ozturk et al. (2012)
Prunus salicina. (Golden Globe)	Total phenolics	846 ± 54 mg/kg	Díaz-Mula et al. (2011)
Prunus salicina (Sun Gold)	Total phenolics	1,562 ± 129 mg/kg	Díaz-Mula et al. (2011)
Plum skin			
Prunus salicina (Black Amber)	Anthocyanins	18.89 ± 0.67 mmol/ kg	Díaz-Mula et al. (2008)
Prunus salicina (Black Diamond)	Anthocyanins	5.93 ± 0.22 mmol/ kg	Díaz-Mula et al. (2009)
Prunus salicina (Black Amber)	Anthocyanin	4,370 ± 180 mg/kg	Díaz-Mula et al. (2009)
Prunus salicina (Black Diamond)	Anthocyanin	1,310 ± 100 mg/kg	Díaz-Mula et al. (2009)
Prunus salicina (Larry Ann)	β-carotene	99 ± 7 mg/kg	Díaz-Mula et al. (2009)
Prunus salicina (Black Amber)	β-carotene	62 ± 5 mg/kg	Díaz-Mula et al. (2009)
Prunus salicina (Golden Globe)	Total phenolics	1,267±19 mg/kg	Díaz-Mula et al. (2011)
Prunus domestica (Casselman)	Anthocyanins	N.I.	Medina-Meza et al. (2015)
Prunus salicina (Black Amber)	Total phenolics	4,584±87 mg/kg	Díaz-Mula et al. (2011)
Plum kernel oil			
Prunus domestica	γ-Tocopherol	76.2–182 mg/100 g oil	Górnaś et al, (2016)
Prunus cerasifera	γ-Tocopherol	60.5–170.6 mg/100 g oil	Górnaś et al. (2016)
Prunus salicina (Black Amber)	β-carotene	188 µg/g oil	Popa et al. (2011)
Prunus salicina (Black Amber)	B + γ -Tocopherols	1057.20 l µg/100 g	Popa et al. (2011)
Prunus domestica (Cacanska Rodna)	Tocopherol	87.4 mg/100g oil	Rabrenović et al. (2021)

oxidation as a result of the various phenolic compounds present in the plum pulp extract. The health-promoting aspect of plum kernels for humans is credited to them as being an indispensable source of phenolic compounds, as well as being a cheap source of protein and oil and hence used for human consumption in place of costly protein sources (Górnaś et al., 2015). Plum seed is also utilized for the production of oil containing phenolic compounds with vanillic acid in the highest concentration that bestows plum seed oil with significant antioxidant properties with satisfactory results up to 65°C, thereby making plum seed oil an important candidate for the food, cosmetics, and pharmaceutical industries (Savic et al., 2020). Plum by-products, such as skin, pulp, kernel, etc can be used to treat various degenerative diseases like cancer and cardiovascular diseases, mainly due to the presence of anthocyanins which could help in the prevention of neuronal diseases (Dulf et al., 2016). The phenolic compounds present in the plum are not uniformly distributed, differing in concentration in various parts of the plum. In this regard, out of the various parts of plum fruit, it has been found that the total phenolic content is highest in the skin as compared with the plum pulp. Moreover, the color of the plum fruit's outer surface decides the total phenolic content in the plum skin; for example, the skin of 'Alutus', a dark violet variety of plum, has a higher total phenolic content than the skin of 'Tuleu timpuriu' and 'Oltenal' plum varieties (Cosmulescu et al., 2015). Hence, it can be concluded that the practical value of plum by-products as a dietary intake source of phenolic antioxidants depends on the cultivar and the fruit parts. The antioxidant activity associated with the plum by-products thereby helps to ensure their application in different industries. Plum pomace (the residue remaining after juice extraction from the fruit) is also believed to be an important source of valuable nutrients and is also helpful in the process of production of different valued compounds, which are also exploited as the raw material to support the growth of different kinds of microorganisms. Plum pomace also contains a significant quantity of procyanidins and polyphenols like proanthocyanidins, hydroxycinnamic acids, quercetin glycosides, and anthocyanins (Milala, et al., 2013). These authors also reported high concentrations of carbohydrates such as fructose, glucose, and sorbitol in plum pomace.

Plum products/by-products help in the preservation of meat products by inhibiting the process of oxidation, with insignificant changes in their color and flavor attributes (Karre et al., 2013). The skin and flesh of Japanese (or Chinese) plum (*Prunus salicina*) have been used for the production of fiber microparticles which are utilized as antioxidant agents for the preservation of different food products. The antioxidant potential of fiber microparticles can be attributed to polyphenolic compounds, mainly proanthocyanidins (Basanta et al., 2016). Moreover, the polyphenolic compounds (cyanidins, quercetin derivatives, pentameric proanthocyanidins) present make the skin and pulp of Japanese plum potential candidates for preservation of meat products such as chicken meat products, as observed by Basanta et al. (2018) in raw chicken breast patty. The authors reported that thiobarbituric acid-reactive substances (TBARS) were inhibited up to 50% in raw patties in a 10-d storage period at 4°C. The samples with added microparticles from plum skin resulted in a higher (77–157%) ferric reducing power (FRAP), concluding that plum skin microparticles were effective additives in chicken patties. Plum seeds are also confirmed to be effective antioxidant materials as demonstrated by Savic and Gajic (2021) through the identification of various phenolic compounds, such as rutin, epi-gallocatechin, gallic acid, ferulic acid, syringic acid, epicatechin, caffeic acid, and coumaric acid, in their study of plum seeds using ultrasound-assisted extraction. The oil obtained from the plum kernel is also used as an antioxidant agent in various food products and has been shown to contain a considerable amount of polyphenols. Popa et al. (2011) reported the FRAP antioxidant activity and the polyphenol concentration in plum kernel oil in the ranges of 0.423–1.895 mM Fe^{2+}/L and 0.605–2.85 mM gallic acid/L, respectively, although the number and concentrations of tocopherols were comparatively low. Even the plum seed consists of a very high concentration of proteins and lipids with an acceptable level of antioxidant activity confirmed through the ABTS radical scavenging capacity assay and lipid peroxidation inhibition capacity after digestion by various enzymes, alcalase in particular (González-García et al., 2014).

14.3 PROCESSING AND UTILIZATION OF PLUM WASTES AND BY-PRODUCTS

To improve the sustainability of the food chain, the focus of researchers has switched toward the valorization of waste and by-product streams produced from the processing of agricultural commodities (Sheikh et al., 2020). Given the indispensable role of fruits and vegetables in our diet and the changing consumer dietary habits, their utilization along with that of wastes and by-products from them are in high demand nowadays (Vilariño et al., 2017). The active agents present in by-products, like organic acids, polyphenols, flavonoids, vitamins, minerals, etc., help in the management of various degenerative diseases in humans (Table 14.2). Unfortunately, the by-products obtained from horticultural products like fruits and vegetables have not previously been accorded sufficient importance but the scenario is changing nowadays because the recovery of valuable biomolecules from these waste materials is now possible. Plum processing industries generate a huge amount of residues in the form of skin, pomace, and stones, which comprise about 10–25% of the raw material (Gonzalez-Garcia et al., 2014). The processing of fruits results in the production of byproducts on a large scale that in turn will lead to the loss of nutrients and energy. Moreover, the byproducts cause several environmental problems like pollution of water and soil (Damiani, et al., 2012). Although the food processing industries have grown rapidly during the past few decades the losses and wastes during processing are hard to be ignored by the researchers. The problem related to the generation of byproducts has attracted scientific, political, economic as well as commercial interest and thus has forced the European Union towards a zero-waste economy by the year 2025. The fruit and vegetable wastes produced in the industries can be used as a source to produce energy and thus has the potential to be used as a substitute for the substrates used as conventional sources of energy. Any business can reduce its fixed cost by the valorization of the wastes produced by the industries which need innovative approaches concerned with their storage and disposal in a mannered way (Shalini and Gupta, 2010). Since the production of plums is increasing throughout the world as a result of the byproducts during the processing of the fruit are also produced at an increased rate as the industrial utilization of the fruit results in the production of byproducts like skin, pomace, and pits (Table 14.2). The by-products like skin and pomace are rich in dietary fiber but are either discarded or used to feed animals. These by-products are rich in phytochemicals associated with human health promotion that have attracted attention from the scientific world. To convert the by-products into value-added products, it is necessary to have comprehensive knowledge about the components responsible for their quality, like pigments, pectin content, and carbohydrates, etc. Table 14.2).

14.3.1 Pomace

Pomace is a by product obtained during the production of juice and the dried pomace is known to contain about 40–70% of dietary fiber. In the case of plums, the quantity of dietary fiber (DF) is about 64.5% which depends on the method of juice extraction as well as on the variety of plum. Pectin obtained from fruit pomace is used in food industries where it provides gelling and stabilizing properties in addition to being used as a thickening agent. The proximate composition of plum pomace has 38–49% dry matter total DF (7–13% soluble fraction), 10–17.4 mM TEAC g^{-1} dry matter of antioxidant activity, in addition to having an appreciable amount of polyphenols (Table 14.2). This makes plum pomace a potential candidate for the manufacture of different supplements beneficial for human health rather than being discarding as waste (Gil et al., 2002). The skin proportion in plums is in the range 10–25% of the total weight of fruit. Plum contains higher amounts of soluble dietary fiber as a result of having higher pectin concentrations. Among the various health benefits of soluble dietary fiber, the reduction in the risk of cardiovascular diseases along with lowering LDL cholesterol levels is of particular interest. Insoluble dietary fiber reduces the risk of occurrence of colorectal cancer. Moreover, an increased intake of dietary fiber helps to reduce human bodyweight. Dietary fiber improves the physical, structural, sensory properties as well as texture of various food preparations (Elleuch et al., 2011). Plum pomace is also a good source of organic micronutrients (Table 14.3).

Table 14.3   The Percentage and Weight of Waste Generated in the Processing of Plums During Sorting and Stoning

Processed Weight of Fruits	24.7	Lipiński et al. (2018)
Sorting		
Percentage (%)	100	Dulf et al. (2016);
Weight of fruits	24.7	Kostić et al. (2016);Górnaś et al. (2017)
Percentage of waste (%)	0.5	
Weight of waste	1.2	
Stoning		
Percentage (%)	100	Amah and Okogeri (2019); Dulf et al. (2016);
Weight of fruits	23.4	Kostić et al. (2016);Górnaś et al. (2017)
Percentage of stones (%)	16	
Weight of stones	3.8	

Figure 14.1   Plum (*Prunus domestica*) pits.

Figure 14.2   Plum (*Prunus domestica*) kernels.

14.3.2 Plum Kernel

Generally, fruits, including plums, produce a high proportion of waste in the form of stones/pits (Figure 14.1). Plum pits are wastes of the plum canning and juicing industries. The plum pits/stones left after processing are discarded as waste. An efficient and environmentally friendly approach toward plum pit disposal would reduce the harmful effects on the environment and could help to valorize the stone biomass (otherwise lost) (Dulf et al., 2016). Moreover, industries can benefit from the utilization of plum pits in various value-added fields (Kostić et al., 2016). Plum pits contain the plum kernels (Figure 14.2). Plum kernels, the dry and mature seeds of plum, represent an essential component in a non-traditional source of nutrition. Plum kernels are also used for the manufacture of value-added products because they contain sizable amounts of lipids, proteins, vitamins, minerals, and fibers (Górnaś et al., 2015). In addition, dietary fibers, proteins, oils, minerals, and phenolics, along with cyanogenic compounds, are also found in significant amounts in plum kernels (Dulf et al., 2016). A cyanogenic glycoside (amygdalin), with health-promoting effects, is the cyanogenic compound in plum kernels (Zhang et al., 2019). The oil from the plum seed and that of the apricot kernel have various characteristics in common. In some cases, fruit stones have been employed in the manufacturing of cosmetics, and the potential use of plum kernels as a source of protein, polyphenols, and oil with a high content of unsaturated fatty acids could be advantageous.

14.3.2.1 Chemical Composition of Plum Kernels

The varietal variation within the plum species results in differences in their compositional characteristics. Plum kernels contain relatively high amounts of oils, dietary proteins, fibers, vitamins, minerals, and bioactive compounds, constituting a source of readily available energy (Table 14.4) (Górnaś et al., 2017).

14.3.2.1.1 Oils

Regardless of species and varieties, plum kernels contain a considerable amount of oils. The oil content ranges from 36.52 to 50.0%, which could be extracted and used for human consumption. The oil obtained from plum kernels mostly contains lipophilic antioxidants, phytosterols, and unsaturated fatty acids which often furnish nutty flavor to foods when incorporated (Górnaś et al., 2017). The major fatty acids are oleic acid, linoleic acid, palmitic acid, stearic acid, palmitoleic acid, and eicosanoic acid (Kamel and Kakuda, 1992; Kostić et al., 2016; Górnaś et al., 2017). Plum kernel oil is quite stable and has low risks of rancidity due to the presence of greater proportions of unsaturated fatty acids, such as oleic and linoleic acids (Table 14.5). Higher concentrations of oleic and linoleic acid in plum kernel oil make it an important constituent of edible oils, for the preparation of cosmetics and moisturizing creams, and could be used in the production of biodiesel in the future. It has been reported that plum kernels contain high levels of essential fatty acids (linoleic acid) that are vital for human metabolism and cannot be synthesized by the human body (Table 14.2).

Table 14.4 Chemical Composition of Plum (*Prunus domestica*) Kernels

Moisture content (%)	4.75–8.25	Kamel and Kakuda (1992); Sheikh et al. (2021c); Sheikh et al. (2021)
Carbohydrates (%)	7.3–17.64	
Crude oil (%)	36.52–50.0	El Abd Aal et al. (1987); Sheikh et al. (2021c); Sheikh et al. (2021); Górnaś et al. (2015); Amah and Okogeri, (2019)
Crude protein (%)	27.57–35.91	
Crude fiber (%)	1.9–4.84	
Total ash (%)	2.2–5.18	
Total phenolic content (mg/100 g)	170	Górnaś et al. (2015)
Total flavonoids (mg/100 g)	3.75	
Amygdalin (mg/g)	0.1–17.5	Sheikh et al. (2021c)

Table 14.5 Fatty Acid Composition of Plum (*Prunus domestica*) Kernel Oil (%, w/w)

Oleic acid, C18:1	46.2–68.4	Stacewicz-Sapuntzakis et al. (2001)
Linoleic acid, C18:2	23.9–45.1	Veličković et al. (2016)
Palmitic acid, C16:0	4.5–7.5	Górnaś et al. (2017)
Stearic acid, C18:0	1.0–2.3	
Palmitoleic acid, C16:1	0.4–1.4	
Eicosanoic acid, C20:0	0.1–0.3	

Various diseases such as cardiovascular, atherosclerosis, inflammatory disorders, etc are prevented by high concentrations of unsaturated fatty acid-containing foods which are also helpful in carrying out functions (both physiological and biological) in the human body (El AbdAal et al., 1987; Zhao et al., 2007). Therefore, plum kernel oil is nutritionally attractive and considered to be an important ingredient in different value-added products used for food and nutraceutical supplements due to the potential accruing from the high concentrations of unsaturated fatty acids, its superior stability, and the presence of antioxidant constituents (Matthäus and Özcan, 2009).

14.3.2.1.2 Proteins

Bioactive compounds are present in significant amounts in plant proteins which are attracting increasing interest nowadays due to their digestible and nutritional potential (Gull et al., 2022). The extraction of proteins from plant sources is a tedious process as they contain vacuoles and rigid cell walls while soluble proteins are present at low concentrations (Garcia et al., 2013). Demand for relatively cheap sources of protein is increasing globally and the plum kernels contain a substantial amount of dietary proteins that have not usually been utilized or characterized. Plum kernel meal contains a high level of protein that varies from 28.0 to 41.3 g/kg (Gonzalez-Garcia et al., 2014; Xue et al., 2018). The use of high-intensity focused ultrasound in the extraction process of protein from plum kernel exhibited a yield of 40% (on a dry weight basis) in less than 1 hour (Gonzalez-Garcia et al., 2014). The plum kernel protein presents a relatively full array of essential amino acids (Table 14.6) which indicates that the kernels could play a vital role in health maintenance and as a supplement designed to maintain muscle mass. Plum kernel proteins might be an attractive source of bioactive peptides that have a positive influence on body function (Gonzalez-Garcia et al., 2014). Plum kernel protein can be considered to be a useful agent with which to reduce degenerative diseases, and more than twenty-one bioactive peptides with natural antioxidant and antihypertensive features have been identified (González-García et al., 2014; Vásquez-Villanueva et al., 2015). Among the identified proteins were several histones and seed storage proteins.

14.3.2.1.3 Amygdalin

Many plants can synthesize cyanogenic glycosides, which are related to bitterness as an anti-herbivore defense (Anjum et al., 2022). Cyanogenic glycosides are water-soluble, heat-stable, secondary metabolites. Cyanogenic glycosides are not toxic when intact but they become toxic when enzymes (endogenous or exogenous, such as β-glycosidase and α-hydroxynitrilelyases) come in contact with them, releasing hydrogen cyanide, which causes tissue damage in response to bruising or chewing damage (Sheikh et al., 2021c). The predominant cyanogenic glycoside in the Rosaceae species is amygdalin ($C_{20}H_{27}NO_{11}$) and the plum kernels contain relatively high amounts (0.1–17.5%) of amygdalin (Gonzalez-Garcia et al., 2014). Amygdalin (D-mandelonitrile-β-D- gentiobioside), also known as vitamin B_{17}, purasin, or nitriloside, is derived from phenylalanine and is composed of

Table 14.6 Amino Acid Composition of Plum (*Prunus domestica*) Kernel (mmoles/100 g protein)

Essential Amino Acids		Non-Essential Amino Acids		References
Leucine (Leu)	20.4	Glutamic acid (Glu)	62.2	Kamel and Kakuda (1992);
Valine (Val)	13.9	Aspartic acid (Asp)	33.1	Stacewicz-Sapuntzakis et al. (2001);
Phenylalanine (Phe)	11.9	Glycine (Gly)	29.8	Veličković et al. (2016);Górnaś et al. (2017)
Isoleucine (Ile)	9.1	Alanine (Ala)	20.1	
Threonine (Thr)	9.0	Serine (Ser)	16.5	
Histidine (His)	7.4	Proline (Pro)	15.3	
Lysine (Lys)	6.7	Tyrosine (Tyr)	5.2	
Methionine (Met)	1.6	Arginine (Arg)	27.4	

two molecules of glucose, one of hydrogen cyanide and one of benzaldehyde (Zhang et al., 2019). Amygdalin is classified as a cyanogenic glycoside because it contains a cyanide group between a glycoside and a benzene ring that can be released after hydrolysis, and each molecule includes a nitrile group, which releases hydrogen cyanide (Cho et al., 2006). Depending upon variety and environmental conditions for growth, plum seeds contain different levels of amygdalin,with kernels from green, black, purple, yellow, and red plum containing 17.5, 10.0, 2.16, 1.54, and 0.44 mg/g of amygdalin, respectively (Gonzalez-Garcia et al., 2014). Consumption of cyanogenic seeds can lead to anxiety, dizziness, and confusion along with headaches which are the symptoms of sub-acute cyanide poisoning by the generation of hydrogen cyanide (HCN) through hydrolysis that could be dangerous to humans (Shi et al., 2019). Amygdalin can be preventative of neurodegenerative diseases and has health benefits such as immuno-regulation, anti-fibrosis, auxiliary anticancer, anti-atherosclerosis, anti-inflammation analgesia which have been explored by different studies (Shi et al., 2019; He et al., 2020). To be specific, the lower dose is beneficial to the health, while the higher dose may lead to cyanide poisoning-like symptoms (Shi et al., 2019). To prevent cyanide toxicity, processing methods like roasting, soaking, peeling, pounding, fermenting, grating, grinding, boiling, and drying of the kernels have been used over the years to reduce the cyanide content before consumption through the loss of the water-soluble glycoside (Gonzalez-Garcia et al., 2014). The conventional methods,, including soaking, autoclaving, fermenting, and boiling have been used to reduce the potential for toxicity (Bolarinwa et al., 2016). β-glycosidase activity causes satisfactory degradation of amygdalin during grinding and sequential soaking (Tuncel et al., 1990). Finer particles in general result in faster degradation of glycosides. Thus, finely ground seeds contain no glycosides after 0.5 h of soaking. The reduction of amygdalin levels by soaking showed no side-effects on the essential amino acids (Sheikh et al., 2022). Heat treatments, such as autoclaving, extrusion, and microwave roasting, are commonly used for the reduction of cyanogenic glycoside levels (Wu et al., 2008). Heating with microwaves has recently become the most adaptable method for reducing anti-nutritional factors (Sheikh et al., 2021c). Microwave heating at 500W for 9 minutes after soaking for 1 hour is sufficient for complete degradation of toxic compounds, while microwave heating effectively reduces levels of some anti-nutrients (tannins, phytic acid, and trypsin inhibitor activity) and improves protein quality in selected common beans (Sheikh and Saini, 2022; Sheikh et al., 2021d). Microwave power of 400 W for 4 min and 50 s caused a reduction in hydrogen cyanide content to below the allowed limits (230 mg/kg of linseed) (Ivanov et al., 2012). Various methods of heating, including microwaving, have been used to improve the nutritional value of beans (Hernande-Infante et al., 1998; Sheikh et al., 2021d).

14.4 CONCLUSIONS

A significant part of the harvested plum is processed, resulting in a substantial amount of waste and by-products like skins, pomace, pits/kernels, etc., and there is no definite use for these residues. Plum wastes and by-products are nutritionally attractive and are regarded as a cheap source of unsaturated fatty acids, proteins, fibers, and bioactive constituents that are irretrievably lost if the wastes are incinerated or sent to landfills, wasting their huge potential. The utilization of these wastes and by-products could be useful for the food, cosmetic and pharmaceutical industries. The incorporation of detoxified plum kernels in food systems could result in economic benefits and reduce waste disposal problems. Therefore, it is worthwhile to explore the wastes and by-products for exploitation of the nutritional and bioactive constituents.

ACKNOWLEDGMENT

The financial support to authors in the form of Senior Research Fellowships (SRF) by the Indian Council of Medical Research–New Delhi, India is duly acknowledged.

REFERENCES

Abd Aal, M. E., Gomaa, E. G., & Karara, H. A. (1987). Bitter almond, plum and mango kernels as sources of lipids. *Lipid/Fett*, *89*(8), 304–306.

Amah, U. J., & Okogeri, O. (2019). Nutritional and phytochemical properties of Wild Black Plum (*Vitex doniana*) seed from Ebonyi state. *International of Journal of Horticulture*, *3*(1), 32–36.

Anjum, N., Sheikh, M. A., Saini, C. S., Hameed, F., Sharma, H. K., & Bhat, A. (2022). Cyanogenic glycosides. In G.A. Naik and J. Kour (eds), *Handbook of plant and animal toxins in food* (pp. 191–202). Boca Raton, FL: CRC Press.

Basanta, M. F., Marin, A., De Leo, S. A., Gerschenson, L. N., Erlejman, A. G., Tomás-Barberán, F. A., & Rojas, A. M. (2016). Antioxidant Japanese plum (*Prunus salicina*) microparticles with potential for food preservation. *Journal of Functional Foods*, *24*, 287–296.

Basanta, M. F., Rizzo, S. A., Szerman, N., Vaudagna, S. R., Descalzo, A. M., Gerschenson, L. N., … & Rojas, A. M. (2018). Plum (*Prunus salicina*) peel and pulp microparticles as natural antioxidant additives in breast chicken patties. *Food Research International*, *106*, 1086–1094.

Bolarinwa, I. F., Olaniyan, S. A., Olatunde, S. J., Ayandokun, F. T., & Olaifa, I. A. (2016). Effect of processing on amygdalin and cyanide contents of some Nigerian foods. *Journal of Chemical and Pharmaceutical Research*, *8*(2), 106–113.

Chang, T. S., Siddiq, M., Sinha, N. K., & Cash, J. N. (1994). Plum juice quality affected by enzyme treatment and fining. *Journal of Food Science*, *59*(5), 1065–1069.

Cho, A. Y., Yi, K. S., Rhim, J. H., Kim, K. I., Park, J. Y., Keum, E. H., & Oh, S. (2006). Detection of abnormally high amygdalin content in food by an enzyme immunoassay. *Molecules & Cells (Springer Science & Business Media BV)*, *21*(2), 208–213.

Correia, I. A. S., Borsato, D., Savada, F. Y., Pauli, E. D., Mantovani, A. C. G., Cremasco, H., & Chendynski, L. T. (2020). Inhibition of the biodiesel oxidation by alcoholic extracts of green and black tea leaves and plum pulp: Application of the simplex-centroid design. *Renewable Energy*, *160*, 288–296.

Cosmulescu, S., Trandafir, I., Nour, V., & Botu, M. (2015). Total phenolic, flavonoid distribution and antioxidant capacity in skin, pulp and fruit extracts of plum cultivars. *Journal of Food Biochemistry*, *39*(1), 64–69.

Damiani, C., Silva, F. A., Rodovalho, E. C., Becker, F. S., Asquieri, E. R., Oliveira, R. A., & Lage, M. E. (2012). Aproveitamento de resíduos vegetais para produção de farofa temperada. *Alimentos e Nutrição*, *22*, 657–662.

Díaz-Mula, H. M., Zapata, P. J., Guillén, F., Castillo, S., Martínez-Romero, D., Valero, D., & Serrano, M. (2008). Changes in physicochemical and nutritive parameters and bioactive compounds during development and on-tree ripening of eight plum cultivars: A comparative study. *Journal of the Science of Food and Agriculture*, *88*(14), 2499–2507.

Díaz-Mula, H. M., Zapata, P. J., Guillén, F., Martínez-Romero, D., Castillo, S., Serrano, M., & Valero, D. (2009). Changes in hydrophilic and lipophilic antioxidant activity and related bioactive compounds during postharvest storage of yellow and purple plum cultivars. *Postharvest Biology and Technology*, *51*(3), 354–363.

Díaz-Mula, H. M., Zapata, P. J., Guillén, F., Valverde, J. M., Valero, D., & Serrano, M. (2011). Modified atmosphere packaging of yellow and purple plum cultivars. 2. Effect on bioactive compounds and antioxidant activity. *Postharvest Biology and Technology*, *61*(2–3), 110–116.

Dreher, M. L. (2018). Whole fruits and fruit fiber emerging health effects. *Nutrients*, *10*(12), 1833.

Dulf, F. V., Vodnar, D. C., & Socaciu, C. (2016). Effects of solid-state fermentation with two filamentous fungi on the total phenolic contents, flavonoids, antioxidant activities and lipid fractions of plum fruit (*Prunus domestica* L.) by-products. *Food Chemistry*, *209*, 27–36.

Elleuch, M., Bedigian, D., Roiseux, O., Besbes, S., Blecker, C., & Attia, H. (2011). Dietary fibre and fibre-rich by-products of food processing: Characterisation, technological functionality and commercial applications: A review. *Food Chemistry*, *124*(2), 411–421.

FAOSTAT. (2020). FAO Statistics, Food and Agriculture Organization of the United Nations. FAOSTAT. Available online at: http://www.fao.org/faostat/en/#data/qc/visualize.

Garcia, M. C., Puchalska, P., Esteve, C., & Marina, M. L. (2013). Vegetable foods: A cheap source of proteins and peptides with antihypertensive, antioxidant, and other less occurrence bioactivities. *Talanta*, *106*, 328–349.

Gil, M. I., Tomas-Barberan, F. A., Hess-Pierce, B., Kader, A. A. (2002). Antioxidant capacities, phenolic compounds, carotenoids, and vitamin C contents of nectarine, peach, and plum cultivars from California. *Journal of Agricultural and Food Chemistry, 50*, 4976–4982.

González-García, E., Marina, M. L., & García, M. C. (2014). Plum (*Prunus domestica* L.) by-product as a new and cheap source of bioactive peptides: Extraction method and peptides characterization. *Journal of Functional Foods, 11*, 428–437.

Górnaś, P., Misina, I., Gravite, I., Lacis, G., Radenkovs, V., Olsteine, A., Seglina, D., Kaufmane, E., & Rubauskis, E. (2015). Composition of tocochromanols in the kernels recovered from plum pits: The impact of the varieties and species on the potential utility value for industrial application. *European Food Research and Technology, 241*, 513–520.

Górnaś, P., Rudzinska, M., Raczyk, M., Misina, I., Soliven, A., Lacis, G., & Seglina, D. (2016). Impact of species and variety on concentrations of minor lipophilic bioactive compounds in oils recovered from plum kernels. *Journal of Agricultural and Food Chemistry, 64*(4), 898–905.

Górnaś, P., Rudzi'nska, M., & Soliven, A. (2017). Industrial by-products of plum *Prunus domestica L.* and *Prunus Cerasifera Ehrh* as potential biodiesel feedstock: Impact of variety. *Industrial Crops and Products, 100*, 77–84.

Gull, A., Mohd, A. S., Jasmeet, K., Beenish, Z., Imtiyaz, A. Z., Altaf, A. W., Surekha, B., & Mushtaq, A. L. (2022). Anthocyanins. In J. Kour and G.A. Naik (eds), *Nutraceuticals and health care* (pp. 317–329). Cambridge, MA: Academic Press.

He, X. Y., Wu, L. J., & Wang, W. X. (2020). Amygdalin - A pharmacological and toxicological review. *Journal of Ethnopharmacology, 254*, 112717.

Hernande-Infante, M., Sousa, V., Montalvo, I., & Tena, E. (1998). Impact of microwave heating on hemagglutinins, trypsin inhibitors and protein quality of selected legume seeds. *Plant Foods for Human Nutrition, 52*, 199–208.

Ivanov, D., Kokic, B., Brlek, T., Colovic, R., Vukmirovic, D., Levic, J., & Sredanovic, S. (2012). Effect of microwave heating on content of cyanogenic glycosides in linseed. *Ratarstvo i Povrtarstvo, 49*, 63–68.

Kamel, B. S., & Kakud, Y. (1992). Characterization of the seed oil and meal from apricot, cherry, nectarine, peach, and plum. *Journal of American Oil Chemists Society, 69*, 492–494.

Karre, L., Lopez, K., & Getty, K. J. (2013). Natural antioxidants in meat and poultry products. *Meat Science, 94*(2), 220–227.

Ki, H. S., & Jong, B. E. (2000). Total dietary fiber contents in processed fruit, vegetable, and cereal products. *Food Science and Biotechnology, 9*, 1–4.

Kim, D. O., Jeong, S. W., & Lee, C. Y. (2003). Antioxidant capacity of phenolic phytochemical from various cultivars of plums. *Food Chemistry, 81*, 321–326.

Kostić, M. D., Veličković, A. V., Joković, N. M., Stamenković, O. S., & Veljković, V. B. (2016). Optimization and kinetic modeling of esterification of the oil obtained from waste plum stones as a pretreatment step in biodiesel production. *Waste Management, 48*, 619–629.

Lipiński, A. J., Lipiński, S., & Kowalkowski, P. (2018). Utilization of post-production waste from fruit processing for energetic purposes: Analysis of Polish potential and case study. *Journal of Material Cycles and Waste Management, 20*(3), 1878–1883.

Matthäus, B., Özcan, M. M. (2009). Fatty acids and tocopherol contents of some *Prunus spp.* kernel oils. *Journal of Food Lipids, 16*, 187–199.

Medina-Meza, I. G., & Barbosa-Cánovas, G. V. (2015). Assisted extraction of bioactive compounds from plum and grape peels by ultrasonics and pulsed electric fields. *Journal of Food Engineering, 166*, 268–275.

Milala, J., Kosmala, M., Sójka, M., Kołodziejczyk, K., Zbrzeźniak, M., & Markowski, J. (2013). Plum pomaces as a potential source of dietary fibre: Composition and antioxidant properties. *Journal of Food Science and Technology, 50*(5), 1012–1017.

Milosevic, T., & Milosevic, N. (2012). Factors influencing mineral composition of plum fruits. *Journal of Elementology, 3*, 454–463.

Ozturk, B., Kucuker, E., Karaman, S., & Ozkan, Y. (2012). The effects of cold storage and aminoethoxyvinylglycine (AVG) on bioactive compounds of plum fruit (Prunus salicina Lindell cv.'Black Amber'). *Postharvest Biology and Technology, 72*, 35–41.

Pashova, S. (2006). Chemical composition of plum fruits. *Journal of Mountain Agriculture on the Balkans, 9*, 239–249.

Popa, V. M., Bele, C., Poiana, M. A., Dumbrava, D., Raba, D. N., & Jianu, C. (2011). Evaluation of bioactive compounds and of antioxidant properties in some oils obtained from food industry by-products. *Romanian Biotechnological Letters*, *16*(3), 6234–6241.

Rabrenović, B. B., Demin, M. A., Basić, M. G., Pezo, L. L., Paunović, D. M., & Sovtić, F. S. (2021). Impact of plum processing on the quality and oxidative stability of cold-pressed kernel oil [Impacto del procesamiento de ciruelas en la calidad y la estabilidad oxidativa del aceite del hueso prensado en frío]. *Grasas y Aceites*, *72*(1), e395.

Savic, I., Savic Gajic, I., & Gajic, D. (2020). Physico-chemical properties and oxidative stability of fixed oil from plum seeds (*Prunus domestica* Linn.). *Biomolecules*, *10*(2), 294.

Savic, I. M., & Gajic, I. M. S. (2021). Optimization study on extraction of antioxidants from plum seeds (*Prunus domestica* L.). *Optimization and Engineering*, *22*(1), 141–158.

Shalini. R., & Gupta, D. K. (2010). Utilization of pomace from apple processing industries: A review. *Journal of Food Science and Technology*, *47*, 365–371.

Sheikh, M. A., Anjum, N., Gull, A., & Saini, C. S. (2020). Turnip. In G. A. Nayik & A. Gull (Eds.), *Antioxidants in Vegetables and Nuts-Properties and Health Benefits* (pp. 143–158). Springer, Singapore.

Sheikh, M. A., Rather, M. A., & Anjum, N. (2021a). To study the effect of apricot kernel flour (by-product) on physico-chemical, sensorial and antioxidant properties of biscuits. *Proceedings 2021*, 68, x. In Presented at the 2nd International Electronic Conference on Foods (Vol. 15, p. 30).

Sheikh, M. A., Saini, C. S., & Sharma, H. K. (2021b). Computation of design-related engineering properties and fracture resistance of plum (*Prunus domestica*) kernels to compressive loading. *Journal of Agriculture and Food Research*, *3*, 100101.

Sheikh, M. A., Saini, C. S., & Sharma, H. K. (2021c). Analyzing the effects of hydrothermal treatment on antinutritional factor content of plum kernel grits by using response surface methodology. *Applied Food Research*, *1*, 100010.

Sheikh, M. A., Saini, C. S., & Sharma, H. K. (2021d). Antioxidant potential, anti-nutritional factors, volatile compounds, and phenolic composition of microwave heat-treated plum (*Prunus domestica* L.) kernels: An analytical approach. *British Food Journal*. https://doi.org/10.1108/BFJ-06-2021-0649.

Sheikh, M. A., & Saini, C. S. (2022). Combined effect of microwave and hydrothermal treatment on anti-nutritional factors, antioxidant potential and bioactive compounds of plum (*Prunus domestica* L.) kernels. *Food Bioscience*, *46*, 101467.

Sheikh, M. A., Saini, C. S., & Sharma, H. K. (2022). Synergistic effect of microwave heating and hydrothermal treatment on cyanogenic glycosides and bioactive compounds of plum (*Prunus domestica* L.) kernels: An analytical approach. *Current Research in Food Science*, *5*, 65–72.

Shi, J., Chen, Q., Xu, M., Xia, Q., Zheng, T., Teng, J., Li, M., & Fan, L. (2019). Recent updates and future perspectives about amygdalin as a potential anticancer agent: A review. *Cancer Medicine*, *8*, 3004–3011.

Stacewicz-Sapuntzakis, M., Phyllis, E., Bowen, E. A., Hussain, B. I., Wood, D., & Farnsworth, N. R. (2001). Chemical composition and potential health effects of prunes: A functional food. *Critical Reviews in Food Science and Nutrition*, *41*, 251–286.

Tuncel, G., Nout, M. J., Brimer, L., & Goktan, D. (1990). Toxicological, nutritional and microbiological evaluation of tempe fermentation with *Rhizopus oligosporus* of bitter and sweet apricot seeds. *International Journal of Food Microbiology*, *11*, 337–344.

Vásquez-Villanueva, R., Marina, M. L., & García, M. C. (2015). Revalorization of a peach (*Prunus persica* (L.) Batsch) byproduct: Extraction and characterization of ACE-inhibitory peptides from peach stone. *Journal of Functional Foods*, *18*, 137–146.

Veličković, D., Ristić, M., Karabegović, I. T., Stojičević, S. S., Nikolić, N. Č., & Lazić, M. L. (2016). Plum (*Prunus domestica*) and walnut (*Juglans regia*): Volatiles and fatty oils. *Advanced Technologies*, *5*, 10–16.

Vilariño, M. V., Franco, C., & Quarrington, C. (2017). Food loss and waste reduction as an integral part of a circular economy. *Frontiers in Environmental Science*, *5*, 1–5.

Walkowiak-Tomczak, D. (2008). Characteristics of plums as a raw material with valuable nutritive and dietary properties – A review. *Polish Journal of Food and Nutrition Sciences*, *58*, 401–405.

Wu, M., Li, D., Zhou, Y. G., Brooks, M. S. L., Chen, X. D., & Mao, Z. H. (2008). Extrusion detoxification technique on flaxseed by uniform design optimization. *Separation and Purification Technology*, *61*, 51–59.

Xue, F., Zhu, C., Liu, F., Wang, S., Liu, H., & Li, C. (2018). Effects of high-intensity ultrasound treatment on functional properties of plum (*Pruni domesticae* semen) seed protein isolate. *Journal of the Science of Food and Agriculture*, *98*, 5690–5699.

Zhang, N., Zhang, Q., Yao, J., & Zhang, X. (2019). Changes of amygdalin and volatile components of apricot kernels during the ultrasonically-accelerated debitterizing. *Ultrasonics-Sonochemistry*, *58*, 104614.

Zhao, X., Wang, H., You, J., & Suo, Y. (2007). Determination of free fatty acids in bryophyte plants and soil by HPLC with fluorescence detection and identification by online MS. *Chromatographia*, *66*, 197–206.

CHAPTER 15

Lychee Wastes and By-Products
Chemistry, Processing, and Utilization

Gurwinder Kaur, Barinderjit Singh, Yashi Srivastava, Manpreet Kaur, and Yogesh Gat

CONTENTS

15.1 INTRODUCTION

Lychee is a non-climacteric, subtropical, and perishable fruit that belongs to the family Sapindaceae. It originated in China but is distributed throughout many subtropical countries and continents (Thailand, Brazil, India, Madagascar, and South Africa). China contributes 90% of lychee production. India produces 585,300 million tonnes of lychee from 84,170 hectares yearly; due to climatic restrictions, lychees can be cultivated in only a few states (Bihar, Uttaranchal, West

Bengal, Assam, Punjab, Uttar Pradesh, Jharkhand, and Tripura) of India (Singh et al. 2011). Bihar alone contributes 40% of Indian lychee production in 37.4% of the area with 6.95 t/ha productivity (Mitra 2001). The leading lychee-producing states are Bihar, West Bengal, and Jharkhand, producing 2,34,200 tons, 93,900 tons, and 58,240 tons per year, respectively, while Tripura, Punjab, Uttarakhand, and Odisha produce smaller volumes of lychees (Ghosh 2001). This lychee export is restricted to Nepal, Bangladesh, the Maldives, Bhutan, and UAE, while Gulf states have enough scope for production to meet local demand for lychees. Lychees have been exported from India rather than from Thailand or China on the basis of higher quality and lower price (Mitra 2001).

There are numerous Indian lychee cultivars, such as 'Shahi', 'China', 'Rose Scented', 'Mandraji', 'Purbi', 'Bombai', 'Dehradun', 'Early Seedless', 'Ajhauli', 'Gandaki Sampada', 'Gandaki Yogita', and 'Gandaki Lalima', and selection is based on various quality parameters (Table 15.1) (ICAR-National Research Centre on Litchi 2020). Many commercial lychee varieties, like 'Muzaffarpur', 'Dehradun', 'Early Seedless', 'Late Bedana', 'Bombay Green', 'Rose Scented', 'Kalyani Selection', 'Calcutta', 'Shahi', 'China', 'Kasba', 'Purbi, and 'Early Bedana' are cultivated in Bihar, Uttar Pradesh, West Bengal, and Punjab. Most lychees produced are consumed fresh, while 25% of productivity suffer post-harvest losses. The perishable nature of the fruit leads to lychees being immediately harvested, sold, and processed. The fruits are oval, oblong, thin, and brittle, containing a white, juicy, fleshy, translucent aril for consumption. Lychee fruit has three different parts, i.e., pulp, seed, and peel, the lychee seed and peel contributing 30% and 15% of total fruit fresh weight, respectively. Each lychee contains a dark brown seed, and a semi-translucent or white aril covered by indehiscent peel (Jiang 2001). The fruit pulp is rich in several bioactive components, enhancing its nutritional value, like vitamins, amino acids, carbohydrates, fiber, protein, fat, lipids, and minerals. In addition, many phytochemicals like cyanidin-3-rutinoside, gallic acid, quercetin-3-rutinoside epicatechin, ascorbic acid, procyanidin B2, and procyanidin B4 are also found in different lychee parts (Bhoopat et al. 2011; Jiang et al. 2013).

The lychee is well known for being used for the manufacture of different food products such as jam, jelly, squash, nectar, candy, juice, syrup, cordial, and ready-to-serve drinks (Pujari et al. 2018). Considerable waste and by-products are generated during the manufacturing of these products, and these can be used to produce bioethanol, wine, vinegar, biochar, biogas, sorbitol, and microbiological growth media. The lychee waste and by-products can also be used for non-food applications, such as producing fabric dye and corrosion inhibitors (Srivastava, Priyanka, and Parmar 2018; Varzakas, Zakynthinos, and Verpoort 2016). Lychee has the potential to be used in traditional Chinese herbal medicine due to its therapeutic properties such as anticancer, antihyperlipidemic, anti-inflammatory, antifungal, antiviral, antioxidant, anticoagulant, antidiabetic, hepatoprotective, antihyperglycemic, and cardioprotective activities (Bhoopat et al. 2011; Jiang et al. 2013).

15.2 CHEMISTRY OF LYCHEE WASTES AND BY-PRODUCTS

The different lychee fruit parts are known to be rich in protein, fat, carbohydrates, vitamins such as ascorbic acid, thiamine, riboflavin, and niacin, and minerals like potassium, nitrogen, phosphorus, calcium, magnesium, sodium, and iron (Hibbert 2007; Pareek 2016; Yang et al. 2006). Lychee peel contains high concentrations of polysaccharides like mannose (65.6%), galactose (33.0%), and arabinose (1.4%) on a dry weight basis. These polysaccharides represent substantial antioxidant activities (Yang et al. 2006). The lychee seeds and peel are rich sources of bioactive components, e.g., phenolic acids, flavonoids, lignans, carotenoids, sesquiterpenes, and sterols (Queiroz et al. 2015). Lychee fruit processing residues have many therapeutic effects, with benefits for hemostasis, treating dysentery, prostate cancer, bladder cancer, renal carcinoma urologic, and neoplasms. In addition to this, lychees can also help maintain good liver, brain, spleen, and heart health and function (Zeng et al. 2019; Bhoopat et al. 2011; Guo et al. 2017).

Table 15.1 Characteristics of Different Cultivars of Lychee

S. No.	Parameters	'Shahi'	'China'	'Rose Scented'	'Mandraji'	'Purbi'	'Bombai'	'Dehradun'	'Gandaki Lalima'
1	Season	This is an early-season cultivar	This is a mid- to late-season cultivar	Popular in Uttarakhand and Uttar Pradesh	Matures in the last week of May to first week of June	Matures in the third week of May Peel is thick, very rough with attractive bright red color Pulp is soft, juicy with pleasant flavor			This is a selection from 'Ranchi'. A late-maturing cultivar which ripens in the second week of June
2	Tree	Trees are vigorous with regular bearing	Plant is vigorous but tends to be alternate bearing Heavy bunch-bearing (20–25 fruits per bunch)	Trees are very vigorous	Trees are vigorous		The trees are vigorous. Pulp is greyish-white, sweet, soft and juicy	This is a moderately vigorous tree	Trees are moderately vigorous
3	Fruit shape	Fruits are globose-heart or obtuse shape		The fruits have distinct aroma		Fruits are egg-round to lopsided heart-shaped with uneven shoulder and fruit tip is distinctly pointed	The color of fruit is carmine red.	Fruits possess oblique-heart to conical shape with bright rose pink colour	
4	Cracking	Moderately prone to cracking (8.24%)	Less prone to cracking (4.17%)	Fruits are moderately susceptible to sunburn and cracking		Fruits are less susceptible to cracking		Highly susceptible to sunburn and cracking	
5	Fruit size			Fruits are medium to large,	Fruits are large	Fruits are large		Fruits are medium	
6	Fruit weight	22–26 g	22–26 g	20–22 g, sweet	22–26 g and produced in clusters Peel is thick, very rough, and has attractive bright red color	23–25 g and produced in large clusters		15–20 g	28–32 g
7	Total Sugar						11% total sugar	10.4% total sugar	
8	Acidity				0.43%	0.44%	0.45%	0.44%	

(Continued)

Table 15.1 (Continued)   Characteristics of Different Cultivars of Lychee

S. No.	Parameters	'Shahi'	'China'	'Rose Scented'	'Mandraji'	'Purbi'	'Bombai'	'Dehradun'	'Gandaki Lalima'
9	Pulp							The pulp is greyish-white, soft, and moderately juicy	Pulp is creamy white in color, sweet, soft, and juicy with good flavor
10	Pulp recovery	65–70%	60–67%						>60 per cent It has good shelf-life
11	Total soluble solids (TSS)	19–21°B		21–23°B	19°B	19–20°B	17–20°B	18°B	18–19°B
12	Yield (kg per tree)	140–150	150–160	80–90			80–90	80–90	130–140
13	Seed		Seeds are bold in size				Seeds are elongated, smooth with 2.3 cm length, 1.6 cm width, and weigh 3.4 g		

Source: https://www.nrclychee.org/cultivars-of-lychee (accessed on December 30, 2021)

15.2.1 Phenolic Acids

Lychee seeds, peel, and pomace contain significant concentrations of phenolic acids capable of scavenging free radicals. These compounds play an essential role in preventing the initiation or development of different diseases such as atherosclerosis, inflammatory injury, cancer, diabetes, hypoglycemia, and cardiovascular diseases in the human body. The principal phenolic acids in lychee seeds are protocatechuic acid, coumaric acid, gallic acid, and butylated hydroxytoluene (Upadhyaya and Upadhyaya 2017). The concentrations of these phenolic compounds are mainly affected by geographical distribution, physiological condition, agriculture setup, and processing techniques (Wang et al. 2011). The processing temperature can also affect the release of phenolic compounds in different parts of the lychee. High-temperature extraction at 121°C can help release more phenolic compounds but has adverse effects on the texture and color of the lychee tissue compared with a low-temperature extraction at 70°C that is sufficient for lychee fruit processing (Wang et al. 2020). However, a higher antioxidant effect can be observed at high temperature (121°C) due to the extraction of more heat-resistant phenolic compounds. Although lychee peel and seeds are valuable sources of bioactive compounds, they cannot be consumed directly. Encapsulation of these additives has to be carried out to use these functional ingredients (Rosales et al. 2019).

15.2.2 Flavonoids

The flavonoids are major polyphenolic compounds present in lychee fruit parts, including seeds and peel. The flavonoids can be classified into six subclasses: flavones, flavanols, flavanones, isoflavones, flavan-3-ols, and anthocyanins. Kaempferol-7-O-neohesperidoside and tamarixetin-3-O-rutinoside are the two major flavanol compounds present in lychee seeds (Ren et al. 2013). The lychee fruit peel accounts for 15 percent of the whole fresh fruit by weight and contains a significant concentration of flavonoid. The principal flavanol present in the peel of the lychee fruit is kaempferol (Lal et al. 2018). Kaempferol-3-O-β-D-glucoside and kaempferol-3-O-alpha-rhamnoside are found in the leaves. Lychee seeds are the primary source of flavanones containing pinocembrin-7-O-(6-O-alpha-L-rhamnopyranosyl-β-D-glucopyranoside). The flavan-3-ol epicatechin is found in lychee peel, and seeds. However, gallo-catechin is found only in the peel and seeds, whereas epicatechin-3-gallate is found only in seed (Ma et al. 2014). However, both as defensive components and as chemical signals, flavonoids play a crucial role in adapting plants to their biological environments. Flavonoids exhibit good potential antioxidant activity, which could partly be attributed to their inhibition of proliferation and induction of apoptosis in cancer cells by up-regulation and down-regulation of multiple genes.

Lychee anthocyanins are additional potent free radical scavengers, like several other flavonoids, and exhibit strong antioxidant activity in lipid environments, such as emulsified methyl linoleate, low-density human lipoprotein, and liposomes. The three main anthocyanins in lychee peel are cyanidin-3-glucoside, cyanidin-3-rutinoside, and malvidin-3-glucoside (Ma et al. 2014; Hu et al. 2010). The monosaccharides like glucose, galactose, and arabinose (thw most common sugar components of anthocyanins) are usually connected to the anthocyanidin molecule *via* the C-ring C-3 hydroxyl group (Upadhyaya and Upadhyaya 2017). The main function of anthocyanin pigments is imparting color to specific plant products (Wen et al. 2014).

15.2.3 Lignans

Lychee peel and pomace are good sources of lignans (Upadhyaya and Upadhyaya 2017). There is an increasing interest in plant lignans, which are polyphenolic compounds in which dimers of phenylpropanoid units are connected by their side chains' central carbons in the form of C8-C8 linkages. These compounds are known as significant secondary plant metabolites (Umezawa 2003). They are obtained by dimerization of substituted cinnamic alcohols, known as monolignols, from phenylalanine. In the plant cell wall, they form the building blocks for the formation of lignin. Like other phenolic compounds, lignans help maintain good health in humans due to their excellent antioxidant, anticancer, antimicrobial, antiviral, and immunomodulatory activities (Wen et al. 2014a; Jing et al. 2014).

15.2.4 Carotenoids

The lychee fruit contains many different carotenoids, such as α-carotene, β-carotene, β-cryptoxanthin, and lycopene. The lychee peel contains the highest concentration of β-carotene. These compounds are known for their excellent antioxidant properties (Maoka 2020). Some carotenoids are converted into vitamin A, a critical component of growth and human health. However, processing methods such as drying decrease the amount of β-carotene in the peel, so processing is the vital factor affecting carotenoid degradation (Queiroz et al. 2015). The carotenoids are present in photosynthetic tissue to protect them (and chlorophyll) from photo-oxidative damage (Kumari, Rani, and Nahakpam 2018). The health benefits of carotenoids present in lychees reduce the risk of disease, especially certain cancers and eye diseases (Johnson 2002). Lychee pulp contains eight vitamin E components (α, β, γ, δ tocopherols and tocotrienols) (Cabral, Cardoso, and Pinheiro-Sant'Ana 2014).

15.2.5 Sesquiterpenes and Triterpenes

Many hydrocarbons, alcohols, and related metabolites of sesquiterpenes in lychees are also highly valued for their desirable odor and flavor characteristics (Fiedor, and Burda 2014). Sesquiterpenes are a very diverse class of terpenes, that can be monocyclic, bicyclic, or tricyclic but are less reactive than terpenes. Lycheeoside A, pumilaside A, funingensin A, lycheeoside B, and pterodontriol-D-6-O-β-D-glucopyranoside are major sesquiterpenes present in lychee fruit seeds (Upadhyaya and Upadhyaya 2017). They have attracted considerable attention due to their role in biological systems and their usefulness for human use (Buckle 2015). These compounds are known for their soothing properties and can also aid the immune system to defend us from harmful bacteria, function as antioxidants, and help regenerate cells (Chadwick et al. 2013).

Triterpenes are precursors of sterols in both plants and animals. Active triterpene biosynthesis occurs when sterol formation has been achieved during exponential cell development by relaxing the control mechanism of sterol precursor production (Flores-Sánchez et al. 2002). Triterpenes are natural components of human diets derived from vegetable oils, cereals, and fruits (Siddique and Saleem 2011). The lychee fruit's pomace, peel, and seed contain triterpenes such as lupeol, β-sitosterol, botulin, betulinic acid, and lup-12 20-diene-3-diols, stigmasterol, lycheeol-A, lycheeol-B, and 3-oxotrirucalla-7,24-diene-21-oic acids. (Upadhyaya and Upadhyaya 2017).

15.3 PROCESSING OF LYCHEE WASTE AND BY-PRODUCTS

The perishable nature and limited shelf-life of lychees leads to the immediate harvesting, processing, and selling of lychees. The peel and seeds are the waste materials produced during the pulping process and preparing the different products from the pulp. Apart from peel and seed, pomace is the significant waste produced during juice processing and preparation of the different products from the juice. Therefore, the above discussion shows the huge waste, in the form of peel, seed, and pomace, generated during processing of lychee pulp and juice and the manufacture of their products. Lychee processing waste is a rich source of energy, protein, minerals, and vitamins, as discussed earlier; therefore, it has excellent potential as feed for livestock, poultry, and fish. In addition to macronutrients, lychee waste is an excellent source of bioactive compounds which is in great demand in pharmaceutical, food processing, and other allied industries due to its anticancer, antimicrobial, antioxidative, and immuno-modulatory effects (Zhu et al. 2019; Yang et al. 2011; Kanlayavattanakul et al. 2012). Pomace, peel, and seeds are also good sources of dietary fiber that can be used as supplements or functional ingredients in the production of processed food products with health-beneficial effects. Therefore, fruit processing residues can be utilized to produce peel flour, chutney, bioethanol, wine, biochar, biogas, microbiological growth media, and colorant agents (Sposito et al. 2007; Rojo-Gutiérrez et al. 2020).

15.3.1 Peel Flour

Lychee peel flour is produced by the separation of peel from seed and pulp. The collected peels are dried at 45°C and converted to a powder form. Lychee peel flour has a moderate concentration of phenolic acid, flavonoids, anthocyanins, and ascorbic acid. The flour made from lychee peel can inhibit lipase, which is an important enzyme for converting dietary triacylglycerols into monoacylglycerol and free fatty acids for facilitating absorption by the enterocyte. Lychee peel flour contains 46% soluble and 54% insoluble fiber. The insoluble fiber can increase satiety and intestinal transit, whereas the soluble fiber helps to absorb cholesterol (Sposito et al. 2007).

15.3.2 Chutney

Lychee pomace remaining after juice extraction is used for the preparation of chutney. The pomace is heated in a steel vessel over a low fire with an appropriate amount of sugar, spices, and vinegar. During heating, the mixture is continuously stirred until a thick consistency is obtained. The chutney is finally poured into the sterilized bottles and held in a cool place (Chakraborty, Chaurasiya, and Saha 2010).

15.3.3 Wine

The wine prepared from lychees is light yellow, acidic, rose-enhanced, tannic, and with a low ethanol concentration of about 6.5% (Swami, Thakor, and Divate 2014). Lychee juice delivers a delicacy for the conversion into wine due to its high nutritional and non-nutritional value. Traditionally, lychee wine is prepared from lychee pulp from which juice has been extracted and further converted to wine by fermentation using yeast (*Saccharomyces cerevisiae*) at low temperature. The wine prepared from lychee pulp is pale yellow or golden. Lychee peel and seeds are treated as waste or can be used as by-products. Lychee wine with high dietary benefits is produced by aging and utilizing wine yeast (*Saccharomyces cerevisiae* var. *bayanus*). Lychee and its parts can be utilized in different aspects such as medicinal and aromatic properties due to its short-term availability and problems with storage conditions. Wine can also be prepared by the partial osmotic dehydration technique by which the nutrients and flavor become concentrated, giving a unique body to the wine. The end product results in a low-alcohol, high-flavor beverage. Wine from reconstituted lychee juice results in 11.6% (v/v) ethanol, 0.78% (v/v) titratable acidity, 92 mg/L esters, and 124 mg/L total aldehydes. The light amber-colored wine has an attractive aroma. Another wine preparation method involves lixiviation and fermentation of lychee fruit. The conversion of whole fruit, i.e., pulp, shell and seeds (by-products from the pulper), is achieved to obtain a red-colored lychee wine.

The method involves the cleaning and washing of mature, sound fruits followed by draining. The drained fruits are further passed through mechanical crushers, which fragment the fruits into peel, seeds, and pulp. Crushing is followed by the addition of food-grade sulfurous acid at 40–90 ppm. The acidity (5.0 g/L) and pH (3.33.8) are regulated with tartrate. The yeast culture (*S. cerevisiae*) is added at 100–200 ppm, followed by the addition of liquid polygalacturonase at 40–60 ppm. The alcoholic fermentation process is done at the low temperature of 11–16°C for 10–14 days with the addition of shell (for color) and stone (for bitterness) in lychee juice extracted from the pulp. The natural color is extracted from the lychee peel during the early stages of fermentation.

During fermentation, karusen (froth on top of beer during fermentation) is formed that carries the natural red color of the fruit. The circulation of karusen is regulated regularly to strengthen the lixiviate natural color of the lychee exocarp. After fermentation, the karusen is removed by screen filtration, and fermented lychee juice is collected at the bottom of the fermenter. Clarification of fermented juice is done with the addition of bentonite followed by filtration using the diatomic filter. After partial filtration, the filtered lychee wine is again purified by the membrane filtration process using 0.45 μm pore size filters. The final finished product is then filled in bottles under sterile conditions and aged. This technique produces red-colored lychee wine that contains 12±1% (v/v) alcohol, 6.5 g/L acid (as tartrate), and 20 g/L total reducing sugars (Figure 15.1).

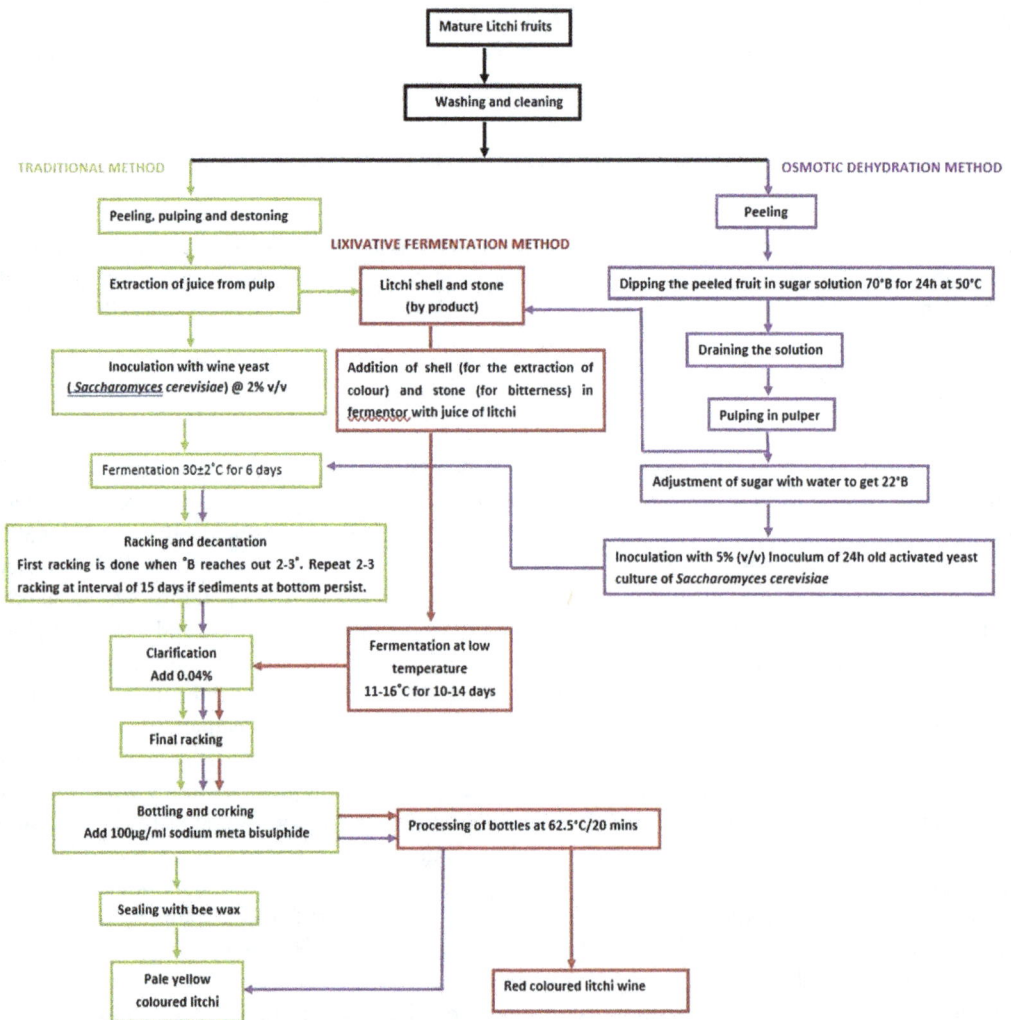

Figure 15.1 Wine production from lychee by-products.

15.3.4 Enzymes

Polyphenol oxidase and peroxidase are the most common enzymes which can be isolated from lychees. These two enzymes can be extracted and purified by fractionating phenyl Sepharose, pyrogallol, catechol, 4-methyl catechol, Sephadex G-100, Q-Sepharose, catechol, and ammonium sulfate (Jiang 2001). After harvesting, the polyphenol oxidase is purified by sequential precipitation in 80% saturated ammonium sulfate in extracts of ripe lychee fruits. Polyphenol oxidase and peroxidase can be utilized for modification of proteins. These enzymes are temperature stable in comparison to others. The processes of browning and staining related to these two enzymes (peroxidase and polyphenol oxidase), say, in sliced banana, are majorly affected by pH and temperature. The activity of these enzymes at 65°C is better than at 85°C for 9 min. Activity of both enzymes was largely lost at 85°C (Jiang 2001), which were utterly inactive at 90°C. These enzymes are pH sensitive and have the greatest activity at pH 6.5 and 7.0.

15.3.5 Lychee-Based Microbiological Media

Lychee-based microbiological media, produced from lychee peel, can be utilized for the production of cellulase from the fungus *Trichoderma reesei* (Ben-Othman, Jõudu, and Bhat 2020). The composition of different lychee-based microbiological growth media is shown in Table 15.2. The microbiological media used in industry, i.e., cetrimide agar, nutrient agar, and MacConkey agar, are very costly. Microorganisms are cultured using appropriate culture media with an ideal environment. Different microorganisms are cultured using different substrates and media. Lychee peel contains basic and complex sugars processed by microorganisms and has attracted much attention for use in animal feed, bio-ethanol, and biogas production (Ben-Othman, Jõudu, and Bhat 2020).

15.3.6 Antioxidant Agents

The therapeutic effect of lychee peel is due to the presence of polyphenolic compounds responsible for enhancing antioxidant activity (Queiroz et al. 2015). The different polyphenolic compounds in lychee peel, such as epicatechin, procyanidin A2, and proanthocyanidin are present at concentrations of 1.7 mg/g, 0.7 mg/g, and 0.4 mg/g, respectively. The whole lychee fresh fruit contains many flavonoids, i.e., approximately 15% by dry weight. The radical scavenging activity of epicatechin is higher than those of procyanidin B4 and procyanidin B2. This scavenging activity showed

Table 15.2 Composition of Lychee-Based Microbiological Growth Medium

Compound	Amount
Lychee strip bed splashed with urea	0.3 g/L
$(NH_4)_2SO4$	1.4 g/L
KH_2PO_4	2.0 g/L
$MgSO_47H_2O$	0.3 g/L
Peptone	1 g/L
Tween 80	0.2 g/L
$FeSO_47H_2O$	0.005 g/L
$MnSO_47H_2O$	0.0016 g/L
$ZnSO_47H_2O$	0.0014 g/L
$CaCl_22H_2O$	0.4 g/L
$CoCl_26H_2O$	0.02 g/L

Source: Ben-Othman, Jõudu, and Bhat 2020

a repressive effect on human breast cancer cell lines. In addition, the lychee fruit pulp contains the volatiles isobutyl acetate, *cis*-Rose oxide, 2-geraniol, isovaleric acid, guaiacol, vanillin, 2-ace-tyl-2-thiazosine, and 2-phenyl ethanol (Sarni-Manchado et al. 2000; Ong and Acree 1998). The different extraction methods, such as conventional extraction, ultrasonic-assisted extraction, and high-pressure extraction, can be used to extract polyphenolic compounds from lychee peel (Prasad 2012; Zhang et al. 2017). High-pressure extraction is the most efficient tool for extraction yield and retention of components (Zhang et al. 2017). The phenolic constituents from lychee peels can be increased by enzyme-assisted extraction (Kessy et al. 2018).

As discussed earlier, the lychee antioxidant properties are due to the high concentrations of polyphenols in the fruits. However, the activity of these compounds can be reduced during fruit storage, although encapsulation of these compounds can help to resolve this problem. The differ-ent encapsulation methods for polyphenolic compounds are described in Table 15.3. Among them, spray drying is the most efficient technique used for the encapsulation of polyphenols. Poly(D, L-lactide-co-glycolide) (PLGA), chitosan, threonine-based peptide, chitosan beta-cyclodextrin xan-than, zein, liposomes (egg phosphatidylcholine and cholesterol), polycaprolactone (PCL), chitosan-tripolyphosphate (CS-TPP), gelatine, bovine β-lactoglobulin (β-lg), kafirin, niosomes (Span 60 and CH in organic solvent), maltodextrin and gum arabic can each be used as a matrix for encapsulation of different polyphenols (Nayak and Rastogi 2010). The encapsulation technique and coating mate-rial are considered to be the most crucial components of this process, so these are selected based on the final product to be encapsulated. The processing conditions, such as temperature, air, and metal ions, affect the activity of the final product.

For specific applications, the particle properties, such as composition, density, size, degrada-tion, and the release mechanism and kinetics, in the final product can be varied by changing the processing parameters. Other factors, such as the cost of the process and optimum concentration of the active core, should also be considered. Oxidation and thermal degradation are reduced by this process, leading to an increase in the shelf-life of the encapsulated bioactive components. Moreover, this method likewise appears to mask an undesirable taste, smell, or flavor, control the delivery, despite there being no changes in the actual properties of the underlying compounds, and improve the bioavailability of the polyphenolic compound. Encapsulation has solved the biggest problem of changing the most remarkable polyphenols into usable mixtures (Munin and Edwards-Lévy 2011).

15.3.7 Colorant Agent

A number of anthocyanin compounds are developed during the ripening of lychee fruit, respon-sible for providing the darker (light green to yellow-green to red to dark red) color to the lychee peel. The amount of anthocyanin accumulation determines the intensity of the red color of the lychee peel. The variation in the anthocyanin content depends on many factors, such as lychee variety, environmental conditions, and physiological development stage of the fruit. A cyanidin and pelargo-nidin mixture is responsible for the red color in lychee peel during the first stage of ripening (Prasad, and Jha 1978), when the lychee fruit contains malvinidin-3-acetylglucoside, cyanidin-3-rutinoside

Table 15.3 Microencapsulation Methods for Polyphenols

Physical Methods	Physiochemical Methods	Chemical Methods
1. Spray-drying	1. Spray-cooling	1. Interfacial polycondensation
2. Fluid bed coating	2. Hot melt coating	2. *In-situ* polymerization
3. Extrusion-spheronization	3. Ionic gelation	3. Interfacial polymerization
4. Centrifugal extrusion	4. Solvent evaporation- extraction	4. Interfacial cross-linking
	5. Simple or complex coacervation	

(>75%), cyanidin-3-glucoside (<17%), and malvinidin-3-acetylglucoside (<9%) (Loapez, Ordorica-Falomir, and Ebeling 1999). The majority of anthocyanins (94.3% of total) come from lychee peel. The anthocyanins can be extracted from lychee peel with 0.5 M HCl and must be purified before analysis (Zhang et al. 2004). The ultra-high pressure-assisted extraction process can increase the extraction yield and bioactive activity of procyanidins from lychee peels (Zhang et al. 2017). The four anthocyanins, i.e., cyanidin glucoside, cyanidin-3-rutinoside, quercetin-3-rutinoside, and quercetin glucoside, are present in lychee peel. However, cyanidin-3-rutinoside (724.6 to 7,706.9 mg/kg) is the major anthocyanin in the lychee peel (Lee, and Wicker 1991; Sarni-Manchado et al. 2000). During storage, the browning index of lychee peel increased while anthocyanin concentrations decreased. Catechol and the anthocyanidin-degrading product have a similar structure that may act as a substrate for polyphenol oxidase; subsequently, the presence of the substrate can accelerate the process of enzymatic browning. The enzyme anthocyanase can catalyze the hydrolysis of anthocyanin and form the aglycone anthocyanidin, which can be extracted from lychee peel (Underhill and Critchley 1994). The microencapsulation technique can reduce the degradation of anthocyanins and the microencapsulated product can be further used to fortify these compounds and provide humans with the associated health benefits.

15.4 UTILIZATION OF WASTES AND BY-PRODUCTS

The valuable product or functional component produced from lychee waste or by-products can be further used in different food formulations for either improving their nutritional value or their enhancing functional properties such as antioxidants or color. The lychee peel is also a good source of different polyphenolic compounds that have good antioxidant potential. These compounds can further be used to reduce the rancidity reaction in fats/oils and high-fat food products. As discussed earlier, lychee peel flour is a very good source of soluble (46%) and insoluble fiber (54%), which help to absorb cholesterol, increase satiety, and accelerate intestinal transit (Sposito et al. 2007). Therefore, lychee peel flour can be used in fiber-enriched bread and biscuits as a dietary fiber source. In addition, the lychee peel also contains a significant concentration of anthocyanin that can be isolated and incorporated into white bread for enhancing the antioxidant activity and color of the bread. Lychee by-products can be further utilized for different non-food applications, such as the production of bioethanol, biochar, biogas, and dye for natural fabrics and to remove metal corrosion.

15.4.1 Bioethanol

Bioethanol can be made from unrefined materials like sugarcane bagasse, corn straw, wheat straw grain, oats, rice straw, sorghum, lychee, etc., depending upon their availability and regional conditions like climate, storage space, and economic viability of transportation. *Saccharomyces cerevisiae* is a yeast used mainly in ethanol production because of its high ethanol productivity, tolerance of high ethanol levels, activity over a vast range of pH levels, and the ability to develop various sugars. Bioethanol production is a three-stage process (Rojo-Gutiérrez et al. 2020):

a. Hydrolysis, by which starch from biomass feedstocks, like cereal grains, or lignocellulose, or large-scale macroalgae, are converted into fermentable monosaccharide sugars, at which point:
b. Fermentation, where these monosaccharide sugars are metabolized into alcohol by the action of a microorganism in a closed-oxygen or anaerobic climate, and eventually:
c. Distillation, a thermochemical isolating cycle whereby ethanol is concentrated to 95%.

The ethanol production from lignocellulosic biomass has probably the greatest potential because these sources are in great abundance around the world, so that the cost of obtaining the material is relatively modest. During lychee juice extraction, about 15% of the organic fruit products are disposed of as waste. A few significant lychee peel attributes, such as the high cellulose content and the largely organic crop production, make it an ideal substrate for bioethanol production. The cellulose should be promptly accessible to cellulase enzymes to change the biomass into ethanol. Subsequently, pre-treatment with dilute acids and bases is needed to eliminate the lignin. This treatment will expand the surface zone for enzymatic activity and frees the hemicellulose sugars, typically arabinose and xylose (Rojo-Gutiérrez et al. 2020). Therefore, the design process permits the yeast to transform the glucose into ethanol during aging.

15.4.2 Biochar

Lychee peel is used as a substrate for the production of biochar. Biochar is a carbon-rich material produced by pyrolysis at high temperature under anaerobic conditions, which leads to thermochemical decay of the natural feedstock. The various types of food waste can be used to produce biochar, affecting its yield and physicochemical properties. For biochar production, hydrothermal carbonisation (HTC) is carried out at 180°C for 12 hours. HTC is a coalification measure that changes crude biomass into a coal-like item, biochar. During HTC, the biomass goes through dehydration, decarboxylation, and the decarbonylation process. The biochar can be utilized for various applications such as fuel, to decontaminate harmful metals from polluted water bodies, to improve soil fertility, to act as an adsorbent, for hydrogen stockpiling, or electrochemical energy (Ben-Othman, Jõudu, and Bhat 2020).

15.4.3 Biogas

The waste produced during lychee processing, such as peel, pomace, and seeds, is a substrate for biogas production. Biogas is generated by the anaerobic digestion process through the biodegradation of organic matter. The main components of biogas are methane, carbon dioxide, hydrogen, nitrogen, and hydrogen sulfide. The biogas produced is further utilized as a renewable energy source, which is very efficient and reasonable in cost, to improve energy security and reduce dependency on less-sustainable sources (Prasad 2012). Biogas production is done by the batch system using digesters by interfacing with a chamber to achieve gas assortment, gas estimating, and a feed gulf. It is fixed by utilizing an elastic plug with a line to separate biogas. Mixing is done manually twice a day in each digester for 45 days. The temperature for mesophilic digestion is maintained at 37°C (Unpaprom, Saetang, and Tipnee 2019).

15.4.4 Dye for Natural Fabrics

Lychee peel can be used to dye fabric in the textile industry as natural pigments have an extraordinary business potential (Srivastava, Priyanka, and Parmar 2018). Lychee fruit peel is collected, dried, and made into powder form. The fabrics are treated with a solution of 0.5 mL mellow cleanser per 100 mL of water and warmed to 50°C before the dyeing process. These fabrics, after pre-treatment, are dipped into the dye solution (lychee peel powder). In addition, ferrous sulfate, aluminum potassium sulfate, and tannic acid are added to the solution and heated up to 90°C for 45 min. The fabrics are washed and dried after the whole process (Srivastava, Priyanka, and Parmar 2018).

15.4.5 Corrosion Inhibitors

Lychee peel strips and extract can be used to remove metal corrosion. The advantage of using lychee peel strips and the extract is that it is an environmentally friendly method. The corrosion inhibitory property of lychee is utilized to inhibit the rate of corrosion of the metal or composite (Singh, Gupta, and Gupta 2019). Lychee strips are dried in an oven at 60°C and crushed to a fine powder. The powder form is refluxed in twice-distilled water for four hours. The refluxed solution is used to eliminate any tainting.

15.5 CONCLUSIONS

Lychee processing extends fruit availability, creates employment, increases income, and helps to minimize usually substantial post-harvest losses. The wastes and by-products generated during lychee processing contain abundant bioactive components and exhibit significant benefits for humans. The residues can be utilized to prepare value-added products, such as peel flour, chutney, bioethanol, wine, enzymes, lychee-based microbiological media, and dyeing agents. The conversion of these wastes and by-products into value-added products helps the lychee processing industries to increase their competitiveness. Furthermore, these residues have applications in the food, pharmaceutical, cosmetic, and chemical industries.

REFERENCES

Ben-Othman, S., I. Jõudu, and R. Bhat. 2020. "Bioactives from agri-food wastes: Present insights and future challenges." *Molecule* 25(3): 1–34.

Bhoopat, L., S. Srichairatanakool, D. Kanjanapothi, T. Taesotikul, H. Thananchai, and T. Bhoopat. 2011. "Hepatoprotective effects of lychee (*Litchi chinensis Sonn*.): A combination of antioxidant and anti-apoptotic activities." *Journal of Ethnopharmacology* 136(1): 55–66.

Buckle, J. 2015. "Basic plant taxonomy, basic essential oil chemistry, extraction, biosynthesis, and analysis." In *Clinical Aromatherapy*, edited by J. Buckle, 37–72. London: Churchill Livingstone.

Cabral, T. A., L. D. M. Cardoso, and H. M. Pinheiro-Sant'Ana. 2014. "Chemical composition, vitamins and minerals of a new cultivar of lychee (*Litchi chinensis* cv. *Tailandes*) grown in Brazil." *Fruits* 69(6): 425–434.

Chadwick, M., H. Trewin, F. Gawthrop, and C. Wagstaff. 2013. "Sesquiterpenoids lactones: Benefits to plants and people." *International Journal of Molecular Sciences* 14(6): 12780–12805.

Chakraborty, I., A. K. Chaurasiya, and J. Saha. 2010. "Litchi delicacies – Few value added items par excellence." *Acta Horticulturae* 863: 637–644.

Fiedor, J., and K. Burda. 2014. "Potential role of carotenoids as antioxidants in human health and disease." *Nutrients* 6(2): 466–488.

Flores-Sánchez, I. J., J. Ortega-López, M. del C. Montes-Horcasitas, and A. C. Ramos-Valdivia. 2002. "Biosynthesis of sterols and triterpenes in cell suspension cultures of *Uncaria tomentosa*." *Plant and Cell Physiology* 43(12): 1502–1509.

Ghosh, S. P. 2001. "World trade in litchi: Past, present and future." *Acta Horticulturae* 558: 23–30.

Guo, H., H. Luo, H. Yuan, Y. Xia, P. Shu, X. Huang, and J. Zhang. 2017. "Litchi seed extracts diminish prostate cancer progression via induction of apoptosis and attenuation of EMT through Akt/GSK-3beta signaling." *Scientific Reports* 7: 41656.

Hibbert, L. 2007. "Proximate analysis, nutritive value, total phenolic content and antioxidant activity of *Litchi Chinensis sonn*." *Professional Engineering* 20(21): 20–21.

Hu, Z. Q., X. M. Huang, H. B. Chen, and H. C.Wang. 2010. "Antioxidant capacity and phenolic compounds in litchi (*Litchi chinensis Sonn*.) peel." *Acta Horticulturae* 863: 567–574.

ICAR-National Research Centre on Litchi. 2020, December 30. Cultivars of litchi. https://www.nrclitchi.org/cultivars-of-litchi.

Jiang, G., S. Lin, L. Wen, Y. Jiang, M. Zhao, F. Chen, K. N. Prasad, X. Duan, and B. Yang. 2013. "Identification of a novel phenolic compound in litchi (*Litchi chinensis Sonn*.) peel and bioactivity evaluation." *Food Chemistry* 136: 563–568.

Jiang, Y. M. 2001. "Properties of litchi polyphenol oxidase." *Acta Hortic* 558: 367–373.

Jing, Y., L. Huang, W. Lv, H. Tong, L. Song, X. Hu, and R. Yu. 2014. "Structural characterization of a novel polysaccharide from pulp tissues of *Litchi chinensis* and its immunomodulatory activity." *Journal of Agricultural and Food Chemistry* 62: 902–911.

Johnson, E. J. 2002. "The role of carotenoids in human health." *Nutrition in Clinical Care* u5(2): 56–65.

Kanlayavattanakul, M., D. Ospondpant, U. Ruktanonchai, and N. Lourith. 2012. "Biological activity assessment and phenolic compounds characterization from the fruit pericarp of *Litchi chinensis* for cosmetic applications." *Pharmaceutical Biology* 50: 1384–1390.

Kessy, H. N. E., K. Wang, L. Zhao, M. Zhou, and Z. Hu. 2018. "Enrichment and biotransformation of phenolic compounds from litchi pericarps with angiotensin I-converting enzyme (ACE) inhibition activity." *LWT—Food Science and Technology* 87: 301–309.

Kumari, N., R. Rani, and S. Nahakpam. 2018. "Changing pattern of chlorophyll content and carotenoid in different flushes of five litchi varieties." *Journal of Pharmacognosy and Phytochemistry* 7(1): 719–722.

Lal, N., S. K. Pandey, V. Nath, V. Agrawal, A. S. Gontia, and H. L. Sharma. 2018. "Total phenol and flavonoids in byproduct of Indian litchi : Difference among genotypes." *Journal of Pharmacognosy and Phytochemistry* 7(3): 2891–2894.

Lee, H. S., and L. Wicker. 1991. "Anthocyanin pigments in the skin of lychee fruit." *Journal of Food Science* 56: 466–483.

Loapez, J. R., C. Ordorica-Falomir, and P. W. Ebeling. 1999. "Changes in anthocyanin concentration in lychee (*Litchi chinensis* Sonn.) pericarp during maturation." *Food Chemistry* 65: 195–200.

Ma, Q., H. Xie, S. Li, R. Zhang, M. Zhang, and X. Wei. 2014. "Flavonoids from the pericarps of *litchi chinensis*." *Journal of Agricultural and Food Chemistry* 62(5): 1073–1078.

Maoka, T. 2020. "Carotenoids as natural functional pigments." *Journal of Natural Medicines* 74(1): 1–16.

Mitra, S. K. 2001. "Overview of lychee production in the Asia-Pacific region." https://www.fao.org/3/ac684e/ac684e04.htm (accessed on 03/06/2022).

Munin, A., and F. Edwards-Lévy. 2011. "Encapsulation of natural polyphenolic compounds: A review." *Pharmaceutics* 3: 793–829.

Nayak, C., and N. K. Rastogi. 2010. "Effect of selected additives on microencapsulation of anthocyanin by spray drying." *Drying Technology* 28(12): 1396–1404.

Ong, P. K. C., and T. E. Acree. 1998. "Gas chromatography/olfactory analysis of lychee (*Litchi chinesis* Sonn.)." *Journal of Agricultural and Food Chemistry* 46(6): 2282–2286.

Pareek, S. 2016. "Nutritional and biochemical composition of lychee (*litchi chinensis sonn.*) cultivars." In *Nutritional Composition of Fruit Cultivars*, edited by Monique S. J. Simmonds, Victor R. Preedy, 395–418. USA: Academic Press.

Prasad, R. D. 2012. "Empirical study on factors affecting biogas production." *ISRN Renewable Energy* 2012: 1–7.

Prasad, U. S., and O. P. Jha. 1978. "Changes in pigmentation patterns during litchi ripening: Flavonoid." *The Plant Biochemical Journal* 5: 44–49.

Pujari, K. H., A. P. Save, and P. P. Relekar. 2018. "Utilization of litchi (*Litchi chinensis*) fruits "Bedana" for the preparation of squash." *Acta Horticulturae* 1211: 191–197.

Queiroz, E. R., C. M. P. Abreu, K. S. Oliveira, V. O. Ramos, and R. M. Fráguas. 2015. "Bioactive phytochemicals and antioxidant activity in fresh and dried lychee fractions." *Revista Ciência Agronômica* 46: 163–169.

Ren, S., D. Xu, Y. Gao, Y. Ma, and Q. Gao. 2013. "Flavonoids from litchi (*Litchi chinensis* Sonn.) seeds and their inhibitory activities on α -glucosidase." *Chemical Research in Chinese Universities* 29(4): 682–685.

Rojo-Gutiérrez, E., J. Buenrostro-Figueroa, L. López-Martínez, D. Sepúlveda, and R. Baeza-Jiménez. 2020. "Biotechnological potential of cottonseed, a by-product of cotton production." In *Valorisation of Agro-Industrial Residues–Volume II: Non-Biological Approaches*, edited by Z. Zakaria, C. Aguilar, R. Kusumaningtyas, P. Binod, 63–82. Applied Environmental Science and Engineering for a Sustainable Future. Basel, Switzerland: Springer Nature.

Rosales, M. P., R. E. Jiménez, M. O. A. Ramos, B. M. T. Hernández, O. F. García, C. M. P. Salgado, and C. M. S. López. 2019. "Phenolic compounds in the pulp, peel and seed of litchi (*litchi chinensis sonn.*) in different stages of maturation." *Nutrition & Food Science International Journal* 8(3): 1–8.

Sarni-Manchado, P., E. L. Roux, C. L. Guerneve, Y. Lozano, and V. Cheynier. 2000. "Phenolic composition of litchi fruit pericarp." *Journal of Agricultural and Food Chemistry* 48: 5995–6002.

Siddique, H., and M. Saleem. 2011. "Beneficial health effects of lupeol triterpene: A review of preclinical studies." *Life Sciences* 88(7–8): 285–293.

Singh, A., S. K. Singh, and S. D. Pandey. 2011. *Vision 2030; Litchi scenario in India.* National Research Center for Litchi. https://nrclitchi.icar.gov.in/uploads/download/Vision-2030_NRCL.pdf

Singh, M. R., P. Gupta, and K. Gupta. 2019. "The litchi (*Litchi Chinensis*) peels extract as a potential green inhibitor in prevention of corrosion of mild steel in 0.5 M H_2SO_4 solution." *Arabian Journal of Chemistry* 12(7): 1035–1041.

Sposito, A. C., B. Caramelli, F. Fonseca, and M. C. Bertolami. 2007. "IV Diretriz Brasileira sobre Dislipidemias e Prevenção da Aterosclerose: Departamento de Aterosclerose da Sociedade Brasileira de Cardiologia." *Arq Bras Cardiol* 28: 2–19.

Srivastava, A., Priyanka, and M. S. Parmar. 2018. "Eco-friendly dyeing of natural fabrics using discarded litchi fruit peel." *Fibre2Fashion* 51(1): 58–65.

Swami, S. B., N. J. Thakor, and A. D. Divate. 2014. "Fruit wine production: A review." *Journal of Food Research and Technology* 2(3): 93–100.

Umezawa, T. 2003. "Diversity in lignan biosynthesis." *Phytochemistry Reviews* 2: 371–390.

Underhill, S., and C. Critchley. 1994. "Anthocyanin decolorisation and its role in lychee peel browning." *Australian Journal of Experimental Agriculture* 34: 115–122.

Unpaprom, Y., N. Saetang, and S. Tipnee. 2019. "Evaluation of mango, longan and lychee trees prunung leaves for the production of biogas via anaerobic fermentation." *Maejo International Journal of Energy and Environmental communication* 1(3): 20–26.

Upadhyaya, D. C., and C. P. Upadhyaya. 2017. "Bioactive compounds and medicinal importance of *Litchi chinensis*." In *The Lychee Biotechnology*, edited by M. Kumar, V. Kumar, R. Prasad, and A. Varma, 333–361. Singapore: Springer.

Varzakas, T., G. Zakynthinos, and F. Verpoort. 2016. "Plant food residues as a source of nutraceuticals and functional foods." *Foods* 5(4): 88.

Wang, H. C., Z. Q. Hu, Y. Wang, H. B. Chen, and X. M. Huang. 2011. "Phenolic compounds and the antioxidant activities in litchi pericarp: Difference among cultivars." *Scientia Horticulturae* 129: 784–789.

Wang, Z., G. Wu, B. Shu, F. Huang, L. Dong, R. Zhang, and D. Su. 2020. "Comparison of the phenolic profiles and physicochemical properties of different varieties of thermally processed canned lychee pulp." *RSC Advances* 10(12): 6743–6751.

Wen, L., D. Wu, Y. Jiang, K. N. Prasad, S. Lin, G. Jiang, J. He, M. Zhao, W. Luo, and B. Yang. 2014. "Identification of flavonoids in litchi (*Litchi chinensis Sonn.*) leaf and evaluation of anticancer activities." *Journal of Functional Foods* 6(1): 555–563.

Wen, L., J. He, D. Wu, Y. Jiang, K. N. Prasad, M. Zhao, S. Lin, G. Jiang, W. Luo, and B. Yang. 2014a. "Identification of sesquilignans in litchi (*Litchi chinensis Sonn.*) leaf and their anticancer activities." *Journal of Functional Foods* 8(1): 26–34.

Yang, B., J. Wang, M. Zhao, Y. Liu, W. Wang, and Y. Jiang. 2006. "Identification of polysaccharides from pericarp tissues of litchi (*Litchi chinensis Sonn.*) fruit in relation to their antioxidant activities." *Carbohydrate Research* 341(5): 634–638.

Yang, B., Y. Jiang, J. Shi, F. Chen, and M. Ashraf. 2011. "Extraction and pharmacological properties of bioactive compounds from longan (Dimocarpus longan Lour.) fruit — A review." *Food Research International* 44: 1837–1842.

Zeng, Q., Z. Xu, M. Dai, X. Cao, X. Xiong, S. He, and D. Su. 2019. "Effects of simulated digestion on the phenolic composition and antioxidant activity of different cultivars of lychee pericarp." *BMC Chemistry* 13(1): 27.

Zhang, R., D. Su, F. Hou, L. Liu, F. Huang, L. Dong, Y. Y. Deng, Z. Wei, and M. Zhang. 2017. "Optimized ultra-high-pressure-assisted extraction of procyanidins from lychee pericarp improves the antioxidant activity of extracts." *Bioscience, Biotechnology, and Biochemistry* 81: 1–10.

Zhang, Z., P. Xuequnb, C. Yang, Z. Ji, and Y. Jiang. 2004. "Purification and structural analysis of anthocyanins from litchi peel." *Food Chemistry* 84: 601–604.

Zhu, X. R., H. Wang, J. Sun, B. Yang, X. W. Duan, and Y. M. Jiang. 2019. "Peel and seed of litchi and longan fruits: Constituent, extraction, bioactive activity, and potential utilization." *Journal of Zhejiang University: Science B* 20(6): 503–512.

CHAPTER 16

Tomato Wastes and By-Products
Chemistry, Processing, and Utilization

Priyanka Suthar, Rafia Rashid, Shafiya Rafiq, Asmat Farooq,
Yogesh Gat, Nazmin Ansari, and Aparna Bisht

CONTENTS

16.1 INTRODUCTION

The Solanaceae or potato family consists of 98 genera and about 2,700 species (Olmstead and Bohs, 2006). *Solanum lycopersicum* L., commonly known as tomato, is from the genus *Solanum*, which is the largest genus in the family (Aflitos et al., 2014). It originated in South America and was classified by Linnaeus in 1753 (Da Silva et al., 2008). Tomato holds the second position in terms of production of vegetables worldwide with around 100 million tons every year from 144 countries (Kalogeropoulos et al., 2012). Over-production of tomatoes in the market results in a fall in the price of tomatoes, causing losses to the sellers as well as to the producers; in order to prevent these losses, the glut of tomato production is exported to other, neighbouring countries.

Tomato is a crop which is cultivated for local as well as export use. Almost five million hectares of land were under cultivation of tomatoes worldwide in 2014 which resulted in the production of 171 million tonnes of tomatoes. China is the leading producer of tomatoes, followed by India (FAOSTAT, 2017). Tomatoes are used raw or in a cooked form or as a basic ingredient in many dishes, so it is one of the most heavily consumed vegetables. Geographical zones where temperatures range between 18 to 28°C are suitable for the cultivation of tomatoes and it can be grown in open fields or in green-houses using methods like seeding, soil preparation, transplanting, pruning, chemical fertilizers and pest/disease control techniques (Da Silva et al., 2008). Breeding of tomatoes has played an important role in increasing and improving the production of the crop. One of the important reasons for breeding tomatoes is to increase the total soluble solids (TSS) of the fruit to achieve higher °Brix in the products. In addition to increasing TSS, breeding also helped to improve productivity per unit land area, quality of the fruit, and to make the crop resistant to biotic and tolerant to abiotic stresses (Grandillo et al., 1999). Three-quarters of the dry matter of tomato is soluble in nature and seeds and

Table 16.1 Nutritional Composition of Tomato

Nutritional Composition		Concentration (g/100 g)
Macronutrients	Sugar (g)	3.92
	Dietary fibre (g)	1.20
	Fat (g)	0.20
	Protein (g)	0.88
	Ash (g)	0.50
	Water (%)	94.50
Vitamins	Vitamin C, total ascorbic acid (mg)	12.7
	Thiamine (mg)	0.037
	Riboflavin (mg)	0.019
	Niacin (mg)	0.594
	Beta-carotene (μg)	449
	Alpha-carotene (μg)	101
	Lycopene (μg)	2573
	Lutein+zeaxanthin (μg)	123
Minerals	Calcium (mg)	10
	Iron (mg)	0.27
	Magnesium (mg)	11
	Phosphorus (mg)	24
	Potassium (mg)	237
	Sodium (mg)	5
	Zinc (mg)	0.17
	Copper (mg)	0.059
	Manganese (mg)	0.114

peel represent about 1–3% of the total fresh weight. The dry matter is only 5–10% of the total fresh weight, half of which is reducing sugars, whereas 10% is organic acid (with citric and malic acids being present in significant amounts). In the market, tomatoes are available in the form of juice, paste, purée, ketchup and sauce which represent 80% of the processed tomato market (Shi et al., 2008).

Reasons like improper processing result in the large-scale production of tomato waste and by-products from the tomato processing industries. The tomato waste and by-products can be categorized into peel waste, seed waste, and pomace which generally remains unutilized and which is produced from either households, industries, the food sector (ready-to-eat food, catering and restaurants), losses within the distribution chain, etc. The total statistical quantification of tomato pomace yield is difficult to collect but is estimated to be in the range of 5.4 to 9.0×10^6 tonnes per year (Lu et al., 2019). The waste generated is a good source of nutrients and minerals and can be used as food supplements (Mirabella et al., 2014). Chlorine and sulphur are present in ample amounts in this waste. The chlorine and sulphur can be helpful in detoxifying body waste by stimulating liver, while sulphur also prevents cirrhosis of the liver (Bhowmik et al., 2012). Utilization of these wastes and by-products for the production of secondary production is an emerging research area, so recent research mainly focuses on the utilization of these wastes and by-products in nutraceutical and pharmaceutical product industries as well as for the generation of energy in the form of biogas, hydrogen gas, bio-ethanol, etc. (Mirabella et al., 2014) (Table 16.1).

16.2 CHEMISTRY OF TOMATO WASTES AND BY-PRODUCTS

Tomato pomace, skin (peel) and seeds are major wastes and by-products generated by the food processing industries. The study by Previtera et al. (2016) showed that the addition of tomato peel to tomato paste resulted in higher nutritional value in terms of lutein and lycopene content compared with other marketed tomato pastes. The nutritional analysis by Reboul et al. (2005) of tomato peel-enriched tomato paste at a concentration of 6% led to higher levels of lycopene and beta carotenoids, in comparison with addition of chylomicron lipoproteins as a control, by 34.1% and 74.0%, respectively. According to the literature, the type of food products and the concentration of tomato peel and its components play a significant role in properties of prepared foods. These results were concluded from the low solubility of tomato peel which could be less compatible with food products. In this regard, tomato peel generally undergoes certain modifications before its actual application. From a nutritional point of view, tomato peel is an excellent source of lycopene, dietary fibre and phenols. The amount of potassium is reported to be the mineral present at the highest concentration, i.e., 1.1 g per 100 g while sodium was the mineral at the lowest concentration (70 mg per 100 g). This low sodium-to-potassium ratio makes tomato peel helpful for patients suffering from cardiovascular diseases (Elbadrawy and Sello, 2016). Zuorro and Lavecchia (2013) reported the lycopene content in tomato peel to 387 ± 25 mg/100 per 100 g dry weight (DW). In another study by Knoblich et al. (2005), the estimated level of lycopene in tomato peel was reported to be nearly 73.4 mg per 100 g. Total phenolics in tomato peel were determined by Navarro-González et al. (2011). The study showed 158.1 mg total phenolic content as gallic acid-equivalents per 100 g in tomato peel. However, the type of phenolic acids in the tomato peel were identified as vanillic, caffeic, gallic, catechein and protocatechuic acids, and the amino acids threonine, aspartic acid, valine, serine, methionine, glutamic acid, isoleucine, proline, leucine, glycine, phenylalanine, alanine, histidine, cystine, lysine, tyrosine and arginine, as well as ammonia (Elbadrawy and Sello, 2016). The types of flavonoids were reported by Valdez-Morales et al. (2014) to be quercetin glucoside, myricetin, rutin, quercetin-3-O-β-glucoside, kaempferol, isorhamnetin, naringenin and apigenin.

Tomato seeds are mostly discarded in the waste from tomato processing plants such as in the production of tomato sauce, tomato juice and paste. Tomato seeds as a tomato waste are an excellent source of both proteins and oils with smaller amounts of dietary fibres as well as the important functional component lycopene. Tomato seeds are rich sources of fatty acids and essential oils. Tomato seeds contain nearly 20 to 36.9 g oil per 100 g (on dry weight basis). It was predicted that tomato seeds could lead to the production of nearly 0.14 million tons of oil every year worldwide. The tomato seed oil content range was reported to be between 17.8 and 24.5 g per 100 g (Giuffrè et al., 2017; Mechmeche et al., 2017; Yilmaz et al., 2015) which is significantly higher than that of other fruit seed oils like grapes that lie in the range 4.53–11.13 g per 100 g) (Al Juhaimi et al., 2017). Oils from tomato seeds are considered to be an edible oil with a high nutritional composition. The composition of tomato seed oil includes various fatty acids, such as stearic, oleic, arachidic, palmitic and linoleic acids which constitute nearly 80% of the total oil composition. Generally, the range of linoleic acid falls between 37.6 and 72.7 g per 100 g, making it the major fatty acid in tomato-seed-based oil. Followed by linoleic acid, oleic acid is the second most prominent fatty acid with a range of 15.5 to 29.7 g per 100 g (Botineştean et al., 2015; Cantarelli et al., 1993; El-Tamimi et al., 1979; Ahmadi et al., 2014; Roy et al., 1996; Yilmaz et al., 2015). These fatty acids are beneficial in the treatment of atherosclerosis, thrombosis; dilate blood vessels and high cholesterols (Shao et al., 2015) so that tomato seeds have the potential to be used as a novel vegetable oil. However, a study published by Ma et al. (2014) reported a high level of cycloeucalenol (25.67%) which was much higher than the concentrations of both oleic and linoleic acids (16.7% and 7.85%, respectively). The differences between the various published results may be due to the different types of methods and quantifications used. In terms of amino acids in tomato seed protein, glutamic acid is the most prominent with a concentration range between 19.44 and 24.37%. After glutamic acid, aspartic acid stands second with a concentration range between 8.82 and 10.32% (Latlief and Knorr, 1983). The amino acid profile for tomato seed protein was developed by Brodowski and Geisman (1980) and stated that the concentration of amino acids is not dependent on different maturity levels (unripe to ripe) in fruit but, in contrast, a trend of decreasing protein concentration was observed as the fruits approached maturity.

16.3 PROCESSING OF TOMATO WASTES AND BY-PRODUCTS

The tomato wastes, such as seeds and peel, and by-products like pomace show major differences in their chemical composition so that it becomes necessary to separate tomato seeds and peel from the pomace waste as an initial step in recovering the important bioactive compounds and to effective use tomato pomace in value-added products (Sogi et al., 2000). Usually, the separation of seeds and peels from the tomato pomace is categorized into two types: dry and wet separation. During wet separation, water is mixed with the pomace in a container or tank. This separation of peel and seeds is based on their difference in density as the low-density peel floats on top of the water tank whereas the higher-density seeds tend to settle at the bottom. On the other hand, dry separation is carried out by drying pomace with the dried pomace being kept in a cyclone separator with an air flow in which the seeds drop downwards and settle at the bottom whereas peel is carried upward by air flow to escape from the topmost outlet at the feed rate of 6.4 m/s air velocity, representing 40 kg peel per hour. When the moisture content of dried pomace was kept at 8%, the peel separation efficiency of pomace was recorded at 68.56% and the purity of the peel and seed fractions were 82.20% and 86.11%, respectively (Shao et al., 2015). In overview, both separation techniques, i.e., wet and dry separation of peel and seeds from pomace, have their own advantages and disadvantages, hence it is difficult to conclude which is superior. In this regard, wet separation can be directly subjected to separation but may consume more water than the dry separation. However, during dry separation, the consumption of energy is high. Dry separation is recommended for producing high peel purity, whereas wet separation is suitable to achieve high seed purity. Other notable comparisons between wet and dry separation are with regard to the soluble components. The wet separation method results in greater losses of soluble components from peels and seed. Lastly, wet separation results in sewage wastes whereas dry separation leads to dust productions, both involving issues of environmental and health problems.

Lycopene is highly soluble in lipids and organic solvents, which include hexane, petroleum ether, chloroform, ethyl acetate and acetone. The total extraction yield of lycopene depends on the type of solvent and sample used, the extraction conditions and quantification methods. However, the use of organic solvents for extraction purposes is harmful for both nature and humans, therefore a number of alternative eco-friendly and safe extraction techniques have been developed in order to reduce the consumption of these solvents. In this regard, several alternative strategies have been used by researchers for extraction of lycopene from tomato, such as microwave- (Ho et al., 2015), ultrasound- (Yilmaz et al., 2017) and enzyme-assisted extraction methods (Choudhari and Ananthanarayan, 2007), as well as high press (Strati and Oreopoulou, 2014) and supercritical extraction (Rozzi et al., 2002).

The production of pectin from tomato peel is cheap and cost effective as well as being a solution to the management of waste generated from tomato processing industries. A study by Grassino et al. (2016) showed pectin extraction from tomato peel waste from two different batches of extraction with yields of 32.6% and 31.9%. However, the degree of esterification was reported to be 82% with high-methoxy pectin. The process of pectin extraction from tomato peel was developed by Grassino et al. (2016). During extraction, tomato peel is treated with ammonium oxalate and oxalic acid for 24 hours at 90°C, followed by the first filtration. The tomato peel is then re-extracted with ammonium oxalate and oxalic acid, followed by the second filtration. Precipitation of pectin was carried out by adding 96% ethanol, with subsequent filtration and washing of the precipitate with ethanol and acetone. Finally, oven drying at 40°C was carried out to yield the tomato pectin. Recently, novel extraction methods like ultrasound assisted-extraction and high hydrostatic pressure-extraction (HHPE) were developed by Grassino et al. (2020) in order to improve recovery of pectin and polyphenols from tomato peel waste. The study revealed 15% increased recovery of pectin by using HHPE after 45 min of recycling when compared with the convectional extraction method for 180 min. Both HHPE and the convectional method showed similar mass fractions of

10 g of tomato seed meal and deionised water (10:1 to 30:1)

↓

pH(7.5 -11.5) maintained during extraction at 30 to50°C in water bath

↓

Suspension was centrifuged at 2600×g for 20 min and supernatant was collect at isoelectric point of 3.9

↓

Protein precipitate was separated by centrifugation for 25 min at 2600×g followed by freeze drying

↓

The solid residue was collected and vacuum dried at 60°C

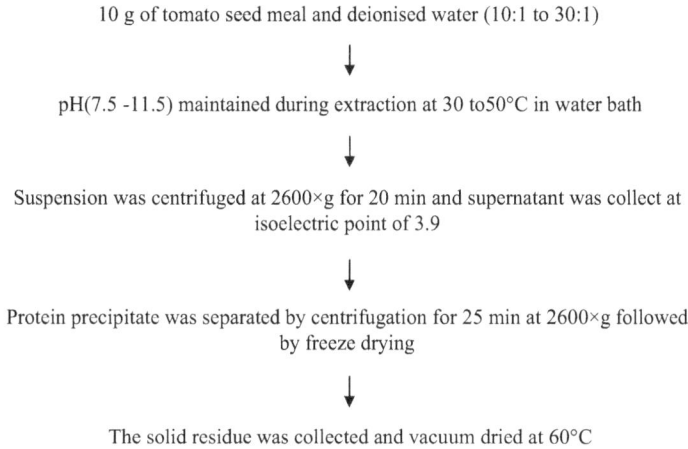

Figure 16.1 Extraction of protein from tomato seeds.

total sugars, anhydrouronic acid and total phenols. The results showed the improved efficiency of pectin and polyphenols extraction by using both HHPE and ultrasound-assisted extraction. Also, these novel methods allowed reduced time to achieve functional component extraction.

However, extraction yield plays a significant role in the commercialization of any new product; therefore, it becomes important to discuss the extraction techniques used to enhance the composition of tomato seed oils. Eller et al. (2010) reported the highest tomato seed oil yield by using hot ethanol and then hot hexane whereas supercritical carbon dioxide extraction gave the lowest tomato seed oil yield. However, the supercritical carbon dioxide treatment also reported high concentrations in the oil of individual phytosterols as well as total phytosterols. The phytosterols which were extracted in the highest concentrations were cycloartanol, sitosterol and stigmasterol. The extraction yield was highly dependent on two factors, i.e., the pressure and temperature used during tomato seed oil extraction. It was suggested that a lower temperature and higher pressure with carbon dioxide showed favourable results by enhancing the oil solubility in the solvent. The solubility of tomato seed oil was highest at 313 K temperature and 24.5 MPa pressure (Roy et al., 1996). Tomato seed oil yield was nearly 35% on a dry weight basis. The oil extracted from tomato seeds is a mixture of both saturated and unsaturated fatty acids. Among the saturated fatty acids, palmitic acid is highest (12.26%) in concentration and then stearic acid (5.15%). Under the class of unsaturated fatty acids, linoleic acids is at the highest concentration (56.12%) and then oleic acid (22.17%). These data show that tomato seed oil is a rich source of fatty acids (Fahimdanesh and Erfan Bahrami, 2013). Tomato seed protein isolation was reported by Liadakis et al. (1995) and the extraction process is represented in a flow chart (Figure 16.1). The protein in tomato seeds showed a total extraction yield of 66.1%, a total protein yield of product of 43.6% and a protein content product of 72.0%.

16.4 UTILIZATION OF TOMATO WASTES AND BY-PRODUCTS

In tomato, polysaccharides are the most important component of pomace due to their physical properties like water uptake and absorption by the prepared pomace powder. These properties can be used as a cheap as well as natural substitute for expensive food hydrocolloids for food preparations. The addition of tomato waste pomace into different food formulations also led to the question of changes to properties like colour and flavour in the final prepared products. Therefore, for food products which require large amounts of tomato paste in their production, such as tomato ketchups, this issue is of great importance. To date, no study has reported on the impact of tomato waste powder on physicochemical properties of tomato-based products.

Tomato waste powder is a key component used in tomato-based formulations like ketchup to improve or maintain the consistency and texture of the final product. According to several reports, the total dietary fibre content in the tomato pomace powder ketchup was higher than that in the ketchup prepared from fresh tomato pomace due to the processing technology in which seeds were not removed. These differences in the processing led to the higher specific surface area of the powder than that of fresh tomato pomace particles and consequently affected their bonding affinity towards polysaccharides. The pomace powder ketchup was also reported to have a significant impact on yield stress and thixotropy when compared with fresh tomato pomace ketchup packed in tubes. Also, the thermal stability of tomato pomace powder-based ketchup was greater and its use has been recommended as being useful in bakery fillings. In spite of promising applications, the entire process requires costly processing equipment as well as higher consumption of energy than for processed fresh tomato pomace (Belović et al., 2018). The tomato pomace powder can be used in various value-added products like ketchup to enhance its natural fibre content. The fresh tomato pomace can be added with other food ingredients, such as water, salt, syrup or sugar, glucose, xanthan gum, vinegar, or guar gum, followed by homogenization at 30°C and heated to 60°C, packed and pasteurized.

Tomato peel is the most important type of tomato by-product which is usually obtained during the preparation of tomato paste. However, the generation of tomato by-products in European countries was discussed by Navarro-González et al. (2011) who reported up to 200,000 tons of tomato by-product from a total of 16 million metric tons tomatoes in 2005. In Spain, about 30,000 tons of tomato solid wastes were produced from 2,300,000 tons of fresh tomatoes in 2010. Waste management in the EU is focused on the prevention of waste generation, so that recycling of tomato peel becomes very important and highly recommended. In this regard, dried and lyophilized tomato peel powders have been introduced into tomato paste at a rate of 2%. The incorporation of tomato peel powder into tomato paste does not alter the sensory characteristics of the paste in terms of taste and appearance, which is therefore comparable with the market-available tomato paste (Torbica et al., 2016). The introduction of vitamin B_{12} can be carried out by utilizing tomato pomace powder as a substrate. In this regard, the tomato pomace is subjected to hydrolysis by using the fungus *Trichoderma reesei* and followed by sugar fermentation by the bacterium *Propinobacterium shermanii* to achieve the final yield of vitamin B_{12}. According to the study report, the aeration of the culture of *Trichoderma reesei* resulted in improvement in the cellulase activity as the higher concentration of nitrogen fell rapidly due to pH inhibition. The maximum level of degraded cellulose was observed at day 14 i.e., 34.4% of total available and reducing sugar achieved was 15 g per litre. Subsequently, inoculation of *Propinobacterium shermanii* into the concentrated reducing sugar substrate led to a yield of 11.1 mg/L of vitamin B_{12} under optimized conditions. It was suggested that increasing the degree of hydrolysis of cellulose will lead to the generation of a significant amount of vitamin B_{12} generation to justify the efficient extraction. However, the purification is still not economically feasible at this point. The fermentation yield of vitamin B_{12} from tomato pomace is 50 to 55 mg/kg so it could be useful as a feedstuff for animals (Haddadin et al., 2001).

In food processing units, the by-products of tomato, such as tomato pomace, are usually utilized in their powder form due to their nutritional richness mainly in the form of lycopene, dietary fibre, phenolics, essential amino acids and unsaturated fatty acids. The pomace and its constituents have been studied in order to develop value-added products such as cookies, pasta, crackers, sausages, bread, hamburgers, tomato paste and many more. The incorporation of tomato pomace in the food may result in either desirable or undesirable effects on the developed food products. Mostly, tomato pomace is added in new products to enhance the nutrient value in terms of lycopene, protein and fibres along with antioxidant properties but sometimes leads to unfavourable textural and sensory properties of the foods. The food products based on wheat flour have most potential application with the incorporation of tomato pomace and its components. Of these, baked products are the most widely studied. Addition of tomato pomace into the wheat flour showed an increased concentration of dietary fibres, proteins and lycopene in the foods. A study by Isik and Topkaya (2016) showed that the incorporation of tomato pomace powder at 12% into crackers improved protein by 6.4%, insoluble dietary fibre by 620%, soluble fibres by 145.1%, total dietary fibre by 359%, total antioxidant activity by 152% and total phenolics by 143%. On the other hand, the authors also reported a little bitterness due to the presence of furostanol saponins which are present in tomato pomace. The textural properties of tomato pomace-supplemented flatbread were reported by Majzoobi et al. (2011), who concluded that the addition of tomato pomace led to increased moisture content and softer textural properties of the prepared flatbreads. The incorporation of tomato pomace into the flatbreads also showed increased loaf volume, although the opposite effect was reported by Nour et al. (2015), who showed a decreased loaf volume of tomato pomace-supplemented flatbread. The increase in volume was due to the strengthening of cell walls trapping air bubbles from the hydrocolloids present in the tomato pomace, whereas the decreased volume was attributed to the higher fibre content caused by the tomato pomace which led to a dilution effect on gluten protein, restricting gluten hydration by water; the study even showed disrupted interaction between starch and gluten. In order to summarize this, the addition of tomato pomace or its components into the flour-based edibles is feasible but other important effects depend on the composition of the tomato by-product and the selected food type. Rizk et al. (2014) prepared a tomato peel lycopene-based ice cream. The results showed that 1%, 4% and 5% scored lowest in terms of sensory characteristics. However, 2% and 3% showed the highest sensory scores. The effect of tomato pomace on the textural and sensory properties and the consumer acceptance of these effects were reported to be diverse. The study by Nour et al. (2015) showed that the incorporation of 10% tomato pomace did not cause different effects on the acceptability of the prepared bread. However, another study by Bhat and Hafiza (2016) argued that, whereas the addition of 5% tomato pomace did not have a significant effect on the overall acceptability of cookies, unfavourable effects were reported with the addition of 25% tomato pomace. In another study, by Altan et al. (2008), blends of barley and tomato pomace flour were studied to check the effect on the extrusion process. The tomato pomace powder was added at the 2–10% level, with the die temperature maintained between 140 and 160°C and the screw speed at 150 to 200 rpm. The samples were extruded and 20 samples were analysed in terms of texture, colour, odour, taste and overall acceptability. The study showed that the parameters and product response were significantly affected by cooking temperature, pomace level and screw speed. The preferred treatments were 2% and 10% pomace level extruded at 160°C and at 200 rpm speed in terms of colour, taste, texture and overall acceptability. Hence, the prepared barley/tomato pomace-based extruded snacks proved to be both nutritional and acceptable (Altan et al., 2008). In a recent report by Karthika et al. (2016), blends of rice flour, cornflour and tomato pomace (seed and peel separately) were used to develop a ready-to-eat snack by using a co-rotating twin-screw extruder and the effect of tomato pomace addition on final product was investigated. The developed formulation was then processed in an extruder (twin-screw) on various parameters such as water feed (14%), solid feed, processing temperature (30 to 140°C) and screw speed (300 to 350 rpm). The results, using a D-optimal mixture, suggested that 25% rice flour, 40% corn flour, 5% tomato seed and 25% tomato peel gave the greatest desirability of the extruded product. The extruded product was high in nutritious fibre. Another study, by Carlson et al. (1981), added tomato seeds (dried ground seed powder) to wheat flour with replacement levels of 5%, 10%, 15% or 20%. The effects of tomato seed powder on the amino acid profile, staling rate and loaf volume were reported. It was reported that replacement of 20% tomato seed powder into a bread formula did not change the staling rate but loaf volume was increased by 20.4%. As mentioned before, the tomato pomace is rich in phytonutrients and an excellent source of fibre, fat and protein. Utilization of tomato pomace will ensure extra income for the tomato processing industries as well as reducing the problems associated with waste disposal. Hence tomato pomace is a good choice for the development of functional products and can be added to flour in different cereal-based products (de Valle et al., 2006). Tomato pomace-based bread and muffins were recently reported by Mehta et al. (2018), and showed softer texture than the control sample; the study also showed improved shelf-life of tomato pomace-based muffins and bread of up to 12 days and 8 days, respectively.

The results of incorporation of tomato pomace into meat products and their analogues, such as ham, hamburgers and sausages, had already been published. In this regard, the added dietary fibres and antioxidants into the meat products could be developed to meet the consumer demand to lower energy density with extra protection from antioxidants. In general, tomato pomace and tomato peels are preferred over tomato seed in meat products. The tomato pomace has the advantage of its rich source of natural red colourant and flavour which makes it highly suitable for red meats (Yadav et al., 2016). The report by Savadkoohi et al. (2014) showed that addition of 7% tomato pomace reduced by 6.4% the hardness of meat-free sausages developed from soy protein isolates, starch, egg albumen and egg white powder as filler. Injection of brine solution containing 3% tomato pomace into the beef ham lowered the chewiness and hardness by 42.5% and 6.2%, respectively, in comparison with the control sample. In a study by Viuda-Martos et al. (2014), the addition of dry tomato peel (0 to 6% w/w) to either cooked or raw hamburgers was described. The authors reported that addition of lycopene from tomato peel at 4.5% showed good acceptance and retained a final lycopene content of 4.9 mg per 100 g in the hamburger (cooked). Other observations, like increased colour characteristics (a* and b*), modified textural properties and lowered pH values, were attributed to addition of dry tomato peel. Also, the hardness value of cooked samples was higher in the presence of 6% dried tomato peel than in the control batch. Owing to the nutritional profile of tomato pomace and its components, mainly proteins, they are considered to be a nutrient supplement in to the feed mix. Therefore, the tomato pomace could act as a good source of daily ration for animals. So far, the potential utilization of tomato pomace and its other components in the feed mix has been studied with sheep, cows, cattle, rabbits, goats, hens, pigs, roosters, ewes, carp, hamsters and chickens. In general, the addition of tomato pomace into the feed of animals led to significant changes in terms of increased weight gain and yield which ultimately enhanced the nutritional profile of foods of animal origin like milk, meat and eggs. The pomace generated by fresh tomato-based food processing industries could be used directly in the feed but mostly it is subjected to dehydration in order to extend the shelf-life of the pomace. It is important to note that the efficiency of feeding the tomato pomace was highly varied for each and every case and the effects were not consistent. Recently, a study by Arco-Pérez et al. (2017) noted that the addition of tomato pomace at the rate of 20% into the feed diet showed negligible effects on goat milk composition. In contrast, the study by Abbeddou et al. (2015) showed reduced sheep's milk yield and protein content when 10% and 3.2% tomato pomace were added although up to 6.6% increase in fat % was reported in response to incorporation of 29.8% dried tomato pomace into the ewe milk. The authors related these results to both the varied total energy and the fibre content. The incorporation of tomato pomace into animal feed is supported by many authors, reporting an improved unsaturated fatty acids profile in both sheep as well as goat milk (Denek and Can, 2006; Razzaghi et al., 2015). Correia et al. (2017) showed that the addition of 5% dried tomato pomace into the feed did not change the quality of pork in terms of cholesterol, fatty acids, lipid oxidation stability and colour. On the other hand, the feeding of tomato pomace showed an observable increase in lycopene content in the meat with improved α-tocopherol in both the liver and the meat. Regarding rabbits, the study by Peiretti et al. (2013) showed that the addition of tomato pomace at a rate of 6% resulted in increased consumer preference with regard to rabbit meat.

16.5 CONCLUSIONS

Tomato is the most in-demand vegetable crop which is consumed all over the world, making tomatoes the most important vegetable crop. The wastes and by-products generated from tomato plants are peel, pomace and seeds which are nutritionally rich and therefore can be utilized in other functional products. The seeds of tomatoes possess important essential oil components like linoleic, oleic, palmitic, stearic and arachidic acids. In general, tomato by-products are added to functional products to improve a number of nutritional parameters such as fibre content, polyphenol concentration or protein content. Bioactive compounds like polyphenols, ascorbic acid and lycopene have strong free radical scavenging activity. Lycopene is the most useful bioactive compound as well as being a natural colourant present in tomato. Many novel technologies have been introduced to extract and utilize these functional ingredients in various products in order to improve the extract efficiency and to save time. The presence of these bioactive substances in tomato by-products and the associated extraction technologies widen the scope of their utilization in the food and pharmaceutical industries.

REFERENCES

Abbeddou, S., Rischkowsky, B., Hilali, M. E. D., Haylani, M., Hess, H. D., & Kreuzer, M. (2015). Supplementing diets of Awassi ewes with olive cake and tomato pomace: On-farm recovery of effects on yield, composition and fatty acid profile of the milk. *Tropical Animal Health and Production*, *47*(1), 145–152.

Aflitos, S., Schijlen, E., de Jong, H., de Ridder, D., Smit, S., Finkers, R., Wang, J., Zhang, G., Li, N., et al. (2014). Exploring genetic variation in the tomato (Solanum section Lycopersicon) clade by whole-genome sequencing. *The Plant Journal*, *80*(1), 136–148.

Ahmadi, K. N., Tavakolipour, H., Hasani, M., & Amiri, M. (2014). Evaluation and analysis of the ultrasound-assisted extracted tomato seed oil. *Journal of Food Biosciences and Technology*, *4*(2), 57–66.

Al Juhaimi, F., Geçgel, Ü., Gülcü, M., Hamurcu, M., & Özcan, M. M. (2017). Bioactive properties, fatty acid composition and mineral contents of grape seed and oils. *South African Journal of Enology and Viticulture*, *38*(1), 103–108.

Altan, A., McCarthy, K. L., & Maskan, M. (2008). Evaluation of snack foods from barley–tomato pomace blends by extrusion processing. *Journal of Food Engineering*, *84*(2), 231–242.

Arco-Pérez, A., Ramos-Morales, E., Yáñez-Ruiz, D. R., Abecia, L., & Martín-García, A. I. (2017). Nutritive evaluation and milk quality of including of tomato or olive by-products silages with sunflower oil in the diet of dairy goats. *Animal Feed Science and Technology*, *232*, 57–70.

Belović, M., Torbica, A., Lijaković, I. P., Tomić, J., Lončarević, I., & Petrović, J. (2018). Tomato pomace powder as a raw material for ketchup production. *Food Bioscience*, *26*, 193–199.

Bhat, M. A., & Hafiza, A. (2016). Physico-chemical characteristics of cookies prepared with tomato pomace powder. *Journal of Food Processing and Technology*, *7*(1), 34–45.

Bhowmik, D., Kumar, K. S., Paswan, S., & Srivastava, S. (2012). Tomato-a natural medicine and its health benefits. *Journal of Pharmacognosy and Phytochemistry*, *1*(1), 33–43.

Botineştean, C., Gruia, A. T., & Jianu, I. (2015). Utilization of seeds from tomato processing wastes as raw material for oil production. *Journal of Material Cycles and Waste Management*, *17*(1), 118–124.

Brodowski, D., & Geisman, J. R. (1980). Protein content and amino acid composition of protein of seeds from tomatoes at various stages of ripeness. *Journal of Food Science*, *45*(2), 228–229.

Cantarelli, P. R., Regitano-d'Arce, M. A. B., & Palma, E. R. (1993). Physicochemical characteristics and fatty acid composition of tomato seed oils from processing wastes. *Scientia Agricola*, *50*(1), 117–120.

Carlson, B. L., Knorr, D., & Watkins, T. R. (1981). Influence of tomato seed addition on the quality of wheat flour breads. *Journal of Food Science*, *46*(4), 1029–1031.

Choudhari, S. M., & Ananthanarayan, L. (2007). Enzyme aided extraction of lycopene from tomato tissues. *Food Chemistry*, *102*(1), 77–81.

Correia, C. S., Alfaia, C. M., Madeira, M. S., Lopes, P. A., Matos, T. J. S., Cunha, L. F., … & Freire, J. P. B. (2017). Dietary inclusion of tomato pomace improves meat oxidative stability of young pigs. *Journal of Animal Physiology and Animal Nutrition*, *101*(6), 1215–1226.

da Silva, D. J. H., Abreu, F. B., Caliman, F. R. B., Antonio, A. C., & Patel, V. B. (2008). Tomatoes: Origin, cultivation techniques and germplasm resources. In *Tomatoes and tomato products-nutritional, medicinal and therapeutic properties* (pp. 3–25). Science Publishers.

Del Valle, M., Cámara, M., & Torija, M. E. (2006). Chemical characterization of tomato pomace. *Journal of the Science of Food and Agriculture*, 86(8), 1232–1236.

Denek, N., & Can, A. (2006). Feeding value of wet tomato pomace ensiled with wheat straw and wheat grain for Awassi sheep. *Small Ruminant Research*, 65(3), 260–265.

Elbadrawy, E., & Sello, A. (2016). Evaluation of nutritional value and antioxidant activity of tomato peel extracts. *Arabian Journal of Chemistry*, 9, S1010–S1018.

Eller, F. J., Moser, J. K., Kenar, J. A., & Taylor, S. L. (2010). Extraction and analysis of tomato seed oil. *Journal of the American Oil Chemists' Society*, 87(7), 755–762.

El-Tamimi, A. H., Morad, M. M., Raof, M. S., & Rady, A. H. (1979). Tomato seed oil I fatty acid composition, stability and hydrogenation of the oil. *Fette, Seifen, Anstrichmittel*, 81(7), 281–284.

Fahimdanesh, M., & Bahrami, M. E. (2013). Evaluation of physicochemical properties of Iranian tomato seed oil. *Journal of Nutrition and Food Science*, 3(206), 76–89.

FAOSTAT. (2017). Database of food and agriculture organization of the United Nations. http://www.fao.org/faostat/en/.

Giuffrè, A. M., Capocasale, M., & Zappia, C. (2017). Tomato seed oil for edible use: Cold break, hot break, and harvest year effects. *Journal of Food Processing and Preservation*, 41(6), e13309.

Grandillo, S., Zamir, D., & Tanksley, S. D. (1999). Genetic improvement of processing tomatoes: A 20 years perspective. *Euphytica*, 110(2), 85–97.

Grassino, A. N., Djaković, S., Bosiljkov, T., et al. (2020). Valorisation of tomato peel waste as a sustainable source for pectin, polyphenols and fatty acids recovery using sequential extraction. *Waste and Biomass Valorization*, 11(9), 4593–4611.

Grassino, A. N., Halambek, J., Djaković, S., Brnčić, S. R., Dent, M., & Grabarić, Z. (2016). Utilization of tomato peel waste from canning factory as a potential source for pectin production and application as tin corrosion inhibitor. *Food Hydrocolloids*, 52, 265–274.

Haddadin, M. S. Y., Abu-Reesh, I. M., Haddadin, F. A. S., & Robinson, R. K. (2001). Utilisation of tomato pomace as a substrate for the production of vitamin B12–a preliminary appraisal. *Bioresource Technology*, 78(3), 225–230.

Ho, K. K., Ferruzzi, M. G., Liceaga, A. M., & San Martín-González, M. F. (2015). Microwave-assisted extraction of lycopene in tomato peels: Effect of extraction conditions on all-trans and cis-isomer yields. *LWT-Food Science and Technology*, 62(1), 160–168.

Isik, F., & Topkaya, C. (2016). Effects of tomato pomace supplementation on chemical and nutritional properties of crackers. *Italian Journal of Food Science*, 28(3), 525–534.

Kalogeropoulos, N., Chiou, A., Pyriochou, V., Peristeraki, A., & Karathanos, V. T. (2012). Bioactive phytochemicals in industrial tomatoes and their processing byproducts. *LWT-Food Science and Technology*, 49(2), 213–216.

Karthika, D. B., Kuriakose, S. P., Krishnan, A. V. C., Choudhary, P., & Rawson, A. (2016). Utilization of by-product from tomato processing industry for the development of new product. *Journal of Food Processing and Technology*, 7, 608–616.

Knoblich, M., Anderson, B., & Latshaw, D. (2005). Analyses of tomato peel and seed byproducts and their use as a source of carotenoids. *Journal of the Science of Food and Agriculture*, 85(7), 1166–1170.

Latlief, S. J., & Knorr, D. (1983). Tomato seed protein concentrates: Effects of methods of recovery upon yield and compositional characteristics. *Journal of Food Science*, 48(6), 1583–1586.

Liadakis, G. N., Tzia, C., Oreopoulou, V., & Thomopoulos, C. D. (1995). Protein isolation from tomato seed meal, extraction optimization. *Journal of Food Science*, 60(3), 477–482.

Lu, Z., Wang, J., Gao, R., Ye, F., & Zhao, G. (2019). Sustainable valorisation of tomato pomace: A comprehensive review. *Trends in Food Science & Technology*, 86, 172–187.

Ma, Y., Ma, J., Yang, T., Cheng, W., Lu, Y., Cao, Y., & Feng, S. (2014). Components, antioxidant and antibacterial activity of tomato seed oil. *Food Science and Technology Research*, 20(1), 1–6.

Majzoobi, M., Ghavi, F. S., Farahnaky, A., Jamalian, J., & Mesbahi, G. (2011). Effect of tomato pomace powder on the physicochemical properties of flat bread (Barbari bread). *Journal of Food Processing and Preservation*, 35(2), 247–256.

Mechmeche, M., Kachouri, F., Chouabi, M., Ksontini, H., Setti, K., & Hamdi, M. (2017). Optimization of extraction parameters of protein isolate from tomato seed using response surface methodology. *Food Analytical Methods*, 10(3), 809–819.

Mehta, D., Prasad, P., Sangwan, R. S., & Yadav, S. K. (2018). Tomato processing byproduct valorization in bread and muffin: Improvement in physicochemical properties and shelf life stability. *Journal of Food Science and Technology*, 55(7), 2560–2568.

Mirabella, N., Castellani, V., & Sala, S. (2014). Current options for the valorization of food manufacturing waste: A review. *Journal of Cleaner Production*, 65, 28–41.

Navarro-González, I., García-Valverde, V., García-Alonso, J., & Periago, M. J. (2011). Chemical profile, functional and antioxidant properties of tomato peel fiber. *Food Research International*, 44(5), 1528–1535.

Nour, V., Ionica, M. E., & Trandafir, I. (2015). Bread enriched in lycopene and other bioactive compounds by addition of dry tomato waste. *Journal of Food Science and Technology*, 52(12), 8260–8267.

Olmstead, R. G., & Bohs, L. (2006, July). A summary of molecular systematic research in Solanaceae: 1982–2006. In *VI International Solanaceae Conference: Genomics Meets Biodiversity 745* (pp. 255–268).

Peiretti, P. G., Gai, F., Rotolo, L., Brugiapaglia, A., & Gasco, L. (2013). Effects of tomato pomace supplementation on carcass characteristics and meat quality of fattening rabbits. *Meat Science*, 95(2), 345–351.

Previtera, L., Fucci, G., De Marco, A., Romanucci, V., Di Fabio, G., & Zarrelli, A. (2016). Chemical and organoleptic characteristics of tomato purée enriched with lyophilized tomato pomace. *Journal of the Science of Food and Agriculture*, 96(6), 1953–1958.

Razzaghi, A., Naserian, A. A., Valizadeh, R., Ebrahimi, S. H., Khorrami, B., Malekkhahi, M., & Khiaosa-ard, R. (2015). Pomegranate seed pulp, pistachio hulls, and tomato pomace as replacement of wheat bran increased milk conjugated linoleic acid concentrations without adverse effects on ruminal fermentation and performance of Saanen dairy goats. *Animal Feed Science and Technology*, 210, 46–55.

Reboul, E., Borel, P., Mikail, C., Abou, L., Charbonnier, M., Caris-Veyrat, C., Goupy, P., Portugal, H., Lairon, D., & Amiot, M. J. (2005). Enrichment of tomato paste with 6% tomato peel increases lycopene and b-carotene bioavailability in men. *Journal of Nutrition*, 135, 790–794.

Rizk, E. M., El-Kady, A. T., & El-Bialy, A. R. (2014). Characterization of carotenoids (lyco-red) extracted from tomato peels and its uses as natural colorants and antioxidants of ice cream. *Annals of Agricultural Sciences*, 59(1), 53–61.

Roy, B. C., Goto, M., & Hirose, T. (1996). Temperature and pressure effects on supercritical CO_2 extraction of tomato seed oil. *International Journal of Food Science & Technology*, 31(2), 137–141.

Rozzi, N. L., Singh, R. K., Vierling, R. A., & Watkins, B. A. (2002). Supercritical fluid extraction of lycopene from tomato processing byproducts. *Journal of Agricultural and Food Chemistry*, 50(9), 2638–2643.

Savadkoohi, S., Hoogenkamp, H., Shamsi, K., & Farahnaky, A. (2014). Color, sensory and textural attributes of beef frankfurter, beef ham and meat-free sausage containing tomato pomace. *Meat Science*, 97(4), 410–418.

Shao, D., Venkitasamy, C., Li, X., Pan, Z., Shi, J., Wang, B., et al. (2015). Thermal and storage characteristics of tomato seed oil. *LWT-Food Science and Technology*, 63(1), 191–197.

Shao, D., Venkitasamy, C., Shi, J., Li, X., Yokoyama, W., & Pan, Z. (2015). Optimization of tomato pomace separation using air aspirator system by response surface methodology. *Transactions of the ASABE*, 58(6), 1885–1894.

Shi, J., Dai, Y., Kakuda, Y., Mittal, G., & Xue, S. J. (2008). Effect of heating and exposure to light on the stability of lycopene in tomato purée. *Food Control*, 19(5), 514–520.

Sogi, D. S., Bawa, A. S., & Garg, S. K. (2000). Sedimentation system for seed separation from tomato processing waste. *Journal of Food Science and Technology*, 37(5), 539–541.

Strati, I. F., & Oreopoulou, V. (2014). Recovery of carotenoids from tomato processing by-products–A review. *Food Research International*, 65, 311–321.

Torbica, A., Belović, M., Mastilović, J., Kevrešan, Ž., Pestorić, M., Škrobot, D., & Hadnađev, T. D. (2016). Nutritional, rheological, and sensory evaluation of tomato ketchup with increased content of natural fibres made from fresh tomato pomace. *Food and Bioproducts Processing*, 98, 299–309.

Valdez-Morales, M., Espinosa-Alonso, L. G., Espinoza-Torres, L. C., Delgado-Vargas, F., & Medina-Godoy, S. (2014). Phenolic content and antioxidant and antimutagenic activities in tomato peel, seeds, and byproducts. *Journal of Agricultural and Food Chemistry*, 62(23), 5281–5289.

Viuda-Martos, M., Sanchez-Zapata, E., Sayas-Barberá, E., Sendra, E., Pérez-Álvarez, J. A., & Fernández-López, J. (2014). Tomato and tomato byproducts. Human health benefits of lycopene and its application to meat products: A review. *Critical Reviews in Food Science and Nutrition*, 54(8), 1032–1049.

Yadav, S., Malik, A., Pathera, A., Islam, R. U., & Sharma, D. (2016). Development of dietary fibre enriched chicken sausages by incorporating corn bran, dried apple pomace and dried tomato pomace. *Nutrition & Food Science*, *46*(1), 16–29.

Yilmaz, E., Aydeniz, B., Güneşer, O., & Arsunar, E. S. (2015). Sensory and physico-chemical properties of cold press-produced tomato (*Lycopersicon esculentum* L.) seed oils. *Journal of the American Oil Chemists' Society*, *92*(6), 833–842.

Yilmaz, T., Kumcuoglu, S., & Tavman, S. (2017). Ultrasound-assisted extraction of lycopene and β-carotene from tomato-processing wastes. *Italian Journal of Food Science*, *29*(1), 186–194.

Zuorro, A., & Lavecchia, R. (2013). Optimization of enzyme-assisted lycopene extraction from tomato processing waste. *Advanced Materials Research*, *800*, 173–176.

CHAPTER 17

Mango Wastes and By-Products
Chemistry, Processing, and Utilization

Seherpreet Kaur Flora, Falak Grover, Vidisha Tomer, and Ashwani Kumar

CONTENTS

17.1 INTRODUCTION

Mangifera indica (mango) is a member of the family Anacardiaceae, consisting of about 30 species of tropical fruiting trees which are mainly grown in tropical regions. Mango trees attain a height of 10–40 m, with a crown radius of approximately 10 m, are perennials with a symmetrical, rounded canopy with an umbrella-like form (Jahurul et al., 2015). It takes about three months for

DOI: 10.1201/9781003164463-17

the fruit to ripen fully. Once ripened, the fruit has a characteristic, distinctively resiny sweet smell (Parvez, 2016). India produces over 90 million tonnes of various fruits. In 2019, the global yield of mango was 56 million tons, of which India is the largest producer with 40.48% of total world production (Figure 17.1 a). Its uniqueness makes it the national fruit of India and the Philippines and it is the national tree of Bangladesh (Parvez, 2016). Amongst tropical fruits, mango is the second highest (Figure 17.1 b). Some varieties of mango of commercial importance include 'Totapuri', 'Alphonso', 'Dasheri', 'Kesar', 'Banganpalli', 'Langra' and 'Chausa' (Yadav and Singh, 2017). The popularity of *M. indica* is not only because of its unusual flavor but also due to its good nutritional value. The composition of the fruit changes from cultivar to cultivar (Asif et al., 2016). According to the nutrient report given by the United States Department of Agriculture (USDA), the ripened mango pulp contains 83.4 g/100 g of water for fresh fruit. Mature fruits of mango are rich in carbohydrates such as starch, pectin and sugars like glucose, fructose and sucrose (Vergara-Valencia et al., 2007). Mango pulp contains lipids in only minute quantities, although seeds are considered to be a good source of fatty acids (Jahurul et al., 2015). On dry basis, mango contains 0.34–0.52 g ash, 0.3–0.53 g lipid, 0.36–0.4 g protein, 16.2–17.18 g carbohydrate and 0.85–1.06 g dietary fibre per 100 fruit on dry basis. The fruit contains vitamins as well. Hundered gram of edible portion of mango fruit contains 13.2–92.8 mg ascorbic acid, 0.16–0.24 mg pantothenic acid, and 0.79–1.02 mg vitamin E (Maldonado-Celis et al., 2019). In nations lying in the regions of Southeast Asia and Africa,

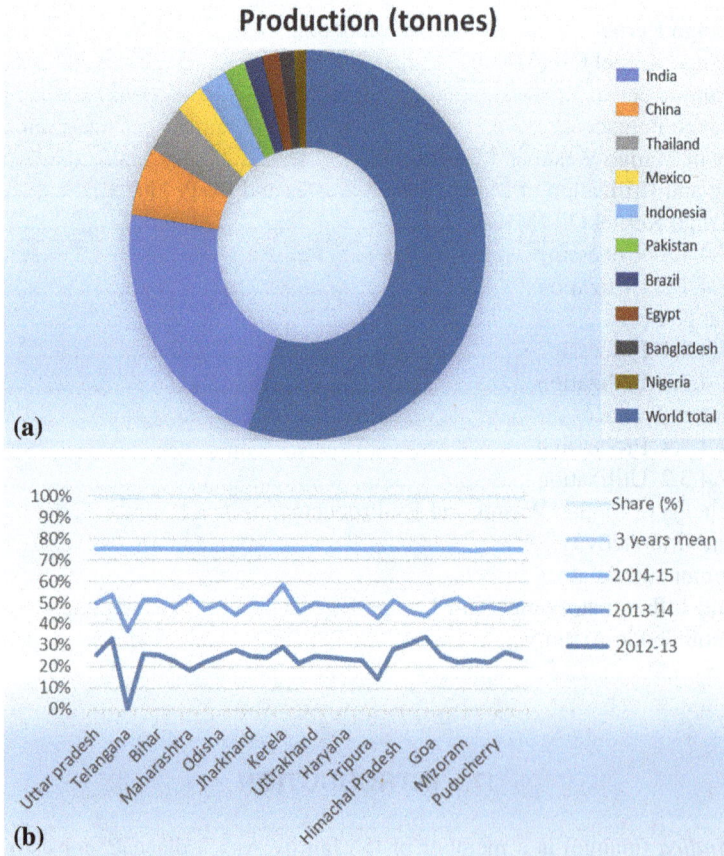

(a)

(b)

Figure 17.1   Production shares of mango as (a) global mango production (metric tonnes) (Parvez, 2016) and (b) statewise production of mango fruit in India (per thousand tonnes) for 2012–13 to 2014–15.

mango and its constituents have been used for ailments under the traditional medicinal therapies since ancient times. In modern times, ripe mango has found applications in the production of various processed products like frozen, canned, dehydrated mango products and beverages (Ediriweera et al., 2017). Other components of the fruit, like the seed and peel, make up a significant proportion of the fruit but are not directly consumed and are therefore discarded. Waste generated from mango processing is generally composed of peel and seeds (stones). Seeds occupy nearly half of the fruit mass. This chapter covers different by-products of mango and their possible utilization.

17.2 MANGO WASTES AND BY-PRODUCTS

A whole fruit or part of a fruit which, although fit to eat, is instead discarded is in general termed waste and the waste which is generated as a result of processing is known as a by-product (Sagar et al., 2018). Unprocessed or processed mango generates peel and seed (stone) as the major waste (Figure 17.2). It has been observed that the seed constitutes nearly 20–60% of the whole fresh fruit weight, whereas the kernel present within the seed accounts for approximately 45–75% of the seed's mass. Mango peel constitutes 7–24% of the total weight of the fruit (Kittiphoom, 2012; Fowomola, 2010).

17.2.1 Mango Kernel

The mango fruit has one oblong seed that differs in size from cultivar to cultivar and has a hairy outer surface. Seed weight varies from 7–24% of fruit weight among different mango cultivars. The seed coat is 1–2 mm thick and is 4–7 cm in length and 3–4 cm in width. The seed coat covers the soft embryo/kernel. In the seed, there is the seed nucleus, which accounts for 45–75% of the entire seed. The mango kernel mainly consists of carbohydrate (32–77%), largely in the form of starch and lipids (6–15%). It is also rich in protein (6–10%) and crude fiber (0.26–4.6%). Moderate mineral content is indicated by an ash content of 1.46% to 3.71% on a dry weight basis. Mango seed kernels contain low protein. The major amino acids are leucine, valine, and lysine. In addition to the nutrients, mango seed kernels also contain a high concentration of phytochemicals like polyphenols

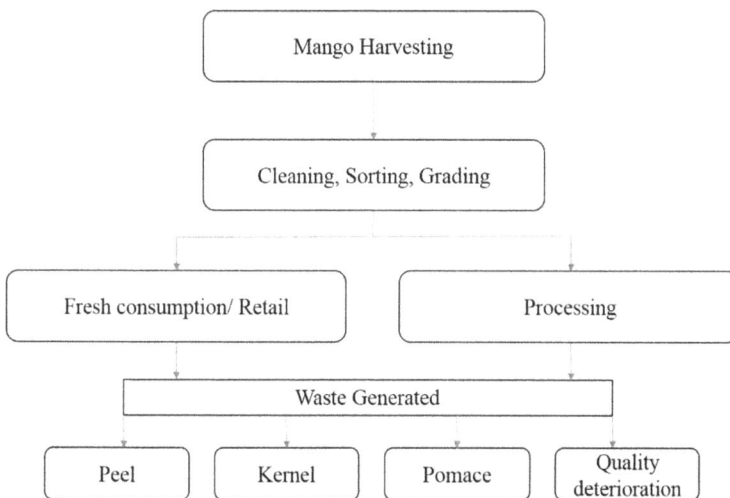

Figure 17.2 Mango wastes generated from mango processing (Source: Mwaurah et al., 2020).

(gallotannins, flavonols, benzophenone derivatives, mangiferin, homomangiferin, isomangiferin, anthocyanins, kaempferol and quercetin) and phytosterols such as campesterol, and tocopherols. Apart from macronutrients and vitamins, the mango kernel is also rich in minerals. It consists of calcium (170 mg), magnesium (210 mg), and potassium (368 mg) per 100 g of kernel on a dry weight basis (Yatnatti et al., 2014). However, the composition as determined by Nzikou et al. (2010) is much different. The author noted less potassium (158 mg), phosphorus (20 mg), magnesium (22.34 mg,) calcium (10.21 mg) and 2.7 mg sodium per 100 g mango kernel on a dry weight basis. Mango kernel is rich in fat-soluble vitamins and their precursors like beta-carotene (15.27 IU), tocopherol (1.30 mg/100 g dry basis, phylloquinone (0.59 mg/100 g dB) and ascorbic acid (0.56 mg/100 g dB). The kernel also contains other vitamins like thiamine (0.08 mg), riboflavin (0.03 mg), biotin (0.19 mg) and cobalamin (0.12 mg) per 100 g kernel weight on a dry weight basis (Fowomola, 2010).

17.2.2 Mango Kernel Oil (MKO)

The mango kernel contains about 8% to 15% oil which is generally termed mango kernel oil. The content of oil in the kernel is similar to that of soybean and cottonseed which contain 18–20% oil content (Nadeem et al., 2016). Sikdar et al. (2017) optimized the conditions for oil recovery from mango kernels using the method of Soxhlet extraction and noted a maximum oil yield of 17.3%. Physical properties of MKO were assessed in various studies. It is solid at room temperature and has a melting point of 32–36°C. The odour of MKO is sweet and nut-like. When in pure form, it appears to be smooth and creamy in color and, like olive oil, it has a mild fragrance. The soft-yellow-colored MKO has a stronger odour than butter, with a lower melting point of around 23–27°C, which indicates that the soft yellow oil should melt on coming into contact with skin. This property makes it highly desirable for cosmetic products. Another category, semi-solid MKO, can be produced while refining the butter produced from mango kernel (Table 17.1).

Generally, processing is necessary for commercial vegetable oils, but since the concentration of free fatty acids in MKO is much lower, it can be used without much processing. MKO contains 42–44% unsaturated and 52–56% saturated fatty acids (Soong and Barlow, 2004). Chemical characterization of the oil reveals oleic acid to be the major fatty acid in the kernel followed by palmitic and

Table 17.1 Nutritional Composition of Mango Waste Parts

Part of Mango	Carbohydrate (g/100 g)	Protein (g/100 g)	Fat (g/100 g)	Fiber (g/100 g)	References
Mango (whole)	16.20–17.18	0.36–0.40	0.30–0.53	0.85–1.06	Maldonado-Celis et al., 2019, Mohammed, Edna and Siraj, 2020
Mango pomace	78.77	10.59	1.74	6.14	Mahmoud et al., 2020
Mango peel	22.25–88.87	1.04–6.16	0.14–8.22	24.35–50.94	Lasano et al., 2019 Kaur and Srivastav, 2018 Torres-León et al., 2018 Onuh et al., 2017 Ojokoh, 2007 Kok and Mohamad, 2020
Mango kernel	43.31–76.81	2.62–26.13	2.76–27.56	24.75	Lasano et al., 2019 Awolu et al., 2018 Richa, Ranu and Anisha, 2018 Kaur and Brar, 2015 Yatnatti et al., 2014 Das et al., 2019
Mango kernel oil	69.20	7.53	11.00	2.45	Abdelaziz, 2018

stearic acids. It contains high concentrations of oleic acid (44.89%) and stearic acid (36.57%) with minor amounts of palmitic (10.06%), linoleic (6.0%) and linolenic (2.48%) acids (Abdelaziz, 2018). The most commonly reported triglycerides of MKO are 1,3-distearoyl-2-oleoyl-glycerol (SOS), which constitutes about 29.4 to 40%, 1-stearoyl-2,3-dioleoylglycerol (SOO) (approximately 14.6 to 23%), 1-palmitoyl-2-oleoyl-3-stearoylglycerol (POS) and 1,3-dipalmitoyl-2-oleoylglycerol (POP) which comprise about 5.7–14.76% and 6.89–8.9% of the triglyceride portion, respectively. The oil also possesses high antioxidant capacity which can be attributed to its high phenolic concentration (9.87 mg/g). This is much higher than other common nut oils like chia and almond (Kittiphoom and Sutasinee, 2013). A mango variety from Pakistan contained 15.5% free fatty acids. The melting point of the oil was found to be 35.2°C, whereas the iodine value and saponification values were reported to be 55.2 and 195 respectively. A low peroxide value (0.22 meqO$_2$/kg indicated good stability (Nadeem et al., 2016). The iodine value changes based on the place of origin, with a variety from Thailand having the lowest iodine value whereas the highest values were obtained from varieties from India, Bangladesh and Malaysia (Muchiri et al., 2012). MKO is used in foods as it does not require partial hydrogenation. Numerous studies have also been conducted to check the efficacy of MKO as opposed to *trans*-fat options. Another reason why MKO finds its use in the food industry is because it is extremely resistant to auto-oxidation (Kittiphoom and Sutasinee, 2013). The physicochemical properties of MKO are shown in Table 17.2.

17.2.3 Mango Peel

Mango peel contains various types of compounds which are essential not just nutritionally or on health grounds but also from an economical perspective. Being a potential source of bioactive compounds, mango peel can ameliorate various lifestyle-related disorders. Additionally, the usage of mango peel not only nurtures value-addition but also helps in the reduction of generated waste and consequent environmental pollution. Reduced waste generation has an additional benefit of the reduced cost of waste disposal. Peel comprises approximately 7–24% of the weight of the mango fruit and contains appreciable quantities of soluble dietary fiber, insoluble dietary fiber and phytochemicals. Moreover, for mango peel, the total fiber concentration accounts for 28% to 78% involving 14 to 50% of insoluble dietary fiber and 13–28% of soluble dietary fiber, neutral sugar and uronic acid. The concentration of dietary fiber present in peel is higher in mature skins. The moisture content ranges from 8.13 to 9.30%, while the ash content of mango peel ranges from 1.38–1.57%. The overall concentration of polyphenols in green peel of raw mango ranged from 90 mg to 110 mg/g peel in 80% acetone extract, whereas that in ripe peel ranged between 55 and 100 mg/g, depending on the cultivar. On extraction of mango peel constituents, the carotenoid content ranged from 74–436 g, with the carotenoids content being found to be higher in peels from ripened fruits. The concentration of ascorbic acid (vitamin C) and tocopherol (vitamin E) in mango peel lies in the ranges 188–392 and 205–509 µg/g dry peel, respectively. It was observed that the concentration of

Table 17.2 Physicochemical Properties of Mango Kernel Oils Extracted with Different Solvents

Physicochemical Property	Ethanol Extract	Hexane Extract	Petroleum Ether Extract	Methanol Extract	Water Extract
Total oil yield (g/g dry weight)	6.96	8.46	8.04	13	2.53
Acid value (mg KOH/g oil)	27.55	0.10	0.15	5.35	-
Peroxide value (mg/g oil)	26.35	8.72	8.82		-
Iodine value (mg/100 g oil)	59.17	38.50	37.25	39.5	-
Saponification value (mg KOH/g oil)	206.0	207.5	190.2	207.5	-
Total phenolic content (mg gallic acd-equivalents/g)	53.5	98.7	77.0	117	118.1

Sources: Adapted from Kittiphoom and Sutasinee (2013), Nzikou et al. (2010), Khammuang and Sarnthima (2011)

these vitamins increased as the fruits ripened. Mango peel (raw or ripe) exhibit significant enzymatic activity and contains a large number of enzyme activities like protease, peroxidase, polyphenol oxidase, xylanase and amylase activities.

Due to the high pectin content, the peel act as a great thickening, gelling and emulsifying agent which finds applications for the manufacture of marmalades, beverages, dairy and meat products (Koubala et al., 2008). It also contains significant amounts of minerals like potassium, copper, zinc, manganese, iron and selenium. In general, dietary fiber extracted from other sources possesses an important trait in that it has high oil-holding capacity. The oil-holding capacity of dietary fiber is associated with the structure of the polysaccharide and depends upon surface properties, overall charge density, thickness, and the hydrophobicity of the fiber particle. Mango peel has comparatively low oil-holding capacity which makes dietary fiber extracted from mango peel useful in fried products where its use can avoid a greasy impression. The mango strips of Indian varieties 'Badami' and 'Raspuri' are a great source of bioactive components like polyphenols, carotenoids, and dietary fiber. In a study from Korea conducted on the peel of 'Irwin' mangos, it contained more polyphenols and bioactive compounds than pulp and thus exhibited effective antioxidant activity. The phenolic compounds extracted from peel are gallic acid, syringic acid, mangiferin, ellagic acid, protocatechuic acid, and quercitin (Kim et al., 2010). More bioactive compounds were detected in thr peel, i.e., the carotenoids β-carotene, violaxanthin and lutein. Other important compounds found in the mango peel are terpenoids (α-pinene, β-pinene, β-myrcene, limonene, *cis*-ocimene, *trans*-ocimene, terpinene, α-guaiene, canfene, fenchene, α-humulene), lactones, aldehydes, acids, sesquiterpenes, esters, aliphatic alcohols and compounds involved in the synthesis of vitamins A, K and E. In another study on the peel of the 'Haden' mango variety, significant concentrations of minerals like calcium (4.445 mg/kg), potassium (2910 mg/kg), manganese (950 mg/kg), iron (175 mg/kg), and zinc (32.5 mg/kg) were observed (Saby et al., 2002).

17.2.4 Mango Pomace

Mango pomace is another by-product of the processing industry obtained after extracting juice from the pulp (Quiles et al., 2018). Mango pomace represents more than half of the total fruit weight and is composed of the pulp, peel and kernel (Mahmoud et al., 2020). In the same study, the nutritional compositions of mango pomace were determined. On analysis, the pomace was found to have a moisture content of 68.57%. The nutritional evaluation on dry weight basis revealed high concentrations of organic matter (97.24%), crude protein and crude fiber (10.59% and 6.14%, respectively). The fat content was low (1.74%) but the pomace contained 78.77% carbohydrates and 2.76% ash in the analyzed samples. In a study carried out by Mahmoud et al. (2020), the effect of solid-state fermentation on chemical analysis, cell wall constituents, energetic and nutritive values of unfermented (UFMP) and fermented (FMP) mango pomace were determined. The moisture concentration of UFMP (68.57%) varied significantly from FMP (75.92%), indicating that solid-state fermentation increased the moisture content of the pomace. Chemical analyses on dry matter basis (%) were carried out wherein the concentrations of organic matter (97.24%), crude protein (10.59%), crude fiber (6.14%), fat (1.74%), carbohydrate (78.77%) and ash (2.76%) of UFMP were compared with those of FMP. Fermentation resulted in an increase in the concentrations of crude protein, fiber, fat, carbohydrate and ash in the fermented pomace, with an exception of organic matter concentration, where the value decreased. A similar increase in trend was observed in the energetic and nutritive values and in the cell constituents, concluding that fermentation (processing) enhanced the nutritional quality of the mango pomace. Furthermore, neutral detergent fiber, acid detergent fiber, acid detergent lignin, hemicelluloses and celluloses represented 32.41%, 15.03%, 2.56%, 17.38% and 12.47%, respectively for fermented powder. The energy composition presented the following values: gross energy (4286 kcal/kg dry mass), digestible energy (3257 kcal/kg dry mass), metabolizable energy (2671 kcal/kg dry mass) and net energy (1496 kcal/kg dry mass).

17.3 CHEMISTRY OF MANGO WASTE AND BY-PRODUCTS

Natural antioxidants provide protection against oxidative damage caused by free radicals, reactive oxygen species, etc. (Ma et al., 2011). In order to determine the antioxidant capacity in mango peel extract, Uma Mahesh et al. (2018) conducted a study of 'Sindhura' mango cultivar peel extracts. 'Sindura' peel extract (89.24±0.57%) was detected to have a high radical scavenging activity at 500 μg/mL. The antioxidant capacity of mango peel is higher in comparison with several other fruit peels. The antioxidant potential of peels of dragon fruit, banana and quince are reported to be 50.1%, 52.1%, and 28%, respectively. Because of the fact that mango fibers possess a large amount of fiber-linked phenolic compounds, it is also termed antioxidant dietary fiber. The interaction between fiber and bioactive components produces positive nutritional effects. This is a significant finding as antioxidant fiber has recently been shown to have a beneficial effect against colon cancer. Antioxidant fiber may also be used in fatty foods for stabilization purposes, as it can limit deterioration and improve product shelf-life. (Ma et al., 2011). Phytochemicals found in mango waste and their health benefits are shown in Table 17.3.

The high DPPH scavenging activity has potential in the counteraction of oxidative stress-related illnesses. Polyphenol-rich parts of peel concentrate could be utilized as useful food/feed supplements (Berardini et al., 2005). Peel strips mainly contain mangiferin (C-glucosyl xanthone), a heat-stable and medicinally potent phytochemical (Ajila et al., 2007). Through studies, it was concluded that mangiferin exerts antioxidant, anti-inflammation, immunomodulation and anti-apoptosis effects in the body. Mangiferin is a compound sparingly soluble in ethanol and insoluble in n-hexane and diethyl ether. It is sparingly soluble in water with a solubility of only 0.111 mg/mL. The antioxidant capacity and polyphenolic concentration, however, encounter a change each time the peel is subjected to processing and also in the presence of different extracts. In a study carried out by Ajila and Rao (2013), the effect of heat treatment on the total phenolic concentration of mango peel against standard ascorbic acid was determined. Results stated that heat treatment delivered a significant decrease ($p<0.05$) in concentration of ascorbic acid (18.24%), polyphenols (4.7%), tannins (76.44%) and flavonoids (7.38%) in contrast with untreated flour while a significant decrease ($p<0.05$) in concentrations of ascorbic acid (13.1%), polyphenols (3.47%), tannins (63.73%) and flavonoids (21.76%) was observed. The estimation of ascorbic acid concentration (112–137.33 mg/100 g) acquired was more than that reported by Sogi et al. (2013), who detected ascorbic acid in the range of 61.22–74.48 mg/100 g in mango peel. The drying technique similarly affects the vitamin C content of mango fruit parts (Sogi et al., 2013).

17.4 PROCESSING AND UTILIZATION OF MANGO-BASED WASTES AND BY-PRODUCTS

This section covers the processing and utilization of mango wastes and by-products. The therapeutic applications of the products will also be discussed.

17.4.1 Mango Kernel Oil (MKO)

Mango kernel oil consists of essential oils and the type of lipids obtained from fruit pulp, fruit peel are usually different from that obtained from the kernel. Most fruit oils are derived from fruit seeds and are used as an energy and carbon storage form when the seeds sprout.

17.4.1.1 Processing

The extraction processes for oil and fat, as well as methods like drying, grinding, dehydration and deodorizing, impact the main and minor components. Furthermore, various procedures applied

Table 17.3 Phytochemical Properties of Compounds Found in Mango Waste Parts

Compound	Structure	Molecular Formula	Biological Activities	References
Gallic acid		$C_7H_6O_5$	Radical scavenger, preventive and therapeutic effects for cardiovascular diseases, cancer, antifungal, neurodegenerative disorders and in aging.	Pal, Avneet and Siddhraj, 2018
m-Digallic acid		$C_{14}H_{10}O_9$	Wound-healing, anti-depressant, antiparkinsonian	Nayeem et al., 2016
p-Digallic acid		$C_{14}H_{10}O_9$	Antioxidant activity, anti-inflammatory, hepatoprotective	Onuh et al., 2017
Ellagic acid		$C_{14}H_6O_8$	Relaxation of blood vessels, oxygen free radical scavenging, lipid-lowering effect, anti-inflammatory and anti-carcinogenic effect	Usta et al., 2013
Mangiferin		$C_{19}H_{18}O_{11}$	Inhibitory activity, anticancer, antioxidant activity	Peng et al., 2015
Penta-galloyl glucose		$C_{41}H_{32}O_{26}$	Free radical sink, an anti-inflammatory agent, antidiabetic agent, enzymatic resistant properties	Patnaik et al., 2019
Quercetin		$C_{15}H_{10}O_7$	Antioxidant, anti-inflammatory, antiviral and anticancer as well as the ability to ease some cardiovascular diseases	Liu et al., 2014

after processing (heating and storing) of oils can diminish or degrade essential bioactive components. Extraction of fruit oils can be done in various ways, like cold or hot pressing, extraction using solvents or a blend of both methods. These techniques can be made more efficient by amalgamating more techniques like microwaves, enzymes and inert gases (Sultana and Ashraf, 2019). In order to extract mango seed oil, the optimal moisture content should be 12–15%. For this, the seeds are separated, washed and dried. After dehydration, the dried kernels are then roasted which ensures better mechanical separation of the hull. Separated kernels are crushed and ground into a fine powder. This fine powder is then used for extraction of the oil. Solvents like n-hexane can be used for extraction with high efficiency (Kittiphoom and Sutasinee, 2013). Different steps involved in the processing of MKO are:

1. Drying and grinding. After separation of the peel and pulp of the mango, the seeds are dried either in the sun or in an oven. For the former, the seeds are usually spread in open spaces to optimize drying using sunlight. Drying is done to reduce the moisture content to 15% which facilitates proper

separation of the hull. Once optimum dehydration has been carried out, the kernels are separated from the endocarp using drum pressing and then can then be ground using different types of grinders. The powder is then used for immediate extraction. If not used immediately, the powder can be stored in airtight jars under refrigeration.

2. MKO extraction. As already discussed, various methods can be used for extracting MKO. The most common methods used are solvent extraction and hydraulic press extraction. A general set-up of a Soxhlet extractor includes a round-bottomed flask containing the solvent. In the neck of the flask, a glass siphon is fitted for extraction. On the upper part of the glass compartment is placed a condenser which is cooled by continuous running water. The powder for oil extraction is kept in a thimble which is made up of either glass or cellulose. Once the set-up is arranged, the solvent is heated which then converts into vapors. The vapors are washed back over the solvent and collect back into the round-bottomed flask. The process is repeated several times until complete extraction is ensured. Continuous washing of the sample allows the extracted material to collect in the flask along with the solvent. For MKO, solvents like n-hexane, chloroform, ethanol, etc. can be used and have shown high efficiency. Another method used is extraction by pressing. Mechanical pressing is one of the oldest methods used for extraction of oil from seeds. The benefit of this method is that it gives better quality of the oil and the cake residue in comparison to other methods like a screw press or extraction by solvents (Babaria, 2011). As in other methods, modifications in the conditions and process parameters can be made to optimize extraction efficiency of MKO (Sultana and Ashraf, 2019). In order to extract MKO efficiently using a solvent, certain advances have been tried in various studies, one of which is the use of sealed containers where extraction takes place under high pressure. Such methods are called pressurized solvent extraction methods. Amongst the many advances, another method is the accelerated solvent extraction system which is employed for removal of oil from solid and semi-solid food samples with the help of conventional liquid solvents and blends at high temperature (up to 200°C) and pressure. Increased temperature and pressure enhance the effectiveness of the extraction method. In another study, it was found that use of microwaves prior to the extraction using various solvents was quite beneficial in improving the oil yield and also in reducing the final extraction time. For this, appropriate pre-treatment conditions involved exposure to microwaves for 60 s at 110 W. The combination successfully reduced the time taken for extraction five-fold in comparison with any conventional methods (Kittiphoom and Sutasinee, 2013). Balacuit et al. (2021) reported microwave-assisted extraction as another advanced method which employs the use of high pressure and temperature (150°C). However, this method possessed a major limitation that some solvents, such as the pure non-polar solvents (e.g., alkanes like hexane, which is used for extraction of MKO), are transparent to microwave radiation and so do not produce the desired results. The use of supercritical fluid extraction is another alternative technique. At temperatures and pressures above the critical point of the material, the solvent becomes a supercritical fluid. The very high density of such fluids (comparable to the gas phase of a supercritical fluid) permits non-volatile organic molecules to dissolve. Carbon dioxide is one of the materials from which it is easy to obtain a supercritical fluid. It has a critical temperature and pressure of 31.3°C and 72.9 atm, respectively. These conditions are readily attainable, making supercritical carbon dioxide straightforward. Supercritical CO_2 is used as a solvent for extraction and can dissolve certain organic compounds, which enables it to substitute for a number of different solvents. Supercritical CO_2 has low toxicity, is cost effective, non-flammable and can be disposed of quickly. As soon as the extraction is complete, the pressure returns back to atmospheric conditions and, with this, the CO_2 is instantaneously converted back into a gas and escapes. The pure extracts are thus left behind which can be further used as desired. Another method employed for the extraction of MKO from the kernel is hydrodistillation. This method is popularly used to extract essential oils from herbs and other medicinal plants in various systems. In this method, the kernel powder is placed in a tank and filled with a generous volume of water. The tank is then heated from the bottom and converts water into vapor. This steams rises, extracts the oil and carries it toward the condenser. On reaching the condenser, it is separated from the solvent. The powder-water mixture left behind in the tank is separated into two layers using the gravity separation method and the two layers are collected separately.

17.4.1.2 Utilization

The mango seed preparations find application in the beauty and cosmetics industries. Seed powder and oil can be stored for a period of more than 3–4 years if stored under refrigerated conditions. MKO can find application as a substitute for cocoa butter in chocolate and confectioneries. On being cold-pressed, mango kernel oil renders out a butter instead of an oil, which is an exceptional source of essential fatty acids, minerals and vitamins, and hence acts as a good alternative to cocoa butter. Mango kernel butter also shows valuable creaming properties for skin lotions and lubricants (Nadeem et al., 2016). Uses of MKO as a skin treatment and its ability to act as an antidepressant were studied by Nadeem et al. (2016). The study reported that MKO could effectively be used in the treatment of dry skin. On topical application, it provides softness and also provides protection of the skin from frostbite. Rashes and itching are severe ailments in hot and humid weather conditions especially for infants and babies. MKO contains several phenolic compounds and other bioactives which can possibly play an important role in alleviating these conditions. The same compounds in MKO possess the power to eradicate dark spots on the skin and the phenolic compounds also fight against aging. MKO also finds use in sunscreens and shaving cream owing to its fatty acid composition and can also be used to avoid and remove stretch marks developing during pregnancy. Due to the presence of phenolics, MKO acts as an antidepressant and relieves muscle fatigue and cramps. Since MKO possesses very low quantities of free fatty acids but significant amounts of phenolic components, this leads to lower incubation periods for MKO than any other commercially available vegetable oil. Hence, it can effectively be used to replace artificial antioxidants used to preserve fats and oils. Mangiferin, chlorogenic acid, quercetin and caffeic acid are the major phenolic compounds present in MKO (Nadeem et al., 2016). Functional properties of MKO can be enhanced by using methods like fractionation, *trans*-esterification and inter-esterification, which can make it more suitable for industrial applications (Nadeem et al., 2016). Apart from its multiple benefits to the body, owing to its antioxidant and profile rich in total phenolics, another advantage of mango kernel oil is its utilization in active packaging. Adilah et al. (2020) attempted to develop a film by making use of by-products obtained from food industries. The authors used mango kernel extracts in various proportions (1–5%) as a natural antioxidant in films which were made using soy protein isolate and fish gelatine. The results showed that incorporation of MKO made the films thicker and more translucent. An enhancement in the antioxidant concentration was observed in both the films in association with an increase in extract concentration. During the process of refining, the medicinal efficacy of MKO can be adversely affected.

17.4.2 Mango Peel

17.4.2.1 Processing

The processing of mangos involves the collection of the fruits of the mango variety to be processed, followed by removal of the by-products (seed, pulp, peel). The by-products are then subjected to drying in order to reduce the initial moisture content below 12–15%. This is followed by grinding of the by-product (in this case, peel) to obtain a powdered form which can be further incorporated as needed (Keshwani and Mishra, 2018). As stated in the previous section, bioactive compounds like phenolic compounds and carotenoids are present in generous amounts in mango peel. The efficacy of phenolic compounds depends upon the efficiency of extraction. There are various methods for extraction of the compounds from mango peel involving the opening of the complex matrix and solubilizing the compounds using various solvents. Solvents like hexane, ethanol and acetone have been tested for mango peel extraction. These solvent-assisted extractions, if coupled with other techniques like microwaves or ultrasound, give higher yields and greater efficiency, in addition to the supercritical fluid extraction method which has already been tested and proved beneficial. Nadeem et al. (2016) successfully extracted non-polar flavonoids and carotenoids using supercritical carbon dioxide extraction from mango peel. Pressurized ethanol was found to be useful for extracting polar polyphenolic compounds. (Nadeem et al., 2016). For storage of mango peel, drying has been considered to be an effective process. Drying makes the final product shelf-stable, not only by reducing moisture content in the final product but also by leading to inactivation of enzymes capable of the destruction of beneficial compounds and reducing the rate of microbial growth. However, since drying processes involve raised temperatures, care must be taken to control the temperature as most of the bioactive compounds can attain significant damage from exposure to high temperatures (Kittiphoom and Sustainee, 2013). Several methods for drying mango peel were compared in a study, such as hot air, infra-red, vacuum, low-temperature (freeze-) drying, etc. Maximum retention of total phenolic compounds was observed in freeze-drying (3,185 mg/100 g) whereas the lowest retention was reported by infrared drying (3,049 mg/100 g). Similar patterns were observed for carotenoids (highest retention for freeze-drying (4.05) and lowest for infrared drying (1.8 mg/100 g) and ascorbic acid (highest for freeze-drying (75.4) and lowest for infrared drying (71.4 mg/100 g) on a dry weight basis. However, research is needed on alternative drying methods for mango peel, like heat pumps or refractory windows (already used in mango pulp production). These alternative methods are currently used only in research studies and their impact on the final phytochemical and nutrient composition is yet to be studied for mango peel. Phenolic mixtures from mango peel strips were effectively removed by utilizing subcritical water, with this dissolvable being viewed as harmless to the ecosystem for extracting bioactive mixes (Tunchaiyaphum et al., 2013).,

17.4.2.2 *Utilization*

Mango peels can be used for the development of useful components for diverse food uses, thanks to their rich nutrient profile (e.g., dietary fiber and polyphenols), as several researchers have documented. Mango peel meat is currently being used in several food items, such as bread, sponges, biscuits and other baking items, as a versatile ingredient (Aziz et al., 2012). Mango peels also contain reasonable concentrations of fat (2.16–2.66%) (Ajila et al., 2007). Owing to their rich levels of bioactive compounds, which are responsible for playing a major part in prevention of diseases, a study was carried out to determine the importance of mango peel in the treatment/avoidance of diseases. The research focused on the influence on rheological, physical, sensory and antioxidant properties of biscuits of incorporation at various substitute levels of mango peel (5, 15 or 20%) and mango kernels powders at 20, 30, 40, or 50%. The findings showed that peel powder possessed reasonable concentrations of ash, fiber and water, while mango kernel powder contained more fat and protein than mango peel powder. Farinographic results of mango peel integrated into wheat flour showed an improvement from 60.4% to 67.6% in water absorption, whereas the value declined in response to kernel powder. The concentration of phenolics in response to incorporation of different amounts of peel and kernel powder rose from 3.84 to 24.37 mg/g biscuits. On fortification of the biscuits with mango peel powder, the biscuits exhibited increased free radical scavenging properties. Incorporation of mango peel or kernel powder up to 10% and 40%, respectively, produced satisfactory mango-flavored biscuits. The study therefore concluded successful incorporation of mango flavor and enhanced antioxidant potential using mango peel or kernel powder. Further research conducted on the same grounds by Aslam et al. (2014) attempted the incorporation of mango peel and kernel powder to enhance antioxidant properties and dietary fiber content of biscuits. Incorporation of mango peel enhanced the content of crude fiber and bioactive components, whereas the use of kernel powder enhanced the concentrations of total phenolics, protein and ash in comparison with control or peel-fortified biscuits. The best sensorial results were obtained by supplementing with 10% peel powder and 5% kernel powder. Another important finding drawn from this study was that the fortified biscuits were shelf-stable and no quality changes were perceived at room temperature for up to thirty days. Ajila and Rao (2013) assessed the effect of incorporation of mango peel powder on the quality of macaroni. The authors studied quality parameters like cooking parameters, firmness, bioactive composition and sensorial traits. On supplementation with mango peel powder, the dietary fiber increased by more than 50 percent, the polyphenols showed a more than three-fold increase in concentration and carotenoid (84 µg/g) concentration increased significantly. Mango peel contains distinct enzyme profiles like proteases, peroxidases, amylases, xylanases and polyphenol oxidases. Proteases are well known for their industrial applications and have a plethora of uses in biotechnology. Mango peel contains serine proteases as the major type, which can be purified simply as compared with other proteases as there are only a few substances which show interference with the process of purification of these enzymes. Furthermore, amylases extracted from mango peel find applications in various industries like the fruit and beverage, confectionary, detergent, medicinal and biotechnology sectors. Encapsulation has recently been used as a novel technique for conservation and delivery of these enzymes (Nadeem et al., 2016). Yang and Li (2013) showed a new biological path for the production of silver nanoparticles as a derivative of mango peel extract. They have been shown to have outstanding antibacterial activity in non-woven fabrics lined with silver nanoparticles. These authors suggested that such nanoparticles could be used for various medicinal applications.

17.4.3 Mango Pomace

17.4.3.1 Processing

Fruit pulp is rich in fiber and, in order to extract fiber concentrates, fruit pomaces are exposed to a series of steps, consisting of desugaring, grinding and bleaching (Quiles et al., 2018). As the name suggests, the desugaring step reduces the sugar content from the pulp. In this step, the free monosaccharides are removed from the pulp which generates a concentrated mixture of dietary fiber. After this, washing is performed slowly so that soluble fibers are not lost. The mixture obtained after desugaring and washing are subjected to filtration through a 100-µm pore size nylon filter. This step segregates the insoluble residues. The above process is repeated to ensure better extraction. The desugared fruit residues are then dried in a hot-air oven at a temperature of 70°C for 7 hours to acquire a fiber concentrate. Ultrafine, fine and coarse fiber concentrate powder are obtained after grinding the fiber concentrate, followed by bleaching. The categories are divided on the basis of particle size achieved by passing through sieves of different mesh sizes (Hu et al., 2018). Bleaching is performed to separate colored molecules like phenols and tannins, and hydrogen peroxide is used as the bleaching agent. In this process, dried desugared powder is mixed with 2.5 litres of 3% hydrogen peroxide and the pH is adjusted to 11.5 using 7 M sodium hydroxide solution for a duration of 60 minutes approximately. The insoluble portion is then neutralized using 7 M hydrochloric acid and separated by filtering through a metallic sieve. The fraction so obtained is washed three times, oven dried and ground in a mill. The powder obtained is called the bleached fiber concentrate (Gould and Sharp, 1990).

17.4.3.2 Utilization

As stated earlier, pomace processing is mainly carried out to extract the fiber-rich polysaccharides and bioactive compounds present and to utilize them for human benefit. Therefore, the benefits provided by the pomace are attributable to the polysaccharides and bioactive compounds present in the pomace. Plant polysaccharides (here, obtained from mango) have attracted considerable attention due to their antioxidant, antitumor, antidiabetic, anticancer, anti-viral and immunomodulating properties. Hu et al. (2018) carried out a study wherein the ultrafine particle extraction method was used to extract the polysaccharides and the anticancer bioactives. Ultrasonic extraction can be used to extract polysaccharides from mango pomace. The best results (3.89% yield) were obtained under the following conditions: temperature 74°C, ultrasonic power 170 W, for 100 minutes with a raw material-to-water ratio of 1:40 ing mL^{-1}. DEAE-52 and Sephadex G-100 column chromatography were used to purify three novel polysaccharide fractions (MG-1, MG-2, and MG-3) from the crude polysaccharide extract. The high performance gel permeation chromatography findings indicated that the molecular weights of MG-1, MG-2, and MG-3 were 1.78×10^4, 3.26×10^4 and 2.79×10^6 Da, respectively. Mannose, rhamnose glucuronic acid, galacturonic acid, glucose, galactose, xylose, arabinose and fucose constituted the monosaccharide composition of MG-1, MG-2 and MG-3. In addition, Fourier Transform-Infrared spectroscopy, ^{1}H Nuclear Magnetic Resonance and Scanning Electron Microscopy have been performed for tentative structural characterizations of mango pomace (MGP). The MGPs were also examined *in vitro* for anticancer activity (Hu et al., 2018).

17.5 HEALTH BENEFITS OF MANGO WASTES AND BY-PRODUCTS

Mango wastes and by-products are known for their pharmacological properties. These find applications in alleviating and protecting against different diseases and ailments like cancer, cardiovascular diseases and allergies to name a few. The current section deals with the pharmacological properties of mango by-products.

17.5.1 Antiviral Activity

The antiviral activity of mangiferin has been proven for some viruses. The activity of mangiferin was studied against Herpes simplex virus type 2 (HSV-2) and the results showed that it did not affect the HSV-2 directly but instead inhibits the later events associated with HSV-2 replication. It also inhibited Herpes simplex virus type 1 multiplication within the cell and was also effective against the cytopathic effects of HIV (Parvez, 2016).

17.5.2 Immunomodulatory Effect

Mangiferin has also been identified as a key constituent in mango which exhibits anti-inflammatory effects. The compound mediated downregulation and suppression of NF-κB, otherwise activated by inflammatory agents like tumor nuclear factor. Mangiferin also increases the intercellular content of the natural antioxidant glutathione and possibly might play a role in chemotherapeutic agent-mediated cell death. This might make mangiferin an important constituent in combination treatment for cancer. (Sarkar et al., 2004). The immunomodulatory effect of mangiferin was assessed on thioglycollate-elicited mouse macrophages that were stimulated with lipopolysaccharide and gamma interferon (IFN-γ). Mangiferin (10 μM) was found to inhibit NF-κB-mediated signaling pathways. It also inhibited Toll-like receptors and a number of pro-inflammatory cytokines such as IL-1α, IL-1, IL-6, IL-12, TNF-α, granulocyte and macrophage colony-stimulating factors (Leiro et al., 2003). Moreover, in addition to controlling reactive oxygen species scavenging, mangiferin has also been found to be responsible for modulating the expression of many genes critical for regulating apoptosis, viral multiplication, development and propagation of tumors, inflammation and many autoimmune conditions. The studies propose that mangiferin can act as a crucial component not just for treating but also for preventing cancer and inflammatory diseases and may provide protection against oxidative stress damage and mutagenesis (Ediriweera et al., 2017).

17.5.3 Anti-Inflammatory and Anti-Cancer Effects

The impact of Vimang and mangiferin in vascular smooth muscle cells of mesenteric normal-tensive and spontaneously hypertensive arteries (Wistar Kyoto rats) on the expression of COX-2

Table 17.4   Health Benefits of Mango Waste and By-Products

Health Disorder	Mango Part	Compound	Outcomes	References
Antioxidant activity	Peel	3,4-Dihydroxybenzoic acid (protocatechuic acid)	Mangiferin is effective in gut protection against gastric injury due to its antisecretory and antioxidant mechanisms	Baretto et al., 2008
	Fruit pulp	Kaempferol		
		Linalool		
		Mangiferin		
		Quercetin		
		β-Carotene		
	Peel	β-Carotene		
Anti-inflammatory	Peel	5-(11Z-Heptadecenyl)-resorcinol and 5-(8,Z,11,Z-heptadecadienyl)-resorcinol	Inhibits inflammatory cellular activities, regulates expression of genes responsible for inflammation, increases cellular resistance against injuries caused by inflammation	Knödler et al., 2007, Jyotshna, Khare and Shanker, 2016, Telang et al., 2013
	Fruit pulp	Kaempferol		
		Mangiferin		
		Shikimic acid		
Antimicrobial	Fruit pulp	Kaempferol	Proven antibacterial activity of both Gram-positive and -negative bacteria and the yeast *Candida albicans.*	Moskaug et al., 2004
		Mangiferin		
		Quercetin		
	Peel	3,4-Dihydroxybenzoic acid (protocatechuic acid)		
Antidiabetic	Fruit pulp	Kaempferol	An oral dosage of 250 mg/kg bodyweight is effective in diabetes mellitus in rats.	Bhowmik et al., 2009
	Peel	3,4-Dihydroxybenzoic acid (protocatechuic acid)		
	Fruit pulp	Mangiferin		
		β-Carotene		
		Quercetin		
Cytotoxic and apoptotic	Fruit pulp	Kaempferol	It is effective against cyclophosphamide-induced immune depression, like lymphoid organ atrophy, poor cellular response, low antigen-specific immunoglobulin M; it reduces lipid peroxidation, and decreases superoxide dismutase activities.	Schieber, Berardini and Carle, 2003
		Linalool		
		Mangiferin		
		Quercetin		
		β-Carotene		
	Peel	3,4-Dihydroxybenzoic acid(protocatechuic acid)		
	Seed	Mangiferin		
		Gallic acid		

(Continued)

Table 17.4 (Continued)   Health Benefits of Mango Waste and By-Products

Health Disorder	Mango Part	Compound	Outcomes	References
Neuroprotective	Fruit pulp	Linalool Mangiferin	Stimulates cell propagation and induces an increase in the level of nerve growth factor and tumor necrosis factor (TNF)-α. Mangiferin is effective in enhancing recognition memory *via* a mechanism which might lead to enhanced neurotrophin and cytokine levels.	Park et al., 2012
Cardio-protective	Fruit pulp	Mangiferin	Decreases production of lipid peroxides and retains the myocardial marker enzyme activity at near-normal level.	Soong and Barlow, 2006
Anticoagulant	Fruit pulp	Shikimic acid	Reduced yeast-induced hyperpyrexia	Gonzalez-Aguilar et al., 2008
Blood pressure lowering activity	Fruit pulp	Quercetin	Significantly reduced total serum cholesterol, triglycerides, low-density lipoprotein, very-low-density lipoprotein and enhanced high density lipoproteins in rats.	Park et al., 2012
Gastro-protective	Seed	Mangiferin	Mangiferin significantly improved gastro-intestinal tract movement at oral doses of 30 mg/kg and 100 mg/kg by 89% and 93%, respectively.	Sandhu and Lim, 2008

and iNOS and noradrenaline-induced vasoconstriction was examined (SHR rats) in rats with and without interleukin-1 (1 ng mL^{-1}; 24 h) stimulation. In the absence of IL-1, Vimang (0.5 mg/ml) and mangiferin (0.025 mg/ml) had no effect on iNos or COX-2 vascular expression in either strain; however, they may suppress IL-1β-induced enzymes, which suggests a strong anti-inflammatory function. The elimination of contractions of noradrenalin (0.1–30 μM), both in WKY and SHR rats, Vimang (1, 0.5 and 0.25 mg mL^{-1}), but not mangiferin (1 and 0.5 mg mL^{-1}). This indicates that various compounds will mediate the inhibitor effect of Vimang on vasoconstrictors' reactions and on COX-2 and iNOS expression (Mahalik et al., 2015).

17.5.4 Antidiabetic Activity

Mangiferin is very effective in treating streptozotocin-induced diabetic rats (streptozotocin damages pancreatic β-cells and induces chronic effects on hyperglycemia, tendency to develop atheroschlerosis and oxidative damage to heart and kidney tissues). Diabetic experimental rats were given mangiferin and the diabetic control rats were supplied with mangiferin or insulin (positive control) every day for 28 days from 30 days onward. As predicted, streptozotocin treatment resulted in a significant decline in catalase and superoxide dismutase function in the kidney. The functions increased in the heart muscles probably due to some alternative compensatory mechanism and were unchanged in the red blood corpuscles. It also increased the levels of malondialdehyde (MDA) (a marker of ROS-induced damage), glycosylated Hb, creatine phosphokinase, glucose, total cholesterol, low-density lipoproteins, and diminished high-density lipoproteins in both the tissues (Muruganadan et al., 2002, 2005). Mangiferin also achieved antidiabetic effect through pathways other than pancreatic ß-cell insulin release/secretion. The mechanisms may include stimulating the utilization of glucose by the peripheral tissues, enhancing glycogenesis and glycolysis pathways and reduction in glycemic response by inhibiting glucose intake (Saxena and Vikram, 2004; Kumar et al., 2021). As a result of these properties, it can be said that mango by-products are highly beneficial in a number of ways. However, their benefits are yet to be exploited for the use of mankind (Table 17.4).

17.6 CONCLUSIONS

Mango is a major fruit crop known for its unusual flavor. It is one of the most grown fruits in the world after banana. The pulp is highly nutritious and used in many food products. The wastes and by-products have no less potential. The peel and seed are generally considered as the wastes of mango. The nutritional content and phytochemical profile of mango-based wastes and by-products have led to their incorporation for the development of functional foods. The processed products from mango-based wastes and by-products increase their value addition and introduce them into the circular economy of waste utilization.

BIBLIOGRAPHY

Abdelaziz, S.A., 2018. Physico chemical characteristics of mango kernel oil and meal. *Middle East Journal of Applied Sciences*, 8(1):01–06.

Abdul Aziz, N.A., Wong, L.M., Bhat, R., & Cheng, L.H., 2012. Evaluation of processed green and ripe mango peel and pulp flours (*Mangifera indica* var. Chokanan) in terms of chemical composition, antioxidant compounds and functional properties. *Journal of the Science of Food and Agriculture*, 92(3):557–563.

Adilah, A.N., Noranizan, M.A., Jamilah, B., & Hanani, Z.N., 2020. Development of polyethylene films coated with gelatin and mango peel extract and the effect on the quality of margarine. *Food Packaging and Shelf Life*, 26:100577.

Ajila, C.M., Aalami, M., Leelavathi, K., & Rao, U.P., 2010. Mango peel powder: A potential source of anti-oxidant and dietary fiber in macaroni preparations. *Innovative Food Science & Emerging Technologies*, *11*(1):219–224.

Ajila, C.M., Naidu, K.A., Bhat, S.G., & Rao, U.P., 2007. Bioactive compounds and antioxidant potential of mango peel extract. *Food Chemistry*, *105*(3):982–988.

Ajila, C.M., & Rao, U.P., 2013. Mango peel dietary fibre: Composition and associated bound phenolics. *Journal of Functional Foods*, *5*(1):444–450.

Asif, A., Farooq, U., Akram, K., Hayat, Z., Shafi, A., Sarfraz, F., Sidhu, M.A.I., Rehman, H.U., & Aftab, S., 2016. Therapeutic potentials of bioactive compounds from mango fruit wastes. *Trends in Food Science & Technology*, *53*:102–112.

Aslam, H.K.W., Raheem, M.I.U., Ramzan, R., Shakeel, A., Shoaib, M., & Sakandar, H.A., 2014. Utilization of mango waste material (peel, kernel) to enhance dietary fiber content and antioxidant properties of biscuit. *Journal of Global Innovations in Agricultural and Social Sciences*, *2*(2):76–81.

Awolu, O.O., Sudha, L.M., & Manohar, B., 2018. Influence of defatted mango kernel seed flour addition on the rheological characteristics and cookie making quality of wheat flour. *Food Science & Nutrition*, *6*(8):2363–2373.

Babaria, M.P.M., 2011. *Extraction of oil from mango kernel by hydraulic pressing.* Junagadh: JAU.

Balacuit, J.N.G., Guillermo, J.D.A., Buenafe, R.J.Q., & Soriano, A.N., 2021. Comparison of microwave-assisted extraction to soxhlet extraction of mango seed kernel oil using ethanol and n-hexane as solvents. *ASEAN Journal of Chemical Engineering*, *21*(2), 158–169.

Barreto, J.C., Trevisan, M.T., Hull, W.E., Erben, G., De Brito, E.S., Pfundstein, B., Würtele, G., Spiegelhalder, B., & Owen, R.W., 2008. Characterization and quantitation of polyphenolic compounds in bark, kernel, leaves, and peel of mango (*Mangifera indica* L.). *Journal of Agricultural and Food Chemistry*, *56*(14):5599–5610.

Berardini, N., Fezer, R., Conrad, J., Beifuss, U., Carle, R., & Schieber, A., 2005. Screening of mango (*Mangifera indica* L.) cultivars for their contents of flavonol O-and xanthone C-glycosides, anthocyanins, and pectin. *Journal of Agricultural and Food Chemistry*, *53*(5):1563–1570.

Bharti, R.P., 2013. Studies on antimicrobial activity and phytochemical profile of *Mangifera indica* leaf extract. *IOSR Journal of Environmental Science, Toxicology and Food Technology*, *7*(3):74–78.

Bhowmik, A., Khan, L.A., Akhter, M., & Rokeya, B., 2009. Studies on the antidiabetic effects of *Mangifera indica* stem-barks and leaves on nondiabetic, type 1 and 2 diabetic model rats. *Bangladesh Journal of Pharmacology*, *4*(2):110–114.

Can Karaca, A., Baskaya, H., Guzel, O., & Ak, M.M., 2020. Characterization of some physicochemical properties of spray-dried and freeze-dried sour cherry powders. *Journal of Food Processing and Preservation*, *44*(12):e14975.

Che Rahim, A., 2019. *Anti-proliferative effect of methyl gallate isolated from mangifera pajang kosterman in selected cancer cell lines* (Doctoral dissertation, Universiti Tun Hussein Onn Malaysia).

Coelho, E.M., de Souza, M.E.A.O., Corrêa, L.C., Viana, A.C., de Azevêdo, L.C., & dos Santos Lima, M., 2019. Bioactive compounds and antioxidant activity of mango peel liqueurs (*Mangifera indica* L.) produced by different methods of maceration. *Antioxidants*, *8*(4):102.

Das, P.C., Khan, M.J., Rahman, M.S., Majumder, S., & Islam, M.N., 2019. Comparison of the physico-chemical and functional properties of mango kernel flour with wheat flour and development of mango kernel flour based composite cakes. *NFS Journal*, *17*:1–7.

Dorta, E., Lobo, M.G., & González, M., 2012. Using drying treatments to stabilise mango peel and seed: Effect on antioxidant activity. *LWT-Food Science and Technology*, *45*(2):261–268.

Ediriweera, M.K., Tennekoon, K.H., & Samarakoon, S.R., 2017. A review on ethnopharmacological applications, pharmacological activities, and bioactive compounds of *Mangifera indica* (Mango). *Evidence-Based Complementary and Alternative Medicine*, *2017*:6949835.

Ellong, E.N., Adenet, S., & Rochefort, K., 2015. Physicochemical, nutritional, organoleptic characteristics and food applications of four mango (*Mangifera indica*) varieties. *Food and Nutrition Sciences*, *6*(02):242.

Feng, S.T., Wang, Z.Z., Yuan, Y.H., Sun, H.M., Chen, N.H., & Zhang, Y., 2019. Mangiferin: A multipotent natural product preventing neurodegeneration in Alzheimer's and Parkinson's disease models. *Pharmacological Research*, *146*:104336.

Ferreira, S., Araujo, T., Souza, N., Rodrigues, L., Lisboa, H.M., Pasquali, M., Trindade, G., & Rocha, A.P., 2019. Physicochemical, morphological and antioxidant properties of spray-dried mango kernel starch. *Journal of Agriculture and Food Research*, *1*:100012.

Fowomola, M.A., 2010. Some nutrients and antinutrients contents of mango (*Magnifera indica*) seed. *African Journal of Food Science*, 4(8):472–476.

Gonzalez-Aguilar, G.A., Celis, J., Sotelo-Mundo, R.R., De La Rosa, L.A., Rodrigo-Garcia, J., & Alvarez-Parrilla, E., 2008. Physiological and biochemical changes of different fresh-cut mango cultivars stored at 5°C. *International Journal of Food Science & Technology*, 43(1):91–101.

González-Aguilar, G.A., Wang, C.Y., Buta, J.G., & Krizek, D.T., 2001. Use of UV-C irradiation to prevent decay and maintain postharvest quality of ripe 'Tommy Atkins' mangoes. *International Journal of Food Science & Technology*, 36(7):767–773.

Gould, W.P., & Sharp, J.L., 1990. Caribbean fruit fly (*Diptera: Tephritidae*) mortality induced by shrink-wrapping infested mangoes. *Journal of Economic Entomology*, 83(6):2324–2326.

Gumte, S.V., Taur, A.T., Sawate, A.R., & Kshirsagar, R.B., 2018. Effect of fortification of mango (*Mangifera indica*) kernel flour on nutritional, phytochemical and textural properties of biscuits. *Journal of Pharmacognosy and Phytochemistry*, 7(3):1630–1637.

Hannan, A., Asghar, S., Naeem, T., Ullah, M.I., Ahmed, I., Aneela, S., & Hussain, S., 2013. Antibacterial effect of mango (*Mangifera indica* Linn.) leaf extract against antibiotic sensitive and multi-drug resistant Salmonella typhi. *Pakistan Journal of Pharmaceutical Sciences*, 26(4):715–719.

Hu, K., Dars, A.G., Liu, Q., Xie, B., & Sun, Z., 2018. Phytochemical profiling of the ripening of Chinese mango (*Mangifera indica* L.) cultivars by real-time monitoring using UPLC-ESI-QTOF-MS and its potential benefits as prebiotic ingredients. *Food Chemistry*, 256:171–180.

Jahurul, M.H.A., Zaidul, I.S.M., Ghafoor, K., Al-Juhaimi, F.Y., Nyam, K.L., Norulaini, N.A.N., Sahena, F., & Omar, A.M., 2015. Mango (*Mangifera indica* L.) by-products and their valuable components: A review. *Food Chemistry*, 183:173–180.

Joyce, O.O., Latayo, B.M., & Onyinye, A.C., 2014. Chemical composition and phytochemical properties of mango (*mangifera indica*.) seed kernel. *International Journal of Advanced Chemistry*, 2:185–187.

Jyotshna, P., & Shanker, K. 2016. Mangiferin: A review of sources and interventions for biological activities. *BioFactors*, 42(5):504–514. doi:10.1002/ biof.1308

Kabir, Y., Shekhar, H.U., & Sidhu, J.S., 2017. Phytochemical compounds in functional properties of mangoes. In M. Siddiq, J.K. Brecht, & J.S. Sidhu (eds.), *Handbook of mango fruit: Production, postharvest science, processing technology and nutrition* (pp. 237–254). Hoboken, NJ: John Wiley & Sons Ltd.

Kalita, P., 2014. An overview on *mangifera indica*: Importance and its various pharmacological action. *PharmaTutor*, 2(12):72–76.

Kaur, A., & Brar, J.K., 2015. Use of mango seed kernels for the development of antioxidant rich biscuits. *International Journal of Scientific Research*, 78(96):535–538.

Kaur, B., & Srivastav, P.P., 2018. Effect of cryogenic grinding on chemical and morphological characteristics of mango (*Mangifera indica* L.) peel powder. *Journal of Food Processing and Preservation*, 42(4):e13583.

Keshwani, D., & Mishra, S., 2018. Utilization of mango and its by-products by different processing methods. *Aslan Journal of Science and Technology*, 9(10):8896–8091.

Khammuang, S., & Sarnthima, R., 2011. Antioxidant and antibacterial activities of selected varieties of Thai mango seed extract. *Pakistan Journal of Pharmaceutical Sciences*, 24(1):37–42.

Khare, P., & Shanker, K., 2016. Mangiferin: A review of sources and interventions for biological activities. *BioFactors*, 42(5):504–514.

Khurana, R.K., Kaur, R., Lohan, S., Singh, K.K., & Singh, B., 2016. Mangiferin: A promising anticancer bioactive. *Pharmaceutical Patent Analyst*, 5(3):169–181.

Kim, H., Moon, J.Y., Kim, H., Lee, D.S., Cho, M., Choi, H.K., Kim, Y.S., Mosaddik, A., & Cho, S.K., 2010. Antioxidant and antiproliferative activities of mango (*Mangifera indica* L.) flesh and peel. *Food Chemistry*, 121(2):429–436.

Kittiphoom, S., & Sutasinee, S., 2013. Mango seed kernel oil and its physicochemical properties. *International Food Research Journal*, 20(3):1145.

Kittiphoom, S., 2012. Utilization of mango seed. *International Food Research Journal*, 19(4):1325.

Knödler, M., Wenzig, E.M., Bauer, R., Conrad, J., Beifuss, U., Carle, R., & Schieber, A., 2007. Cyclooxygenase inhibitory 5-alkenylresorcinols isolated from mango (*Mangifera indica* L.) peels. *Planta Medica*, 73(9):031.

Kok, J.W., & Mohamad, T.R.T., 2020. Nutritional composition and antioxidant properties of golden lily and chokanan mango (mangifera indica) peels. *Universiti Malaysia Terengganu Journal of Undergraduate Research*, 2(3):35–44.

Koubala, B.B., Kansci, G., Mbome, L.I., Crépeau, M.J., Thibault, J.F., & Ralet, M.C., 2008. Effect of extraction conditions on some physicochemical characteristics of pectins from "Améliorée" and "Mango" mango peels. *Food Hydrocolloids*, 22(7):1345–1351.

Kumar, N., Petkoska, A.T., AL-Hilifi, S.A., & Fawole, O.A., 2021. Effect of chitosan–pullulan composite edible coating functionalized with pomegranate peel extract on the shelf life of mango (*Mangifera indica*). *Coatings*, 11(7):764.

Larrauri, J.A., Rupérez, P., & Saura-Calixto, F., 1997. Mango peel fibres with antioxidant activity. *Zeitschrift für Lebensmitteluntersuchung und-Forschung A*, 205(1):39–42.

Lasano, N.F., Hamid, A.H., Karim, R., Pak Dek, M.S., Shukri, R., & Ramli, N.S., 2019. Nutritional composition, anti-diabetic properties and identification of active compounds using UHPLC-ESI-orbitrap-MS/MS in *Mangifera odorata* L. peel and seed kernel. *Molecules*, 24(2):320.

Lawson, T., Lycett, G.W., Ali, A., & Chin, C.F., 2019. Characterization of Southeast Asia mangoes (*Mangifera indica* L) according to their physicochemical attributes. *Scientia Horticulturae*, 243:189–196.

Leiro, J., García, D., Escalante, M., Delgado, R., Ubeira, F.M., 2003. Anthelminthic and anti-allergic activities of *Mangifera indica* L. stem bark components Vimang and Mangiferin. *Phytotherapy Research*, 17:1203–1208.

Li, L., Wang, S., Chen, J., Xie, J., Wu, H., Zhan, R., & Li, W., 2014. Major antioxidants and in vitro antioxidant capacity of eleven mango (*Mangifera indica* L.) cultivars. *International Journal of Food Properties*, 17(8):1872–1887.

Liu, F.X., Fu, S.F., Bi, X.F., Chen, F., Liao, X.J., Hu, X.S., & Wu, J.H., 2013. Physico-chemical and antioxidant properties of four mango (*Mangifera indica* L.) cultivars in China. *Food Chemistry*, 138(1):396–405.

Liu, F., Wang, Y., Li, R., Bi, X., & Liao, X., 2014. Effects of high hydrostatic pressure and high temperature short time on antioxidant activity, antioxidant compounds and color of mango nectars. *Innovative Food Science & Emerging Technologies*, 21, 35–43.

Ma, X., Wu, H., Liu, L., Yao, Q., Wang, S., Zhan, R., Xing, S., & Zhou, Y., 2011. Polyphenolic compounds and antioxidant properties in mango fruits. *Scientia Horticulturae*, 129(1):102–107.

Mahalik, G., Jali, P., Sahoo, S., & Satapathy, K.B., 2015. Ethnomedicinal, phytochemical and pharmacological properties of *Mangifera indica* L: A review. *Structure*, 29:31.

Mahmoud, A.E.E.D., Omer, H.A.A.A., Mohammed, A.T., & Ali, M.M., 2020. Enhancement of chemical composition and nutritive value of some fruits pomace by solid state fermentation. *Egyptian Journal of Chemistry*, 63(10):3713–3720.

Maldonado-Celis, M.E., Yahia, E.M., Bedoya, R., Landázuri, P., Loango, N., Aguillón, J., Restrepo, B., & Guerrero Ospina, J.C., 2019. Chemical composition of mango (*Mangifera indica* L.) fruit: Nutritional and phytochemical compounds. *Frontiers in Plant Science*, 10:1073.

Melo, P.E., Silva, A.P.M., Marques, F.P., Ribeiro, P.R., Brito, E.S., Lima, J.R., & Azeredo, H.M., 2019. Antioxidant films from mango kernel components. *Food Hydrocolloids*, 95:487–495.

Mirza, B., Croley, C.R., Ahmad, M., Pumarol, J., Das, N., Sethi, G., & Bishayee, A., 2020. Mango (*Mangifera indica* L.): A magnificent plant with cancer preventive and anticancer therapeutic potential. *Critical Reviews in Food Science and Nutrition* 1:27.

Mohammed, S., Edna, M., & Siraj, K., 2020. The effect of traditional and improved solar drying methods on the sensory quality and nutritional composition of fruits: A case of mangoes and pineapples. *Heliyon*, 6(6):e04163.

Moskaug, J.Ø., Carlsen, H., Myhrstad, M., & Blomhoff, R., 2004. Molecular imaging of the biological effects of quercetin and quercetin-rich foods. *Mechanisms of Ageing and Development*, 125(4):315–324.

Muchiri, D.R., Mahungu, S.M., & Gituanja, S.N., 2012. Studies on mango (*Mangifera indica*, L.) kernel fat of some Kenyan varieties in Meru. *Journal of the American Oil Chemists' Society*, 89(9):1567–1575.

Muruganandan, S., Srinivasan, K., Gupta, S., Gupta, P.K., & Lal, J., 2005. Effect of mangiferin on hyperglycemia and atherogenicity in streptozotocin diabetic rats. *Journal of Ethnopharmacology*, 97(3):497–501.

Mutua, J.K., Imathiu, S., & Owino, W., 2017. Evaluation of the proximate composition, antioxidant potential, and antimicrobial activity of mango seed kernel extracts. *Food Science & Nutrition*, 5(2):349–357.

Mwaurah, P.W., Kumar, S., Kumar, N., Panghal, A., Attkan, A.K., Singh, V.K., & Garg, M.K., 2020. Physicochemical characteristics, bioactive compounds and industrial applications of mango kernel and its products: A review. *Comprehensive Reviews in Food Science and Food Safety*, 19(5):2421–2446.

Nadeem, M., Imran, M., & Khalique, A., 2016. Promising features of mango (*Mangifera indica* L.) kernel oil: A review. *Journal of Food Science and Technology*, 53(5):2185–2195.

Nayeem, N., Asdaq, S.M.B., Salem, H., & AHEl-Alfqy, S., 2016. Gallic acid: A promising lead molecule for drug development. *Journal of Applied Pharmacy*, 8(2):1–4.

Nwaokobia, K., Ogboru, R.O., & Idibie, C.A., 2018. Extraction of edible oil from the pulp of Persea americana (Mill) using cold process method. *World News of Natural Sciences. An International Scientific Journal*, 17:130–140.

Nzikou, J.M., Kimbonguila, A., Matos, L., Loumouamou, B., Pambou-Tobi, N.P.G., Ndangui, C.B., Abena, A.A., Silou, T., Scher, J., & Desobry, S., 2010. Extraction and characteristics of seed kernel oil from mango (*Mangifera indica*). *Research Journal of Environmental and Earth Sciences*, 2(1):31–35.

Ojha, P., Raut, S., Subedi, U., & Upadhaya, N., 2019. Study of nutritional, phytochemicals and functional properties of mango kernel powder. *Journal of Food Science and Technology Nepal*, 11:32–38.

Ojokoh, A.O., 2007. Effect of fermentation on the chemical composition of mango (*Mangifera indica* R) peels. *African Journal of Biotechnology*, 6:1979–1981.

Onuh, J.O., Momoh, G., Egwujeh, S., & Onuh, F., 2017. Evaluation of the nutritional, phytochemical and antioxidant properties of the peels of some selected mango varieties. *American Journal of Food Science and Technology*, 5(5):176–181.

Pal, S.M., Avneet, G., & Siddhraj, S.S., 2018. Gallic acid: Pharmacological promising lead molecule: A review. *International Journal Pharmacognosy and Phytochemical Research*, 10:132–138.

Park, S.N., Lim, Y.K., Freire, M.O., Cho, E., Jin, D., & Kook, J.K., 2012. Antimicrobial effect of linalool and α-terpineol against periodontopathic and cariogenic bacteria. *Anaerobe*, 18(3):369–372.

Parvez, G.M., 2016. Pharmacological activities of mango (*Mangifera Indica*): A review. *Journal of Pharmacognosy and Phytochemistry*, 5(3):1.

Patnaik, S.S., Simionescu, D.T., Goergen, C.J., Hoyt, K., Sirsi, S., & Finol, E.A., 2019. Pentagalloyl glucose and its functional role in vascular health: Biomechanics and drug-delivery characteristics. *Annals of Biomedical Engineering*, 47(1):39–59.

Peng, Z.G., Yao, Y.B., Yang, J., Tang, Y.L., & Huang, X., 2015. Mangiferin induces cell cycle arrest at G2/M phase through ATR-Chk1 pathway in HL-60 leukemia cells. *Genetics and Molecular Research*, 14(2):4989–5002.

Pierson, J.T., Monteith, G.R., Roberts-Thomson, S.J., Dietzgen, R.G., Gidley, M.J., & Shaw, P.N., 2014. Phytochemical extraction, characterisation and comparative distribution across four mango (*Mangifera indica* L.) fruit varieties. *Food Chemistry*, 149:253–263.

Ponis, S.T., Papanikolaou, P.A., Katimertzoglou, P., Ntalla, A.C., & Xenos, K.I., 2017. Household food waste in Greece: A questionnaire survey. *Journal of Cleaner Production*, 149:1268–1277.

Quiles, A., Campbell, G.M., Struck, S., Rohm, H., & Hernando, I., 2018. Fiber from fruit pomace: A review of applications in cereal-based products. *Food Reviews International*, 34(2):162–181.

Rakholiya, K., Kaneria, M., & Chanda, S., 2016. Physicochemical and phytochemical analysis of different parts of Indian Kesar mango–a unique variety from Saurashtra region of Gujarat. *Pharmacognosy Journal*, 8(5):502–506.

Ravani, A., & Joshi, D., 2013. Mango and its by product utilization – A review. *Energy (kcal)*, 74(44):2013.

Ribeiro, S.M.R., Queiroz, J.H., de Queiroz, M.E.L.R., Campos, F.M., & Sant'Ana, H.M.P., 2007. Antioxidant in mango (*Mangifera indica* L.) pulp. *Plant Foods for Human Nutrition*, 62(1):13–17.

Richa, M., Ranu, P., & Anisha, V., 2018. Nutritional composition, anti-nutritional factors and antioxidant activity of waste seeds (mango seeds and tamarind seeds). *Journal of Pharmacognosy and Phytochemistry*, 7(3):1078–1080.

Rivera-Vargas, L.I., Lugo-Noel, Y., McGovern, R.J., Seijo, T., & Davis, M.J., 2006. Occurrence and distribution of *Colletotrichum* spp. on mango (*Mangifera indica* L.) in Puerto Rico and Florida, USA. *Plant Pathology Journal*, 5(2):192–198.

Rumainum, I.M., Worarad, K., Srilaong, V., & Yamane, K., 2018. Fruit quality and antioxidant capacity of six Thai mango cultivars. *Agriculture and Natural Resources*, 52(2):208–214.

Rummun, N., Pires, E., McCullagh, J., Claridge, T.W., Bahorun, T., Li, W.W., & Neergheen, V.S., 2021. Methyl gallate – Rich fraction of *Syzygium coriaceum* leaf extract induced cancer cell cytotoxicity via oxidative stress. *South African Journal of Botany*, 137:149–158.

Saby, J.K., Bhat, S.G., & Prasada, R.J., 2002. Involvement of peroxidase and polyphenol oxidase in mango sap-injury. *Journal Food Biochemistry*, 26:403–414.

Sagar, N.A., Pareek, S., Sharma, S., Yahia, E.M., & Lobo, M.G., 2018. Fruit and vegetable waste: Bioactive compounds, their extraction, and possible utilization. *Comprehensive Reviews in Food Science and Food Safety*, 17(3):512–531.

Sandhu, K.S., & Lim, S.T., 2008. Structural characteristics and in vitro digestibility of mango kernel starches (*Mangifera indica* L.). *Food Chemistry, 107*(1):92–97.

Sarkar, S.K., Gautam, B., Seethambram, Y., & Vijaya, N., 2004. Effect of intercropping sequence with vegetables in young mango orchard under Deccan plateau. *Indian Journal of Horticulture, 61*(2):125–127.

Saxena, A., & Vikram, N.K., 2004. Role of selected Indian plants in management of type 2 diabetes: A review. *The Journal of Alternative and Complementary Medicine, 10*:369–378.

Schieber, A., Berardini, N., & Carle, R., 2003. Identification of flavonol and xanthone glycosides from mango (*Mangifera indica* L. Cv. "Tommy Atkins") peels by high-performance liquid chromatography-electrospray ionization mass spectrometry. *Journal of Agricultural and Food Chemistry, 51*(17):5006–5011.

Schneider, F., 2013. Review of food waste prevention on an international level. *Proceedings of the Institution of Civil Engineers-Waste and Resource Management, 166*(4):187–203.

Serna-Cock, L., García-Gonzales, E., & Torres-León, C., 2016. Agro-industrial potential of the mango peel based on its nutritional and functional properties. *Food Reviews International, 32*(4):364–376.

Shah, K.A., Patel, M.B., Patel, R.J., & Parmar, P.K., 2010. *Mangifera indica* (mango). *Pharmacognosy Reviews, 4*(7):42.

Sikdar, D.C., Hegde, S., Swamynathan, V., Varsha, S., & Rakesh, R., 2017. Solvent extraction of mango (*Mangifera indica* L.) seed kernel oil and its characterization. *International Journal of Technical Research and Applications, 5*(4):43–47.

Singh, J.P., Kaur, A., Shevkani, K., & Singh, N., 2016. Composition, bioactive compounds and antioxidant activity of common Indian fruits and vegetables. *Journal of Food Science and Technology, 53*(11):4056–4066.

Sogi, D.S., Siddiq, M., Greiby, I., & Dolan, K.D., 2013. Total phenolics, antioxidant activity, and functional properties of 'Tommy Atkins' mango peel and kernel as affected by drying methods. *Food Chemistry, 141*(3):2649–2655.

Solís-Fuentes, J.A., del Carmen Durán-de-Bazúa, M., 2011. Mango (Mangifera indica L.) seed and its fats. In V.R. Preedy, R.R. Watson, V.B. Patel (eds.), *Nuts & seeds in healthand disease prevention* (1st ed., pp. 741–748). London, Burlington, San Diego: Academic Press.

Soong, Y.Y., & Barlow, P.J., 2004. Antioxidant activity and phenolic content of selected fruit seeds. *Food Chemistry, 88*(3):411–417.

Soong, Y.Y., & Barlow, P.J., 2006. Quantification of gallic acid and ellagic acid from longan (Dimocarpus longan Lour.) seed and mango (*Mangifera indica* L.) kernel and their effects on antioxidant activity. *Food Chemistry, 97*(3):524–530.

Schack, A.S., Stubbe, J., Steffensen, L.B., Mahmoud, H., Laursen, M.S., & Lindholt, J.S., 2020. Intraluminal infusion of Penta-Galloyl glucose reduces abdominal aortic aneurysm development in the elastase rat model. *PLoS One, 15*(8):e0234409.

Sultana, B., & Ashraf, R., 2019. Mango (*Mangifera indica* L.) seed oil. In M.F. Ramadan (ed.), *Fruit oils: Chemistry and functionality* (pp. 561–575). Cham: Springer.

Telang, M., Dhulap, S., Mandhare, A., & Hirwani, R., 2013. Therapeutic and cosmetic applications of mangiferin: A patent review. *Expert Opinion on Therapeutic Patents, 23*(12):156.

Thitikornpong, W., Palanuvej, C., & Ruangrungsi, N., 2019. In vitro antidiabetic, antioxidation and cytotoxicity activities of ethanolic extract of *Aquilaria crassna* leaves and its active compound; mangiferin. *Indian Journal of Traditional Knowledge, 18*:144–150.

Torres-León, C., Ramírez-Guzman, N., Londoño-Hernandez, L., Martinez-Medina, G.A., Díaz-Herrera, R., Navarro-Macias, V., Alvarez-Pérez, O.B., Picazo, B., Villarreal-Vázquez, M., Ascacio-Valdes, J., & Aguilar, C.N., 2018. Food waste and byproducts: An opportunity to minimize malnutrition and hunger in developing countries. *Frontiers in Sustainable Food Systems, 2*:52.

Torres-León, C., Rojas, R., Contreras-Esquivel, J.C., Serna-Cock, L., Belmares-Cerda, R.E., & Aguilar, C.N., 2016. Mango seed: Functional and nutritional properties. *Trends in Food Science & Technology, 55*:109–117.

Tunchaiyaphum, S., Eshtiaghi, M.N., & Yoswathana, N., 2013. Extraction of bioactive compounds from mango peels using green technology. *International Journal of Chemical Engineering and Applications, 4*(4):194.

Umamahesh, K., Ramesh, B., Kumar, B.V., & Reddy, O.V.S., 2019. In vitro anti-oxidant, anti-microbial and anti-inflammatory activities of five Indian cultivars of mango (*Mangifera indica* L.) fruit peel extracts. *Journal of Herbmed Pharmacology, 8*(3):238–247.

Usta, C., Ozdemir, S., Schiariti, M., & Puddu, P.E., 2013. The pharmacological use of ellagic acid-rich pomegranate fruit. *International Journal of Food Sciences and Nutrition, 64*(7):907–913.

Vergara-Valencia, N., Granados-Pérez, E., Agama-Acevedo, E., Tovar, J., Ruales, J., & Bello-Pérez, L.A., 2007. Fibre concentrate from mango fruit: Characterization, associated antioxidant capacity and application as a bakery product ingredient. *LWT-Food Science and Technology, 40*(4):722–729.

Vijay, K.V., Rahaman, S.M., Wadhwani, M.K., Kumari, S., Kumari, M., Kumar, S., Homa, F., Sengupta, S., & Bairwa, S.L., 2019. Diagnosis on strengths and threats of mango in the context of Bihar: A case study. *Journal of Crop and Weed, 15*(2):94–99.

Vilela, C., Santos, S.A., Oliveira, L., Camacho, J.F., Cordeiro, N., Freire, C.S., & Silvestre, A.J., 2013. The ripe pulp of *Mangifera indica* L.: A rich source of phytosterols and other lipophilic phytochemicals. *Food Research International, 54*(2):1535–1540.

Vyas, A., Syeda, K., Ahmad, A., Padhye, S., & Sarkar, F., 2012. Perspectives on medicinal properties of mangiferin. *Mini Reviews in Medicinal Chemistry, 12*(5):412–425.

Wall-Medrano, A., Olivas-Aguirre, F.J., Ayala-Zavala, J.F., Domínguez-Avila, J.A., Gonzalez-Aguilar, G.A., Herrera-Cazares, L.A., & Gaytan-Martinez, M., 2020. Health benefits of mango by-products. In *Food wastes and by-products: Nutraceutical and health potential.* (pp. 159–191). Hoboken, NJ: John Wiley & Sons Ltd.

Wauthoz, N., Balde, A., Balde, E.S., Van Damme, M., & Duez, P., 2007. Ethnopharmacology of *Mangifera indica* L. bark and pharmacological studies of its main C-glucosylxanthone, mangiferin. *International Journal of Biomedical and Pharmaceutical Sciences, 1*(2):112–119.

Yadav, D., & Singh, S.P., 2017. Mango: History origin and distribution. *Journal of Pharmacognosy and Phytochemistry, 6*(6):1257–1262.

Yang, N., & Li, W.H., 2013. Mango peel extract mediated novel route for synthesis of silver nanoparticles and antibacterial application of silver nanoparticles loaded onto non-woven fabrics. *Industrial Crops and Products, 48*:81–88.

Yatnatti, S., Vijayalakshmi, D., & Chandru, R., 2014. Processing and nutritive value of mango seed kernel flour. *Current Research in Nutrition and Food Science Journal, 2*(3):170–175.

Zafar, T.A., & Sidhu, J.S., 2017. Composition and nutritional properties of mangoes. In *Handbook of mango fruit: Production, postharvest science, processing technology and nutrition* (pp. 217–236). Hoboken, NJ: John Wiley & Sons Ltd.

CHAPTER 18

Cashew Wastes
Chemistry, Processing, and Utilization

Barinderjit Singh, Gurwinder Kaur, Reetu, and Shafiya Rafiq

CONTENTS

DOI: 10.1201/9781003164463-18

18.1 INTRODUCTION

Cashew (*Anacardium occidentale*), a tropical evergreen tree native to Northeastern Brazil, is grown for its economic potential in several countries around the world. (Nam et al. 2014). Brazil has long been by far the top producer of cashews with 95 percent of the world's production of cashew fruit (Kouassi et al. 2018). The leaves of the tree are oval, alternating, yellowish green to dark green-brownish in color, and the petiole is pinnate and has a smooth surface, with small flowers that are initially pale green before turning reddish and fragrant. The cashew fruit consists of two parts, the nut and the cashew apple. The cashew apple is the primary waste generated during the recovery of cashew nuts. The cashew apple is the fleshy portion of the cashew fruit connected to the cashew nut (Figure 18.1). Cashew apples are categorized as red, orange, yellow, or greenish-yellow cultivars based on color. In the yellow varieties, the total soluble solids (TSS) content is greater than in the red varieties. The cashew apple is 9–10 times heavier than the nut. The cashew apple is a rich source of carbohydrates, amino acids, minerals, vitamins, organic acids, tannins, fibres, and water. In addition, the cashew apple is also rich in vitamin C and carotenoids (Lopes et al. 2012). The cashew apple is a light reddish to yellow fruit, the juice of which can be distilled into a sweet, astringent fruit drink or liquor (Brijyog, Singh, and Maiti 2017). The astringency of the fruit is due to the presence of polyphenols, tannins, and oily substances in the skin's waxy layer (Devi, Naveen, and Kannan 2015; Talasila, Vechalapu, and Shaik 2012). The cashew apple has tremendous potential to be processed into jam, ice cream, juice, candy, syrup, chutney, pickle, and other products such as wine, mannitol, and vinegar (Antony, Kunhiraman, and Abdulhameed 2020). The presence of a high sugar concentration in cashew apples can also indicate a vast potential to produce ethanol or bioethanol.

18.2 PARTS OF THE CASHEW APPLE

The cashew apple is an accessory fruit or pseudo-fruit produced from the swollen pedicel and receptable of the flower and is the part of the cashew fruit that is attached to the cashew nut, the tree's real fruit. The cashew nut, a double-hulled shell that contains a kidney-shaped, green seed, is attached to the bottom of cashew apple. About 90% of the total fresh weight of a cashew apple is water. The cashew apple is comprised of two parts: the skin and the flesh. The flesh of the cashew apple is very juicy, but the skin is thin and delicate. Owing to its fragility, cashew apple is not a

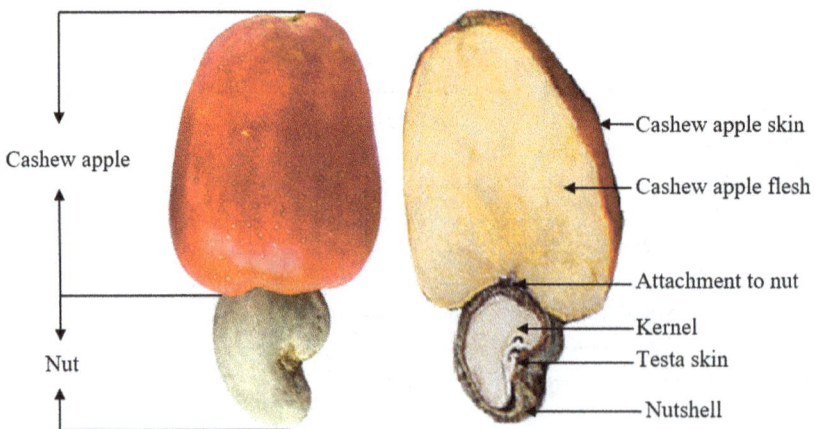

Figure 18.1 Parts of the cashew fruit.

commonly available fruit. It can be found at fruit stalls near cashew-growing regions. The cashew apple is small to medium in size, about 7 cm long, with white flesh. The ripe fruit is sweet, crispy, and juicy with a faint, rose fragrance. The skin of the cashew apple is waxy and smooth. It has glossy skin that turns from green to bright red, orange, or yellow as it matures. The skin is one of the most important parts of the cashew apple because this outer shell skin protects the cashew apple flesh. The flesh of the cashew apple is spongy, fibrous, and delicate, yet it is strong. The flesh is somewhat astringent and is not palatable to customers. The flesh is usually cut into very fine pieces to reduce its fibrous texture and sprinkled with salt before eating to eliminate the astringent taste.

18.3 CHEMISTRY OF THE CASHEW APPLE

The composition of the cashew apple is 84.4–88.7% moisture, 0.7–0.9% protein, 0.1% fat, 12.3% carbohydrate, 0.9% crude fibre, 200 mg minerals/100 g fruit fresh weight, 0.02 mg thiamine/100 g, 0.05 mg riboflavin/100 g, 0.4 mg nicotinic acid/100 g, and 23 µg carotene/100 g (Nam et al. 2014; Sobhana 2019). The main sugars found in cashew apples are glucose, fructose, and sucrose. The cashew apple is considered to be a good substrate for the manufacture of some fermented products, such as wine, alcohol, and vinegar processing due to the presence of a high sugar concentration (Msoka et al. 2017). The various mineral compounds in cashew apples are calcium, magnesium, potassium, sodium, phosphorus, copper, zinc, and iron. The major mineral in cashew apple is potassium (76.0 mg/100ml) followed by calcium (43.0 mg/100 mL), magnesium (10.92 mg/100 mL), phosphorus (0.40 mg/100 mL), sodium (0.33–0.45 mg/100 mL), and copper (0.07–0.12 mg/100 mL). Potassium contributes to cell organization and permeability maintenance; calcium plays an important role in skeletal tissue development; magnesium is involved in nervous system stabilization and muscle contraction in the metabolic regulations of biomolecules such as coenzymes, and iron is the most important element in anaemia prevention, being a central element of red blood cells (Msoka et al. 2017). The minor minerals are zinc, copper, and iron (0.05–0.08 mg/100 mL) (Lowor and Agyeute-Badu 2009). In addition to nutritional components, cashew apple also contains many bioactive components that are protective against many diseases. The fruit antioxidant activity is linked to vitamins and phenolics, which play an important role in scavenging free radicals and decreasing the body's oxidative stress damage. The fruit is also a rich source of fibre that could decrease the risk of different chronic diseases (Nam et al. 2014).

18.3.1 Ascorbic Acid

Ascorbic acid is typically present in plant tissues that undergo active growth and development. Fruit and vegetables constitute about 90% of the ascorbic acid in the human diet. The cashew apple is a good source of ascorbic acid that supports better health, being present in all maturity stages of the cashew apple, from unripe fruits (1038 mg/100 g dry matter), medium ripe fruits (1392 mg/100 g dry matter), to ripe fruits (1731 mg/100 g dry matter) (Gordon et al. 2012). In certain fruits, the ascorbic acid content increases at ripening, whereas in others, the ascorbic acid content decreases. Therefore, ascorbic acid production during fruit ripening seems to rely on the fruit variety. The cashew apple is also a rich source of vitamin A precursors. The high ascorbic acid content in cashew apple juice strongly supports the need for its exploitation and increased use. Cashew apple steaming before juice extraction results in a 19.9% decrease in ascorbic acid content. The drop in ascorbic acid level during processing may be attributed to inevitable losses resulting from the vitamin's heat-labile nature (Inyang and Abah 1997). As an endogenous plant defence mechanism, natural antioxidants play a significant role. Their intake also confers protection against some human degenerative diseases, as the production of oxidant radicals in biological systems such as animals and plants is a natural phenomenon.

18.3.2 Phenolic Acids

The phenolic acids are one of the important groups of polyphenols which exhibit high levels of antioxidant activity. These compounds are classified into two subgroups, hydroxybenzoic acids and hydroxycinnamic acids, derived from benzoic and cinnamic acid non-phenolic molecules, respectively (Queiroz et al. 2011). Unlike other phenolic compounds, due to the presence of one carboxylic group in the molecule, hydroxybenzoic and hydroxycinnamic acids show an acidic character. Phenolic acids found in cashew apples include gallic acid, protocatechuic acid, ellagic acid, caffeic acid, *p*-coumaric acid, ferulic acid, and conjugated and free cinnamic acid (Michodjehoun-Mestres et al. 2009).

18.3.3 Flavonoids

Around two-thirds of the dietary phenolics are accounted for by flavonoids and are mainly present as glycosides and partly as esters, rather than as the free compounds. The food plate's flavonoid content has been documented to offer biological benefits, such as reducing cancer and cardiovascular disease. The cashew apple contains different glycosylated flavonols and anthocyanins. The glycosylated flavonols in the cashew apple are myricetin-3-*O*-galactoside, myricetin-3-*O*-glucoside, myricetin-3-*O*-xylopyranoside, myricetin-3-*O*-arabinopyrannoside, myricetin-3-*O*-arabinofuranoside, myricetin-3-*O*-rhamnoside, quercetin-3-*O*-galactoside, quercetin-3-*O*-glucoside, quercetin-3-*O*-xylopyranoside, quercetin-3-*O*-arabinopyrannoside, quercetin-3-*O*-arabinofuranoside, quercetin-3-*O*-rhamnoside, and kaemferol-3-*O*-glucoside (de Brito et al. 2007). In the case of anthocyanin, 5-methylcyanidin 3-*O*-hexoside, has been identified as a methyl derivative of cyanidin in the cashew apple extract. The highest concentration of total anthocyanins is present in the skin of the cashew apple (de Brito et al. 2007).

18.3.4 Tannins

Tannins, a class of polyphenols, are also present in cashew apples. Tannins with varying molecular weights are astringent and bitter compounds. These are a group of oligomers of polyhydroxy-flavan-3-ol and polymers with carbon-carbon linkages between subunits of flavanol. Condensed (0.64 mg/100 g) and hydrolyzable (0.18 mg/100 g) are the two major forms of tannin present in cashew apple (Haminiuk et al. 2012; Emelike and Ebere 2016). In the cashew apple, tannins are the key phenolic compounds that are mostly responsible for the astringency of the cashew apple and its juice, hence its low preference and consumption (Vergara et al. 2010; Msoka et al. 2017).

18.3.5 Lignin

The composition of cashew apple bagasse makes it a potential raw material used to manufacture fuel and chemicals, such as ethanol and xylitol, as an alternative to costly lignocellulosic material. Cashew apple bagasse has a high content of lignin (33–35.5%). Lignin is the most abundant natural polymer on earth, insoluble in water and stable in nature, which serves as a 'glue' to bind cellulose with hemicellulose. Lignin has a number of functional groups that specify the polarity and consistency, such as aliphatic hydroxyls, carbonyls, phenolic hydroxyls, methoxyls, and benzyl alcohol groups (de França Serpa et al. 2020). Pre-treatment measures are required to disrupt the lignocellulosic material's resistant structure to increase the matrix's digestibility before conversion. The structure of cashew apple bagasse is highly recalcitrant to microbial and enzymatic biotransformation. The lignin-containing biomass discarded during pre-treatment can be used as a by-product for the manufacture of various value-added goods (Reis et al. 2017).

18.3.6 Carotenoids

Carotenoids are among the most widely distributed pigments, having red, orange, and yellow colours. Carotenoids are lipid-soluble pigments which are present in cashew apple in significant amounts. In the red-coloured cashew apples, carotenoids are present at higher concentrations than in the yellow-coloured and greenish-yellow cashew apple varieties (Msoka et al. 2017).

The cashew apple carotenoids are α-carotene, auroxanthine, δ-carotene, β-cryptoxanthin, *trans*-, and *cis*-β-carotene. Carotenoids have been associated with reducing the risk of many degenerative diseases, including cancer, because of good radical or singlet oxygen trapping properties (De Abreu et al. 2013). *Cis*-violaxanthin is the primary carotenoid (28.3% of total carotenoids) present in cashew apple, followed by *cis*-neoxanthin (22.0%), lutein (15.3%), and neochrome (10.6%). Minor carotenoid compounds are zeanoxanthin, α-carotene, and 13-*cis*-β-carotene. Cashew apple waste is an interesting source of xanthophylls. The important carotenoids responsible for the bright yellow colour of the cashew apple are auroxanthin and β-cryptoxanthin (De Abreu et al. 2013).

18.3.7 Fibre

Fibre is the part of plant-derived food components that human digestive enzymes cannot fully break down. It is best known for its ability to alleviate constipation, maintain a healthy weight, lower heart disease risk, diabetes, and certain cancer forms. Juice processing from cashew apple produces the condensed pulp that mainly contains lignocellulosic material. The two forms of cashew apple fibres, including industrial and artisanal cashew apple fibre, are present in commercial juice. The yellow and red varieties of cashew apples are picked and sanitized to obtain industrial fibres using a chlorinated water solution. Cashew apple flesh is then crushed and pulped for juice extraction. The fibres are collected in a fine-mesh sieve and stored at 4°C in vacuum-sealed plastic bags. Both red and yellow cashew apple varieties are extracted in the test kitchen to obtain artisanal fibres, using a centrifugal juicer to extract the juice. The resulting fibre is hand cut, and, by pressing through a sieve, the remaining juice is extracted. The artisanal fibres are kept at 4°C using vacuum-sealed plastic bags (Sucupira et al. 2020).

18.4 PROCESSING AND UTILIZATION OF THE CASHEW APPLE

The cashew apple can be processed into different value-added fermented (wine, feni, vinegar, mannitol, and xylitol) and non-fermented products (jam, jelly, syrup, squash, pickle, ready-to-serve drink, fruit bar, candy, and cashew apple powder) (Antony, Kunhiraman, and Abdulhameed 2020).

18.4.1 Cashew Apple Wine

Cashew apple wine is a bright yellow drink with a 7.0% alcohol content. It is mildly acidic with a high tannin content (Mohanty et al. 2006; Gawankar et al. 2018; Ketaki, Agte, and Arora 2020). Undistilled alcoholic wine is prepared by fermentation of cashew apple juice. The cashew apples are washed thoroughly with distilled water and soaked in 5% salt solution to reduce the tannin content. The fruits are steamed for 15 mins at 15 psi of pressure and then pounded to remove the juice in a mixer-cum-grinder. The juice is filtered and treated with sodium metabisulfite to prevent the growth of harmful micro-organisms, such as acetic acid bacteria, wild yeasts, and moulds. Cane sugar and tartaric acid are added to the juice to obtain 17°Brix and pH 3.6. Various microbial strains may be used for production of wine from cashew apple juice (Apine and Jadhav 2015). The starter culture most frequently used is *Saccharomyces cerevisiae* (Preethi et al. 2019). Fermentation is carried out at room temperature (28–32°C) for between six days to six weeks (Mohanty et al. 2006). Wine racking is carried out when total soluble solids (TSS) reach 2–3° Brix. Two or three more rackings are performed at 15-day intervals in order to extract any sediment accumulated in the wine (Mohanty et al. 2006). The wine is clarified with the addition of 0.04% bentonite and analysed after racking. Sodium metabisulfite (100 mg/mL) is introduced as a preservative before bottling (Preethi et al. 2019).

18.4.2 Feni

Feni is a three-fold-distilled liquor made from cashew apples. Feni has been an essential part of the Goan food culture for over 400 years (Ketaki, Agte, and Arora 2020). The strong-smelling liquor feni is a trademark of Goa. Sattari is a small town in Goa, renowned for being the capital of feni production in Goa. Cashew apple feni has recently earned its geographical indication registration as a Goa specialty alcoholic beverage. It is a fruity spirit, with a distinct, astringent, and pungent scent, and a peculiar taste. Ethanol, acetic acid, ethyl acetate, acetaldehyde, furfural, and copper are the hydrocarbons, volatile and mineral components of cashew apple feni (Preethi et al. 2019).

Feni is made from fermented cashew apple juice. The juice is boiled before the distillation process, and the condensed liquid is collected. There are three grades of liquor derived from cashew apple juice: urrac, cazulo, and feni. Urrace is a mildly alcoholic beverage, formed by the first distillation. and contains around 15% alcohol (Ketaki, Agte, and Arora 2020; Ranganaekar 2009). Cazula is a stronger liquor, which is prepared by combining urrac with fermented cashew apple juice, containing 40% alcohol, and is produced by the second distillation. A combination of cazulo and fermented juice is used in the third distillation to produce feni as an end product. It contains about 43–45% alcohol, which makes it very heavy and smelly to drink. The high sugar level (6.7–10.5%) in cashew apple is helpful for the rapid fermentation (Rangnekar 2009). Feni distillation takes approximately eight hours to complete (Ketaki, Agte, and Arora 2020).

18.4.3 Vinegar

Vinegar is produced by fermentation of ethanol or sugars by acetic acid bacteria and contains 5–8% acetic acid by volume. Based on the source materials, there are several vinegar types. Vinegar is used as a flavouring agent, acidic cooking ingredient, or pickling agent in the culinary arts. After discarding the damaged or infected apples, fully ripe, firm, and good-quality cashew apples are washed in water, and the juice is extracted. The preparation of vinegar from cashew apples consists of two stages, including alcoholic fermentation and acidic fermentation (Preethi et al. 2019). Sago starch, ammonium sulfate, citric acid, and yeast are the ingredients used for vinegar preparation. First, the sago starch is boiled in water followed by addition of the juice after cooling. The sugar is added to the clarified juice and strained. Then, it is finished with the addition of citric acid, ammonium sulfate, and yeast, and held for seven days in narrow-mouthed plastic bottles with cotton plugs. The alcoholic fermentation begins, and, after one week, the alcoholic ferment is seen at the top and the tannin below. The fermented supernatant juice is isolated through filtration. The mother vinegar is added for acidic fermentation at this stage. The wide-mouthed glass jar containing the juice after adding the mother vinegar is kept again for two weeks for acid fermentation. A live *Acetobacter aceti* bacterium in the mother vinegar works on the alcoholic ferment and transforms the alcohol into acid. The clear juice is filtered into a clean stainless-steel vessel and pasteurized at 70–80°C for ten minutes. The chemical composition of cashew apple vinegar is pH 4.6, TSS 17 °Brix, titratable acidity(0.36%, reducing sugar 6.44%, phenolics 0.12 g/100 mL, and tannin (2.2 mg/100 mL) (Preethi et al. 2019).

18.4.4 Mannitol and Xylitol

Mannitol is a fructose-derived, six-carbon sugar alcohol that occurs naturally in fruits and vegetables and can be produced by several organisms. D-mannitol is the most abundant polyol in nature.

It is believed that almost 75% of mannitol gets fermented by intestinal flora. The glycemic index of mannitol is zero, which makes it an attractive food. So, mannitol can be used as a low-caloric food for diabetic patients. It can be used in foods that require stability at high humidity, as it is nonabsorptive in nature. The glucose-fructose syrup prepared from cashew apple can be used to produce mannitol and gluconic acid. The yeast *Saccharomyces cerevisiae* consumes NADH for mannitol production. Fructose mixture is converted into gluconate and mannitol using glucose dehydrogenase and malate dehydrogenase. By this method, up to 88% of fructose can be converted to mannitol (Wisselink et al. 2002).

Xylitol, a five-carbon sugar alcohol, has a sweetness equivalent to sucrose. D-xylose is the second most abundant sugar in nature, and it can be converted to xylitol (Antony, Kunhiraman, and Abdulhameed 2020). The metabolism of xylitol is independent of insulin so that it can be used in diabetes treatment. Compared with others, xylitol sugars contribute to lower fat tissue development, and thus can be used by obese individuals. Cashew apple bagasse is a lignocellulosic by-product of the cashew apple juice industry. When cashew apple bagasse is pretreated with dilute acid, a suitable substrate is produced that can be converted to xylitol using *Kluyveromyces marxianus CCA510* (Antony, Kunhiraman, and Abdulhameed 2020). A higher concentration of ethanol and xylitol can be obtained from this microorganism compared with *Saccharomyces cerevisiae* (Wisselink et al. 2002).

18.4.5 Lactic Acid

Lactic acid is an organic acid that is white in the solid state and is water-miscible. It forms a colourless solution in the dissolved state. Lactic acid is used to ferment and preserve food substances (Urbonaviciene et al. 2015). Many cheap materials can be used as raw materials from which to produce lactic acid, such as starchy, cellulosic, and renewable materials, such as agricultural waste. Lactic acid has a wide variety of applications such as in the food, textile, pharmaceutical, cosmetics, and chemical industries. In addition to these uses, it can be used as an acidulant, flavouring agents, and a bacterial inhibitor. It can be converted into useful chemicals such as esters, bio-solvents, and so on due to inclusion of hydroxyl and carboxyl groups in its structure (Miller et al. 2019). It is recognized as GRAS (Generally Regarded as Safe) and is authorized by the United States Food and Drug Administration to be used as a food additive and is commonly used in every food industry sector. Two processes, the chemical method and the fermentative method, primarily produce lactic acid. The key downside of chemical synthesis is that it results in racemic mixtures, whereas fermentative processes produce the single-isomer product. Lactic acid is the primary end product of lactic acid bacteria carbohydrate fermentation (Komesu et al. 2017). The production of lactic acid from cashew apple juice in a low-batch reactor is a low-cost and efficient process. Parameters such as pH, temperature, and reducing sugar concentrations affect the assessment of biomass formation and production of lactic acid. The pH to produce lactic acid at a temperature of 38°C should be 6.5. After incubation for three days and the addition of ammonium sulfate, the yield of lactic acid increases. Owing to its natural sources, the use of lactic acid is considered a strong substitute and more acceptable by consumers than synthetic food additives and preservatives, antioxidants, flavourings, and acidifying agents (Cizeikiene et al. 2013).

18.4.6 Enzymes

Different enzymes, such as tannase and pectinase, can be produced from the cashew apple. The enzyme tannase, also referred to as tannin acyl hydrolase, acts on tannin and catalyzes the hydrolysis of bonds found in gallic acid-producing molecules of hydrolyzable tannins and gallic acid esters. The main product of tannic acid degradation is gallic acid. In the food industry, gallic acid is used as a basis for the chemical synthesis of food preservatives such as gallates and pyrogallol.

Propyl gallate is a very effective antioxidant in food and for the synthesis of antibacterial drugs in the pharmaceutical industry (Srivastava and Kar 2009). It is an inducible enzyme that has a wide range of industrial uses. Tannase is used in the industrial processing of fruit juices as a clarifying agent. It decreases the astringent taste of cashew apple caused by the presence of high tannin concentrations (Barthomeuf, Regerat, and Pourrat 1994). Microorganisms are used to produce tannase, including bacteria, fungi, and yeasts. The fungi include *Aspergillus*, *Mucor*, *Penicillium*, *Fusarium*, *Neurospora*, and *Trichoderma*, while the bacteria include *Pseudomonas*, *Bacillus*, *Streptococcus*, and *Achromobacter*, and a few yeast species, including *Candida*, *Debaryomyces*, and *Mycotorula*, are used as tannase producers (Chávez-González et al. 2012). Tannase production is achieved at the industrial level using *Aspergillus* sp. in a solid-state fermenter. The cashew apple's tannin content is important and can be used as a useful substrate to produce the tannase enzyme. In solid-state fermentation, cashew apple bagasse is used as a substrate for maximizing enzyme productivity. The use of tannase at the industrial level is restricted by crucial factors, i.e., production costs and inadequate knowledge of the basic characteristics, physicochemical properties, catalytic characteristics, control mechanisms, and potential uses (Belmares et al. 2004).

Pectic, pectinase, or pectinolytic enzymes are the complex enzymes that hydrolyze the pectic substances. These enzymes are categorized into three major classes based on their mechanism of action, i.e., protopectinases, esterases, and depolymerases. Protopectinases degrade and transform insoluble protopectin to strongly polymerized soluble pectin, esterase helps to de-esterify pectin, and depolymerase plays a significant role in the breakdown of pectin (Shet, Desai, and Achappa 2018). Pectinase is responsible for reducing pectic compounds such as pectinic acids, propectins, pectins, and pectic acids by a complex enzymatic mechanism. This enzyme plays an essential role in virtually all living organisms' metabolic activities, such as plants, animals, fungi, bacteria, and viruses. In many industrial applications, pectinolytic enzymes are used, primarily in the food industry, in specific operations such as the clarification of fruit juices, wines, the extraction of vegetable oils, the grinding of vegetable fibres, the curing of coffee and cocoa, and the processing of pectin-free starch (Satapathy et al. 2020). Cashew apple bagasse is used in the solid-state fermentation system to produce polygalacturonase in a tray bioreactor. Parameters, such as temperature, thickening of the substrate, moisture content, spore concentration, and ammonium sulfate concentration, affect polygalacturonase activity. The activity of the enzyme is highest at 35°C after 29 hours of incubation. Depending on the micro-organism and fermentation conditions, such as pH, incubation period or cultivation time, temperature, source of carbon and nitrogen, agitation, substrate types and concentration, and the use of different preparations, the capacity of these enzymes to break down the biomolecules varies. The yield, stability, and cost of enzyme development are the major obstacles to exploiting pectinases' commercial potential (Sudeep et al. 2020).

18.4.7 Jam

Cashew apples can be utilized to produce homemade products such as jam. The delicious cashew apple jam can be prepared using conventional methods and the osmotic dehydration method. Furthermore, the different variants of mixed-fruit jam can be produced by including other fruits, such as mango and pineapple, in the cashew apple jam recipe (Preethi et al. 2019). However, the cashew apple contains a high concentration of tannin that needs to be reduced before jam preparation. The tannin content can be reduced by immersing cashew apples in 3–5% salt solution for three days. Salt solution needs to be changed daily for a better reduction of tannin content. After de-tanning, the cashew apple is again cleaned with freshwater followed by steaming for 15–20 minutes to soften the pulp, to facilitate crushing of the cashew apple pulp, and to better mix with sugar. The jam can be prepared by following the recipe described in Table 18.1 (Ketaki, Agte, and Arora 2020). In the conventional method, the cashew apple pulp is heated at a low temperature until the appropriate jam consistency or up to 65° Brix has been achieved. Sometimes, the selected cashew

Table 18.1 Recipes for Cashew Apple Jam Preparation

Ingredient	Cashew Apple Jam Recipe	Mixed Fruit Jam Recipe
Cashew apple pulp	1 kg	500g
Mango pulp	-	500g
Sugar	1 kg	1 kg
Citric acid	2–5 g	2.5 g
Reference	Ketaki, Agte, and Arora (192020)	Kerala Agricultural University (2016)

Table 18.2 Recipes for Cashew Apple Syrup Preparation

Ingredient	Recipe I	Recipe II
Clarified cashew apple juice	1 L	1 L
Sugar	1–1.25 kg	2 kg
Citric acid	20–25 g	15 g
Sodium benzoate	0.08%	-
Lemon-yellow colour	-	0.5 g
Reference	Suganya and Dharshini (2011)	Ketaki, Agte, and Arora (192020); Preethi et al. (2019); Runjala and Kella (2017)

apples have insufficient pectin to attain the right gel consistency in a jam. Therefore, pectin can be added to further enhance the gel consistency. Citric acid is added during the cooling process for a better taste. Now, the jam is packed into pasteurized bottles under aseptic conditions to enhance the jam's shelf-life. However, the conventional methods can reduce the nutritional value, colour, and flavour of the jam due to more prolonged heating at higher temperatures. In the osmotic dehydration method, the cashew apple (one part) is dipped in osmotic solution (50° Brix, eight parts) containing potassium metabisulfite (KMS) (1%) for 4 h at 40°C (Ketaki, Agte, and Arora 2020). After draining away the osmotic solution, the osmotically dehydrated apple is converted into pulp to mix in freshly prepared osmotic solution (50° Brix) containing citric acid and pectin. The mixture is heated until the set at 65° Brix, followed by cooling and packing into pasteurized bottles under aseptic conditions. The high total soluble solids could help to retain maximum vitamins, tannins, total phenols content and their antioxidant activity (Preethi et al. 2019).

18.4.8 Syrup

The syrup is one of the most nutritious commercial products made from unfermented cashew apple juice by adding varying concentrations of sugar, citric acid, and preservatives (Sobhana 2019). Cashew apple has a high tannin content. Tannins are an anti-nutritional component, so cashew apple or cashew apple juice is not consumed directly due to its astringency. Tannins interfere with the assimilation of proteins in the body, resulting in nutrients becoming unavailable (Preethi et al. 2019). Detanning is, therefore, an important step before any product is taken out of cashew apple. Clarification (removal of tannins) is important in the processing of cashew apple juice to increase the juice's acceptability. The different approaches used to reduce tannin content in cashew apple juice are microfiltration, use of tannase enzyme, and clearing agents (sago, gelatin, and starch). These methods help to reduce astringency and increase the shelf-life of the juice for further use.

The different recipes for preparing the cashew apple syrup are given in Table 18.2 (Ketaki, Agte, and Arora 2020; Preethi et al. 2019). Fresh ripe cashew apples are picked and washed in clean water for the syrup preparation. The cashew apples in 2% brine solution for five minutes. The cashew apples are rinsed again with clean water, and the juice is extracted and filtered. The required amount of sugar is added to the juice, stirring continuously at moderate heat, until the sugar is fully dissolved. A small quantity of citric acid is added and mixed well. Sodium benzoate (0.08%) is then added to the juice. After stirring continuously at a low flame for some time, it is taken off the heat and the solution allowed to cool. To maximize the syrup's shelf-life, it is poured into sterilized glass bottles or food-grade PET bottles and sealed airtight and kept cool and dry (Gawankar et al. 2018). To serve, the syrup is diluted five-fold with cold water to be used as a fresh drink.

18.4.9 Jelly

Jelly is a semi-solid product prepared by boiling a clear, strained fruit extract free of pulp after adding the required amount of sugar, citric acid, and pectin. Relationships between the three essential ingredients, i.e., pectin, sugar, and citric acid, are critical for the product's quality. Low pectin or citric acid can prevent gel formation whereas too little sugar results in a tough jelly. The amount of sugar in the jelly depends on the pectin's quantity and consistency when the sugar content of the mixture is increased, or the pectin content is reduced, and the jelly becomes weaker. The concentration of sugar should be that of a saturated sugar solution for the better development of jelly. The citric acid is added to neutralize the charge on the carboxyl groups of pectin and increase the molecules' tendency to associate to form a gel. The pH must be below 3.5 for making jelly (Preethi et al. 2019).

For jelly preparation, cashew apples are washed properly with tap water and then diced into pieces. The cashew apple pieces (250 g) are boiled in 500 mL water for 30 mins in a stainless-steel pan to make a cashew apple extract. The extract is then filtered. The appropriate amount of sugar is added to the cashew apple extract. The other ingredients, such as pectin or some other fruit extract, are mixed into the cashew apple extract. The citric acid is added after the jelly has been cooked, to enhance the flavour and avoid sugar crystallization by converting it to invert sugar. Citric acid reduces the pH and helps to form a gel. The remaining sugar is added to the mixture and boiled until the TSS is 68° Brix. When the mix is sufficiently dense in consistency, the sheet measurement should judge the endpoint. A small amount of jelly is taken in a spoon during the boiling process and then allowed to drop. If the product comes in the form of a sheet or flakes instead of flowing in a continuous stream, the product is ready. The jelly is poured into clean and dry wide-mouth pre-pasteurized PET bottles or hot sterilized bottles. The bottles are packed at least 90% full, leaving no more than 1.25 cm of space at the top of the bottles ("headspace"e). After the bottles have cooled down, they can be sealed for storage (Preethi et al. 2019).

18.4.10 Pickle

Pickle is an innate food product in all societies and cultures around the globe. It is eaten as a delicious, spicy accompaniment to a meal. Pickling requires the preservation of food products at a highly acid formulation, allowing them to be preserved for more than two years without refrigeration. Pickles are sustained by a combination of increased acidity (reduced pH), added salt, reduced moisture, and added spices. Vinegar is the main ingredient in the production of pickles. It is extracted from natural sugars or starches *via* a two-step fermentation process. Starch is converted to sugar, which is then fermented by yeast to form alcohol. Alcohol is exposed to *Acetobacter*, which converts it into vinegar. The vinegar is added for flavour and also prevents the development of unwanted micro-organisms. Salt helps to extract excess water from the fruit and vegetables and to unlock the aromatic juices and flavors. It gives the pickles a solid texture. Spices improve flavour and taste (Ketaki, Agte, and Arora 2020).

Cashew apple pickles can be prepared using any of the recipes given in Table 18.3 (Preethi et al. 2019). Fresh cashew apples are washed and cut to a uniform size for the preparation of the pickle. The cashew apple pieces are marinated in the salt solution for at least 24 hours. Gingelly (sesame) oil is boiled in a pre-heated steel vessel and fenugreek powder, turmeric powder, chilli powder, and

Table 18.3   Recipes for Cashew Apple Pickles Preparation

Ingredient	Recipe I	Recipe II
Detanned cashew apple slices	500 g	1 kg
Gingelly oil	75 mL	100 mL
Chilli powder	75 g	100 g
Fenugreek powder	20 g	10 g
Asafoetida powder	10 g	10 g
Turmeric powder	5 g	5 g
Mustard powder	5 g	5 g
Mustard	10 g	2 g
Garlic paste	-	10 g
Ginger paste	-	10 g
Green chilli paste	-	10 g
Vinegar	100 mL	150 mL
Sodium benzoate	0.75 g	0.75 g
Salt	To taste	To taste
Reference	Preethi et al. (2019)	Runjala and Kella (2017)

mustard powder are added to the boiling gingelly oil (Ketaki, Agte, and Arora 2020; Preethi et al. 2019). After the chilli powder colour changes, garlic paste, ginger paste, and green chilli paste are added. The detanned cashew apple slices are then added to the boiling mixture to cook the mixture for at least 5–10 minutes, followed by mixing the vinegar and salt with constant stirring at low temperature. A small amount (1 pinch or 0.75 g) of sodium benzoate is added (Runjala and Kella. 2017). The mixture is allowed to cool. After cooling, the liquid is poured into dry and sterilized glass pots.

18.4.11 Candy

The cashew apple is a convenient alternative material for adding value to products such as candy and for generating income for cashew farmers because they can be preserved for months without undesirable changes, preserving their organoleptic properties such as aroma, taste, texture, and colour (Ketaki, Agte, and Arora 2020). Candy, also referred to as sweets or lollies, is a confectionery product made from a syrup that is boiled until it begins to caramelize by dissolving desired concentration of sugar. The kind of candy depends on the ingredients and how long the mixture is cooked. Candy has a wide range of textures, ranging from smooth and chewy to rough and brittle. Sugar candies include hard candies, soft candies, caramels, marshmallows, taffy, and other candies whose principal ingredient is sugar. It is possible to categorize sugar candy into non-crystalline and crystalline groups. Non-crystalline candies, such as hard candies, caramels, toffees, and nougats, are homogenous and can be chewy or hard. In their composition, crystalline candies have small crystals, are creamy that melt in the mouth or are easy to chew, such as fondant and fudge.

Cashew apple candy is an economical alternative capable of adding value to raw materials. Cashew apples are selected that are sound and ripe with a soluble solid content between 10.5 and 11.5 and they must not be of the sour kind to make cashew apple candy. Their colour can be red or yellow, and there are no selection criteria in this respect. To minimize the microbial load on the fruits' surface, the cashew apples are washed and surface sterilized by immersion in sodium hypochlorite solution (with 8% of active chlorine). The sugar syrup is prepared to make candy by dissolving the sugar in water and heating it well. In this solution, citric acid and potassium metabisulfite are dissolved (Preethi et al. 2019). In order to fully immerse the apples, the pre-prepared cashew apple pieces are dropped into the boiling sugar syrup. The vessel should be covered with a lid and held as such for one day. On the second day, the fruit should be taken out, more sugar added and the cashew apple pieces dropped back in while heating. After seven days, the sugar syrup amount will be reduced to one-third, which will allow the cashew apples to be kept for 8–10 days. The solution should not have less than the TSS of 70° Brix. The cashew apples from the syrup solution are removed and drained after removal and allowed to dry in an open area by spreading them over a sheet of polythene. After the mixture has cooled down, it is placed on a wooden cutting table. The mixture is broken into pieces or molds are used to make the various candy shapes. Cashew apple candies are packaged in high-density plastic bags, labelled, and stored in a dry and well-ventilated place (Runjala and Kella. 2017).

18.4.12 Squash

Fruit squash is a beverage which is consumed after dilution and which essentially consists of a moderate quantity of fruit pulp/juice (min 25%) to which cane sugar is added for sweetening to raise the TSS above 40° Brix, with citric acid added to 3.5% or less. It should contain class II preservatives such as sulfur dioxide (≤ 350 ppm) or benzoic acid (≤ 600 ppm). Squash is prepared by blending the juice with sugar, citric acid, and water (Preethi et al. 2019).

Different variants of cashew apple squash can be prepared by selecting ingredients highlighted in Table 18.4 (Ketaki, Agte, and Arora 2020; Preethi et al. 2019). Good-quality cashew apples are selected and washed properly, peeled and the juice extracted. The juice is clarified by adding polyvinylpyrrolidone and filtering it through a muslin cloth. The other ingredients, such as sugar, water, and citric acid, are dissolved into the clarified juice which should be stirred well under moderate heat until well mixed and slightly sticky. Preservatives such as KMS are added to increase the shelf-life of the squash and the mixture is allowed to cool at room temperature, before being strained through a muslin cloth to remove any pulp or impurities from the prepared squash. The squash is stored in sterilized bottles in a cool place (Preethi et al. 2019; Runjala and Kella. 2017).

18.4.13 Ready-to-Serve Beverage

Ready-to-serve (RTS) beverage is a type of fruit drink containing at least 10% fruit juice and 10% total soluble solids and about 0.3% acid. Before serving, it does not require dilution. It is prepared by adding the appropriate amount of clarified juice with water and sugar. With storage, the total soluble solids increase, which may be due to polysaccharide hydrolysis into monosaccharides and oligosaccharides. The TSS influences the juice blending ratio, processing temperature, and potassium metabisulfite during storage. Due to the conversion of acids into salts and the production of sugars by enzymes, particularly invertase, the titratable acidity decreases as time passes. The RTS shelf life can be increased by using class II preservatives such as sulfur dioxide (≤ 70 ppm) or benzoic acid (≤ 120 ppm). UHT processing of RTS drinks is quite popular due to the longer shelf-life and the lower loss of nutrients during processing.

Table 18.4 Recipes for Cashew Apple Squash Preparation

Ingredients	Recipe I	Recipe II	Recipe III
Cashew apple juice	8 L	1 L	250 mL
Polyvinylpyrrolidone	10 g	-	-
Sodium benzoate	6 g	-	-
Sugar	3 kg	1.6 kg	400g
Citric acid	100 g	2–5 g	2.5 g
Water	-	1.40 L	350 mL
Lemon-yellow colour	-	1.25 mg	-
Reference	Preethi et al. (2019)	Runjala and Kella (2017)	Ketaki, Agte, and Arora (2020)

Table 18.5   Recipes for Cashew Apple Ready-to-Serve (RTS) Beverage Preparation

Ingredient	Cashew Apple-RTS Beverage	Cashew Apple + Any Other Juice Blended RTS Beverage
Clarified cashew apple juice	150 mL	200 mL
Other fruit juice	-	100 mL
Sugar	120 g	200 g
Water	730 mL	1 L
Citric acid	5 g	5 g
Lemon-yellow colour	Should not exceed 100 ppm	-
Reference	Ketaki, Agte, and Arora (2020)	Preethi et al. (2019)

The cashew apple RTS beverage can be prepared by following the recipe given in Table 18.5 (Ketaki, Agte, and Arora 2020). The requisite amount of sugar is dissolved in water. To prepare the sugar syrup, the sugar should be dissolved fully followed by boiling. The syrup solution is strained to remove any impurities. Clarified hot cashew apple juice is added immediately with continuous stirring and a small quantity of citric acid is added to the solution. Finally, the TSS should be adjusted to 10% and the acidity to 0.3% (Gawankar et al. 2018). The solution is cooled and lemon-yellow colour added to it for better acceptance colour of the product. The mixture is strained and allowed to cool down. For bottling, the bottles are pasteurized by keeping in boiled water for 20 mins to avoid contamination, to increase the shelf-life of cashew apple RTS beverage. The prepared product is poured into pre-pasteurized bottles and then capped with hot-water-sterilized lids to ensure the safety under storage of the prepared RTS beverage. Packing can also be done in food-grade plastic covers using a liquid packaging machine.

18.4.14 Fruit Bar

A fruit bar is a dehydrated and shelf-stable product used as a confectionery with low water activity and low moisture content (15–25%). By combining fruit purées or pulp extracted from ripe pulpy fruit with sugar or other nutritional sweeteners, and other ingredients and additives needed for the product, the fruit bar is prepared and dehydrated to form sheets that can be cut to the desired shape and size. It is suitable for use as an instant food since it is easy to carry. In terms of packaging and transport, the fruit bar is an incredibly convenient product that is easy to eat and tasty, that can be eaten and distributed anywhere (Poornakala et al. 2020). Cashew apple fruit bars are pure and nutritious items, eaten as a snack as a healthier alternative to other sweets, cookies, or cakes, which contain very high concentrations of fat or sugar.

The cashew apple bar is made from purée that can be extracted from the pulp. Sugars, citric acid, pectin, and permissible preservatives are added to the purée. In conventional cashew apple fruit bar preparation, sugar is essential. In a cabinet dryer, the mixture is dried in the form of sheets. Natural ripe fruit pulps, along with other ingredients, result in improved flavour in bars. In order to minimize fruit bar browning and microbial load, before drying, sodium metabisulfite must be applied to the cashew apple pulp. The cashew apple fruit bar is also prepared with a 50:50 ratio of blended ripe banana. In 1000 g of fruit purées, this blended fruit bar contains 7% sugar, 5% date powder, 0.2% sodium metabisulfite, and 0.5% citric acid, and dried for 8 hours at 80°C. Improved nutritional characteristics, sensory characteristics (colour, texture, and flavour), and storage stability during product formulation are the main objectives of selecting one or more fruits for blending. Different types of packaging materials such as laminated aluminium foil, low-density polyethylene, high-density polyethylene, and polypropylene can be used for the wrapping of fruit bars.

18.4.15 Cashew Apple Powder

Fruit powders have economic advantages over their liquid precursors, such as reduced volume/weight, reduced packaging, ease of handling/transportation, and much longer shelf-life. To meet customers' expectations of storage and ease-of-use, powder form can be made from fruit that has health-promoting and nutritional benefits (Poornakala et al. 2020).

For the preparation of powder, high-grade cashew apples are selected. After removing the nuts, the cashew apples are washed in clean, cold water and soaked in 2% salt solution (for 10 mins) for the de-tanning process. The cashew apples are then rinsed in clean water and dried at room temperature. The cashew apple is cut into slices and steamed (2 mins) for blanching. The cashew apple slices are then exposed to sulfur fumes to prevent mould growth and dark fruit colour. The slices are then blended and strained. After straining, the juice and pulp residue are separated. The pulp residue is dried at 60°C and then ground to a powder. The dehydrated cashew apple powder is packed and stored.

This dehydrated powder is used in the preparation of wheat laddu, sponge cake, chocolate, soup, and sweet kadabu recipes at the rate of 10% substitution (Ketaki, Agte, and Arora 2020). Cashew apple biscuits are also prepared using cashew apple powder, ghee, sugar, and maida in required quantities and subjected to further baking. Cashew apple juice blended with skimmed milk in the ratio of 15:85 is subjected to spray drying, and the powder obtained is used in the preparation of halwa and lassi.

18.5 CONCLUSIONS

In the domestic and global forum, cashew apple, a waste from cashew fruit production, has a potential market. The main obstacle to the full potential of using the cashew apple and related products is the lack of knowledge and technology at the storage and processing stage. It is one of the primary areas of indigenous fruit utilization and opens up wider market opportunities and enormous scope for commercialization scope. After harvest, the cashew apple is subjected to several processes that affect its physicochemical characteristics and quality attributes. There are many conventional and industrial ways to eliminate the cashew apple's astringency before the preparation of any product from it. The outstanding qualities of cashew apple offer immense possibilities for its processing into different value-added products. As discussed in the chapter, the consumption of these products can be industrialized. Cashew apple value-added products are thus a valid research area for food technologists, industrialists, and producers, and these products represent alternative, sustainable and nutritious foods or drinks.

REFERENCES

Antony, A. P., S. Kunhiraman, and S. Abdulhameed. 2020. "Bioprocessing with cashew apple and its by-products." In *Valorisation of agro-industrial residues – Volume II: Non-biological approaches. Applied environmental science and engineering for a sustainable future*, edited by Z. Zakaria, C. Aguilar, R. Kusumaningtyas, and P. Binod, 83–106. Cham: Springer.

Apine, O. A., and J. P. Jadhav. 2015. "Research paper fermentation of cashew apple (*Anacardium Occidentale*) juice into wine by different *Saccharomyces cerevisiae* strains: A comparative study." *Biotechnology* 4 (3): 6–10.

Barthomeuf, C., F. Regerat, and H. Pourrat. 1994. "Production, purification and characterization of a tannase from *Aspergillus niger* LCF 8." *Journal of Fermentation and Bioengineering* 77 (3): 320–23.

Belmares, R., J. C. Contreras-Esquivel, R. Rodríguez-Herrera, A. R. Coronel, and C. N. Aguilar. 2004. "Microbial production of tannase: An enzyme with potential use in food industry." *LWT – Food Science and Technology* 37 (8): 857–64.

Brijyog, L. P. Singh, and A. Maiti. 2017. "Pharmacological importance of anacardium occidentale: A review." *Asian Journal of Pharmaceutical Education and Research* 6 (1): 40–51.

Chávez-González, M., L. V. Rodríguez-Durán, N. Balagurusamy, A. Prado-Barragán, R.Rodríguez, J. C. Contreras, and C. N. Aguilar. 2012. "Biotechnological advances and challenges of tannase: An overview." *Food and Bioprocess Technology* 5 (2): 445–59.

Cizeikiene, D., G. Juodeikiene, A.Paskevicius, and E. Bartkiene. 2013. "Antimicrobial activity of lactic acid bacteria against pathogenic and spoilage microorganism isolated from food and their control in wheat bread." *Food Control* 31 (2): 539–45.

de Abreu, F. P., M. Dornier, A. P. Dionisio, M. Carail, C. Caris-Veyrat, and C. Dhuique-Mayer. 2013. "Cashew apple (*Anacardium occidentale L.*) extract from by-product of juice processing: A focus on carotenoids." *Food Chemistry* 138 (1): 25–31.

de Brito, E. S., M. C. P. de Araújo, L. Z. Lin, and J. Harnly. 2007. "Determination of the flavonoid components of cashew apple (*Anacardium occidentale*) by LC-DAD-ESI/MS." *Food Chemistry* 105 (3): 1112–1118.

de França Serpa, J., J. de Sousa Silva, C. L. B. Reis, L. Micoli, L. M. A. e Silva, K. M. Canuto, A. C. de Macedo, and M. V. P. Rocha. 2020. "Extraction and characterization of lignins from cashew apple bagasse obtained by different treatments." *Biomass and Bioenergy* 141: 1–11.

Devi, V. C., J. Naveen, and P. V. Kannan. 2015. "Process optimization of bio-ethanol production through cashew apple juice fermentation." *International Journal of ChemTech Research* 8 (1): 79–81.

Emelike, N. J. T., and C. O. Ebere. 2016. "Effect of treatments on the tannin content and quality Assessment of cashew apple juice and the kernel." *European Journal of Food Science and Technology* 4 (3): 25–36.

Gawankar, M. S., B. R. Salvi, C. D. Pawar, M. H. Khanvilkar, S. P. Salvi, N. V. Dalvi, K. V. Malshe, D. S. Kadam, Y. S. Saitwal, and P. M. Haldankar. 2018. "Technology development for cashew apple processing in Konkan region- A review." *Advanced Agricultural Research & Technology Journal* 2 (1): 40–47.

Gordon, A., M. Friedrich, V. M. da Matta, C. F. H. Moura, and F. Marx. 2012. " Changes in phenolic composition, ascorbic acid and antioxidant capacity in cashew apple (*Anacardium Occidentale L.*) during ripening ." *Fruits* 67 (4): 267–76.

Haminiuk, Charles W. I., G. M. Maciel, M. S. V. Plata-Oviedo, and R. M. Peralta. 2012. Phenolic compounds in fruits - An overview." *International Journal of Food Science and Technology* 47 (10): 2023–44.

Inyang, U. E., and U. J. Abah. 1997. "Chemical composition and organoleptic evaluation of juice from steamed cashew apple blended with orange juice." *Plant Foods for Human Nutrition* 50 (4): 295–300.

Kerala Agricultural University. 2016. *Package of practices recommendations: Crops.* 15th edition. Thrissur: Kerala Agricultural University.

Ketaki, B., V. Agte, and A. Arora. 2020. "*Anacardium occidentale* by-product: A review on sustainable application and added-value." *Journal of Food Nutrition and Metabolism*, 3 (1): 1–6.

Komesu, A., J. A. Ra. de Oliveira, L. H. da Silva Martins, M. R. Wolf Maciel, and R. M. Filho. 2017. "Lactic acid production to purification: A review." *BioResources* 12 (2): 4364–83.

Kouassi, E. K. Appiah, Y. Soro, C. Vaca-Garcia, and K. B. Yao. 2018. "Chemical composition and specific lipids profile of the cashew apple bagasse." *Rasayan Journal of Chemistry* 11 (1): 386–91.

Lopes, M. M. de Almeida, M. R. A. de Miranda, C. F. H. Moura, and J. E. Filho. 2012. "Bioactive compounds and total antioxidant capacity of cashew apples (*Anacardium occidentale* L.) during the ripening of early dwarf cashew clones." *Ciência e Agrotecnologia* 36 (3): 325–32.

Lowor, S. T., and C. K. Agyeute-Badu. 2009. "Mineral and proximate composition of cashew apple (*Anarcadium Occidentale* L.) juice from northern savannah, forest and coastal savannah regions in Ghana." *American Journal of Food Technology* 4 (4): 154–61.

Michodjehoun-Mestres, L., J. M. Souquet, H. Fulcrand, C. Bouchut, M. Reynes, and J. Marc Brillouet. 2009. "Monomeric phenols of cashew apple (*Anacardium Occidentale* L.)." *Food Chemistry* 112 (4): 851–57.

Miller, C., A. Fosmer, B. Rush, T. McMullin, D. Beacom, and P. Suominen. 2019. "Industrial production of lactic acid." *Comprehensive Biotechnology* 3: 179–188.

Mohanty, S., P. Ray, M. R. Swain, and R. C. Ray. 2006. "Fermentation of cashew (*Anacardium Occidentale* L.) "apple" into wine." *Journal of Food Processing and Preservation* 30 (3): 314–22.

Msoka, R., N. Kassim, E. Makule, and P. Masawe. 2017. "Physio-chemical properties of five cashew apple (*Anacardium Occidentale* L.) varieties grown in different regions of Tanzania." *International Journal of Biosciences* 11 (5): 386–95.

Nam, T. Nhat, N. P. Minh, D. Thi, and A. Dao. 2014. "Investigation of processing conditions for dietary fiber production from cashew apple (*Anacardium Occidentale* L.) residue." *International Journal of Scientific & Technology Research* 3 (1): 148–56.

Poornakala, S. J., V. M. Indumathi, and K. Shanthi. 2020. "Value addition in cashew apple: A review." *International Journal of Chemical Studies* 8 (6): 819–825.

Preethi, P., A. D. Rajkumar, M. Shamsudheen, and M. G. Nayak. 2019. "Prospects of cashew apple – A compilation report", *Technical Bulletin No.* 2/2019. ICAR- Directorate of Cashew Research, Puttur, 28.

Queiroz, C., A. J. R. da Silva, M. L. M. Lopes, E. Fialho, and V. L. Valente-Mesquita. 2011. "Polyphenol oxidase activity, phenolic acid composition and browning in cashew apple (*Anacardium Occidentale* L.) after processing." *Food Chemistry* 125 (1): 128–32.

Ranganaekar, D. 2009. "Geographical indications and localisation ndiccase study of Feni Dwijen Rangnekar." *ESRC Project on GI and FENI* 1–64.

Reis, C. L. Borges, L. M. Alexandre e Silva, T. H. S. Rodrigues, A. K. N. Félix, R. S. de Santiago-Aguiar, K. M. Canuto, and M. V. P. Rocha. 2017. "Pretreatment of cashew apple bagasse using protic ionic liquids: enhanced enzymatic hydrolysis." *Bioresource Technology* 224: 694–701.

Runjala, S., and L. Kella. 2017. "Cashew apple (*Anacardium occidentale* L.) therapeutic benefits, processing and product development: An overview." *The Pharma Innovation Journal* 6 (7): 260–264.

Satapathy, S., J. R. Rout, R. G. Kerry, H. Thatoi, and S. L. Sahoo. 2020. "Biochemical prospects of various microbial pectinase and pectin: An approachable concept in pharmaceutical bioprocessing." *Frontiers in Nutrition* 7: 1–17.

Shet, A. R., S. V. Desai, and S. Achappa. 2018. "Pectinolytic enzymes: Classification, production." *Research Journal of Life Sciences, Bioinformatics, Pharmaceutical and Chemical Sciences* 4 (3): 337–48.

Sobhana, A. 2019. "Cashew apple utilization-generating wealth from waste." *Advances in Nutrition & Food Science* 4 (4): 1–5.

Srivastava, A., and R. Kar. 2009. "Characterization and application of tannase produced by *Aspergillus Niger* ITCC 6514.07 on Pomegranate Rind." *Brazilian Journal of Microbiology* 40 (4): 782–89.

Sucupira, N. R., L. B. de Sousa Sabino, L. G. Neto, S. T. Gouveia, R. W. de Figueiredo, G. A. Maia, and P. H. Machado de Sousa. 2020. "Evaluation of cooking methods on the bioactive compounds of cashew apple fibre and its application in plant-based foods." *Heliyon* 6 (11): 1–9.

Sudeep, K. C., J. Upadhyaya, D. R. Joshi, B. Lekhak, D. K. Chaudhary, B. R. Pant, T. R. Bajgai, R. Dhital, S. Khanal, N. Koirala and V. Raghavan. 2020. "Production, characterization and industrial application of pectinase enzyme isolated from fungal strains." *Fermentation* 6 (2): 1–10.

Suganya, P., and R. Dharshini. 2011. "Value added products from cashew apple -an alternate nutritional source." *International Journal of Current Research* 3 (7): 177–80.

Talasila, U., R. R. Vechalapu, and K. B. Shaik. 2012. "Clarification, preservation and shelf life evaluation of cashewapple juice." *Food Science and Biotechnology* 21 (3): 709–14.

Urbonaviciene, D., P. Viskelis, E. Bartkiene, G. Juodeikiene, and D. Vidmantiene. 2015. "The use of lactic acid bacteria in the fermentation of fruits and vegetables — Technological and functional properties." In *Biotechnology*, edited by D. Ekinci, 135–64. London: IntechOpen.

Vergara, C. M. de Araújo Chagas, T. L. Honorato, G. A. Maia, and S. Rodrigues. 2010. "Prebiotic effect of fermented cashew apple (*Anacardium occidentale* L) juice." *LWT - Food Science and Technology* 43 (1): 141–45.

Wisselink, H., R. Weusthuis, G. Eggink, J. Hugenholtz, and G. Grobben. 2002. "Mannitol production by lactic acid bacteria: A review." *International Dairy Journal* 12:151–161.

CHAPTER 19

Pineapple Wastes and By-Products
Chemistry, Processing, and Utilization

Suheela Bhat, Priyanka Suthar, Shafiya Rafiq, Asmat Farooq, and Touseef Sheikh

CONTENTS

19.1 INTRODUCTION

Pineapple (*Ananas comosus* L.) belongs to the family Bromeliaceae and is the only fruit in this family grown commercially for its nutritional and consumer acceptability. Pineapple is the third most important fruit in the world in terms of production, and countries in Asia, Africa and Central America are the main producers (Lobo and Paull, 2017). Being a tropical fruit, pineapple is mainly used for the preparation of juices like ready-to-serve drinks, concentrates and squash, jellies and canned products, with the remaining part (pomace) being discarded. Rapid growth of the

DOI: 10.1201/9781003164463-19

fruit processing industries has led to huge production of fruit wastes and by-products. Due to the high amount of organic matter in fruit by-products and wastes as these have high biological oxygen demand (BOD) and chemical oxygen demand (COD), problems related to disposal, environmental pollution and sustainability are numerous. Moreover, poor transportation and infrastructure also cause wastage. In addition, there is a considerable loss of valuable nutrients and biomass that could otherwise be employed in the development of low-cost value-added products. Such wastage can be reduced by introduction of modern technology and processing techniques to reutilize the wastes, bearing in mind that the net cost of reusing wastes should not exceed that of their safe disposal. In 2017, the world production of pineapple was 25.9 million tons and hence it is considered to be the second most important tropical fruit crop (Casabar et al., 2019). The main pineapple-producing countries are Costa Rica, the Philippines, Brazil, Thailand and India with production in 2017 of 3,056.45, 2,671.71, 2,253.9, 2,123.18 and 1,861 thousand tons, respectively (Casabar et al., 2019).

19.2 PINEAPPLE WASTES AND BY-PRODUCTS

From the enormous production, approximately 40–80% is considered to be by-products and wastes, and is discarded, despite being a good source of many nutrients and bioactive phytochemicals. Tropical and sub-tropical fruit processing generally have much higher proportions of by-product generation than temperate fruits and pineapple is no exception. It is believed that the percentage of by-products and wastes from pineapple is greater than the part that is being commercially utilized. In addition to processing waste, rough handling during transportation and storage results in 55% of fruits being unfit for fresh use (Nunes et al., 2009). The utilization of pineapple-derived wastes and by-products into beneficial products was thought to be one of the processing strategies by which to reduce wastage. However, these wastes are also prone to microbial spoilage and thus lack the potential for further exploitation. Drying, transportation and storage costs of such wastes are more expensive than the real value of these materials so that eco-friendly, cost effective and highly efficient utilization of wastes is becoming essential. The major pineapple wastes generated during processing are peel (30–42%), core (9–10%), stem (2–5%) and crowns (2–4%) (Table 19.1).

Pineapple wastes and by-products consist of very high concentrations of reducing sugars, proteins and cellulosic components. Several studies have been carried out to find possible ways to convert pineapple wastes and by-products into useful products. It is anticipated that the discarded by-products from pineapple processing industries can be utilized for further industrial processes like fermentation, bioactive component extraction, etc.

19.2.1 Peel

Pineapple peel represents 30–42% of total waste generated from the pineapple industry. Peel is one of the major by-products in terms of weight percentage in the pineapple processing industry.

Table 19.1   The Major Constituents Present in Pineapple

Concentration (% Dry Basis)	Fresh	Dry	Ensiled	Skin	Pulp	Crown	Peel
Pectin	6.7	7.1	5.1	-	-	-	-
Cellulose	11.2	12	9	14	14.3	29.6	19.8
Hemicelluloses	7	6.5	4.7	20.2	22.1	23.2	11.7
Protein	3.13	3.3	0.91	4.1	4.6	4.2	-
Reducing sugars	25.8	27.8	5	-	-	-	-
Non-reducing sugars	5.7	4.9	1.7	-	-	-	-
Lignin	11.52	11	9	1.5	2.3	4.5	-

Source: Adapted from Dorta and Sogi (2017)

The relatively higher peel proportion is due to deep-set eyes in pineapple that require additional hand peeling. In certain cultivars of pineapple like 'Phu Lau', peel percentage is greater than other cultivars due to much deeper eyes of clones in the 'Queen' group. Pineapple peel is a major waste and by-product generated from pineapple processing (Ketnewa et al., 2012). Chemically, peel consists of a large number of sugars that can be utilized as a substrate in various fermentation processes like methane production, hydrogen generation, etc. (Choonut et al., 2014).

19.2.2 Core

Core makes up 9–10% of the total wastes generated from the pineapple industry and is the second-largest by-product/biowaste. Core is considered to be the main source of dietary fiber from the whole pineapple. It has relatively high beneficial value when it comes to prevention of important medical issues related to diet, especially in the West. Moreover, the pineapple core is rich in the proteolytic enzyme bromelain, and has the highest concentration of vitamin C in any part of this fruit. Pineapple juice concentrate, alcoholic beverage production and vinegar are some of the main products obtained from the pineapple core (Ketnawa et al., 2012) (Figure 19.1).

19.2.3 Stem and Crown

The stem consists of 2–5% of the total biowastes from the pineapple industry whereas the crown represents 2–4% of the total solid biowastes/by-products; the crown is sometimes used as planting material. Both the crown and stem contain negligible amounts of sugars and are fibrous in nature. Biodegradable wastes and by-products from pineapple processing industries have very high BOD (Biochemical Oxygen Demand) and COD (Chemical Oxygen Demand) values that can cause serious environmental problems. Disposal of such products is not cost effective for industries due to high transportation costs. Moreover, the limited availability of landfill sites and the strict legal restrictions further hinder the process of waste disposal. Recent approaches focus on co-digestion of solid wastes so as to reduce it by 50–60% in order to reduce transportation costs and reduce the space required for its disposal (Upadhyay et al., 2010). Moreover, researchers focus on minimizing the generation of wastes from pineapple canneries by adapting highly efficient processing techniques to make the process economical and beneficial. Another important decision regarding by-product management from the pineapple industries is its utilization due to the fact that the wastes are very rich in many nutrients and bioactive chemicals. The economic conditions also influence the utilization of wastes. The selling price of the final product should not exceed the value of converting wastes and by-products into the final product

19.3 CHEMISTRY OF PINEAPPLE WASTES AND BY-PRODUCTS

The analysis of pineapple wastes and by-products was reported by Roda and Lambri (2019) who stated that concentrations of nearly 5% protein and approximately 5% polyphenols were present in pineapple wastes. The pineapple wastes also contain small amounts of mineral elements like Ca, Zn, Fe, Cd, Mn, Na and Cu. Potassium concentration is reported to be the most abundant of the elements in pineapple waste. Analysis of the liquid waste and by-products showed the presence of important sugars like glucose, sucrose and fructose. The pineapple leaf fibers have high concentrations of cellulose, lignin and hemicelluloses. Lignin and hemicellulose concentrations are 4–15% and 6–19%, respectively. According to Salve and Ray (2020), the pineapple waste and by-products have concentrations (mg/100 g) of ascorbic acid 26.5 mg/100 g, ash 0.04 mg/100 g, crude fiber 0.60 g/100 g fresh weight, moisture 91.35%- non-reducing sugars 8.8%, protein 10 mg/100 g, reducing sugars 8.2%, titratable acidity 1.86%, total soluble solids 10.2% and total sugars 9.75%.

Liquid byproducts	Major products	Solid byproducts

```
                              ┌──────────────┐
                              │  Receiving   │
                              └──────┬───────┘
                                     │
    ┌──────────────┐          ┌──────┴───────┐
    │ Waste water  │◄─────────│   Washing    │
    └──────────────┘          └──────┬───────┘
                                     │
                              ┌──────┴───────┐       ┌──────────────┐
                              │ Size sorting │──────►│  Cull fruits │
                              └──────┬───────┘       └──────────────┘
                                     │
                              ┌──────┴───────┐       ┌──────────────┐
                              │  Decrowing   │──────►│    Crown     │
                              └──────┬───────┘       └──────────────┘
                                     │
                              ┌──────┴───────┐       ┌──────────────┐
                              │   Peeling    │──────►│     Peel     │
                              └──────┬───────┘       └──────────────┘
                                     │
                              ┌──────┴───────┐       ┌──────────────┐
                              │    Coring    │──────►│     Core     │
                              └──────┬───────┘       └──────────────┘
                                     │
                              ┌──────┴───────┐       ┌──────────────┐
                              │   Trimming   │──────►│  Eyes, flesh │
                              └──────┬───────┘       └──────────────┘
                                     │
                              ┌──────┴───────┐       ┌──────────────┐
                              │   Slicing    │──────►│Damaged slices│
                              └──────┬───────┘       └──────────────┘
                                     │
                              ┌──────┴───────┐
                              │   Filling    │
                              └──────┬───────┘
                                     │
                              ┌──────┴───────┐
                              │ Adding sugar │
                              └──────┬───────┘
                                     │
                              ┌──────┴───────┐
                              │  Exhausting  │
                              └──────┬───────┘
                                     │
                              ┌──────┴───────┐
                              │   Seaming    │
                              └──────┬───────┘
                                     │
                              ┌──────┴───────────┐
                              │ Thermal processing│
                              └──────┬───────────┘
                                     │
                              ┌──────┴───────┐
                              │   Cooling    │
                              └──────┬───────┘
                                     │
                              ┌──────┴───────┐
                              │   Labeling   │
                              └──────────────┘
```

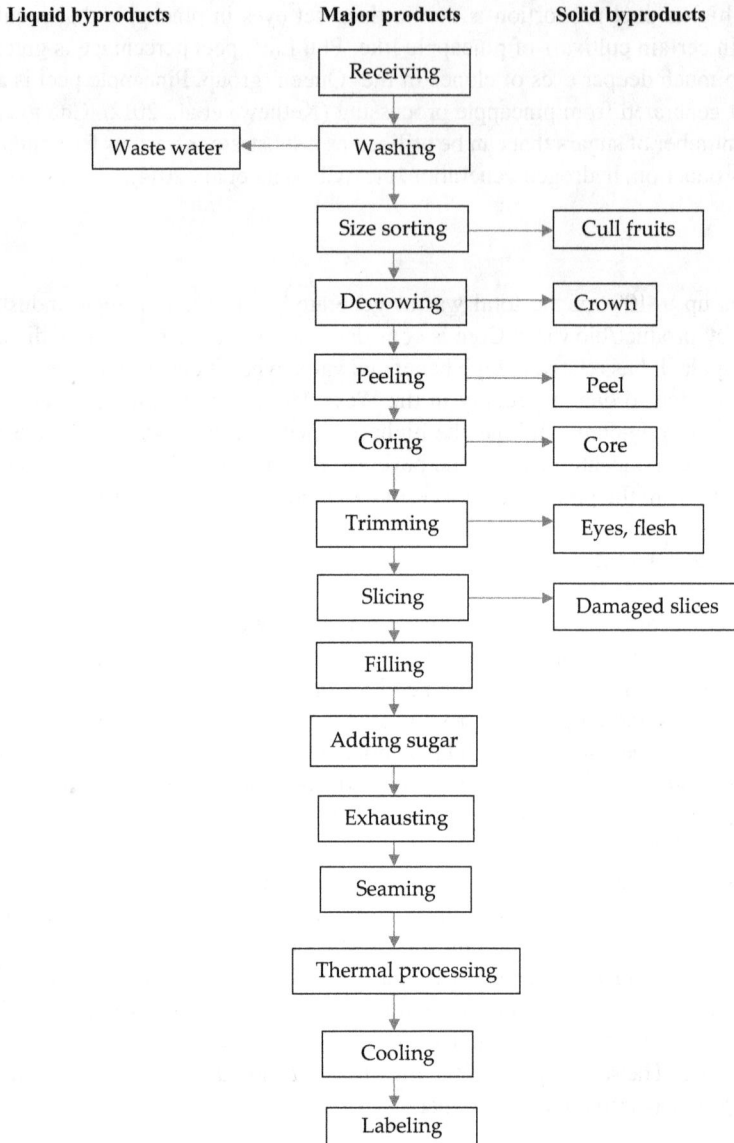

Figure 19.1 Major by-products during different stages of pineapple processing.

Antioxidants are bioactive compounds that play important roles in the food and chemical industries. Antioxidants like butylated hydroxyanisole and butylated hydroxytoluene can be synthesized chemically but are not considered safe due to many reports of induced tumor growths by such antioxidants (Saraswaty et al., 2017). The presence of antioxidants in pineapple wastes and by-products has been reported by many researchers. Gorinstein et al. (1999) reported the presence of nearly 40.40 mg/100 g of antioxidants in pineapple fruit in terms of gallic acid-equivalents. Sun et al. (2002) explored the presence of natural antioxidants in wastes of pineapple as a potential commercial source of antioxidants. Different researchers reported concentrations of different antioxidants in different parts of pineapple wastes. One of the most promising applications of pineapple wastes was found to be pineapple extract as a reducing agent in the biosynthesis of metal nanoparticles.

Table 19.2 Antioxidant Properties of Different By-Products of Pineapple

Pineapple Waste	Antioxidant Activity	References
Fresh peel	0.6 mg/mL	Saraswaty et al. (2017).
Dried peel	1.2 mg/mL	
Residue	1.3 mg/g GAE	Oliveira et al. (2009)
Peel	386 µg/mL	Azizan et al. (2020)
Crown	341 IC_{50}(µg/mL)	
Core	ND	

Note: GAE means Gallic Acid Eqvivalent; IC_{50} means the concentration of sample required to scavenge 50% DPPH free radicals.

Several extraction solvents and different techniques were used by researchers, with ethanol/water at a ratio of 55: 45% (v/v) reported to provide the highest yield of antioxidants and avoid microbial contamination. Lubaina et al. (2020) evaluated the polyphenol compounds in pineapple peel wastes. The study reported that gallic acid, ferulic acid, kaempferol and myricetin were present in pineapple wastes (Table 19.2).

19.4 PROCESSING OF PINEAPPLE WASTES AND BY-PRODUCTS

Discarded fruits, as well as wastes and by-products from the fruit and vegetable industries, are anticipated to be utilized for further industrial processes like fermentation, extraction of bioactive compounds and other valued processed products, etc. There are many reports on the utilization of wastes and by-products from different plant processing industries to enhance the economic value of fruits and their by-products. Pineapple canneries generate a lot of solid wastes that can be utilized as a substrate for production of bromelain, organic acids, ethanol, etc. As a matter of fact, these wastes and by-products are potential sources of many essential sugars, vitamins and growth factors (Larrauri et al., 1997; Dacera et al., 2009). Many attempts have been used over the years to explore different possibilities of using wastes and by-products. In early times, ion exchange was considered to be one of the most successful methods for obtaining sugars from pineapple effluent. This sugar was used in the syruping process in pineapple canneries for preservation of pineapple slices (Beohner and Mindler, 1949).

19.4.1 Extraction of Bromelain

Extraction of bromelain from pineapple wastes is considered to be one of the most successful uses in food processing for product development. Pineapple-based wastes are a very rich source of proteolytic enzymes like bromelain and other cysteine proteases that are present in different parts of pineapple wastes (Ketnawa et al., 2010; Rolle, 1998; Schieber et al., 2001). Bromelain is used for many commercial applications in the food, dietary and cosmetics industries (Uhlig, 1998; Walsh, 2002). It is used for meat tenderizing, brewing, baking, as well as for the production of protein hydrolysates and peptides (Ketnawa et al., 2011; Walsh, 2002). Many other important applications of bromelain include in detergent formulations and skin softening agent and for leather tanning and as a wool softener, etc. (Subhabrata and Mayura, 2006). Bromelain has also been used as a traditional medicine to treat many ailments, like wound healing, as an anti-inflammatory agent and as a digestive aid due to its protease nature (Bitange et al., 2008; Koh et al., 2006). Due to its broad-spectrum applications, ranging from the food industry to the textiles industry, commercial bromelain is very expensive. Different methods have been used to extract bromelain from pineapple wastes and by-products in order to obtain a cost-effective product (Upadhyay et al., 2010). Ketnawa et al., (2012) analyzed peel, core, stem and crown tissues for extraction of bromelain and measured their pro-teolytic activity with casein. Extracted bromelain from crown waste, with a molecular weight of 28 kDa, showed highest activity with casein. Pineapple peel is the most promising by-product for extraction of bromelain as it has been reported that the concentration of bromelain in pineapple peel is about 10 to 30% (Zhou et al., 2021). Ketnawa et al. (2011) isolated and characterized bromelain from core, stem and crown of pineapple with high levels of enzyme yield reported in peel (29–40%), followed by core (9–10%), stem (2–5%) and crown (2–4%), by weight. Ketnawa et al. (2011) reported that an increase in production of pineapple increased the waste and production of bromelain, and for the by-products increased their added value, economic value and also decreased environmen-tal problems due to waste disposal. Papain-based bromelain does not disappear in over-ripened pineapple fruit and is obtained from waste streams of fruit that are mostly considered to be wastes, making it more cost effective for commercial production. The enzyme bromelain is extracted from by-products of pineapple in crude form and needs further purification for commercial applica-tion. Many techniques have been used for purification of crude bromelain, including liquid–liquid extraction procedures, polyethyleneglycol or ammonium sulfate precipitation, ethanol and polishing techniques, such as affinity chromatography/ion-exchange (Bresolin et al., 2013). Processing under severe conditions of precipitation, sterilization and auto-digestion minimizes the proteolytic char-acteristic of bromelain, thereby inhibiting its medicinal properties. Gupta et al., (2007) successfully studied the stem bromelain immobilization using histidine *via* a metal affinity support; such an immobilized bromelain had better thermal stability over a wide range of temperatures. In addition, combining immobilized bromelain with tea polyphenols enhanced their thermal stability further (Liang et al., 1999). Extracted bromelain showed natural stability without the use of a preservative and retained 80% of its activity during refrigerated storage temperature for 180 days (Bhattacharya and Bhattacharya, 2009).

19.4.2 Production of Ethanol

Next to bromelain production, alcohol production from renewable pineapple wastes as substrate has particularly attracted interest over the years. At an industrial scale, ethanol production was obtained from biomass using sucrose and starch as fermentation enhancers but such production was considered to pose concerns with respect to the food supply chain (Hahn-Hagerdal et al., 2006; Field et al., 2008). Additional conversion of natural sources like forest and grassland into biomass also poses concerns regarding land use changes and gas emissions (Farrell et al., 2006; Searchinger et al., 2008), these problems focusing attention on crop residues such as wastes and by-products as substrates (Table 19.3).

In addition to other wastes and byproducts from fruit industries, pineapple wastes and by-prod-ucts are available in large amounts and their feasibility for producing ethanol was much greater than for other industrial wastes. Pineapple wastes mainly consist of pineapple trimmings from pineapple canneries that are produced in vast quantities throughout the world. Pineapple wastes contain nearly 10% of dry matter that is composed of 96% organic matter (Abdullah, 2007). The dry matter from pineapple wastes and by-products has very high BOD and COD, making it one of the most serious

Table 19.3 Various Food Applications of Bromelain

Food Sector	Use	Benefits
Bakery and confectionary	Baking agent	Improves dough relaxation. Helps to produce hypoallergenic flour
Fruits and vegetables	Anti-browning agent	Inhibits browning of sliced fruits. Prevents phenol oxidation
Alcoholic beverages	Antimicrobial agent	Prevents growth of bacteria. Enhances protein stability of beer. Prevents haze formation

Table 19.4 Production of Alcohol from Pineapple Waste

Organism	Yield (%)	Productivity (g/L/h)	References
Zymomonas mobilis	92.4	2.81	Tanaka et al. (1999)
Saccharomyces cerevisiae	92.5	3.75	Nigam (1999a)
Saccharomyces cerevisiae	86.3	42.8	Nigam (2000)

concerns regarding disposal (Ban-Koffi and Han, 1990). Pineapple wastes and by-products contain abundant sugar making it economically feasible for the production of ethanol. The main sugars present in pineapple biomass were found to be glucose and xylose. Other important sugars present in by-products were galactose, mannose, arabinose and uronic acid followed by smaller quantities of rhamnose. Such analysis revealed that pineapple wastes and byproducts are especially rich in cellulosic, hemicellulosic and pectic compounds (Huang et al., 2011). Nigam (1999b) reported the use of the anaerobic yeast strain *Saccharomyces cerevisiae* ATCC 24553 for industrial production of bioethanol using pineapple canneries-based wastes. The production of ethanol from pineapple by-products not only solves problems related to its disposal but also provides renewable biofuels which can represent an additional income stream. The risks of fossil fuels and their impact on environment and their rapid depletion have encouraged scientists and industrialists to find fuel alternatives, especially from wastes generated every year from different fruit industries. Pineapple wastes serve as a good for production of biofuels due to their high economic value (Table 19.4).

19.4.3 Production of Organic Acids

Traditionally, the process of fermentation involves conversion of free sugar groups into different organic acids (Ueno et al., 2003). Many wastes and by-products that are rich in sugars like pineapple wastes and by-products not only cause disposal issues but also are more prone to bacterial growth and hence spoilage, causing loss of valuable nutrients. Pineapple wastes, especially peel and core, are rich in sugars and hence make a perfect substrate for conversion of sugars into organic acids. Easy availability of pineapple by-products has increased demand for this material as a good substrate for solid-state fermentation in food and other sectors. Many researchers report production of acids like lactic acid, succinic acid, acetic acid and citric acid from fermentation processes using pineapple peel as substrate under controlled conditions. Despite the widespread use of all organic acids in food sector, lactic acid is one the most important acid due to its global use in the food and biopolymer sectors. In food processing industries, lactic acid is used as a food preservative and in different food formulations whereas in other sectors, especially in the cosmetic industry, it is an important ingredient in many products. The basic technique for its production involves use of the lactic acid bacterium *Lactobacallus lactis* (Ueno et al., 2003) and a carbohydrate-rich substrate. Different researchers have reported the use of pineapple wastes as a rich substrate for lactic acid production due to easy hydrolyzation of pineapple carbohydrates into simple sugars like glucose and fructose by the action of invertase. Among other acids, production of ferulic acid has also been reported by Tilay et al. (2008) from pineapple wastes and by-products using an alkali extraction method. In addition to its acidic properties, ferulic acid is used in the food industry and other sectors for its antioxidant properties.

Another important acid produced from pineapple wastes and byproducts is citric acid. It is used in the food, pharmaceutical, cosmetic and many other industrial sectors. Acetic acid produced from pineapple by-products acts as an acid flavor enhancer and preservative in the food processing industry. Production of acetic acid as a preservative was achieved from pineapple by-products through solid-state fermentation using the fungus *Aspergillus niger* and pineapple residues as substrate (Kumar et al., 2003) (Table 19.5).

Table 19.5   Production of Organic Acids by Fermentation of Pineapple Wastes

Organic Acid	Micro-Organism	Yield	References
Lactic acid	*Rhizopus arrhizus*	19.3 g/L	Jin et al. (2003)
	Rhizopus oryzae	14.7 g/L	
	Lactobacillus delbrueckii	0.78–0.82 g/g glucose	Idris and Suzana (2006)
	Lactococcus lactis	92 g/L	Ueno et al. (2003)
Citric acid	*Yarrowia lipolytica*	20.24 %	Imandi et al. (2008)
	Aspergillus foetidus	16.0%	Tilay et al. (2008)
	Aspergillus niger	11.3%	Kumar et al. (2003)
Succinic acid	*Escherichia coli*	6.2 g/L	Jusoh et al. (2014)

The maximum yield was found to be 54.2% which was much higher than with other methods and substrates used. In another study, Imandi et al., (2008) used the bacterium *Yarrowia lipolytica* during the solid-state fermentation process using only dried pineapple by-products (pomace) as substrate. The production cost was found to be very inexpensive and total maximum production of citric acid reached 202.35g/kg dried pineapple waste. The use of organic acids in food industries is widespread required characteristics vary from product to product. The most diversified use among all other acids was found to be that of acetic acid. It is the type of organic acid that is used as both a flavor enhancer and a food preservative. It can be used in both concentrated and diluted form; in the case of the diluted form, table vinegar is the most common, containing 18% acetic acid. Among other uses of acetic acid, its ability to kill infectious bacteria makes it a suitable antiseptic in the pharmaceutical industry (Madhusudhan, 2016). Acetic acid is also used in certain clinical diagnostics with the most important being cervical cancer screening tests (Fokom Domgue et al., 2015). Pineapple peel is used as a sugar-rich substrate for production of acetic acid during the fermentation process (Raji et al., 2012).

19.4.4 Production of Carbohydrates

Pineapple wastes and by-products contain a much higher percentage of carbohydrates than other fruit wastes/by-products. The presence of sugars in pineapple wastes and by-products (pomace) makes its use possible to generate different products in the biofuels, food and bio-based chemicals industries (Banerjee et al., 2017). Such a high percentage of carbohydrates makes pineapple wastes a suitable substrate for production of valuable chemicals like xylitol, xylooligosaccharides, etc. that have potential use in the food industry. However, the percentage of sugars varies at different maturity levels, reaching up to nearly 55%; among the wastes, stem-based pineapple waste has the lowest percentage at 35%. The reducing sugar level was found to be higher in pineapple peel. It has been reported that pineapple peel wastes contain nearly 9.75% of total sugars and 8.8% of reducing sugars more than its prime marketed product, pineapple juice. Such a high percentage of carbohydrates makes the peel a suitable substrate for fermentation of yeast. Pineapple refineries produce both solid and liquid waste by-products, and analysis of liquid byproducts revealed the presence of sucrose, glucose and fructose fractions with fructose in a higher percentage than glucose or sucrose. The percentage of sugars was reported to be similar to that in pineapple juice concentrate as reported by Krueger et al. (1992). In contrast, Sasaki et al. (1998) reported that sucrose was the predominant sugar, rather than glucose and fructose. Variation in the composition of by-product sugar levels and their identity may be a result of the process used to collect the wastes and by-products and also due to variation in season, area, variety and the canning industry.

In addition to free sugars, the pineapple by-product pomace, remaining after juice extraction, consists of much higher levels of crude fiber than the actual edible part of the fruit. These crude fibers consist of dietary fibers that cannot be digested in the small intestine although they undergo complete or partial fermentation in the large intestine. In Indian varieties of pineapple, Ramulu and Rao (2003) reported maximum crude fiber of 0.50 g/100 g of fruit by-product. Dietary fiber ensures correct gastrointestinal tract health and good bowel movement, although excess dietary fiber may bind some trace elements.

19.4.5 Production of Cellulase

Cellulase is the main enzyme used for production of bioethanol from pineapple by-product or other industrial cellulosic wastes. Conversion of cellulose to simple sugar form involves a process of hydrolysis that is performed by the enzyme cellulase. Production of bioethanol from pineapple wastes and by-products is largely dependent on the cost of cellulase. The production of enzymes through a fermentative route requires a substrate rich in sugars. Bandikari et al., (2014) suggested that by-products and wastes from different fruit canneries could be the potential substrate for fermentation, although these by-products may require certain pretreatments prior to fermentation. Pineapple wastes and by-products, as discussed earlier, are high in sugars and are easily available in large volumes at very low cost to serve as a rich substrate for the production of cellulase. Pineapple wastes and by-products are considered superior to other fruit or vegetable by-products due to little or no requirement for pretreatment prior to fermentation. Saravanan et al. (2013) reported cellulase production from solid-state fermentation using pineapple wastes as substrate. Pineapple-based cellulase was reported to have a maximum enzyme activity of 8.61 IU/mL. The microbe used for cellulase production was the fungus *Trichoderma reesei* (Banerjee et al., 2018). Cellulase production from pineapple wastes and by-products was also studied by Amaeze et al. (2015) using *Aspergillus niger* and *Saccharomyces cerevisiae*.

19.5 UTILIZATION OF PINEAPPLE WASTES AND BY-PRODUCTS

The wastes and by-products from pineapple are excellent sources of pectin, insoluble fibres, proteins, simple sugars and macronutrients, phenolic compounds, minerals and vitamins (Roda and Lambri, 2019). The utilization of pineapple wastes and by-products in food and non-foods is discussed in this section.

19.5.1 Pineapple-Based Wastes and By-Products in Food Applications

19.5.1.1 Meat and Meat-Based Products

The utilization of bromelain enzyme extracted from pineapple peel, core, stem and crown for tenderization of beef and chicken meat was studied in detail by Singh et al. (2018). The authors reported that, as the concentration of bromelain increases, the TCA- (trichloroacetic acid) soluble peptide contents also increased. The highest level of TCA-soluble peptides was reported to be 798 μmol/g for chicken treated with 20% treated bromelain and 709 μmol/g for beef treated with 20% bromelain. On the other hand, the non-treated beef sample showed the lowest level of TCA-soluble peptides, i.e., 238 μmol/g. Furthermore, the textural properties of bromelain-treated meat samples gradually change as the concentration of the enzyme increases. Recently, Lourenço et al. (2020) fabricated an alginate-based edible film with natural pineapple peel antioxidants in order to control microbial spoilage, to provide a barrier against lipid oxidation and to preserve the color of beef steaks at 4°C for five days.

19.5.1.2 Fermented Food Products

Roda et al. (2017) utilized the pineapple peel and core for the saccharification process to evaluate the total alcohol yield by using physical and enzymatic treatments, followed by microbial fermentation by *Saccharomyces cerevisiae* for 7 to 10 days under aerobic conditions at 25°C. The results showed the production of approximately 7% alcohol content. After 30 days, the alcoholic medium was used for production of pineapple vinegar (5% acetic acid) by A fermentation process using *Acetobacter aceti* at 32°C. Two samples, one taken at the beginning and one after acetification, were analyzed for the presence of volatile compounds. The results indicated the presence of L-lysine, mullein and gallic acid in pineapple vinegar. Higher alcohols, ketones and aldehydes were characterized by the aroma in the final product. In another study by Isitua and Ibeh (2010), pineapple wastes and by-products were used for wine production by a controlled fermentation process for five days. The Mauritius pineapple's core and peel were used for the production of a non-alcoholic wine by Kodagoda and Marapana (2017) and analyzed after 5 and 56 days of fermentation. The results showed alcohol content less than 0.5% (v/v) level. Also, no production of methanol was observed and hence pineapple wastes can be used in the development of non-alcoholic wines.

19.5.2 Pineapple Wastes and By-Products Utilization in Non-Foods

19.5.2.1 Bio-Aerogels

Aerogels are very light, solid materials that are formed by combining a polymer with certain solvents to form a gel and then replacing the liquid with air with an intact structure. Techniques for utilizing wastes and by-products are an important step towards agricultural sustainability and waste management. Moreover, utilizing wastes and by-products before food use provides farmers with additional income due to versatile usage and properties. Over the past decades, synthesis methods have been developed for manufacturing biocompatible cellulose-based aerogels using recycled cellulose fibers of paper, cotton waste and banana peel (Luu et al., 2020). Pineapple fiber aerogels can be potential candidates for several applications such as heat insulation of buildings as well as food preservation (Luu et al., 2020). Many researchers have utilized pineapple leaf fiber for making bio-aerogels that are very versatile. Eco-friendly and cost-effective aerogels are successfully developed from pineapple leaf fiber and the cross-linker polyvinyl alcohol (PVA). The pineapple-based aerosol has the benefit of heat and sound insulation applications due to low density, high porosity, good mechanical strength, low thermal conductivity, moderate thermal stability, and excellent acoustic insulation (Do et al., 2021).

19.5.2.2 Absorbents

After certain pretreatments, pineapple wastes and by-products are widely used in food industries as effective absorbents for certain toxic metals, as observed by Senthilkumaar et al. (2000). Anaerobic digestion of sewage sludge was carried out to trap certain heavy metals like copper, chromium, nickel and zinc by using raw fermented liquid waste from pineapple. The process of fermentation was carried out by the fungus *Aspergillus niger*. Pineapple liquid waste acts as a source of citric acid for the fermentation processes. Efficiency of metal removal was largely dependent on process parameters during the digestion process. Important process parameters include pH, time of contact and type and form of metal in the sludge. Certain metals, like chromium and zinc, were completely removed from sludge at pH 4 and a contact time of 9–11 days using *Aspergillus niger*-fermented liquid as the absorbent. The condition of the metal present also affected the process of its removal, as metals in the oxidizable form and chemical exchangeable stage seem to be removed more easily by using the fermented pineapple by-product liquid. Pineapple by-product liquid was also used as an inexpensive alternative nutrient resource for microbes that was further used in the removal of metal contaminations from sludge digestion processes. *Acinetobacter haemolyticus* is one of the bacterial strains used in the sludge digestion process to reduce chromium contamination. The pineapple wastewater is used as a source of nutrient substratum for microorganisms acting in the sludge digestion processes due to the rich source of sugars utilized by the bacteria (Zakaria et al., 2007).

19.5.2.3 Fibers

Pineapple leaf fiber is one of the by-products of pineapple that possesses high mechanical strength due to its high cellulosic content. The fiber from pineapple leaf is used in certain parts of the world to make coarse textiles and threads (Tran, 2006). Pineapple pulp by-products are also utilized for fiber thread production (Sreenath et al., 1996). The isolation of cellulosic nanocrystals is nowadays an important option that can be processed to produce value-added products for non-food use. Different extrusion techniques have been used to produce reinforced polymeric composites, for which pineapple by-products are very suitable due to their abundant nature, ease of availability and low cost (Arib et al., 2006). The pineapple-based fibers from the crown leaf have the important characteristic of greater mechanical strength than other plant-based fibers, as reported by several researchers (Gorinstein et al., 1999; Lund and Smoot, 1982). The production of extruded fibers from pineapple wastes and by-products is very inexpensive but most of the methods used locally are not yet considered suitable for industrial-scale production. The limitation of the use of pineapple leaf fiber obtained from pineapple by-products is due to its large size. Multiple efforts have been attempted to reduce the size of the fibers produced from pineapple leaves in order to make the extraction process more efficient for industrial use. The fiber obtained is very suitable for usage in plastic and rubber industries due to its high reinforcement properties. Use of such pineapple leaf fibers is not limited to the textile industry but can be combined with other suitable adhesion enhancers and used with other cellulosic fibers (Kengkhetkit and Amornsakchai, 2012).

19.6 CONCLUSIONS

Lack of information on promoting innovations and sustainable developments regarding conversion of wastes and by-products from the pineapple industry into value-added foods has led to huge losses each year. Processing of pineapple-based wastes and by-products can generate high economic returns for pineapple processing industries, achieve nutrient prevention and prevent environmental degradation and issues of waste management. This has led to research, much of it successful, into the utilization of pineapple-derived wastes and by-products and their conversion into valuable products for the development of value-added products. Pineapple wastes and by-products can be converted into valuable products for the food, pharmaceutical and other industries because of the increased demand for natural products.

BIBLIOGRAPHY

Abdullah, A. (2007). Solid and liquid pineapple waste utilization for lactic acid fermentation using *Lactobacillus delbrueckii. Reaktor, 11*(1), 50–52.

Amaeze, N. J., Okoliegbe, I. N., & Francis, M. E. (2015). Cellulase production by *Aspergillus niger* and *Saccharomyces cerevisiae* using fruit wastes as substrates. *International Journal of Applied Microbiology and Biotechnology Research, 3,* 36–44.

Arib, R. M. N., Sapuan, S. M., Ahmad, M. M. H. M., Paridah, M. T., & Zaman, H. K. (2006). Mechanical properties of pineapple leaf fibre reinforced polypropylene composites. *Materials & Design, 27*(5), 391–396.

Azizan, A., Lee, A. X., Abdul Hamid, N. A., Maulidiani, M., Mediani, A., Abdul Ghafar, S. Z., Zolkeflee, N. K. Z., & Abas, F. (2020). Potentially bioactive metabolites from pineapple waste extracts and their antioxidant and α-glucosidase inhibitory activities by 1H NMR. *Foods, 9*(2), 173.

Babu, B. R., Rastogi, N. K., & Raghavarao, K. S. M. S. (2008). Liquid–liquid extraction of bromelain and polyphenol oxidase using aqueous two-phase system. *Chemical Engineering and Processing: Process Intensification*, *47*(1), 83–89.

Bandikari, R., Poondla, V., & Obulam, V. S. R. (2014). Enhanced production of xylanase by solid state fermentation using Trichoderma koeningi isolate: Effect of pretreated agro-residues. *3 Biotech*, *4*(6), 655–664.

Banerjee, J., Singh, R., Vijayaraghavan, R., MacFarlane, D., Patti, A. F., & Arora, A. (2018). A hydrocolloid based biorefinery approach to the valorisation of mango peel waste. *Food Hydrocolloids*, *77*, 142–151.

Banerjee, R., Chintagunta, A. D., & Ray, S. (2017). A cleaner and eco-friendly bioprocess for enhancing reducing sugar production from pineapple leaf waste. *Journal of Cleaner Production*, *149*, 387–395.

Ban-Koffi, L., & Han, Y. (1990). Alcohol production from pineapple waste. *World Journal of Microbiology and Biotechnology*, *6*(3), 281–284.

Beohner, H. L., & Mindler, A. B. (1949). Ion exchange in waste treatment. *Industrial & Engineering Chemistry*, *41*(3), 448–452.

Bhattacharya, R., & Bhattacharyya, D. (2009). Preservation of natural stability of fruit "bromelain" from *Ananas comosus* (pineapple). *Journal of Food Biochemistry*, *33*(1), 1–19.

Bitange, N. T., Zhang, W., Shi, Y. X., & Wenbin, Z. (2008). Therapeutic application of pineapple protease (bromelain). *Pakistan Journal of Nutrition*, *7*, 513–520.

Bresolin, I. R. A. P., Bresolin, I. T. L., Silveira, E., Tambourgi, E. B., & Mazzola, P. G. (2013). Isolation and purification of bromelain from waste peel of pineapple for therapeutic application. *Brazilian Archives of Biology and Technology*, *56*, 971–979.

Casabar, J. T., Unpaprom, Y., & Ramaraj, R. (2019). Fermentation of pineapple fruit peel wastes for bioethanol production. *Biomass Conversion and Biorefinery*, *9*(4), 761–765.

Choonut, A., Saejong, M., & Sangkharak, K. (2014). The production of ethanol and hydrogen from pineapple peel by Saccharomyces cerevisiae and Enterobacter aerogenes. *Energy Procedia*, *52*, 242–249.

Conesa, C., Seguí, L., Laguarda-Miró, N., & Fito, P. (2016). Microwaves as a pretreatment for enhancing enzymatic hydrolysis of pineapple industrial waste for bioethanol production. *Food and Bioproducts Processing*, *100*, 203–213.

Dacera, D. D. M., Babel, S., & Parkpian, P. (2009). Potential for land application of contaminated sewage sludge treated with fermented liquid from pineapple wastes. *Journal of Hazardous Materials*, *167*(1–3), 866–872.

de Oliveira, A. C., Valentim, I. B., Silva, C. A., Bechara, E. J. H., de Barros, M. P., Mano, C. M., & Goulart, M. O. F. (2009). Total phenolic content and free radical scavenging activities of methanolic extract powders of tropical fruit residues. *Food Chemistry*, *115*(2), 469–475.

Do, N. H., Tran, V. T., Tran, Q. B., Le, K. A., Thai, Q. B., Nguyen, P. T., Duong, H. M. & Le, P. K. (2021). Recycling of pineapple leaf and cotton waste fibers into heat-insulating and flexible cellulose aerogel composites. *Journal of Polymers and the Environment*, *29*(4), 1112–1121.

Dorta, E., & Sogi, D. S. (2017). Value added processing and utilization of pineapple by-products. In Maria Gloria Lobo and Robert E. Paull (Eds), *Hand-book of pineapple technology, production, postharvest science, processing and nutrition*. 1st ed. Chichester: John Wiley & Sons, (pp. 196–220).

Farrell, A. E., Plevin, R. J., Turner, B. T., Jones, A. D., O'hare, M., & Kammen, D. M. (2006). Ethanol can contribute to energy and environmental goals. *Science*, *311*(5760), 506–508.

Field, C. B., Campbell, J. E., & Lobell, D. B. (2008). Biomass energy: The scale of the potential resource. *Trends in Ecology & Evolution*, *23*(2), 65–72.

Fokom-Domgue, J., Combescure, C., Fokom-Defo, V., Tebeu, P. M., Vassilakos, P., Kengne, A. P., & Petignat, P. (2015). Performance of alternative strategies for primary cervical cancer screening in sub-Saharan Africa: Systematic review and meta-analysis of diagnostic test accuracy studies. *BMJ*, *351*, h3084.

Gil, L. S., & Maupoey, P. F. (2018). An integrated approach for pineapple waste valorisation. Bioethanol production and bromelain extraction from pineapple residues. *Journal of Cleaner Production*, *172*, 1224–1231.

Gorinstein, S., Zemser, M., Haruenkit, R., Chuthakorn, R., Grauer, F., Martin-Belloso, O., & Trakhtenberg, S. (1999). Comparative content of total polyphenols and dietary fiber in tropical fruits and persimmon. *The Journal of Nutritional Biochemistry*, *10*(6), 367–371.

Grigelmo-Miguel, N., Gorinstein, S., & Martín-Belloso, O. (1999). Characterisation of peach dietary fibre concentrate as a food ingredient. *Food Chemistry*, *65*(2), 175–181.

Gupta, P., Maqbool, T., & Saleemuddin, M. (2007). Oriented immobilization of stem bromelain via the lone histidine on a metal affinity support. *Journal of Molecular Catalysis B: Enzymatic, 45*(3–4), 78–83.

Hahn-Hägerdal, B., Galbe, M., Gorwa-Grauslund, M. F., Lidén, G., & Zacchi, G. (2006). Bio-ethanol–the fuel of tomorrow from the residues of today. *Trends in Biotechnology, 24*(12), 549–556.

Huang, Y. L., Chow, C. J., & Fang, Y. J. (2011). Preparation and physicochemical properties of fiber-rich fraction from pineapple peels as a potential ingredient. *Journal of Food and Drug Analysis, 19*(3), 318–323.

Idris, A., & Suzana, W. (2006). Effect of sodium alginate concentration, bead diameter, initial pH and temperature on lactic acid production from pineapple waste using immobilized Lactobacillus delbrueckii. *Process Biochemistry, 41*(5), 1117–1123.

Imandi, S. B., Bandaru, V. V. R., Somalanka, S. R., Bandaru, S. R., & Garapati, H. R. (2008). Application of statistical experimental designs for the optimization of medium constituents for the production of citric acid from pineapple waste. *Bioresource Technology, 99*(10), 4445–4450.

Isitua, C. C., & Ibeh, I. N. (2010). Novel method of wine production from banana (*Musa acuminata*) and pineapple (*Ananas comosus*) wastes. *African Journal of Biotechnology, 9*(44), 7521–7524.

Jin, B., Huang, L. P., & Lant, P. (2003). Rhizopus arrhizus–a producer for simultaneous saccharification and fermentation of starch waste materials to L (+)-lactic acid. *Biotechnology Letters, 25*(23), 1983–1987.

Jusoh, N., Othman, N., Idris, A., & Nasruddin, A. (2014). Characterization of liquid pineapple waste as carbon source for production of succinic acid. *Jurnal Teknologi, 69*(4). 11–13.

Kengkhetkit, N., & Amornsakchai, T. (2012). Utilisation of pineapple leaf waste for plastic reinforcement: 1. A novel extraction method for short pineapple leaf fiber. *Industrial Crops and Products, 40*, 55–61.

Ketnawa, S., Chaiwut, P., & Rawdkuen, S. (2011). Extraction of bromelain from pineapple peels. *Food Science and Technology International, 17*(4), 395–402.

Ketnawa, S., Chaiwut, P., & Rawdkuen, S. (2012). Pineapple wastes: A potential source for bromelain extraction. *Food and Bioproducts Processing, 90*(3), 385–391.

Ketnawa, S., Rawdkuen, S., & Chaiwut, P. (2010). Two phase partitioning and collagen hydrolysis of bromelain from pineapple peel Nang Lae cultivar. *Biochemical Engineering Journal, 52*(2–3), 205–211.

Kodagoda, K. H. G. K., & Marapana, R. A. U. J. (2017). Development of non-alcoholic wines from the wastes of Mauritius pineapple variety and its physicochemical properties. *Journal of Pharmacognosy and Phytochemistry.* 6(3). 492–497.

Koh, J., Kang, S. M., Kim, S. J., Cha, M. K., & Kwon, Y. J. (2006). Effect of pineapple protease on the characteristics of protein fibers. *Fibers and Polymers, 7*(2), 180–185.

Kringel, D. H., Dias, A. R. G., Zavareze, E. D. R., & Gandra, E. A. (2020). Fruit wastes as promising sources of starch: Extraction, properties, and applications. *Starch-Stärke, 72*(3–4), 1900200.

Krueger, D. A., Krueger, R. G., & Maciel, J. (1992). Composition of pineapple juice. *Journal of AOAC International, 75*(2), 280–282.

Kumar, D., Jain, V. K., Shanker, G., & Srivastava, A. (2003). Utilisation of fruits waste for citric acid production by solid state fermentation. *Process Biochemistry, 38*(12), 1725–1729.

Larrauri, J. A., Rupérez, P., & Calixto, F. S. (1997). Pineapple shell as a source of dietary fiber with associated polyphenols. *Journal of Agricultural and Food Chemistry, 45*(10), 4028–4031.

Liang, H. H., Huang, H. H., & Kwok, K. C. (1999). Properties of tea-polyphenol-complexed bromelain. *Food Research International, 32*(8), 545–551.

Lobo, M. G., & Paull, R. E. (Eds.). (2017). *Handbook of pineapple technology: Production, postharvest science, processing and nutrition.* John Wiley & Sons.

Lourenço, S. C., Fraqueza, M. J., Fernandes, M. H., Moldão-Martins, M., & Alves, V. D. (2020). Application of edible alginate films with pineapple peel active compounds on beef meat preservation. *Antioxidants, 9*(8), 667.

Lubaina, A. S., Renjith, P. R., & Roshni, A. S. (2020). Identification and quantification of polyphenols from pineapple peel by high performance liquid chromatography analysis. *Advances in Zoology and Botany, 8*(5): 431–438.

Lund, E. D., & Smoot, J. M. (1982). Dietary fiber content of some tropical fruits and vegetables. *Journal of Agricultural and Food Chemistry, 30*(6), 1123–1127.

Luu, T. P., Do, N. H. N., Chau, N. D. Q., Lai, D. Q., Nguyen, S. T., Le, D., Thai, Q. B., Le, P. K., & Duong, H. M. (2020). Morphology control and advanced properties of bio-aerogels from pineapple leaf waste. *Chemical Engineering, 78*, 433–438.

Madhusudhan, V. L. (2016). Efficacy of 1% acetic acid in the treatment of chronic wounds infected with pseudomonas aeruginosa: Prospective randomised controlled clinical trial. *International Wound Journal, 13*(6), 1129–1136.

Mochamad Busairi, A. (2008). Conversion of pineapple juice waste into lactic acid in batch and fed batch fermentation system. *Reaktor, 12*(2), 98–101.

Murachi, T. (1964). Amino acid composition of stem bromelain. *Biochemistry, 3*(7), 932–934.

Nakthong, N., Wongsagonsup, R., & Amornsakchai, T. (2017). Characteristics and potential utilizations of starch from pineapple stem waste. *Industrial Crops and Products, 105*, 74–82.

Nie, H., Li, S., Zhou, Y., Chen, T., He, Z., Su, S., Zhang, H., Xue, Y., & Zhu, L. (2008). Purification of bromelain using immobilized metal affinity membranes. *Journal of Biotechnology, 136*(S), S402–S459.

Nigam, J. N. (1999a). Continuous ethanol production from pineapple cannery waste. *Journal of Biotechnology, 72*, 197–202.

Nigam, J. N. (1999b). Continuous cultivation of the yeast Candida utilis at different dilution rates on pineapple cannery waste. *World Journal of Microbiology and Biotechnology, 15*, 115–117.

Nigam, J. N. (2000). Continuous ethanol production from pineapple cannery waste using immobilized yeast cells. *Journal of Biotechnology, 80*(2), 189–193.

Nunes, M. C. N., Emond, J. P., Rauth, M., Dea, S. & Chau, K. V. (2009). Environmental conditions encountered during typical consumer retail display affect fruit and vegetable quality and waste. *Postharvest Biology and Technology, 51*, 232–241.

Raji, Y. O., Jibril, M., Misau, I. M., & Danjuma, B. Y. (2012). Production of vinegar from pineapple peel. *International Journal of Advanced Scientific Research and Technology, 3*(2), 656–666.

Ramulu, P., & Rao, P. U. (2003). Total, insoluble and soluble dietary fiber contents of Indian fruits. *Journal of Food Composition and Analysis, 16*(6), 677–685.

Roda, A., De Faveri, D. M., Giacosa, S., Dordoni, R., & Lambri, M. (2016). Effect of pre-treatments on the saccharification of pineapple waste as a potential source for vinegar production. *Journal of Cleaner Production, 112*, 4477–4484.

Roda, A., & Lambri, M. (2019). Food uses of pineapple waste and by-products: A review. *International Journal of Food Science & Technology, 54*(4), 1009–1017.

Roda, A., Lucini, L., Torchio, F., Dordoni, R., De Faveri, D. M., & Lambri, M. (2017). Metabolite profiling and volatiles of pineapple wine and vinegar obtained from pineapple waste. *Food Chemistry, 229*, 734–742.

Rolle, R. S. (1998). Enzyme applications for agro-processing in developing countries: An inventory of current and potential applications. *World Journal of Microbiology and Biotechnology, 14*(5), 611–619.

Salve, R. R., & Ray, S. (2020). Comprehensive study of different extraction methods of extracting bioactive compounds from pineapple waste-A review. *The Pharma Innovation Journal, 9*(6): 327–340.

Saraswaty, V., Risdian, C., Primadona, I., Andriyani, R., Andayani, D. G. S., & Mozef, T. (2017). Pineapple peel wastes as a potential source of antioxidant compounds. In *IOP conference series: Earth and environmental science* (Vol. 60, No. 1, p. 012013). IOP Publishing.

Saravanan, P., Muthuvelayudham, R., & Viruthagiri, T. (2013). Enhanced production of cellulase from pineapple waste by response surface methodology. *Journal of Engineering*, Article ID 979547, 8 pages.

Sasaki, K., Tanaka, T., & Nagai, S. (1998). Use of photosynthetic bacteria for the production of SCP and chemicals from organic wastes. In A. M. Martin (Ed.), *Bioconversion of waste materials to industrial products*. Boston, MA: Springer, (pp. 247–292).

Schieber, A., Stintzing, F. C., & Carle, R. (2001). By-products of plant food processing as a source of functional compounds—recent developments. *Trends in Food Science & Technology, 12*(11), 401–413.

Searchinger, T., Heimlich, R., Houghton, R. A., Dong, F., Elobeid, A., Fabiosa, J., Tokgoz, S., Hayes, D & Yu, T. H. (2008). Use of US croplands for biofuels increases greenhouse gases through emissions from land-use change. *Science, 319*(5867), 1238–1240.

Senthilkumaar, S., Bharathi, S., Nithyanandhi, D., & Subburam, V. (2000). Biosorption of toxic heavy metals from aqueous solutions. *Bioresource Technology, 75*(2), 163–165.

Singh, T. A., Sarangi, P. K., & Singh, N. J. (2018). Tenderisation of meat by bromelain enzyme extracted from pineapple wastes. *International Journal of Current Microbiology and Applied Sciences, 7*(9), 3256–3264.

Sreenath, H. K., Sudarshanakrishna, K. R., Prasad, N. N., & Santhanam, K. (1996). Characteristics of some fiber incorporated cake preparations and their dietary fiber content. *Starch-Stärke, 48*(2), 72–76.

Subhabrata, S., & Mayura, D. (2006). Industrial and clinical applications excluding diagnostic clinical. *Enzymology*, *1*, 1–25.

Sun, J., Chu, Y. F., Wu, X., & Liu, R. H. (2002). Antioxidant and antiproliferative activities of common fruits. *Journal of Agricultural and Food Chemistry*, *50*(25), 7449–7454.

Tanaka, K., Hilary, Z. D., & Ishizaki, A. (1999). Investigation of the utility of pineapple juice and pineapple waste material as low-cost substrate for ethanol fermentation by *Zymomonas mobilis*. *Journal of Bioscience and Bioengineering*, *87*(5), 642–646.

Tilay, A., Bule, M., Kishenkumar, J., & Annapure, U. (2008). Preparation of ferulic acid from agricultural wastes: Its improved extraction and purification. *Journal of Agricultural and Food Chemistry*, *56*(17), 7644–7648.

Tochi, B. N., Wang, Z., Xu, S. Y., & Zhang, W. (2008). Therapeutic application of pineapple protease (bromelain): A review. *Pakistan Journal of Nutrition*, *7*(4), 513–520.

Tran, A. V. (2006). Chemical analysis and pulping study of pineapple crown leaves. *Industrial Crops and Products*, *24*, 66–74.

Ueno, T., Ozawa, Y., Ishikawa, M., Nakanishi, K., & Kimura, T. (2003). Lactic acid production using two food processing wastes, canned pineapple syrup and grape invertase, as substrate and enzyme. *Biotechnology Letters*, *25*(7), 573–577.

Uhlig, H. (1998). *Industrial enzymes and their applications*. John Wiley & Sons, pp. 209–210.

Upadhyay, A., Lama, J. P., & Tawata, S. (2010). Utilization of pineapple waste: A review. *Journal of Food Science and Technology Nepal*, *6*, 10–18.

Van Tran, A. (2006). Chemical analysis and pulping study of pineapple crown leaves. *Industrial Crops and Products*, *24*(1), 66–74.

Walsh, G. (2002). *Proteins: Biochemistry and biotechnology*. Wiley, pp. 408–410.

Zakaria, Z. A., Zakaria, Z., Surif, S., & Ahmad, W. A. (2007). Biological detoxification of Cr (VI) using wood-husk immobilized *Acinetobacter haemolyticus*. *Journal of Hazardous Materials*, *148*(1–2), 164–171.

Zhou, W., Ye, C., Geng, L., Chen, G., Wang, X., Chen, W., Sa, R., Zhang, J., & Zhang, X. (2021). Purification and characterization of bromelain from pineapple (Ananas comosus L.) peel waste. *Journal of Food Science*, *86*(2), 385–393.

CHAPTER 20

Pear Wastes and By-Products
Chemistry, Processing, and Utilization

Fozia Hameed, Neeraj Gupta and Rukhsana Rehman

CONTENTS

20.1 INTRODUCTION

Pear (*Pyrus communis* L.) belongs to the family Rosaceae. The name "pear" has its origins from the Latin words *pira* or *pera*, and with some other words, including *poire* in French, *peer* in German and *acras* in Greece for wild and *apios* for cultivated types of pear (Silva et al., 2014). Different vernacular names for the pear fruit are listed in Table 20.1. The lawyer and botanist, Jean-Baptiste de La Quintinie – passionate about pear cultivation – said in a report, "Among all fruits, there is nothing as beautiful and noble as the pear". The fruit is hardy in nature and flourishes widely in temperate and subtropical regions throughout the world (Kapoor and Ranote, 2015). It is one of the world's most popular pome fruits with a characteristic compartmented core, and is native to the coastline and slightly temperate regions of North Africa and Western Europe (Mahammad et al., 2010) and has also been naturalized in Switzerland, Germany, northern Italy, Greece, Moldova, Ukraine, China, Iran, Korea, Bhutan and Uzbekistan (Silva et al., 2014). China, the United States, Italy, Argentina, France, Japan, South Africa, South Korea and Turkey are the world's top pear producers, with Turkey ranking seventh. (Ozturk et al., 2009). The entire area occupied by pears in India was 42 ha, with an output of 313 metric tons (MT) in 2019–20 (National Horticulture Board, 2020). Uttarakhand (38%) has the highest pear production share in India, followed by Jammu and Kashmir (26%), Punjab (21%), Tamil Nadu (10%), Himachal Pradesh (4%) and other states (1%). Commercial pear varieties grown in various Indian states include 'Bartlett', 'Red Bartlett', 'Thumb Pear' or 'Chusni', 'Max Red Bartlett', 'William Bartlett', 'Flemish Beauty', 'Conference', 'Laxton's Superb', 'Shinsui', 'Kosui', 'Pathernakh', 'Gola', 'Hosui', 'Jargnel', 'Winter Nelis', 'Beurre Hardy', 'Victoria' and 'Keiffer'.

DOI: 10.1201/9781003164463-20

Table 20.1   Vernacular Names of Pear Fruit

	Type of Language	Pear Name
Indian names	Tamil	Perikai
	Kannada	Berikkai
	Malayalam	Sabariil
	Hindi	Babbu-ghosha
	Sanskrit	Amritphala
	Telugu	Berikkai
	Kashmiri	Tang, Kishtabahira
	Urdu	Nashpati, batangi
	Punjabi	Kainth, shegal
Other names	Nepali	Chinia naspaati, passi, mayal
	French	Poire
	Japanese	Nashi
	Thai	Luk Phaer
	German	Birne
	Persian	Golabi
	Bhutan	Pashia
	Korean	Bae
	Georgia	Khechechuri

Pear is a table or dessert fruit valued for its sensory attributes and nutritional properties. It contains significant quantity of saccharides, minerals, vitamins, proteins, lipids and fiber (Silva et al., 2014). Apart from the micro- and macronutrients, pear is known to contain a number of secondary metabolites like flavonols, phenolic acids, anthocyanins and the glycosylated hydroquinone arbutin (Liaudanskas et al., 2017). The antioxidant capacity of non-nutritive substances aids the body in scavenging damaging free radicals, which might also protect tissues from oxidation, prevent cancer and promote immunity (Hameed et al., 2020). In the food industry, pear is regarded as a valuable fruit due to its high functional values and associated health-related characteristics. Pear is either consumed raw (in the fresh form) or is used for the preparation of several food products like juices, jams, ice creams, cakes, pies, risotto and so on, although processing of pears produces tons of waste which is generally thrown away. This waste consists of peel, core, seeds, pulp residue and pomace and is believed to contain a plethora of dietary fiber and phenolic compounds.

20.2 CHEMISTRY OF PEAR-DERIVED WASTES AND BY-PRODUCTS

Fruit wastes and by-products include leaves, discarded peel, pulp, seeds and rotten fruits. However, pomace, peel and seeds are the major food wastes produced by the pear processing industry. The concentrations of polyphenolic compounds and other phytochemicals in pear peel, seeds and pomace are higher than in the edible tissues, indicating that these wastes and by-products might be promising resources for extracting bioactive components. Pear wastes and by-products are known to contain a number of phytoconstituents such as flavonoids, phenolic acids, sterols, alkaloids, triterpenoids and carotenoids (Sahiba et al., 2018). Lin et al. (2012) investigated the antioxidant ability, phenolic composition and their interactions of water- and fat-soluble extracts of pear organics and nutraceuticals, identifying a large number of compounds with substantial antioxidant characteristics. The positive relationship between antioxidant potential and phenolic content indicated that phenolics might be one of the most important contributors to the antioxidant capacity of pear. These antioxidants (polyphenolic as well as other phytochemicals) and some other bioactive substances have anticancer, antimicrobial, antioxidative and immune-modulatory properties. Furthermore, they lower the risk of cardiovascular disease and capillary fragility, limit blood clotting and inhibit thrombosis and diabetes in vertebrates (Lin and Harnly, 2008).

Large volumes of seeds are generated during the processing of pear fruits globally. Pear seeds are a rich source of both micro- and macronutrients. Mahammad et al. (2010), describing the composition of pear (*Pyrus communis*) fruit, reported that pear pulp and seeds include important levels of essential amino acids with values higher than recommended when compared with the reference standard, underlining their high biological value. Pear fruit seeds are thought to contain a high concentration of oil and significant nutritional composition, which can be used to augment the diet of domestic animals. The oil production in pear seeds was 17.34% to 32.6% (w/w) dw, with palmitic acid (6.14–9.64%), oleic acid (28.40–39.18%), and linoleic acid (50.73–63.78%) being the chief fatty acids found in seed pear oil, accounting for 96–99% of the fatty acids measured. Other fatty acids reported in pear seed oil, include palmitoleic acid, stearic acid, arachidic acid, gondoic acid, eicosadienoic and lignoceric acid. The seed oil consists of 77.846 g/100 g oil of unsaturated fatty acids with additional fatty acids (Yukui et al., 2009). The fatty acid composition of pear seed oil is shown in Table 20.2. Pear seed oil contains 9.73 g of saturated fatty acids and 77.85 g of unsaturated fatty acids (20.67 g of monounsaturated fatty acids) per 100 g of oil (Yukui et al., 2009). While analyzing the chemical make-up of pear seed oils from several pear varieties, Górnás et al. (2016) discovered that the oil is high in lipophilic substances such as tocopherols and carotenoids, as well as essential fatty acids. Total tocochromanols were found to be in the range of 120.5 to 216.1 mg/ 100 g of oil. Carotenoid and squalene concentrations were 0.69–2.99 mg/100 g oil and 25.6–40.8 mg/100 g oil, respectively. The fatty acid composition of pear seed oil is found to be rich in saturated fatty acids, with palmitic acid (44.31%) and stearic acid (8.08%) being the most abundant saturated fatty acids, and oleic acid (42.45%) being the most prevalent unsaturated fatty acid (Isaac et al, 2014). If this oil is appropriately handled, it might help meet the challenge to obtain vegetable oils for surface coatings and other new industrial purposes. Pear seed oil has a greater concentration of the anti-corrosion fatty acid linoleic acid than other commonly used edible oils, according to Mushtaq et al. (2019). The amino acids in pear seeds are threonine, methionine, tyrosine, leucine, isoleucine, lysine, valine, glutamic acid, glycine, aspartic acid, histidine, arginine and serine (Mahammad et al., 2010).

20.2.1 Peel

The health-promoting and nutritional value of a natural plant-based commodity is usually better assessed by examining their anticancer agents and bioactive composition, which depend on the type of natural product and its growth conditions (Scalzo et al., 2005). The peel of the fruit is usually

Table 20.2   Fatty Acid Composition of Pear Seed Oil

Type of Fatty Acid	Content (g/100g)
C16:0	6.388
C16:1	0.119
C18:0	1.746
C18:1	20.281
C18:2	56.801
C18:3	0.320
C20:0	1.251
C20:1	0.275
C20:2	0.050
C22:0	0.238
C24:0	0.109

discarded during processing operations. However, it is considered to be nutritious and can be a tremendous source of carbohydrates, proteins and antistress compounds. Fruit peel contains significant quantities of bioactive compounds, especially polyphenols. Manzoor et al. (2013) observed the antioxidant properties and phenolic concentration of the peel and pulp of two pear cultivars viz, Nakh and Nashpati. The peel extracts of both cultivars showed high total phenolic concentration (619.80 and 601.50 mg gallic acid-equivalents (GAE)/100 g dry weight (DW)), total flavonoid concentration (543.50 and 561.30 mg CE/100 g) and antioxidant activity (49.71 and 49.94%). Furthermore, either the peel or the pulp of the pear fruit showed a significant reduction in linoleic acid peroxidation, reflecting antioxidant activity. The peel extract inhibited peroxidation by 60.32 to 60.60%, whereas the pulp extract inhibited peroxidation by only 34.15 to 34.45%. This study revealed that pear skin had much greater antioxidant activity and phenolic content than did pear pulp, and so might be exploited as a feasible resource of natural antioxidants for functional food products and medicinal purposes. Li et al. (2014) found that the pear peel is a good source of phenolic compounds and triterpenes, with levels 6–21 times higher than those found in the pear core. Different pear cultivars growing in the Ardahan region of Turkey are rich in antioxidant activity, according to Abaci et al. (2015). The 'Bal' cultivar had the highest concentration of phenolic compounds and the highest antioxidant activity. The skin of the different pear genotypes had much higher concentrations of phenolic compounds and higher antioxidant activity than the corresponding pulp tissue. In addition to some well-known phenolic compounds like chlorogenic acid, arbutin, catechin, quercetin and kaempferol, the peel of pear has also been reported to contain hydroxycinnamoyl malates, procyanidins and triterpenes (Ma et al., 2012). Arbutin and chlorogenic acid were identified as being important phenolics in the peel and flesh of pear fruit by Ozturk et al. (2014), whereas rutin hydrate and rutin-tri-hydrate were identified as minor phenolics. Fattouch et al. (2008) compared the phytoconstituents, antioxidants and antimicrobial actions of pome fruit peels (apple, pear, and quince), and found that apple and quince peel extracts were much more potent than pear peel extract in slowing down the growth of *Staphylococcus aureus*, *Pseudomonas aeruginosa*, and *Bacillus cereus*.

20.2.2 Pomace

The pomace has a diverse composition because it comprises peel, core, seed, calyx, leftover juice, stem and soft tissues, Flotation or sieving can be used to isolate pear seeds from pomace, and standard extraction processes are normally used to recover the oil. Despite the fact that global pear productivity is comparable to apple production, there is much less published research on the composition of pear pomace. Pear pomace is rich in carbohydrates, minerals and dietary fiber, as well as polyphenols, the latter acting as antioxidants. Pear pomace has more insoluble fiber but less soluble fiber than apple pomace. According to a study, total dietary fiber in pear pomace was 43.9% on a dry weight basis, with soluble and insoluble dietary fiber of 7.6% and 36.3%, respectively. Pectin (7% DW) and lignin (5.4% DW) were discovered to constitute the majority of the total dietary fiber, whereas protein content was determined to be 3.9% and ash content was 5.5% (Martin-Cabrejas et al., 1995). Pear pomaces are frequently employed in the supplementation of food goods due to their high nutritional value. Pear pomace includes significant quantities of fiber and polyphenols, both of which have been linked to improved human wellbeing (Pascoalino et al., 2021). Because of the presence of pectins, carotenoids and antioxidants, pear pomace is utilized as a dietary fiber supplement and a functional component in processed food items. Martin-Cabrejas et al. (1995) examined dietary fiber products manufactured from pomace residues of kiwi and pear purées and found that pear and kiwi pomace provided 44 and 26% (dry weight) of total dietary fiber, respectively. The impact of superfine grinding and hydrostatic pressure on pear pomace was investigated by Yan et al. (2019) and found that the administration of both treatments resulted in a considerable increase in soluble dietary fiber content from 11% to 17%, with a comparable drop in insoluble dietary fiber amount from 47% to 44%.

20.3 PROCESSING OF PEAR WASTES AND BY-PRODUCTS

The pear processing sector creates a lot of waste (peel, seeds, pomace), which has a detrimental influence on the environment when disposed of indiscriminately, the resulting pollution costing a lot of money to remediate. Several studies have shown that pear-based wastes and by-products are nonetheless high in bioactive chemicals, particularly phenolic compounds, resulting in important chemical, physical and biological qualities. Pear pomace is also used to recover high levels of dietary fiber (Chockchaisawasdee and Stathopoulos, 2017). The wastes and by-products of pear are a useful source for novel ingredients to be incorporated into food items to have a significant impact on intestinal function. In several publications on the utilization of bioactive components from fruit wastes in food items, antioxidant activity has been shown to have a synergistic impact with the inclusion of dietary fiber (Saura-Calixto et al., 2010). Hemicelluloses are the second most prevalent component in fruit pomace after cellulose, accounting for up to 15–30% of the dry matter (Nawirska and Kwasniewska, 2005). The hemicellulose component of pear pomace is often made up of β-D-xylose, β-D-glucose, α-L-arabinose, α-D-galactose residues and a small fraction of uronic acids from glucuronoxylans, mannans and xyloglucans (Aguedo et al., 2012; Silva et al., 2013). Hemicelluloses are appealing as biopolymers because of their structural complexity, and they can be used in a variety of food and non-food applications, such as films, stabilizing agents, gelling agents, adhesives and viscosity enhancing additives in the nutritional, pharmacological and extra-industrial applications. Rabetafika et al. (2014) devised a method for extracting hemicelluloses from pear pomace. Direct alkaline extraction (sodium hydroxide and hydrogen peroxide) and two-step extraction with a delignification pretreatment were used to remove hemicelluloses from pear pomace. Delignification of the sample (up to 995.4 g kg^{-1} of the original lignins) during the direct alkaline/hydrogen peroxide and two-step acidified sodium chlorite/sodium hydroxide methods enhanced the hemicellulose extraction efficiency to 945.3 g kg^{-1}. Hemicelluloses were mostly made up of xylans (xylose/glucose ratio of 4.7–17.3) and had little lignin (54.5–62.1 g kg^{-1} dry matter). Direct sodium hydroxide extraction produced xylans and glucans with a significant lignin concentration (149.3 g kg^{-1} dry matter). Polymers and oligomers with molecular masses ranging from 1,711 g mol^{-1} to 8,870,000 g mol^{-1} were found in all separated components. The two-step procedure yielded the purest cellulose residue (799.3 g kg^{-1} dry matter).

The production of pectin from pear pomace is economically viable and a solution for waste management generated from the fruit processing industries. In order to meet the increased demand for pectin, recovery is a critical unit activity in the food business (Vriesmann et al., 2012; Vasco-Correa and Zapata, 2017). At the industrial scale, pectin is recovered at extreme temperatures by hydrolyzing protopectin into pectin, although there are new possibilities in pectin synthesis. Polygalacturonase (PGI) was employed to obtain pectin from apple and pear pomace, with excellent yields, indicating that the PGI enzyme had a better affinity for this substrate. For apples and pears, pectin recovery was 20 and 60% greater than with chemical extraction, respectively. These findings suggest that pear pomace, rather than apple pomace, is the more suitable substrate for enzymatic pectin recovery (Franchi et al., 2014).

20.4 UTILIZATION OF PEAR WASTES AND BY-PRODUCTS

The processing of pear to obtain juice and various other value-added products, such as pear paste, pear wine (perry) and pear makjeilli, produces considerable amounts of waste (Rhyu et al., 2014). The waste material may be a potential source of fiber and hence a desirable component in the manufacture of functional food items (Ahmad and Khalid, 2018). This waste is known as pomace or bagasse (a by-product) and is mostly made up of seeds, peels and cores. Pear pomace contains a considerable amount of water that could be a significant environmental contaminant. The juice yield is, however, reported to be 80–85% and up to 15–20% is leftover pomace which consists of about 98% dietary fiber, higher than when compared with other fruit pomaces (Nawirska and Kwasniewska, 2005). Research studies have reported pear pomace to be a potential source of nutraceuticals owing to their dietary fiber content such as non-digestible polysaccharides and lignin. Nawirska and Kwasniewska, (2005) evaluated the dietary fiber fractions of the pear pomace and reported 35% cellulose, 34% liginin, 19% hemicelluloses and 13% pectin. In another study, Rabetafika et al. (2014) reported that pear pomace consisted of 32.43% (dry weight) cellulose, 20.22% hemicelluloses and 19.66% lignin, although the estimate of the hemicellulose content of the pear pomace was lower than that obtained by Martin-Cabrejas et al. (1995), who found it to be close to 35%. Several attempts were made by the researchers to enhance the functional characteristics and chemical nature of the dietary fiber in pear pomace. Furthermore, application of both treatments improved the water- and oil-holding efficiencies from 3.5 to 5.9 g/g and 1.9 to 2.9 g/g, respectively. In addition to dietary fiber, pear pomace also includes essential nutrition-beneficial elements such as polyphenols, triterpenes, and phenolic acids, among others (Rhyu et al., 2014). These bioactive phytochemicals are believed to have potential health-beneficial effects on our metabolism, including antioxidant and anti-inflammatory actions (Li et al., 2014). Due to its nutraceutical composition, pear pomace has been studied with respect to other medical objectives. Sharma et al. (2018) studied the combined effect of *Garcinia cambogia* (Malabar tamarind) and pear pomace extracts on adipogenesis in 3T3-L1 cells. Results indicated that compared with individual extracts the mixture exhibited significant reductions in lipid accumulation, suggesting that the mixture had synergistic inhibitory activity on adipogeneis, with high potential utility as an obesity regulator. In another study, You et al. (2017a) found that aqueous extract of pear pomace successfully suppressed hepatic lipid peroxidation (on account of its antioxidant activity) and thus pear pomace acts as a liver defender in rats fed a high-fat diet. The effect of an ethanolic extract of pear pomace to improve the insulin sensitivity on both *in-vitro* and *in-vivo* experimental animal models fed high-fat diets by You et al. (2017b). Results indicated that, among the components of the pear pomace fiber, insoluble dietary fiber content increased glucose uptake by improving gut microbiota, thus preventing fatness in rats fed a diet with a high lipid content. Similar results were obtained from research by Chang et al. (2017).

Despite the extensive research, pear pomace has been scarcely explored with regard to food technology compared with other fruit pomaces. Bchir et al. (2013) investigated the outcome of incorporating pear pomace fiber on the performance of wheat dough and the quality of bread produced. Nutritional and functional parameters, including water-holding capacity, swelling ability and oil-holding capacity, were determined. It was evident from the results that adding pear flesh fiber concentrate to dough (with no fiber) significantly improved techno-functional properties of the wheat dough by inducing increases in the water absorption capacity from 55.5±0.1 to 57.5±0.1%, tenacity from 83.2±2.4 to 106.5±4.2 mmH$_2$O, stability from 4.0±0.0 to 19.0±0.0 min and a decrease in extensibility from 69.1±5.7 to 35.3±3.1 mm, softening from 60.0±1.0 to 20.0±1.0 (BU, Brabender Units), breakdown from 34.0±2.8 (BU) and setback from 103.0±0.0 to 94.0±0.0 (BU). Furthermore, addition of pear fiber concentrate significantly increased crumb a* values from 1.96±0.09 to 3.58±0.01 and decreased b* and L* values from 22.28±0.20 to 17.49±0.18 and 74.19±0.88 to 60.52±0.91, respectively. However, a slight change in the crust L* value was observed, increasing from 53.07±0.70 to 54.39±1.85 while a* and b* values decreased from 17.59±2.13 to 11.01±0.84 and 34.77±3.83 to 29.28±0.63, respectively. Bread enriched with pear fiber concentrate was nutritionally equal to white wheat bread with no marked variation in the specific volume and quality of breads. Thus, pear pomace fiber has the potential to develop fiber-enriched breads, thereby raising the value of the by-product. In another study, Bchir et al. (2017) examined the consequence of incorporation of pear pomace fiber on the nutritional, physicochemical and organoleptic characteristics of cereal bars. Cereal bars were prepared using 10% of the by-product. Results showed that addition of pear pomace fiber showed a slight difference in the nutritional composition of the cereal bars compared with control (cereal bars made with commercial wheat bran fiber). The developed cereal bar showed lower water activity level (0.481 ± 0.001 a_w) and is energy dense (300 Kcal/100 g). No significant difference was found in a^* value (~45) and texture standards (cohesiveness, hardness, springiness and chewiness) among the cereal bars developed with pear byproduct and wheat bran. Sensory investigation found that using 10% pear pomace fiber resulted in a cereal bar with the best consumer panel acceptance when compared with commercial cereal bars. However, appearance of the cereal bar developed with pear pomace becomes a limiting factor. Overall, pear by-products might be successfully utilized as functional food ingredients in the development of novel cereal bar compositions. Rocha-Parra et al. (2019) also investigated the particle shape effect of pear pomace on the batter microstructure, density, stickiness, specific volume, consistency and color of enriched layer and sponge cakes. Three pear pomace powders were prepared with dissimilar particle dimensions, namely fine, average and coarse. When pear pomace was administered to batters, there was reduced consistency in bubble dispersion, especially at larger particle sizes. With increasing volumes of pear pomace, the specific volume of the cake decreased dramatically. For sponge cakes, the finest particle size of pear pomace led to a tremendous drop in specific volume. In general, doubling the volume of pear pomace increased hardness while decreasing elasticity, cohesion and resilience, although the impact was dependent on particle size. Medium and coarse particle sizes produced superior textural qualities. These findings suggest that pear pomace with appropriate particle size might be a potential fiber source for various cake recipes.

20.5 CONCLUSIONS

Industrial processing of pear fruit generates large quantities of waste or by-products which presents a serious problem for the environment. These wastes are high in dietary fiber and polyphenols in addition to micro- and macronutrients that promote intestinal health, lower cholesterol level, achieve weight control and improve control of glycaemic and insulin responses. However, these by-products remain underexploited due to lack of adequate processing techniques pivotal for their effective valorization, particularly for recovery of beneficial health-promoting bioactive components like dietary fibers. Increased use of fiber supplements has been shown in studies to be advantageous to human wellbeing. As a result of its high dietary fiber content, pear waste has incredible promise as a possible food additive or functional food component to satisfy the techno-functional requirements of generating well-being value-added goods. However, there is still a significant gap to be addressed to exploit the possible advantages of dietary fiber in the pharmaceutical and food industries in a sustainable way in order to improve value-added healthful products, functional foods and various biofortified foods and juices. There are also other possibilities to study dietary fiber from residues and by-products in order, say, to generate cattle feed. Such gaps are likely to be resolved in the coming future to allow for the long-term use of pear wastes or by-products.

REFERENCES

Abaci, Z. T., Sevindik, E. and Ayvaz, M. 2015. Comparative study of bioactive components in pear genotypes from Ardahan/Turkey. *Biotechnology & Biotechnological Equipment*. https://doi.org/10.1080/13102818. 2015.1095654.

Aguedo, M., Kohnen, S., Rabetafika, H. N., Vanden Bossche, S., Sterckx, J., Blecker, C., et al. 2012. Composition of by-products from cooked fruit processing and potential use in food products. *Journal of Food Composition and Analysis*, 24(1), 13–25.

Ahmad, A. and Khalid, N. 2018. Dietary fibers in modern food production: A special perspective with β-glucans. In: A. M. Grumezescu & A. M. Holban (eds). *Biopolymers for Food Design*. Cambridge, MA: Academic Press, pp. 125–156.

Bchir, B., Jean-franc, T., Rabetafika, H. N. and Blecker, C. 2017. Effect of pear apple and date fibres incorporation on the physico-chemical, sensory, nutritional characteristics and the acceptability of cereal bars. *Food Science and Technology International*, 24, 198–208.

Bchir, B., Rabetafika, H. N., Paquot, M. and Blecker, C. 2013. Effect of pear, apple and date fibres from cooked fruit by-products on dough performance and bread quality. *Food and Bioprocess Technology*, 7, 1114–1127.

Chang, S., Cui, X., Guo, M., Tian, Y., Xu, W., Huang, K., and Zhang, Y. 2017. Insoluble dietary fiber from pear pomace can prevent high-fat diet-induced obesity in rats mainly by improving the structure of the gut microbiota. *Journal of Microbiology and Biotechnology*, 27, 856–867.

Chockchaisawasdee, S. and Stathopoulos, C. E. 2017. Extraction, isolation and utilization of bioactive compounds from fruit juice industry waste. In: Q. V. Vuong (ed). *Utilisation of bioactive compounds from agricultural and food production waste*. Boca Raton, FL: CRC Press, pp. 272–313.

Fattouch, S., caboni, P., Coroneo, V., Tuberoso, C., Angioni, A., Dessi, S., Marzouki, N. and Cabras, P. 2008. Comparative analysis of polyphenolic profiles and antioxidant and antimicrobial activities of Tunisian pome fruit pulp and peel aqueous acetone extracts. *Journal of Agricultural and Food Chemistry*, 56, 1084–1090.

Franchi, M. L., Marzialetti, M. B., Pose, G. N. and Cavalitto, S. F. 2014. Evaluation of enzymatic pectin extraction by a recombinant polygalacturonase (PGI) from apples and pears pomace of Argentinean production and characterization of the extracted pectin. *Journal of Food Processing and Technology*, 5, 352.

Gornas, P., Rudzinska, M., Raczyk, M., Misina, I., Soliven, A. and Seglina, D. 2016. Chemical composition of seed oils recovered from different pear (*Pyrus communis* L.) cultivars. *Journal of the American Oil Chemists' Society*, 93, 267–274.

Hameed, F., Gupta, N., Rahman, R., Anjum, N. and Nayik, G. A. 2020. Jamun. In: Nayik, G. A. and Gull, A. (eds). *Antioxidants in fruits: Properties and health benefits*. Springer, Singapore.

Isaac, I. O., Ekpa, o. D. and Ekpe, U. J. 2014. Extraction, characterization of African pear (*Dacryodes Edulis*) oil and its application in synthesis and evaluation of surface coating driers. *International Journal of Advanced Research in Chemical Science (IJARCS)*, 1(4), 14–22.

Kapoor, S. and Ranote, P. S. 2015. Antioxidant potentials and quality of blended pear-jamun (*Syzygium cumini* L.) Juice. *International Research Journal of Biological Sciences*, 4(4), 30–37.

Li, X., Wang, T., Zhou, B., Gao, W., Cao, J. and Huang, L. 2014. Chemical composition and antioxidant and anti-inflammatory potential of peels and flesh from 10 different pear varieties (*Pyrus* spp.). *Food Chemistry*, 152, 531–538.

Liaudanskas, M., Zymonè, K., Viškelis, J., Klevinskas, A., and Janulis,V. 2017. Determination of the phenolic composition and antioxidant activity of Pear extracts. *Journal of Chemistry*, 7856521.

Lin, L. and Harnly, J. M. 2008. Phenolic compounds and chromatographic profiles of pear skins (*Pyrus* spp.). *Journal of Agricultural and Food Chemistry*, 56, 9094–9101.

Lin, X., Chen, S., Xu, X. R. and Xia, E. Q. 2012. Potential of fruit wastes as a natural resource of bioactive compounds. *International Journal of Molecular Sciences*, 13(7), 8308–8323.

Ma, J., Wang, S., Zhang, K., Wu, Z., Hattori, M., Chen, G. and Ma, C. 2012. Chemical components and antioxidant activity of the peels of commercial apple-shaped pear (fruit of *Pyrus pyrifolia* cv. pingguoli). *Journal of Food Science*, 77(10), C1097–C1102.

Mahammad, M. U., Kamba, A. S., Abubakar, L. and Bagna, E. A. 2010. Nutritional composition of pear fruits (*Pyrus communis*). *African Journal of Food Science and Technology*, 1(3), 76–81.

Manzoor, M., Anwari, F., Ahmed, I. B. and Jamili, A. 2013. Variation of the phenolics and antioxidant activity between peel and pulp parts of pear (*Pyrus communis*) fruit. *Pakistani Journal of Botany*, 45(5), 1521–1525.

Martın-Cabrejas, M. A., Esteban, R. M., Lopez-Andreu, F. J., Waldron, K. and Selvendran, R. R. 1995. Dietary fiber content of pear and kiwi pomaces. *Journal of Agricultural and Food Chemistry*, 43, 662–666.

Mushtaq, M., Akram, S., Ishaq, S. and Adnan, A. 2019. Pear (*Pyrus communis*) seed oil. In: Ramadan, M. (ed). *Fruit oils: chemistry and functionality*. Springer, Cham, pp. 859–874.

National Horticulture Board (NHB). 2020. Horticultural statistics at a glance 2019, pp. 1–514.

Nawirska, A. and Kwasniewska, M. 2005. Dietary fibre fractions from fruit and vegetable processing waste. *Food Chemistry*, 91, 221–225.

Ozturk, A., Demirsoy, L., Demirsoy, H., Asan, A. and Gul, O. 2014. Phenolic compounds and chemical characteristics of pears (*Pyrus Communis* L.), *International Journal of Food Properties*. https://doi.org/10.1080/10942912.2013.835821.

Ozturk, I., Ercisli, S., Kalkan, F. and Demir, B. 2009. Some chemical and physic-mechanical properties of pear cultivars. *African Journal of Biotechnology*, 8(4), 687–693.

Pascoalino, L. A., Reis, F. S., Prieto, M. A., Barreira, J. C. M., Ferreira, I. C. F. R. and Barros, L. 2021. Valorization of bio-residues from the processing of main Portuguese fruit crops: From discarded waste to health promoting compounds. *Molecules*, 26(9), 2624.

Rabetafika, H. N., Bchir, B., Blecker, C., Paquot, M. and Wathelet, B. 2014. Comparative study of alkaline extraction process of hemicelluloses from pear pomace. *Biomass and Bioenergy*, 61, 254–264.

Rhyu, J., Kim, M. S., You, M. K., Bang, M. A. & Kim, H. A. 2014. Pear pomace water extract inhibits adipogenesis and induces apoptosis in 3T3-L1 adipocytes. *Nutrition Research and Practice*, 8, 33–39.

Rocha-Parra, A. F., Belorio, M., Ribotta, P. D., Ferrero, C. and Gómez, M. 2019. Effect of the particle size of pear pomace on the quality of enriched layer and sponge cakes. *International Journal of Food Science & Technology*, 54(4), 1265–1275.

Sahiba, Ali, M. and Juyal, D. 2018. A review on pharmacognostical and phytochemical evaluation of Pyrus pashia Buch-Ham ex D. Don. *The Pharma Innovation Journal*, 7(5), 186–189.

Saura-Calixto, F., Pérez-Jiménez, J., and Goñi, I. 2010. Dietary fiber and associated antioxidants in fruit and vegetables. In: L. A. De la Rosa, E. Álvarez-Parilla, & G. A. González-Aguilar (eds). *Fruit and vegetable phytochemicals*. Iowa: Wiley-Blackwell, pp. 223–234.

Scalzo, J., Politi, A., Pellegrini, N., Mezetti, B. and Battino, M. 2005. Plant genotype affects total antioxidant capacity and phenolic contents in fruit. *Nutrition*, 21(2), 207–213.

Sharma, K., Kang, S., Gong, D., et al. 2018. Combination of *Garcinia cambogia* extract and pear pomace extract additively suppresses adipogenesis and enhances lipolysis in 3T3-L1 cells. *Pharmacognosy Magazine*, 14, 220–226.

Silva, G., Souza, T. M., Barbieri, R. L. and de Oliveira, A. C. 2014. Origin, domestication and dispersing of pear (*Pyrus* spp.). *Advances in Agriculture*, 4(1), 1–8.

Silva, L., Shahidi, F. and Coimbra, M. A. 2013. Dried pears: Phytochemicals and potential health effects. In: Alasalvar, C., and Shahidi, F. (eds). *Dried fruits: Phytochemicals and health effects*. Oxford: Blackwell Publishing Ltd, pp. 325–56.

Vasco-correa, J. and Zapata, A. D. Z. 2017. Enzymatic extraction of pectin from passion fruit peel (*Assiflora edulis* f. *flavicarpa*) at laboratory and bench scale. *LWT-Food Science and Technology*, 108, 171–177.

Vriesmann, L. C., Teµfilo, R. F. and Petkowicz, C. L. O. 2012. Extraction and characterization of pectin from cacao pod husks (*Heobroma cacao* L.) with citric acid. *LWT-Food Science and Technology*, 49, 108–116.

Yan, L., Li, T., Liu, C. and Zheng, L. 2019. Effects of high hydrostatic pressure and superfine grinding treatment on physicochemical/functional properties of pear pomace and chemical composition of its soluble dietary fibre. *LWT Food Science and Technology*, 107, 171–177.

You, M. K., Jin, R. and Kim, H. A. 2017a. Pear pomace water extract suppresses hepatic lipid peroxidation and protects against liver damage in rats fed a high fat/cholesterol diet. *Food Science and Biotechnology*, 26, 801–806.

You, M. K., Kim, H. J., Rhyu, J. and Kim, H. A. 2017b. Pear pomace ethanol extract improves insulin resistance through enhancement of insulin signaling pathway without lipid accumulation. *Nutrition Research and Practice*, 11, 198.

Yukui, R., Wenya, W., Rashid, F. and Qing, L. 2009. Fatty acids composition of apple and pear seed oils. *International Journal of Food Properties*, 12(4), 774–779.

CHAPTER 21

Quince Wastes and By-Products
Chemistry, Processing, and Utilization

Nadia Bashir, Beena Munaza, Shafiya Rafiq, Monika Sood, and Sushil Sharma

CONTENTS

21.1 INTRODUCTION

Quince (*Cydonia oblonga* Miller) is a temperate pome fruit belonging to the Rosaceae family. The quince is a small tree or shrub bearing edible fruits which has been cultivated since ancient times in Asia, spreading to the eastern areas of the Himalayan mountains and throughout Europe to the west (Monka et al., 2014). The cultivation of quince also touched the Mediterranean region in classical times and the fruit was used by the Romans (Vaughan and Geissler, 2009). Quince is a small, deciduous and spineless tree cultivated in warm climates and reaches up to a height of 8 m and a width of 4 m (Khoubnasabjafari and Jouyban, 2011). Quince is known by different names: '*bahee dana*' in Urdu, '*beh*' in Farsee, '*Strythionas*' in Greek, '*Bihi*' in Hindi and 'Bamchount' in Kashmiri (Mir et al., 2015). The quince fruit is an irregularly shaped pome fruit possessing multiple mucilage-covered seeds and has a large diameter of 10–12 cm. The surface of the quince fruit is covered by thin hair-like structures which usually disappear upon ripening (Fazeenah et al., 2016). Quince fruits are harvested during October to November and the fruit requires a ripening temperature of 20°C because of its climacteric nature. Quince fruit has remarkable nutritional composition (both macro- and micro-nutrients) providing immense benefits to human health. Quince is a rich source of organic acids with citric acid, ascorbic acid, malic acid, aspartic acid and glutamic acid as the dominant ones. Quince fruit also contains significant amounts of dietary fiber and is rich in bioactive compounds such as phenolics and flavonoids (Fazeenah et al., 2016). The proximate composition of quince reported by Rodriguez-Guisado et al. (2009) over different varieties grown in Southeastern Spain showed a water content of 76.72%, a crude fiber content of 5.33% and a low fat content (1.95%). Quince fruit is a rich source of vitamins and minerals with vitamin C, calcium, phosphorus, iron and potassium concentrations of 15.12, 8.25, 16.83, 195.62 and 0.74 mg/100 g, respectively (Gheisari and Abhari, 2014). The pulp of ripe quince fruit has values for total soluble solids (TSS), titratable acidity, total sugars and reducing sugars of 14.10 °Brix, 1.18, 8.13% and 5.0%, respectively (Rasheed et al., 2018).

The quince plant is very easy to cultivate owing to its resistance to pests and diseases and tolerance of harsh climatic conditions (Milic et al., 2010). The average worldwide quince production amounts to 469,325 tonnes, with Turkey being the highest producer (106,716 tonnes) followed by China (100,470 tonnes), Uzbekistan (52510 tonnes), Iran (35,315 tonnes) and Morocco (34,810 tonnes). In India, Quince fruit is mostly cultivated on 470 hectares of arable land in the Kashmir region (Rather et al., 2020). Quince, a valuable fruit, has an astringent taste which limits its consumption in the fresh form. Therefore, quince fruit is mainly used in the preparation of processed products like canned fruit, marmalade, jam, jellies and alcoholic beverages (Elena et al., 2016). The processing operations, like peeling, coring and pitting, generates waste and by-products like seeds, mucilage and peel which can be utilized in different ways. The fruit fresh weight is comprised of 90.6% pulp, 4.4% peel and core and seeds amounting to 5% of the total fruit. The amount of waste generated varies, depending upon the technology employed during processing.

21.2 CHEMISTRY OF QUINCE WASTE AND BY-PRODUCTS

21.2.1 Quince Seeds

Quince seeds, usually considered to be a waste product, are rich in phenolics, antioxidants and dietary fiber. There are approximately ten seeds present in a mature quince fruit (Hakala et al., 2014; Jouki et al., 2014). On a dry weight basis, quince seeds are a good source of protein (35.55%), carbohydrates (29.52%), fat (23.56), ash (3.63%), hemicelluloses (9.58%), cellulose (11.39%) and lignin (2.08%) (Kurt and Alatar, 2018). The seeds of quince also exhibit a rich phenolic profile composed of caffeoylquinic acid, dicaffeoylquinic acid, lucenin-2, vicenin-2, stellarin 2, isoschaftoside, schaftoside, 6-C-pentosyl-8-C-glucosyl chrysoeriol and 6-C-glucosyl-8-C-pentosyl chrysoeriol. Organic acids, such as citric, ascorbic, malic, quinic, shikimic and fumaric acids, have also been reported in quince seeds (Silva, 2005). Glutamic acid, aspartic acid and asparagine are the free amino acids present in quince seeds in abundance (Hanan et al., 2020). Enriched bioactive compounds in quince seeds provide health benefits against diarrhea, dysentery, intestinal colic, constipation and respiratory tract diseases (Hanan et al., 2020). Some fat-soluble bioactive components, such as tocopherols (exhibiting vitamin E activity), phytosterols (campesterol, stigmasterol and sitosterol) and phenolic acids, are also present in quince seeds. These bioactive components present in quince seed can provide immense health benefits including treatment of gastrointestinal disorders such as dysentery, constipation and colitis, as well as respiratory disorders (Aslam et al., 2014). According to Janbaz et al. (2013), the antispasmodic property associated with quince seeds might be due to the presence of components activating the muscarinic receptors and the calcium antagonist mechanism.

21.2.2 Seed Mucilage

Quince seeds exude mucilage as a by-product when steeped in water, and it can act as a biodegradable polymer used for food coatings. The mucilage of quince seed is mainly composed of heteropolysaccharide fractions like rhamnogalacturonan (pectinaceous fraction) and arabinoxylans (hemicellulosic fraction), which are highly branched and are associated with different side chain groups like galactans or galactomannans, imparting them with unique film-forming properties (Soukoulis et al., 2018). Quince seed mucilage, a plant biopolymer with a polysaccharide–protein structure, contains flavonoids, sterols, alkaloids, saponins, resins, phenolics and terpenoids (Moghbel and Tayebi, 2015).

21.2.3 Peel

Quince peel is a valuable waste material resulting from fruit processing operations like the manufacture of jams, marmalades and jellies. Quince peel is rich in antioxidants, pectin, dietary fiber and anti-microbial compounds and finds uses in different industrial sectors. The pre-dominant polyphenols present in the peel include quercetin, rutin, hyperin, isoquercetin, chlorogenic acid, kaempferol and kaempferol 3-O-glucoside (Rather et al., 2020). The synergistic action of these polyphenols can result in a number of health benefits including antioxidant, anti-inflammatory, cardioprotective and hyperglycemic activities (Ashraf et al., 2016). Quince peel contains high concentrations of pectin which can act as a protective shield against colon damage in colitis, thereby conferring health benefits (Minaiyan et al., 2012). The incorporation of this health-promoting by-product into human foods can greatly benefit human health. Quince peel resulting from food processing operations, usually considered to be an agro-waste, can be utilized as a potent anti-microbial agent, for which synergistic activity between polyphenols like chlorogenic acid might be responsible (Hakkinen et al., 1999). The total phenolic content of quince peel, as reported by Magalhães et al. (2009), corresponded to 6.3 g/kg with 5-O-caffeoylquinic acid (29%) as the dominant antioxidant whereas the EC_{50} of DPPH free radical scavenging activity of peel extract was found to be 0.8 mg mL^{-1}. On the basis of these phenolic and antioxidant components, Magalhães et al. (2009) used peel extract for protection of erythrocyte proliferation and reported a significant protection of the erythrocyte membrane from hemolysis. The study carried out by Fattouch et al. (2007) reported that anti-microbial activity of quince peel extract with minimum inhibitory and bactericidal concentrations in the range of 10^2–5×10^3 µg polyphenol mL^{-1}. The greatest inhibition was recorded against the bacteria *Staphylococcus aureus* and *Pseudomonas aeruginosa* in contrast to *Escherichia coli* and the yeast *Candida albicans*. Therefore, quince peel can be exploited for the development of antibiotic or bio-control agents.

21.3 PROCESSING AND UTILIZATION OF QUINCE WASTE AND BY-PRODUCTS

Increased urbanization and enhanced productivity of fruits, with a consequential rise in waste generation, is a serious concern in terms of environmental contamination by degrading the soil and water sources (Panda et al., 2016). However, these waste and by-products resulting from different unit operations, like peel and seeds, are rich in proteins, bioactive components, minerals, fibers and other important components and therefore can act as a functional ingredient in the development of nutraceutical foods (Kodagoda and Marapana, 2017). The utilization of these by-products not only conserves and protects the environment but can also improve the economy.

21.3.1 Quince Seeds

The combination of high nutrient-density and the presence of bioactive components makes quince seeds an excellent ingredient for functional food formulations. The quince seeds, as wastes and by-products from quince-processing industries, are an inexpensive and rich source of protein (21%), as compared with cereals, with good functional properties (Sandovaloliveros and Paredeslópez, 2013). Quince seeds have been used for the extraction of protein isolates, based on alkaline extraction and the isoelectric point method, with a protein content of 87% (Deng et al., 2019). Deng et al. (2020)

used fraction characterization to separate the quince proteins into albumin (88%) and glutelin (85%) fractions with good functional properties. The quince proteins have been used to formulate protein-rich food products. Kurt and Alatar (2018) prepared a quince seed powder with high fiber content which was subjected to air drying at 45°C for 24 h and then milled and sieved into a fine powder as the pre-mix ingredient in ice cream preparations for enhancing melting characteristics and textural characteristics. The study concluded that incorporation of quince seed powder (at 0.5% or 0.75%) reduced the ice-crystallization and imparting a smooth texture to the ice cream. The addition of quince seed powder also enhanced the nutritional composition of the ice cream in terms of protein and fiber contents. Apart from food application, quince seeds have also been used in soaps and in the oil industry. According to Górnás et al. (2014), quince seed oil obtained from the cold-pressing technique showed the highest concentrations of tocopherols, β -carotene and total phenolic compounds (726.20, 10.77 and 64.03 mg kg^{-1}, respectively) and high peroxide value (0.59 mEq O$_2$ kg^{-1}), highlighting the potential use of quince seed oil in food applications in the near future. Wang et al. (2017) reported that oil extracted from underutilized Chinese quince seeds using supercritical fluid extraction exhibited higher concentrations of unsaturated fatty acids (86.37–86.75%) and α-tocopherol (576.0–847.6 mg kg^{-1}), lower acid value (3.97 mg g^{-1}), and lower peroxide value (0.02 meq O$_2$ kg^{-1}). The study also reported that leftover protein meal following oil extraction had the highest nitrogen solubility index (49.64%) and protein dispersibility index (50.80%), meaning that the protein meal could be successfully used in the supplementation and fortification of foods.

21.3.2 Quince Seed Mucilage

Quince seed mucilage exhibits hydrophilic, emulsifying and foaming properties which have shown its use as a bulking agent, or its packaging and coating properties in food products (Rezagholi et al., 2019). The hydrophilic nature of the quince seed mucilage enables it to form hydrogels, a 3D structure which can be utilized in efficient drug delivery systems, tissue engineering and food coatings, by replacing synthetic and non-biodegradable polymers (Hussain et al., 2019). A number of methods have been employed for the preparation of quince seed mucilage involving the basic step of soaking dried quince seeds in water followed by precipitation/filtration of the mucilage. The separated mucilage is then subjected to different drying methods such as freeze-drying, oven-drying and spray-drying, resulting in quince seed mucilage (Tamri et al., 2014; Mughbal and Tayabi, 2015). Jouki et al. (2014) formulated quince seed-based mucilage films with antioxidant and antimicrobial properties for rainbow fish fillets against *Staphylococcus aureus*, *Escherichia coli*, *Staphylococcus putrefaciens*, and *Yersinia enterocolitica*. Kozlu et al. (2020) coated mandarin orange slices with quince seed mucilage film and stored them at 4°C for 10 days. The quince seed mucilage film retarded weight loss, delayed tissue softening and retained the color in the coated mandarin slices in contrast to fruit slices without the film coating. Yousefi et al. (2018) employed quince seed mucilage as a healthy fat replacer in hamburgers and reported a positive effect of quince seed mucilage on the texture of hamburgers owing to the enhanced water-holding capacity of the mucilage. In addition to food coatings, the quince seed mucilage can also be utilized as a wound-healing agent and has been traditionally been used for this purpose in the Iranian medicine system. The rich phytochemical profile of quince seeds, together with amino acids (like glutamic acid, aspartic acid and asparagine) and potent antioxidants like ascorbic acid, suggests that caffeoylquinic acid might be responsible for the wound-healing ability of quince seed mucilage (Dessimond et al., 2002). A study conducted by Tamri et al. (2014) concluded that quince seed mucilage incorporated into eucerin creams was more effective in healing wounds in rabbits than the creams used alone. Kawhara et al. (2017) also reported a positive effect of quince seed mucilage on atopic dermatitis-like symptoms. Shahbaz and Moosavy (2019) formulated a film from quince seed mucilage in collaboration with TiO$_4$ and demonstrated its efficacy against the bacteria *Staphylococcus aureus*, *Bacillus subtilis*, *Bacillus cereus*, *Listeria monocytogenes*, *Salmonella typhimurium*, and *Escherichia coli* strain O157:H7.

The quince seed mucilage is extracted by a hot water method and then incorporated with titanium dioxide and silicon dioxide nanoparticles for preparation of nano-compost films as antimicrobial films to increase the shelf-life of coated fruits and vegetables.

21.3.3 Quince Peel

Quince peel is a valuable waste and by-product resulting from fruit processing operations like the production of jams, marmalades and jellies. The peel is rich in antioxidants, dietary fiber and anti-microbial components and finds use in different purposes. De Escalada Pla et al. (2010) extracted a dietary fiber fraction with valuable properties and composition from quince peel using an ethanol extraction method followed by drying. Pectin, a heterogeneous complex polysaccharide, is mainly found between plant cell walls and is responsible for cellular structure and integrity. Pectin is mainly composed of D-galacturonic acid units linked by α (1-4) linkages along with neutral sugar molecules and (1-2)-linked side chains of L-rhamnose (Hadi et al., 2020). The carboxyl groups in the galacturonic acid residues of pectin are esterified with methyl groups and the degree of methylation determines the degree of gelling (Brown et al., 2014). Pectin is mainly used as an additive in the food industry and it exhibits gelling, emulsifying and thickening properties. Quince peel contains considerable amounts of pectin and can act as a non-traditional but economic source of pectin extraction (Ferreira et al., 2004). Brown et al. (2014) extracted pectin from quince peel using different extraction techniques such as hot air drying and low-pressure super-heated steam drying. The pectin was extracted from dried quince peel by immersing it in nitric acid solution (85 g L^{-1}) in a rotary evaporator followed by filtration of the hot acid extract. The filtrate was concentrated using two techniques, namely ultrafiltration and vacuum evaporation. The pectin was precipitated from the filtrate using ethanol (95%) and allowed to stand overnight. The coagulated pectin was then washed with ethanol (70%) followed by freeze drying. The freeze-dried pectin was then converted into powder using a grinder. The pectin extracted in this way exhibited yield, purity and degree of methylation characteristics indistinguishable from pectin obtained from traditional sources. Hellin et al. (2003) estimated the efficacy of pectin extracted from Japanese quince on quality parameters of white bread and reported a positive impact of quince-based pectin on crumb elasticity and hardness. The incorporation of pectin extracted from quince fruits at levels of 0.5% enhanced bread volume by 7%.

Quince peel is also a rich source of polyphenols and other bioactive components which possesses high antioxidant activities as compared with quince fruit pulp or seeds. Mir et al. (2015) reported higher antioxidant potential of quince peel in terms of DPPH scavenging activity, total phenolic concentration, reducing power and H_2O_2 scavenging activity. Silva et al. (2004) also concluded that there was a higher antioxidant activity and total phenolic concentration in peel than in quince pulp. Riahi-Chebbi et al. (2015) explored anti-tumoral effect of *Cydonia oblonga* peel pulp polyphenolic extracts on both non-tumorigenic cells, NIH 3T3 fibroblasts, HEK-293 embryonic kidney cells and human colon adenocarcinoma LS174 cells. Techniques such as flow cytometry, Western blot analysis and polymerase chain reaction (PCR) were employed to assess cell cycle distribution, cell apoptosis and intra-cellular reactive oxygen species (ROS). The study concluded that the synergistic effect of quince peel polyphenol extract resulted in cell death and reduced formation of reactive oxygen species, thereby suggesting its potential use in the development of a non-toxic anti-colon cancer drug. Therefore, quince peel can be utilized as a functional ingredient in food formulations where it can act as a source of antioxidant components conferring health benefits. Elena et al. (2016) utilized quince peel as a source of natural color and for improving the antioxidant and phenolic content of an alcoholic beverage. Mudura et al. (2016) employed quince peel extract for enhancing the antioxidant and phenolic content of fruit liqueurs. The study reported positive impacts of quince peel extract on the antioxidant and phenolic components as well as on the color of fruit liqueurs. Anvar et al. (2019) utilized quince peel powder at 0–15% levels in the development of sponge cakes. The study revealed an increase in total phenolics, iron and calcium content of sponge cakes along with improved batter consistency and overall acceptability.

21.4 CONCLUSIONS

Quince fruit is rich in antioxidants, polyphenols, dietary fibers and is processed into jams, jellies and marmalades. Quince-derived wastes and by-products can cause environmental degradation and can pollute soil and water resources. Fortunately, these by-products have a rich metabolite profile in terms of bioactive components or other proximate constituents. Currently, people are becoming more health conscious and the demand for functional foods with health benefits is increasing day by day. Wastes and by-products resulting from quince processing can serve as ideal chemical feedstocks for formulation of functional food products with health benefits. There is a need to carry out extensive research on the utilization of quince by-products which could prove useful for reducing waste disposal costs and pollution, as well as providing novel ingredients for the food, pharmaceutical and cosmetic industries.

REFERENCES

Anvar, A., Nasehi, B., Noshad, M., & Barzegar, H. (2019). Improvement of physicochemical & nutritional quality of sponge cake fortified with microwave-air dried quince peel. *Iranian Food Science Technology Research Journal*, 15, 69–78.

Ashraf, U. M., Muhammad, G., Hussain, M. A., & Bukhari, S. N. (2016). A medicinal plant rich in phytonutrients for pharmaceuticals. *Frontiers in Pharmacology*, 7(163), 1–20.

Aslam, M., & Sial, A. A. (2014). Effect of hydroalcoholic extract of *Cydonia oblonga* miller (Quince) on sexual behaviour of wistar rats. *Advances in Pharmacological Science*, 2014, 1–7.

Brown, V. A., Lozano, J. E., & Genovese, D. B. (2014). Pectin extraction from quince (*Cydonia oblonga*) peel applying alternative methods, effect of process variables & preliminary optimization. *Food Science and Technology International*, 20(2), 83–98.

De Escalada Pla, M. F., Uribe, M., Fissore, E. N., Gerschenson, L. N., & Rojas, A. M. (2010). Influence of the isolation procedure on the characteristics of fiber-rich products obtained from quince wastes. *Journal of Food Engineering*, 96, 239–248.

Deng, Y., Huang, L., Zhang, C., & Xie, P. (2020). Chinese quince seed proteins: sequential extraction processing and fraction characterization. *Journal of Food Science and Technology*, 57(2), 764–774.

Deng, Y., Huang, L., Zhang, C., Xie, P., Cheng, J., Wang, X., & Li, S. (2019). Physicochemical and functional properties of Chinese quince seed protein isolate. *Food Chemistry*, 283, 539–548.

Dissemond, J., Goos, M., & Wagner, S. N. (2002). The role of oxidative stress in the pathogenesis and therapy of chronic wounds. *Der Hautarzt; Zeitschrift fur Dermatologie, Venerologie, und verwandte Gebiete*, 53(11), 718–723.

Elina, M., Coldea, T. E., & Fărcaş, A. (2016). Quince peel extract addition to liqueur for improving antioxidant activity and phenolic content. *Hop and Medicinal Plants*, 24(1/2), 63–70.

Fattouch, S., Caboni, P., Coroneo, V., Tuberoso, C. I., Angioni, A., Dessi, S., ... & Cabras, P. (2007). Antimicrobial activity of Tunisian quince (*Cydonia oblonga* Miller) pulp and peel polyphenolic extracts. *Journal of Agricultural and Food Chemistry*, 55(3), 963–969.

Fazeenah, A. A., & Quamri, M. A. (2016). Behidana (*Cydonia oblonga* Miller)–A review. *World Journal of Pharmaceutical Research*, 5(11), 79–91.

Ferreira, I. M., Pestana, N., Alves, M. R., Mota, F. J., Reu, C., Cunha, S., & Oliveira, M. B. P. (2004). Quince jam quality, microbiological, physicochemical and sensory evaluation. *Food Control*, 15(4), 291–295.

Gheisari, H. R., & Abhari, K. H. (2014). Drying method effects on the antioxidant activity of quince (*Cydonia oblonga* Miller) tea. *Acta Scientiarum Polonorum Technologia Alimentaria*, 13(2), 129–134.

Gornas, P., Siger, A., Juhņeviča, K., Lācis, G., Šnē, E., & Segliņa, D. (2014). Cold-pressed Japanese quince (*Chaenomeles japonica* (Thunb.) Lindl. ex Spach) seed oil as a rich source of α-tocopherol, carotenoids and phenolics, A comparison of the composition and antioxidant activity with nine other plant oils. *European Journal of Lipid Science and Technology*, 116(5), 563–570.

Hadi, S. T., Fadhil, N. J., Khalaf, A. S., & Alhadithi, H. J. (2020) Extraction of pectin from quince (*Cydonia Oblonga*) fruit husk and using it in jam industry. *Biochemical and Cellular Archives*, 20(1), 2163–2166.

Hakala, T. J., Saikko, V., Arola, S., Ahlroos, T., Helle, A., Kuosmanen, P., & Laaksonen, P. (2014). Structural characterization and tribological evaluation of quince seed mucilage. *Tribology International*, 77, 24–31.

Hakkinen, S. H., Karenlampi, S. O., Heinonen, I. M., Mykkanen, H. M., & Torronen, A. R. (1999). Content of the flavonols quercetin, myricetin, and kaempherol in 25 edible berries. *Journal of Agricultural Food Chemistry*, 47, 2274–2279.

Hanan, E., Sharma, V., & Ahmad, F. J. (2020). Nutritional composition, phytochemistry and medicinal use of quince (*Cydonia oblonga* Miller) with emphasis on its processed and fortified food products. *Journal of Food Processing & Technology*, 11(6), 1–13.

Hellin, P., Jordán, M. J., Vila, R., Gustafsson, M., Göransson, E., Åkesson, B., Gröön, I., Laencina, J., & Ros, J. M. (2003). Processing and products of Japanese quince (*Chaenomeles japonica*) fruits. *Japanese quince—potential fruit crop for northern Europe. Final report of FAIR-CT97-3894*. Swedish University of Agricultural Sciences, Alnarp, 169–175.

Hussain, M. A., Muhammad, G., Haseeb, M. T., & Tahir, M. N. (2019). Quince seed mucilage, a stimuli-responsive/smart biopolymer. In *Functional biopolymers. Polymers and polymeric composites, a reference series*, Springer, Cham, 127–148.

Janbaz, K. H., Shabbir, A., Mehmood, M. H., & Gilani, A. H. (2013). Insight into mechanism underlying the medicinal use of *Cydonia oblonga* in gut and airways disorders. *Journal of Animal and Plant Science*, 23, 330–36.

Jouki, M., Mortazavi, S. A., Yazdi, F. T., Koocheki, A., & Khazaei, N. (2014). Use of quince seed mucilage edible films containing natural preservatives to enhance physico-chemical quality of rainbow trout fillets during cold storage. *Food Science and Human Wellness*, 3(2), 65–72.

Kawahara, T., Tsutsui, K., Nakanishi, E., Inoue, T., & Hamauzu, Y. (2017). Effect of the topical application of an ethanol extract of quince seeds on the development of atopic dermatitis-like symptoms in NC/Nga mice. *BMC Complementary and Alternative Medicine*, 17(1), 1–8.

Khoubnasabjafari, M., & Jouyban, A. (2011). A review of phytochemistry and bioactivity of quince (Cydonia oblonga Mill.). *Journal of Medicinal Plants Research*, 5(16), 3577–3594.

Kodagoda, K. H. G. K., & Marapana, R. A. U. J. (2017). Utilization of fruit processing by-products for industrial applications: A review. *International Journal of Food Science and Nutrition*, 2(6), 24–30.

Kozlu, A., & Elmacı, Y. (2020). Quince seed mucilage as edible coating for mandarin fruit; determination of the quality characteristics during storage. *Journal of Food Processing and Preservation*, 44(11), 1–8.

Kurt, A., & Atalar, I. (2018). Effects of quince seed on the rheological, structural and sensory characteristics of ice cream. *Food Hydrocolloids*, 82, 186–195.

Magalhães, A. S., Silva, B. M., Pereira, J. A., Andrade, P. B., Valentão, P., & Carvalho, M. (2009). Protective effect of quince (*Cydonia oblonga* Miller) fruit against oxidative hemolysis of human erythrocytes. *Food and Chemical Toxicology*, 47(6), 1372–1377.

Milić, D., Vukoje, V., & Sredojević, Z. (2010). Production characteristics and economic aspects of quince production. *Journal on Processing and Energy in Agriculture*, 14(1), 36–39.

Minaiyan, M., Ghannadi, A., Etemad, M., & Mahzouni, P. (2012) A study of the effects of *Cydonia oblonga* Miller (Quince) on TNBS-induced ulcerative colitis in rats. *Research in Pharmaceutical Sciences*, 7(2), 103–110.

Mir, S. A., Masoodi, F. A., Gani, A., Ganaie, S. A., Reyaz, U., & Wani, S. M. (2015). Evaluation of antioxidant properties of methanolic extracts from different fractions of quince (*Cydonia oblonga* Miller). *Advances in Biomedicine and Pharmacy*, 2(1), 1–6.

Moghbel, A., & Tayebi, M. (2015). Quince seeds biopolymer, extraction, drying methods and evaluation. *Jundishapur Journal of Natural Pharmaceutical Products*, 10(3), 1–7.

Monka, A., Grygorieva, O., Chlebo, P., & Brindza, J. (2014). Morphological and antioxidant characteristics of quince (*Cydonia oblonga* Mill.) and chinese quince fruit (*Pseudocydonia sinensis* Schneid.). *Potravinarstvo*, 8(1), 333–340.

Mudura, E., Coldea, T. E., & Farcas, A. (2016) Quince peel extract addition to liqueur for improving antioxidant activity and phenolic content. *Hop and Medicinal Plants*, 24, 63–70.

Panda, S. K., Mishra, S. S., Kayitesi, E., & Ray, R. C. (2016). Microbial-processing of fruit and vegetable wastes for production of vital enzymes and organic acids, Biotechnology and scopes. *Environmental Research*, 146, 161–172.

Rasheed, M., Hussain, I., Rafiq, S., Hayat, I., Qayyum, A., Ishaq, S., & Awan, M. S. (2018). Chemical composition and antioxidant activity of quince fruit pulp collected from different locations. *International Journal of Food Properties*, 21(1), 2320–2327.

Rather, G. A., Bhat, M. Y., Sana, S. S., Ali, A., Gul, M. Z., Nanda, A., & Hassan, M. (2020). Quince 20. In Gulzar Nayik and Amir Gull (Eds), *Antioxidants in Fruits, Properties and Health Benefits*, Springer, Singapore, 397 –416.

Rezagholi, F., Hashemi, S. M. B., Gholamhosseinpour, A., Sherahi, M. H., Hesarinejad, M. A., & Ale, M. T. (2019). Characterizations and rheological study of the purified polysaccharide extracted from quince seeds. *Journal of the Science of Food and Agriculture*, 99(1), 143–151.

Riahi-Chebbi, I., Haoues, M., Essafi, M., Zakraoui, O., Fattouch, S., Karoui, H., & Essafi-Benkhadir, K.(2015). Quince peel polyphenolic extract blocks human colon adenocarcinoma LS174 cell growth and potentiates 5-fluorouracil efficacy. *Cancer Cell International*, 16(1), 1–15.

Rodríguez-Guisado, I., Hernández, F., Melgarejo, P., Legua, P., Martínez, R., & Martínez, J. J. (2009). Chemical, morphological and organoleptical characterisation of five Spanish quince tree clones (*Cydonia oblonga* Miller). *Scientia Horticulturae*, 122(3), 491–496.

Sandoval-Oliveros, M. R., & Paredes-López, O. (2013). Isolation and characterization of proteins from chia seeds (*Salvia hispanica* L.). *Journal of Agricultural and Food Chemistry*, 61(1), 193–201.

Shahbazi, Y., & Moosavy, M. H. (2019). Physico-mechanical and antimicrobial properties of quince seed mucilage supplemented with titanium dioxide and silicon oxide nanoparticles. *Nanomedicine Research Journal*, 4(3), 157–163.

Silva, B. M., Andrade, P. B., Ferreres, F., Seabra, R. M., Beatriz, M., Oliveira, P. P., & Ferreira, M. A. (2005). Composition of quince (*Cydonia oblonga* Miller) seeds, phenolics, organic acids and free amino acids. *Natural Product Research*, 19(3), 275–281.

Silva, B. M., Andrade, P. B., Valentão, P., Ferreres, F., Seabra, R. M., & Ferreira, M. A. (2004). Quince (*Cydonia oblonga* Miller) fruit (pulp, peel, and seed) and jam, antioxidant activity. *Journal of Agricultural and Food Chemistry*, 52(15), 4705–4712.

Soukoulis, C., Gaiani, C., & Hoffmann, L. (2018). Plant seed mucilage as emerging biopolymer in food industry applications. *Current Opinion in Food Science*, 22, 28–42.

Tamri, P., Hemmati, A., & Boroujerdnia, M. G. (2014). Wound healing properties of quince seed mucilage, In vivo evaluation in rabbit full-thickness wound model. *International Journal of Surgery*, 12(8), 843–847.

Vaughan, J., & Geissler, C. (2009). *The new Oxford book of food plants*. OUP, Oxford.

Wang, L., Wu, M., Liu, H. M., Ma, Y. X., Wang, X. D., & Qin, G. Y. (2017). Subcritical fluid extraction of Chinese quince seed, optimization and product characterization. *Molecules*, 22(4), 528.

Yousefi, N., Zeynali, F., & Alizadeh, M. (2018). Optimization of low-fat meat hamburger formulation containing quince seed gum using response surface methodology. *Journal of Food Science & Technology*, 55(2), 598–604.

CHAPTER 22

Pomegranate Wastes and By-Products
Chemistry, Processing, and Utilization

Shiv Kumar, Poonam Baniwal, Harpreet Kaur, Rekha Kaushik,
Sugandha Sharma and Naseer Ahmed

CONTENTS

22.1 INTRODUCTION

Pomegranate (*Punica granatum* L.) is a deciduous shrub belonging to the Lythraceae family which is adapted to different regions of tropics and subtropics (Kalamara et al., 2015; Dhinesh and Ramasamy, 2016). The name 'pomegranate' is derived from the Latin '*Malum granatum*' which means 'grainy apple'. Pomegranate is a non-climacteric fruit, primarily composed of juice (78%) and seeds (22%) (Kalamara et al., 2015). The edible portion of pomegranate is recognized as the seeds (actually 'arils') which is an adequate source of nutrients and phytochemicals such as polyphenols, organic acids, anthocyanins, anthocyanidins, procyanidins, tannins, polysaccharides, vitamins, minerals and other minor biological constituents (Erkan and Kader, 2011). The physicochemical composition of pomegranate is significantly affected by the variety and the cultivation and prevailing environmental conditions (Erkan and Kader, 2011). The major producers of pomegranate in the world are India, Iran, China, the USA and Turkey, of which India is the world's largest producer of pomegranates (Chandra et al., 2010). The pomegranate aril is generally consumed fresh or can be utilized for the preparation of various fruit preserve products such as jams, jellies, beverages, wine, syrups, anardana, frozen arils, powder and coloring and flavoring agent (Erkan and Kader, 2011). The juice of pomegranates is a rich source of sugars such as glucose and fructose (Viuda Martos et al., 2010). Pomegranate fruit represents excellent nutritive, pharmacological and functional properties. Pomegranate is widely cultivated on a commercial

DOI: 10.1201/9781003164463-22

scale due to containing nutritional and bioactive components which offer numerous therapeutic applications, for which pomegranate can be transformed into different pharmaceuticals and functional food products. The major challenge is maintaining the quality of this fruit for a long time. The post-harvest losses of this fruit are around 20–40% (Dhinesh and Ramasamy, 2016). With the advancement of scientific knowledge, to keep the quality and extend the shelf-life of pomegranates, different techniques were performed and accepted including fast-pre-refrigeration, controlled or modified atmospheric packaging and other smart and active packaging. Each part of the pomegranate, i.e., fruit, leaves, arils, peel, seeds and flowers have profitable value and can be formulated into different value-added products. The fruit is mainly used for the production of juice and further into jams, jellies, etc. The edible portions of fruit correspond to 52% of the total weight which includes juice (78%) and seeds (22%). The juice contains moisture (85.4%), total sugars (10.6%), pectin (1.4%), total acidity (0.1 g/100 mL) and ascorbic acid (0.7 mg/100 mL). Peel and its membranes contribute around 50% of the total fruit weight. Seeds and peel offer unique nutraceutical and pharmaceutical properties. The seeds mainly contribute 85% water and 10% sugar (Caruso et al., 2020) and exhibit numerous functional properties including inhibition of eicosanoid biosynthesis enzymes, immunostimulatory and antioxidant activities and protective effects related to gentamicin-induced nephrotoxicity (Kalamara et al., 2015). The by-products are rich sources of various bioactive components, i.e., phenolics, tannins and flavonoids. They also possess efficient antioxidant and free radical scavenging properties (Sorrenti et al., 2019). Peel and flowers are used as natural dyes. Jam is produced from the arils, which can be dried, ground and used as seasoning (Chandra et al., 2010).

22.2 CHEMISTRY OF POMEGRANATE WASTES AND BY-PRODUCTS

Pomegranate has various positive effects on human health due to the presence of various bioactive components, such as phenolic acids, flavonoids and various tannins which are found in the skins of pomegranate (Bialonska et al., 2009). The pomegranate fruits inhibit inflammatory markers and indicate antioxidant properties owing to the contents of flavonoids and phenolics. (Bae et al., 2010). The fruits are the richest source of punicallagin, and are also rich in quecertin, minerals, glucose, caffeic acid, catechin and ellagic acid, as well as derivatives of cyanidin-3, 5-diglucosides, pelargonidin 3-glucosides and delphinidin (Bagri et al., 2009). The two major polyphenols present in the peel and seeds of pomegranates are punicalagin and punicalin, and punicic acid, sterols and ellagic acid are also present (Ercisli et al., 2011). Various phenolic constituents are present in the peel, such as tannins, gallic acid, punicalagin, catechin, flavones, anthocyanidins and flavones. Pomegranate juice extraction generates huge amounts of peel and seeds as wastes and by-products (Figure 22.1). Pomegranate seed and peel are excellent sources of phytochemicals and bio-active compounds that represent a tremendous potential for the production of cost-delivered products (Andrade et al., 2019). Almost 50% of the fruit weight is the safe-to-eat portion, made up of arils (40%) and seeds (10%), with the remainder being non-edible in nature. The widely used technology for juice extraction includes the peeling of the outer peel and removal of seeds as they impart bitterness to the processed juice (Charalampia and Koutelidakis, 2017). After juice extraction, clarification is achieved by disposing of haze and the associated phenolic compounds. The production of pomegranate juice leads to production of major waste and by-product streams, such as peel and seeds, at respective steps of processing (Andrade et al., 2019). The pomegranate wastes and by-products show a huge potential in the development of nutraceuticals and value-added products in the food industry (Aslam et al., 2006; Durgac et al., 2008). The wastes and by-products can also act as substrates for the manufacture of organic components that can be used as food additives or supplements and as alternatives to synthetic compounds with health benefits (Cam et al., 2009; Davidson et al., 2009; Tezcan et al., 2009; Kumar, et al., 2018).

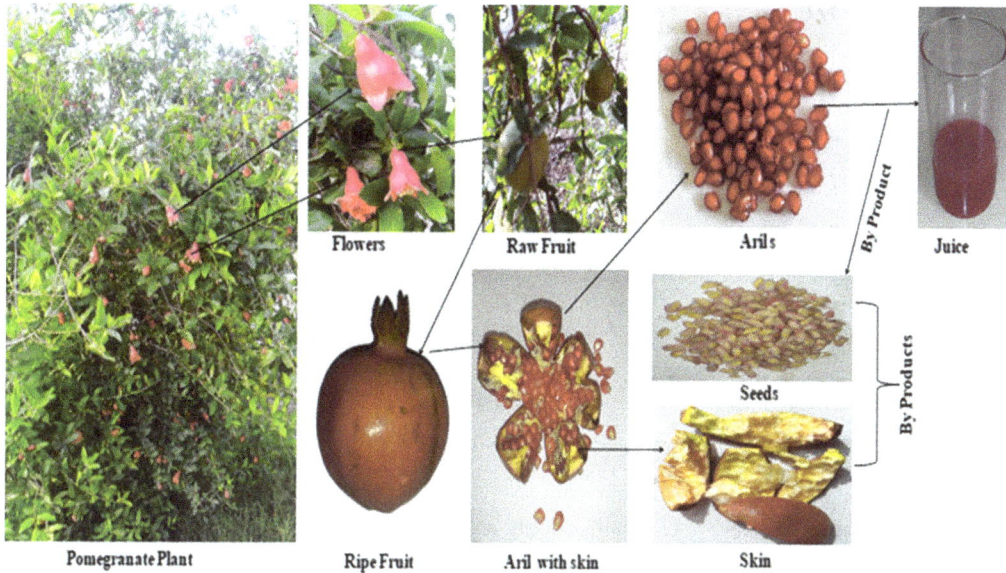

Figure 22.1 Pomegranate plant and its fruit, with by-products.

22.2.1 Pomegranate Peel

Peel refers to the thick, inedible covering around the consumable fruit aril and accounts for almost 50% of the weight of the entire fruit. Peel is a source of polyphenols, minerals and complex polysaccharides, whilst arils consist of 85% water containing sugars, phenolics, pectin, organic acids and flavonoids, particularly anthocyanins. Pomegranate peel is a rich source of polyphenols with nearly double the concentration of antioxidants than is present in the pulp or juice. It is a potential nutrient-dense ingredient for the development of phytochemical-rich food products. Pomegranate peel powder is an effective antioxidant containing flavonoids, phenolics and pro-anythocyanidins as well as high levels of vitamin C (Chidambara et al., 2004; Adhami et al., 2006; Conley, 2008; Rajan et al., 2011) (Table 22.1). The presence of phytochemical compounds in pomegranate peel includes health-affective compounds against cancers, diabetes, atherosclerosis, Alzheimer's disease, nephrotoxicity, hepatotoxicity, chronic pain, and other degenerative diseases (Kulkarni et al., 2007; Zahin et al., 2010; Chidambara et al., 2004). Pomegranate peel is an extraordinary treasure trove of bioactive components including phenolic acids, flavonoids and hydrolyzable tannins that have validated health benefits Pectin is the primary polysaccharide present, accounting for nearly 25% of the peel (Talekar, 2018). The protein content in the peel is quite low, only 3%, although it is higher in pomegranate seeds (14%) (Rowayshed et al., 2013). Polyphenols in peel include gallic acid (14.14%) chlorogenic acid (2.35%), procatechuic acid (14.51%), coumarin (3.53%), vanillic acid (3.85%), oleuropein (0.59%), caffeic acid (2.74%), quercetin (0.95%) and ferulic acid (1.85%) (Charalampia and Koutelidakis, 2017). Their concentration in the by-products varies with respect to the variety (Seeram et al., 2005; Li et al., 2005; Mirdehghan and Rahemi, 2007). Approximately 30% of the phytochemicals in pomegranate peel are comprised of anthocyanidins. Intake of dried pomegranate peel shows positive results in the form of rapid weight gain in bull calves (Shabtay et al., 2008).

22.2.2 Pomegranate Seed

The edible part of pomegranate fruit is composed of hundreds of seeds which are surrounded with a red, juicy and sweet seed casing known as an aril (Ozgul, 2005). The seeds and arils are

Table 22.1   By-Products of Pomegranate and Their Uses

S. No.	Chemical Constituents	Uses	References
		Peel	
1.	Ellagitannins	Cardiovascular health, wound healing properties	Pathak, et al. (2017)
2.	Gallic acid	Antidiabetic, cardiovascular health, anticarcinogenic	Venkitasamy et al. (2019)
3.	Caffeic acid	Anti-inflammatory activity	Jaiswal et al. (2010)
4.	Ellagic acid	Antiviral, antioxidant properties	Rowayshed et al. (2013); Kumar et al. (2018)
5.	Anthocyanidins	Anti-atherogenic and antioxidant activities	Kaplan et al. (2001)
		Seeds	
6.	Sterols	Regulate the cholesterol level	Verardo et al. (2014)
7.	Phytoestrogens	Harmon replacement therapy	Ewies (2002); Glazier and Bowman (2001).
8.	Punicic acid	Anti-obesity, antiproliferative	Aruna et al. (2016)
9.	Isoflavones		
10.	Polyphenols	Antimetastatic, anticancer	Syed et al. (2007); Kumar et al. (2020)

edible. The seeds make up 10% of the fruit whereas the aril accounts for 50% of the fruit weight. The seeds are a rich source of lipids, mainly polyunsaturated fatty acids (PUFA). Pomegranate seeds contain proteins, minerals, crude fibers, vitamins, sugars, pectin, isoflavones and polyphenols (Kandylis and Evangelos, 2020). The seeds contain hydrolyzable tannins, such as ellagitannins, which possess antioxidant activity. Pomegranate seed oil mainly consists of conjugated octadeca-trienoic fatty acids with a considerable concentration of punicic acid (Jurenka, 2008). The oil of the pomegranate seed has a restrictive impact on human skin as well as on cancers related to the breast. Phytoestrogenic compounds are present in pomegranate seed oil and the pomegranate is considered to be rich in phenolics, with robust antioxidant activity. Its juice and seeds are considered to be a medicine for the throat as well as against coronary heart disease. Pomegranate seed oil is a splendid source of fatty acids, punicic acid, sterols, phenyl aliphatic glycosides, ellagic acid and isoflavones (Farag et al., 2014). Pomegranate seeds are also sources of fiber (50%), rich in cellulose and lignin (Akhtar et al., 2015).

22.3 PROCESSING OF POMEGRANATE WASTES AND BY-PRODUCTS

Pomegranate processing to extract the juice as the main product generates wastes and by-products such as peel and seeds. The peel and seeds from processed pomegranate fruits can be converted into a powder as shelf-stable product with nutraceutical potential. The extracted seeds and peels from processed pomegranate are further dried by different methods such as traditional (sun drying) and modern techniques (hot-air oven, tray drying, etc.) for production of dried ground seed powder. Pomegranate seed powder can be incorporated into value-added bakery goods such as cookies, where it increases antioxidant activity, enhances specific volume and increases springiness. The seed powder contains different essential fatty acids which are beneficial for health (Table 22.1). The pomegranate-derived peel is used to produce an antioxidant-rich extract and antimicrobial-rich products. The antioxidant and antimicrobial activities of pomegranate peel extracts, using green solvents, have been demonstrated using *in-vitro* models, HPLC analysis, total phenolic determination, antioxidant assay using β-carotene (Linoleate Model System), radical scavenging activity (DPPH Method), lipid peroxidation measurement (TBA assay) and hydroxyl radical scavenging activity (Singh et al., 2002). The pomegranate seeds have been used for production of oil using solvent extraction and enzymatic methods (Talekar et al., 2018). The pomegranate oils are a rich source of dietary fiber, protein and conjugated fatty acids. The seeds were treated with a protease and centrifuged to recover fibers (insoluble), protein and oil in different phases. The highest recovery of protein was 13.2%, of insoluble fibers was 97.6 g per 100 g seed residue and of oil was 22.9%, when the seeds were incubated with protease at a concentration of 50 U/g for 14 h, pH 7.2 at 45°C. Pomegranate peel and seeds lead to the production of various bioactive components, such as tannins, phenolic compounds, flavonoids, fatty acids, sterols, vitamins, minerals and dietary fibers. The by-products are also used for the production of industrial enzymes, single-cell protein and lovastatin with a number of economic and waste management advantages (Charalampia and Koutelidakis, 2017). Also, the fruit peel is used in pharmaceutical, nutraceutical and cosmeceutical industries because of its value-additive potential (Magangana et al., 2020).

22.4 UTILIZATION OF POMEGRANATE WASTES AND BY-PRODUCTS

Pomegranate is a wonderful fruit with many beneficial properties. The benefits are not only limited to the consumable portion, but are also associated with wastes and byproducts with significant health benefits and potential in value-addition to the food processing industry. The literature details the addition of pomegranate-based byproducts to food and in non-food applications to increase their market value and to be more environmentally friendly. Pomegranate peel extract is rich in polyphenols and its addition to different dairy products increases their shelf-life and nutraceutical potential. Addition of pomegranate peel extract to cheese packaging reduced lipid oxidation and increased its shelf-life (Sandhya et al., 2018). Pomegranate seed powder is a rich source of dietary fiber and its addition to yogurt improved the antioxidant activities, linoleic acid content and decreased the atherogenicity indexes (Van et al., 2019). Pomegranate peel extracts are also used as food additives because they are of value in storage and processing including refrigeration and thermal processing (Charalampia and Koutelidakis, 2017). Novel edible food packaging materials are attracting great attention. Pomegranate peel has been used in the design of packaging materials to enhance the antimicrobial activity and strength of the materials due to incorporation of this rich source of phenolics and polysaccharides. Various coatings have been formulated from different polymers with added pomegranate peel to increase the integrity of coatings and films. Addition of pomegranate peel results in the proliferation of flexible power, phenolic profile, and antioxidant action and prevents elongation where there may be a constant thickness of film. The prepared film has proven to exhibit better packaging stability than regular films. Chitosan is a polysaccharide derived from deacetylated chitin and has been used to coat material for packaging of food materials. Chitosan and pomegranate peel-based coatings have been widely used for the coating of strawberries, white shrimp, rainbow trout, etc., resulting in enhanced sensory characteristics and improved shelf stability (Duran et al., 2016). Peel-based powder has also been integrated with starch for the formation of an edible film which improves the shelf-life of the packaged product, exhibits antimicrobial properties and reinforces the strength of the film. Starch-based film incorporating pomegranate peel exhibits antimicrobial activity against *Staphylococcus aureus* and *Salmonella* spp. and improves the tensile strength and stiffness of the matrix coating/film (Ali et al., 2019). Meat and meat products are perishable in nature and are prone to undesirable changes which causes taste modifications and color loss and hence affects the sensory properties and acceptability among consumers. Incorporation of pomegranate peel into meat products has been used as an additive to prevent spoilage and increase the shelf stability of meat products. Nanoparticles derived from pomegranate peel, when added to meat balls at a concentration of 1.5% extended shelf-life and prevented lipid oxidation during a storage period of 15 days under 4°C. Furthermore, pomegranate peel powder was confirmed to be an efficient natural preservative for beef sausages by reducing their thiobarbituric acid content and achieving improved storage stability. Pomegranate peel powder might be used as natural antimicrobial and antioxidant agents in seafood. The spoilage and biochemical changes in shrimps could be prevented by soaking them in a methanolic extract of pomegranate peel for 15 min (Basiri et al., 2015). Pomegranate wastes and by-products are extensively used in processed cereal products and nuts with improved phenolic profile and textural properties. The shelf-life of hazelnut paste was extended by stopping lipid oxidation by the incorporation of pomegranate peel powder either in encapsulated form or in a crude form (Kaderides et al., 2015). Furthermore, pomegranate peel with orange peel was used to fortify bakery goods. The resulted product was rich in phenolic content and was more stable (Kaderides et al., 2020); the stability of the end- products remained constant throughout the storage period.

In a similar observation, gluten-free bread was prepared by the incorporation of pomegranate peel in order to modify the total phenolic profile and its free radical scavenging activities (Bourekoua et al., 2018). The prepared product had enhanced specific volume and springiness although the chewiness and hardness of the product were reduced due to increased addition of pomegranate powder. The best physical characteristics of the gluten-free bread were achieved with 7.5% incorporation of pomegranate peel (Dib et al., 2018). Subsequently, pomegranate seed powder has also been incorporated into gluten-free cake and pasta, resulting in lower peroxide value and higher fiber, antioxidant and protein values. In the case of pasta, pomegranate peel altered the textural and cooking characteristics. The pasta with an incorporation of 7.5% was the most acceptable (Saeidi et al., 2018).

Dietary fibers are used as functional components and are easily obtained from fruits and vegetables. They have various health benefits including improved glucose tolerance, decreased cholesterol levels, prevention of colon cancer, etc. They also increase water-holding capacity which helps in the digestion process (Koutelidakis, 2016). A study revealed that the incorporation of pomegranate waste and by-products could be used as a potential source of total, insoluble and soluble fiber. Oil-holding capacity of food products was also found to be increased when pomegranate by-products were incorporated, and which retained less oil when fried (De Crizel et al., 2013). Therefore, pomegranate-based wastes and byproducts can be used for the manufacture of fiber-enriched products and also as potential fat replacers by using dietary fiber (Charalampia and Koutelidakis, 2017).

22.5 CONCLUSIONS

Daily consumption of pomegranates, either as juice or as whole fruits, has increased tremendously, resulting in greater production of pomegranate-based wastes and by-products. These materials are rich sources of phytochemicals, antioxidants, tannins, sterols, phenolic compounds, dietary fibers, flavonoids, vitamins, fatty acids and minerals. The wastes and by-products from the pomegranate processing industry can be utilized in various food products such as dairy, meat and fish, cereals and nuts, as well as in the food packaging industry. Incorporation of pomegranate-derived wastes such as peel, seeds, their dried powders, extracts and seed oil increase the nutraceutical activities of the target foodstuff and produce a novel strategy for the formulation of functional food products with health benefits and improved quality. Furthermore, the incorporation of pomegranate-based wastes and byproducts in food product development are one of the models for the development of a circular economy and increased profitability for the farming community.

BIBLIOGRAPHY

Adhami, V. M., & Mukhtar, H. (2006). Polyphenols from green tea and pomegranate for prevention of prostate cancer. *Free Radical Research* 40(10):1095–1104.

Akhtar, S., Ismail, T., Fraternale, D., & Sestili, P. (2015). Pomegranate peel and peel extracts: Chemistry and food features. *Food Chemistry* 174:417–425.

Ali, A., Chen, Y., Liu, H.; Yu, L., Baloch, Z., Khalid, S., Zhu, J., & Chen, L. (2019). Starch-based antimicrobial films functionalized by pomegranate peel. *International Journal of Biological Macromolecules* 129:1120–1126.

Andrade, M. A., Lima, V., Silva, A. S., Vilarinho, F., Castilho, M. C., Khwaldia, K., & Ramos, F. (2019). Pomegranate and grape by-products and their active compounds: Are they a valuable source for food applications. *Trends in Food Science & Technology* 86:68–84.

Aruna, P., Venkataramanamma, D., Singh, A. K., & Singh, R. P. (2016). Health benefits of punicic acid: A review. *Comprehensive Reviews in Food Science and Food Safety* 15(1):16–27.

Aslam, M. N., Lansky, E. P., & Varani, J. (2006). Pomegranate as a cosmeceutical source: Pomegranate fractions promote proliferation and procollagen synthesis and inhibit matrix metalloproteinase-1 production in human skin cells. *Journal of Ethnopharmacology* 103(3):311–318.

Bae, J. Y., Choi, J. S., Kang, S. W., Lee, Y. J., Park, J., & Kang, Y. H. (2010). Dietary compound ellagic acid alleviates skin wrinkle and inflammation induced by UV-B irradiation. *Experimental Dermatology* 19(8):e182–e190.

Baghizadeh, A., Karimi-Maleh, H., Khoshnama, Z., Hassankhani, A., & Abbasghorbani, M. (2015). A voltammetric sensor for simultaneous determination of vitamin C and vitamin B6 in food samples using ZrO2 nanoparticle/ionic liquids carbon paste electrode, *Food Analysis Methods* 8:549–557.

Bagri, P., Ali, M., Aeri, V., Bhowmik, M., & Sultana, S. (2009). Antidiabetic effect of Punicagranatum flowers: Effect on hyperlipidemia, pancreatic cells lipid peroxidation and antioxidant enzymes in experimental diabetes. *Food and Chemical Toxicology* 47(1):50–54.

Basiri, S., Shekarforoush, S. S., Aminlari, M., & Akbari, S. (2015). The effect of pomegranate peel extract (PPE) on the polyphenol oxidase (PPO) and quality of Pacific white shrimp (Litopenaeus vannamei) during refrigerated storage. *LWT-Food Science and Technology* 60(2):1025–1033.

Basu, A., & Penugonda, K. (2009). Pomegranate juice: A heart-healthy fruit juice. *Nutrition Reviews* 67(1):49–56.

Bialonska, D., Kasimsetty, S. G., Schrader, K. K., & Ferreira, D. (2009). The effect of pomegranate (*Punica granatum* L.) byproducts and ellagitannins on the growth of human gut bacteria. *Journal of Agricultural and Food Chemistry* 57(18):8344–8349.

Bourekoua, H., Rozylo, R., Gawlik-Dziki, U., Benatallah, L., Zidoune, M. N., & Dziki, D. (2018). Pomegranate seed powder as a functional component of gluten-free bread. *International Journal of Food Science and Technology* 53:1906–1913.

Çam, M., Hışıl, Y., & Durmaz, G. (2009). Classification of eight pomegranate juices based on antioxidant capacity measured by four methods. *Food Chemistry* 112(3):721–726.

Caruso, A., Barvarossa, A., Tassone, A., Caremella, J., Carocci, A., Catalano, A., Basile, G., Fazio, A., Lacopetta, D., Franchini, C., & Sinicropi, M. S. (2020). Pomegranate: Nutraceutical with promising benefits on human health. *Applied Sciences* 10(19):6915.

Chandra, R., Jadhav, V. T., & Sharma, J. (2010). Global scenario of pomegranate (*Punica granatum* L.) culture with special reference to India. *Fruit, Vegetable and Cereal Science and Biotechnology* 4(Special Issue 2):7–18.

Charalampia, D., & Koutelidakis, A. E. (2017). From pomegranate processing by-products to innovative value added functional ingredients and bio-based products with several applications in food sector. *BAOJ Biotechnology* 3(025):210.

Chidambara Murthy, K. N., Reddy, V. K., Veigas, J. M., & Murthy, U. D. (2004). Study on wound healing activity of *Punica granatum* peel. *Journal of Medicinal Food* 7(2):256–259.

Conley, J. M., Symes, S. J., Kindelberger, S. A., & Richards, S. M. (2008). Rapid liquid chromatography–tandem mass spectrometry method for the determination of a broad mixture of pharmaceuticals in surface water. *Journal of Chromatography A* 34:150–195.

Davidson, M. H., Maki, K. C., Dicklin, M. R., Feinstein, S. B., Witchger, M., Bell, M., & Aviram, M. (2009). Effects of consumption of pomegranate juice on carotid intima–media thickness in men and women at moderate risk for coronary heart disease. *The American Journal of Cardiology* 104(7):936–942.

De Crizel, T. M., Jablonski, A., Rios, A. O., Rech, R., & Flores, S. H. (2013). Dietary fiber from orange by-products as a potential fat replacer. *LWT- Food Science & Technology* 53(1):9–14.

Dhinesh, K. V., & Ramasamy, D. (2016). Pomegranate processing and value addition: A review. *Indian Horticulture Journal* 6(1):1–12.

Dhinesh, K. V., & Ramasamy, D. (2016). Pomegranate processing and value addition. *Journal of Food Processing and Technology* 7(3):31–43.

Dhineshkumar, V., & Ramasamy, D. (2016). Pomegranate processing and value addition: A review. *Indian Horticulture Journal* 6(1):1–12.

Dib, A. Kasprzak, K. Wojtowicz, A. Benatallah, L. Waksmundzka-Hajnos, M. Zidoune, M.N Oniszczuk, T. Karakula-Juchnowicz, H., & Oniszczuk, A. (2018). The effect of pomegranate seed powder addition on radical scavenging activity determined by TLC-DPPH test and selected properties of gluten free pasta. *Journal of Liquid Chromatography and Related Technologies* 41:364–372.

Duran, M., Aday, M. S., Zorba, N. N. D., & Temizkan, R. (2016). 'Potential of antimicrobial active packaging containing natamycin, nisin, pomegranate and grape seed extract in chitosan coating' to extend shelf life of fresh strawberry. *Food and Bioproducts Processing* 98:354–363.

Durgaç, C., Ozgen, M., Simsek, O. Z. H. A. N., Kaçar, Y. A., Kiyga, Y., Çelebi, S., & Serce, S. (2008). Molecular and pomological diversity among pomegranate (*Punica granatum* L.) cultivars in Eastern Mediterranean region of Turkey. *African Journal of Biotechnology* 7(9):294–1301.

El-Nemr, S. E., Ismail, I. A., & Ragab, M. (1990). Chemical composition of juice and seeds of pomegranate fruit. *Die Nahrung* 34(7):601–606.

Ercisli, S., Gadze, J., Agar, G., Yildirim, N., & Hizarci, Y. (2011). Genetic relationships among wild pomegranate (*Punica granatum*) genotypes from Coruh Valley in Turkey. *Genetics and Molecular Research* 10(1):459–464.

Erkan, M., & Kader, A. A. (2011). Pomegranate (*Punica granatum* L.). In E. Yahi (Ed.), *Postharvest biology and technology of tropical and subtropical fruits* (pp. 287–313). Cambridge: Woodhead Publishing.

Ewies, A. A. (2002). Phytoestrogens in the management of the menopause: Up-to-date. *Obstetrical & Gynecological Survey* 57(5):306–313.

Farag, R. S., Latif, M. S. A., Emam, S. S., & Tawfeek, L. S. (2014). Phtochemical screening and polyphenol constituents of pomegranate peels and leave juices. *Agricultural Soil Science* 1(6):86–93.

Ferrazzano, G. F., Scioscia, E., Sateriale, D., Pastore, G., Colicchio, R., Pagliuca, C., & Pagliarulo, C. (2017). In vitro antibacterial activity of pomegranate juice and peel extracts on cariogenic bacteria. *BioMed Research International* 1–7: 376–389.

Fuhrman, B., Volkova, N., & Aviram, M. (2010). Pomegranate juice polyphenols increase recombinant paraoxonase-1 binding to high-density lipoprotein: Studies *in vitro* and in diabetic patients. *Nutrition* 26(4):359–366.

Glazier, M. G., & Bowman, M. A. (2001). A review of the evidence for the use of phytoestrogens as a replacement for traditional estrogen replacement therapy. *Archives of Internal Medicine* 161(9):1161–1172.

Hayouni, E. A., Miled, K., Boubaker, S., Bellasfar, Z., Abedrabba, M., Iwaski, H., & Hamdi, M. (2011). Hydroalcoholic extract based-ointment from *Punica granatum* L. peels with enhanced in vivo healing potential on dermal wounds. *Phytomedicine* 18(11):976–984.

Jaiswal, V., Der Marderoslan, A., & Porter, J. R. (2010). Anthocyanins and polyphenol oxidase from dried arils of pomegranate (*Punica granatum* L.). *Food Chemistry* 118(1):11–16.

Jurenka, J. S., (2008). Therapeutic applications of Pomegranate (*Punica granatum* L.): A Review. *Alternative Medicine Review* 13(2):128–144.

Kaderides, K., Goula, A. M., & Adamopoulos, K. G. (2015). A process for turning pomegranate peels into a valuable food ingredient using ultrasound-assisted extraction and encapsulation. *Innovative Food Science and Emerging Technologies* 31:204–215.

Kaderides, K., Mourtzinos, I., & Goula, A. M. (2020). Stability of pomegranate peel polyphenols encapsulated in orange juice industry by-product and their incorporation in cookies. *Food Chemistry* 310:125849.

Kalamara, E., Goula, A. M., & Adamopoulos, K. G. (2015). An integrated process for utilization of pomegranate wastes—Seeds. *Innovative Food Science & Emerging Technologies* 27:144–153.

Kandylis, P., & Kokkinomagoulos, E. (2020). Food applications and potential health benefits of pomegranate and its derivatives. *Foods* 9(2):122.

Kaplan, M., Hayek, T., Raz, A., Coleman, R., Dornfeld, L., Vaya, J., & Aviram, M. (2001). Pomegranate juice supplementation to atherosclerotic mice reduces macrophage lipid peroxidation, cellular cholesterol accumulation and development of atherosclerosis. *The Journal of Nutrition* 131(8):2082–2089.

Koutelidakis, A., & Dimou, C. (2016). The effects of functional food and bioactive compounds on biomarkers of cardiovascular diseases. In D. Martirosyan (Ed.), *Functional foods text book*, 1st edition (pp. 89–117). Texas: Functional Food Center.

Kulkarni, A. P., Mahal, H. S., Kapoor, S., & Aradhya, S. M. (2007). *In vitro* studies on the binding, antioxidant, and cytotoxic actions of punicalagin. *Journal of Agricultural and Food Chemistry* 55(4):1491–1500.

Kumar, N., Kumar, N., & Kumar, S. (2018). Functional properties of pomegranate (*Punica granatum* L.). *The Pharma Innovation Journal* 7(10):71–81.

Kumar, S., Baniwal, P., Kaur, J., & Kumar, H. (2020). Kachnar (*Bauhinia variegata*). In G. A. Nayik & A. Gull (Eds.), *Antioxidants in fruits: Properties and health benefits* (pp. 365–377). Singapore: Springer.

Lee, C. J., Chen, L. G., Liang, W. L., & Wang, C. C. (2010). Anti-inflammatory effects of *Punica granatum* L. in vitro and in vivo. *Food Chemistry* 118(2):315–322.

Li, Y., Wen, S., Kota, B. P., Peng, G., Li, G. Q., Yamahara, J., & Roufogalis, B. D. (2005). *Punica granatum* flower extract, a potent α-glucosidase inhibitor, improves postprandial hyperglycemia in Zucker diabetic fatty rats. *Journal of Ethnopharmacology* 99(2):239–244.

Magangana, T. P., Makunga, N. P., Fawole, O. A., & Opara, U. L. (2020). Processing factors affecting the phytochemical and nutritional properties of pomegranate (*Punica granatum* L.) peel waste: A review. *Molecules* 25:4690.

Mirdehghan, S. H., & Rahemi, M. (2007). Seasonal changes of mineral nutrients and phenolics in pomegranate (*Punica granatum* L.) fruit. *Scientia Horticulturae* 111(2):120–127.

Morsy, M. K., Mekawi, E., & Elsabagh, R. (2018). Impact of pomegranate peel nanoparticles on quality attributes of meatballs during refrigerated storage. *LWT-Food Science and Technology* 89:489–495.

Mushtaq, M., Gani, A., Punoo, H. A., & Masoodi, F. A. (2018). Use of pomegranate peel extract incorporated zein film with improved properties for prolonged shelf life of fresh Himalayan cheese (kalari/kradi). *Innovative Food Science and Emerging Technologies* 48:25–32.

Newman, R., Lansky, E., & Block, M. (2011). A wealth of phytochemicals. *Pomegranate: The Most Medicinal Fruit*). Sydney: Readhowyouwant, 184.

Olapour, S., Mousavi, E., Sheikhzade, M., Hoseininezhad, O., & Najafzadeh, H. (2009). Evaluation anti-diarrheal effects of pomegranate peel extract. *Journal of the Iranian Chemical Society* 6(Nov):115–43.

Ozgul-Yucel, S. (2005). Determination of conjugated linolenic acid content of selected oil seeds grown in Turkey. *Journal of the American Oil Chemists' Society* 82(12):893–897.

Pan, Z., Zhang, R., & Zicari, S. (Eds.). (2019). *Integrated processing technologies for food and agricultural by-products*. Cambridge: Academic Press.

Pathak, P. D., Mandavgane, S. A., & Kulkarni, B. D. (2017). Valorization of pomegranate peels: A biorefinery approach. *Waste and Biomass Valorization* 8(4):1127–1137.

Prabhu, S., Molath, A., Choksi, H., Kumar, S., & Mehra, R. (2021). Classifications of polyphenols and their potential application in human health and diseases. *International Journal of Physiology, Nutrition and Physical Education* 6:293–301.

Rajan, S., Mahalakshmi, S., Deepa, V. M., Sathya, K., Shajitha, S., & Thirunalasundari, T. (2011). Antioxidant potentials of *Punica granatum* fruit rind extracts. *International Journal of Pharmacy and Pharmaceutical Sciences* 3(3):82–88.

Rowayshed, G., Salama, A., Fadl, M. A., Hamza, S. A., & Emad, A. M. (2013). Nutritional and chemical evaluation of pomegranate (*Punica granatum* L.) fruit peel and seeds powders by products. *Middle East Journal of Applied Sciences* 3(4):169–179.

Saeidi, Z., Nasehi, B., & Jooyandeh, H. (2018). Optimization of gluten-free cake formulation enriched with pomegranate seed powder and trans glutaminase enzyme. *Journal of Food Science and Technology* 55:3110–3118.

Sandhya, S., Khamrui, K., Prasad, W., & Kumar, M. C. T. (2018). Preparation of pomegranate peel extract powder and evaluation of its effect on functional properties and shelf life of curd. *LWT- Food Science and Technology* 92:416–421.

Scappaticci, F. A. (2003). The therapeutic potential of novel antiangiogenic therapies. *Expert Opinion on Investigational Drugs* 12(6):923–932.

Seeram, N. P., Adams, L. S., Henning, S. M., Niu, Y., Zhang, Y., Nair, M. G., & Heber, D. (2005). *In vitro* anti proliferative, apoptotic and antioxidant activities of punicalagin, ellagic acid and a total pomegranate tannin extract are enhanced in combination with other polyphenols as found in pomegranate juice. *The Journal of Nutritional Biochemistry* 16(6):360–367.

Shabtay, A., Eitam, H., Tadmor, Y., Orlov, A., Meir, A., Weingberg, P., Weingberg, Z. G., Chen, Y., Brosh, A., Izhaki, I., & Kerem, Z. (2008). Nutritive and antioxidant potential of fresh and stored pomegranate industrial byproduct as novel beef cattle feed. *Journal of Agricultural and Food Chemistry* 56:10063–10070.

Singh, B., Singh, J. P., Kaur, A., & Singh, N. (2018). Phenolic compounds as beneficial phytochemicals in pomegranate (*Punica granatum* L.) peel: A review. *Food Chemistry* 261:75–86.

Singh, R. P., Chidamba, M. K. N., & Jayaprakash, G. K. (2002). Studies on the antioxidant activity of pomegranate (*Punica granatum*) peel and seed extracts using *in vitro* models. *Journal of Agricultural and Food Chemistry* 50:81–86.

Sorrenti, V., Randazzo, L. C., Caggia, C., Baliistreri, G., Romeo, V. F., Fabroni, S., Timpanaro, N., Marco, R., & Vanella, L. (2019). Benefical effects of pomegranate peel extract and probiotics on pre-adipocyte differentiation. *Frontiers in Microbiology* 10:660. https://doi.org/10.3389/fmicb.2019.00660.

Stowe, C. B. (2011). The effects of pomegranate juice consumption on blood pressure and cardiovascular health. *Complementary Therapies in Clinical Practice* 17(2):113–115.

Sun, W., Yan, C., Frost, B., Wang, X., Hou, C., Zeng, M., & Liu, J. (2016). Pomegranate extract decreases oxidative stress and alleviates mitochondrial impairment by activating AMPK-Nrf2 in hypothalamic paraventricular nucleus of spontaneously hypertensive rats. *Scientific Reports* 6(1):1–12.

Syed, D. N., Afaq, F., & Mukhtar, H. (2007). Pomegranate derived products for cancer chemoprevention. *Seminars in Cancer Biology* 17(5):377–385.

Talekar, S., Patti, A. F., Singh, R. Vijayraghavan, R., & Arora, A. (2018). From waste to wealth: High recovery of nutraceuticals and pomegranate seed waste using a green extraction process. *Industrial Crops and Products* 112:790–802.

Talekar, S., Patti, A. F., Vijayraghavan, R., & Arora, A. (2018). An integrated green biorefinery approach towards simultaneous recovery of pectin and polyphenols coupled with bioethanol production from waste pomegranate peels. *Bioresource Technology* 266:322–334.

Tezcan, F., Gultekin-Ozguven, M., Diken, T., Ozçelik, B., & Erim, F. B. (2009). Antioxidant activity and total phenolic, organic acid and sugar content in commercial pomegranate juices. *Food Chemistry* 115(3):873–877.

Van Nieuwenhove, C. P., Moyano, A., Castro-Gomez, P., Fontecha, J., Saez, G., Zarate, G., & Pizarro, P. L. (2019). Comparative study of pomegranate and jacaranda seeds as functional components for the conjugated linoleic acid enrichment of yogurt. *LWT- Food Science and Technology* 111:401–407.

Venkitasamy, C., Zhao, L., Zhang, R., & Pan, Z. (2019). Pomegranate. In Z. Pan, R. Zhang, & S. Zicari (Eds), *Integrated processing technologies for food and agricultural by-products* (pp. 181–216). Cambridge: Academic Press.

Verardo, V., Garcia-Salas, P., Baldi, E., Segura-Carretero, A., Fernandez-Gutierrez, A., & Caboni, M. F. (2014). Pomegranate seeds as a source of nutraceutical oil naturally rich in bioactive lipids. *Food Research International* 65:445–452.

Viuda-Martos, M., Fernandez-Lopez, J., & PerezAlvarez, J. A. (2010). Pomegranate and its many functional components as related to human health: A review. *Comprehensive Reviews in Food Science and Food Safety* 9(6):635–654.

Vucic, V., Grabez, M., Trchounian, A., & Arsic, A. (2019). Composition and potential health benefits of pomegranate: A review. *Current Pharmaceutical Design* 25(16):1817–1827.

Zahin, M., Aqil, F., & Ahmad, I. (2010). Broad spectrum antimutagenic activity of antioxidant active fraction of *Punica granatum* L. peel extracts. *Mutation Research/Genetic Toxicology and Environmental Mutagenesis* 703(2):99–107.

Index

For Product Safety Concerns and Information please contact our EU
representative GPSR@taylorandfrancis.com
Taylor & Francis Verlag GmbH, Kaufingerstraße 24, 80331 München, Germany

www.ingramcontent.com/pod-product-compliance
Lightning Source LLC
Chambersburg PA
CBHW080910220326
41598CB00034B/5529

9 780367 758950